# Principles and Practices of Nanobiotechnology

# Principles and Practices of Nanobiotechnology

Edited by **Giorgio Salati**

SYRAWOOD
PUBLISHING HOUSE

New York

Published by Syrawood Publishing House,
750 Third Avenue, 9th Floor,
New York, NY 10017, USA
www.syrawoodpublishinghouse.com

**Principles and Practices of Nanobiotechnology**
Edited by Giorgio Salati

International Standard Book Number: 978-1-68286-110-3 (Hardback)

Printed in the United States of America.

# Contents

Preface IX

Chapter 1 **Melanin-templated rapid synthesis of silver nanostructures** 1
George Seghal Kiran, Asha Dhasayan, Anuj Nishanth Lipton, Joseph Selvin,
Mariadhas Valan Arasu and Naif Abdullah Al-Dhabi

Chapter 2 **Investigation of antibacterial effect of Cadmium Oxide nanoparticles on**
***Staphylococcus Aureus* bacteria** 14
Bahareh Salehi, Sedigheh Mehrabian and Mehdi Ahmadi

Chapter 3 **Engineering of near infrared fluorescent proteinoid-poly(L-lactic acid) particles**
**for *in vivo* colon cancer detection** 22
Michal Kolitz-Domb, Igor Grinberg, Enav Corem-Salkmon and Shlomo Margel

Chapter 4 **Development of an insecticidal nanoemulsion with *Manilkara subsericea***
**(Sapotaceae) extract** 35
Caio Pinho Fernandes, Fernanda Borges de Almeida, Amanda Nunes Silveira,
Marcelo Salabert Gonzalez, Cicero Brasileiro Mello, Denise Feder,
Raul Apolinário, Marcelo Guerra Santos, José Carlos Tavares Carvalho,
Luis Armando Cândido Tietbohl, Leandro Rocha and
Deborah Quintanilha Falcão

Chapter 5 **Effects in cigarette smoke stimulated bronchial epithelial cells of a**
**corticosteroid entrapped into nanostructured lipid carriers** 44
Maria Luisa Bondì, Maria Ferraro, Serena Di Vincenzo, Stefania Gerbino,
Gennara Cavallaro, Gaetano Giammona, Chiara Botto, Mark Gjomarkaj and
Elisabetta Pace

Chapter 6 **Investigation of magnetically controlled water intake behavior of Iron Oxide**
**Impregnated Superparamagnetic Casein Nanoparticles (IOICNPs)** 59
Anamika Singh, Jaya Bajpai and Anil Kumar Bajpai

Chapter 7 **Activation of caspase-dependent apoptosis by intracellular delivery of**
**cytochrome c-based nanoparticles** 72
Moraima Morales-Cruz, Cindy M Figueroa, Tania González-Robles,
Yamixa Delgado, Anna Molina, Jessica Méndez, Myraida Morales and
Kai Griebenow

Chapter 8    **Transient extracellular application of gold nanostars increases hippocampal neuronal activity**    83
Kirstie Salinas, Zurab Kereselidze, Frank DeLuna, Xomalin G Peralta and Fidel Santamaria

Chapter 9    **Comparative lung toxicity of engineered nanomaterials utilizing *in vitro*, *ex vivo* and *in vivo* approaches**    90
Yong Ho Kim, Elizabeth Boykin, Tina Stevens, Katelyn Lavrich and M Ian Gilmour

Chapter 10    **Enhanced green fluorescent protein-mediated synthesis of biocompatible graphene**    102
Sangiliyandi Gurunathan, Jae Woong Han, Eunsu Kim, Deug-Nam Kwon, Jin-Ki Park and Jin-Hoi Kim

Chapter 11    **Plants and microbes assisted selenium nanoparticles: characterization and application**    118
Azamal Husen and Khwaja Salahuddin Siddiqi

Chapter 12    **Facile synthesis of fluorescent Au/Ce nanoclusters for high-sensitive bioimaging**    128
Wei Ge, Yuanyuan Zhang, Jing Ye, Donghua Chen, Fawad Ur Rehman, Qiwei Li, Yun Chen, Hui Jiang and Xuemei Wang

Chapter 13    **Novel metal allergy patch test using metal nanoballs**    136
Tomoko Sugiyama, Motohiro Uo, Takahiro Wada, Toshio Hongo, Daisuke Omagari, Kazuo Komiyama, Hitoshi Sasaki, Heishichiro Takahashi, Mikio Kusama and Yoshiyuki Mori

Chapter 14    **Development of antimicrobial biomaterials produced from chitin-nanofiber sheet/silver nanoparticle composites**    142
Vinh Quang Nguyen, Masayuki Ishihara, Jun Kinoda, Hidemi Hattori, Shingo Nakamura, Takeshi Ono, Yasushi Miyahira and Takemi Matsui

Chapter 15    **One pot light assisted green synthesis, storage and antimicrobial activity of dextran stabilized silver nanoparticles**    151
Muhammad Ajaz Hussain, Abdullah Shah, Ibrahim Jantan, Muhammad Nawaz Tahir, Muhammad Raza Shah, Riaz Ahmed and Syed Nasir Abbas Bukhari

Chapter 16    **Interfacial film stabilized W/O/W nano multiple emulsions loaded with green tea and lotus extracts: systematic characterization of physicochemical properties and shelf-storage stability**    157
Tariq Mahmood, Naveed Akhtar and Sivakumar Manickam

Chapter 17    **Exploring cancer metastasis prevention strategy: interrupting adhesion of cancer cells to vascular endothelia of potential metastatic tissues by antibody-coated nanomaterial**    165
Jingjing Xie, Haiyan Dong, Hongning Chen, Rongli Zhao, Patrick J Sinko, Weiyu Shen, Jichuang Wang, Yusheng Lu, Xiang Yang, Fangwei Xie and Lee Jia

Chapter 18   **Invertase-nanogold clusters decorated plant membranes for
fluorescence-based sucrose sensor**                                      **178**
Dipali Bagal-Kestwal, Rakesh Mohan Kestwal and Been-Huang Chiang

Chapter 19   **Multifunctional polymeric nanoparticles doubly loaded with SPION and
ceftiofur retain their physical and biological properties**              **189**
Paula Solar, Guillermo González, Cristian Vilos, Natalia Herrera, Natalia Juica,
Mabel Moreno, Felipe Simon and Luis Velásquez

**Permissions**

**List of Contributors**

# Preface

Nanobiotechnology is a rapidly advancing field with widespread applications. This book focuses on innovative applications and approaches in this field. The various studies by international experts that are constantly contributing towards advancing technologies and evolution of this field are examined in detail. Some of the concepts discussed in this book such as engineered biomaterials, nanoparticles, nanostructures, etc. provide an in-depth knowledge of this field. It aims to serve as a resource guide for students and experts alike and contribute to the growth of the discipline.

The information contained in this book is the result of intensive hard work done by researchers in this field. All due efforts have been made to make this book serve as a complete guiding source for students and researchers. The topics in this book have been comprehensively explained to help readers understand the growing trends in the field.

I would like to thank the entire group of writers who made sincere efforts in this book and my family who supported me in my efforts of working on this book. I take this opportunity to thank all those who have been a guiding force throughout my life.

**Editor**

# Melanin-templated rapid synthesis of silver nanostructures

George Seghal Kiran[1], Asha Dhasayan[2], Anuj Nishanth Lipton[3], Joseph Selvin[3*], Mariadhas Valan Arasu[4] and Naif Abdullah Al-Dhabi[4]

## Abstract

**Background:** As a potent antimicrobial agent, silver nanostructures have been used in nanosensors and nanomaterial-based assays for the detection of food relevant analytes such as organic molecules, aroma, chemical contaminants, gases and food borne pathogens. In addition silver based nanocomposites act as an antimicrobial for food packaging materials. In this prospective, the food grade melanin pigment extracted from sponge associated actinobacterium *Nocardiopsis alba* MSA10 and melanin mediated synthesis of silver nanostructures were studied. Based on the present findings, antimicrobial nanostructures can be developed against food pathogens for food industrial applications.

**Results:** Briefly, the sponge associated actinobacterium *N. alba* MSA10 was screened and fermentation conditions were optimized for the production of melanin pigment. The Plackett-Burman design followed by a Box-Behnken design was developed to optimize the concentration of most significant factors for improved melanin yield. The antioxidant potential, reductive capabilities and physiochemical properties of *Nocardiopsis* melanin was characterized. The optimum production of melanin was attained with pH 7.5, temperature 35°C, salinity 2.5%, sucrose 25 g/L and tyrosine 12.5 g/L under submerged fermentation conditions. A highest melanin production of 3.4 mg/ml was reached with the optimization using Box-Behnken design. The purified melanin showed rapid reduction and stabilization of silver nanostructures. The melanin mediated process produced uniform and stable silver nanostructures with broad spectrum antimicrobial activity against food pathogens.

**Conclusions:** The melanin pigment produced by *N. alba* MSA10 can be used for environmentally benign synthesis of silver nanostructures and can be useful for food packaging materials. The characteristics of broad spectrum of activity against food pathogens of silver nanostructures gives an insight for their potential applicability in incorporation of food packaging materials and antimicrobials for stored fruits and foods.

**Keywords:** Marine *Nocardiopsis*, Melanin, Optimization, Silver nanostructures, Antimicrobial activity

## Background

Silver particles/nanostructures have been used as an effective antimicrobial agent in food and beverage storage for a long time. Silver containing plastics had been incorporated in refrigerator liners and food storage containers [1-3]. FDA has been approved the use of silver based particles for disinfection purpose for the food contacting materials [4].

Silver based nanomaterials and nanocomposite can be devised for the easiest detection of commonly found food adulterants, chemical contaminants, allergens and any changes respond to environmental conditions etc. Silver nanoparticles incorporated cellulose pads are used to control the food pathogens from packed beef meat and reduce the microbial count in fresh cut melon [5]. Apart from this, silver nanoparticles slower the ripening times of stored fruits by catalyzing the destruction of ethylene gas and increase the shelf lives of stored fruits [5]. Several studies have demonstrated the efficacy of silver nanoparticles loaded packaging materials in campaigning against microbial growth in foods [5-8]. Nanostructured antimicrobials have a higher surface area-to-volume ratio than their microscale counterpart and their incorporation in food packaging systems are supposed to be particularly

* Correspondence: josephselvinss@gmail.com
[3]Department of Microbiology, Pondicherry University, Puducherry 605014, India
Full list of author information is available at the end of the article

efficient in their activities against microbial cells [9]. The development of stable, mono dispersible, metallic silver nanostructures synthesis via reliable green synthesis has been an important aspect of current nanotechnology research. The aggregation of silver nanostructures and the insufficient stability of their dispersions lead to loss of their special nanoscale properties. Researchers employ polymer–assisted fabrication routes and various chemical stabilizing agents (surfactants such as CTAB, SDS etc., and polymers such as PVP) for preventing the self-aggregation of nanostructures [10-12]. The use of chemical compounds is toxic and will reduce the biological applicability. The use of natural products such as bio-surfactant, monosaccharides, plant extracts etc. as enhancers and stabilizing agent for silver nanostructures synthesis were extensively studied. The marine glycolipid biosurfactant stabilized silver nanoparticles were synthesized by *Brevibacterium casei* MSA19 under solid state fermentation using agro-industrial and industrial waste as substrate [13]. Apte et al. [14] studied L-DOPA mediated synthesis of melanin by fungi *Yarrowia lipolytica* and the induced melanin has been exploited in the synthesis of silver and gold nanostructures. In this study, rapid reliable approach has been developed to produce uniform silver nanostructures by purified melanin from marine *Nocardiopsis alba* MSA10.

As melanin pigments are used as food colorant and nutritional supplements, which reflects the industrial need to large scale production as natural ingredients. Natural pigment production especially from microorganisms is emerging as an important aspect due to their wide acceptance in various industrial sectors [15] and it replaces the chemically synthesized pigments which cause harmful effects in the natural environment [16]. The microbial pigment, melanin has received considerable attention because of their useful biological activities especially in food and pharmaceutical industries. Melanins are high molecular weight pigments that are produced in microorganisms by oxidative polymerization of phenolic or indolic compounds with free radical generating and scavenging activity [17]. Based on chemical structure, properties and species affiliation, melanins are classified as allo-, pheo-, and eumelanins. The black or brown eumelanins are produced by oxidation of tyrosine through tyrosinase to DOPA (o- dihydroxyphenylalanine) and dopachrome, further the cyclization mediates to form 5,6-dihydroxyindole (DHI) or 5,6-dihydroxyindole-2-carboxylic acid (DHICA) [18]. The yellow-red pheomelanins are synthesized like eumelanins in the first step; the intermediate DOPA undergoes cysteinylation, directly or mediated by glutathione to form various derivatives of benzothiazines [19]. The third types of allomelanins are heterogenous group of polymers synthesized via penta-ketide pathway [20]. Brown pigments may also produce

from L-tyrosine pathway via accumulation and autooxidation of intermediates of tyrosine catabolism [18]). Microbial melanin has a wide range of applications including photo-protective, radioprotective, immuno-modulating, anti-microbial and antitumour activities [21-23]. Actinobacteria were resilient bacteria found among culturable sponge microbes and are current focus on bioactive leads from marine environment [24]. The sponge associated actinomycetes has wide application as antiviral, antibacterial, anti-tumour, anti-helminthic, insecticidal, immuno-modulator, immuno-suppressant and food colorants [25]. Melanin producing microorganisms are ubiquitous in nature; however limited literature is available on actinobacterial melanin production at different cultural conditions. Therefore, this study aims to enhance the production of melanin from marine actinobacterium *N. alba* MSA10, by optimizing various cultural and environmental parameters under submerged conditions as well as melanin mediated synthesis of silver nanostructures.

## Results
### Screening and identification of melanin producers
The strain MSA10 was considered as potential melanin producers among the other isolates obtained from the sponge *Dendrilla nigra*. The MSA10 strain was Gram positive and mycelia appearance under phase contrast microscope, which produce white powdery colonies on the actinomycetes isolation agar. It showed positive results on indole, citrate utilization, urease and triple sugar ion tests and negative results in methyl red, Voges Proskauer and catalase tests. Based on the morphological, biochemical, phylogenetic analysis (UPGMA algorithm) and taxonomic affiliation (RDP-II), the isolate MSA10 was identified as *Nocardiopsis alba* MSA10. The 16S rRNA sequence was deposited in Genbank with an accession number EU563352. It was found that the isolate MSA10 showed clustering exclusively with pigment producing *Streptomyces* strain and also an efficient biosurfactant producer [26]. Melanin production by *N. alba* MSA10 was initiated at 72 h of incubation, the medium changed to light brown, further the color development was increased at 96, 120 and 144 h to light brown, brown and dark brown respectively. The melanin production was depending on the biomass yield and a highest yield of biomass with melanin was obtained at 144 h of incubation (Figure 1).

### Formulation of fermentation media for melanin production
It is evident that different media constituents such as carbon, nitrogen, metal ions, and organic solvents and environmental factors such as pH, temperature, and salinity are

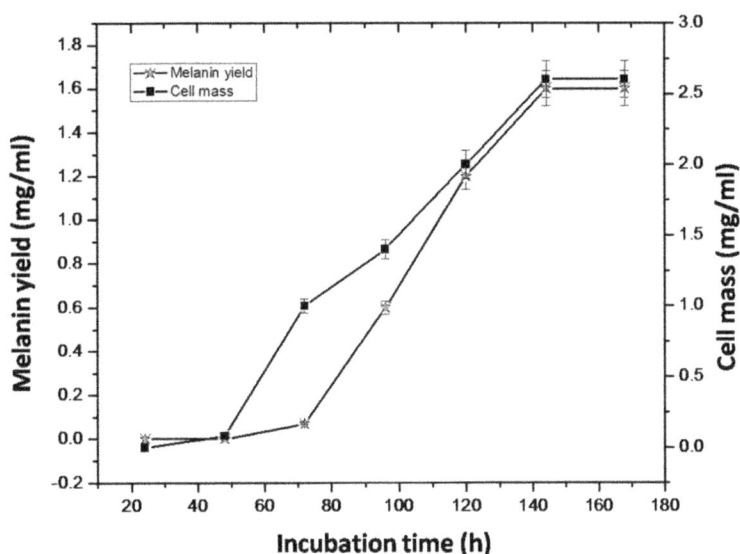

**Figure 1** Melanin production rate at different incubation time by *Nocardiopsis alba* MSA10.

known to play a vital role in the melanin production. The fermentation conditions and media constituents including sucrose, tyrosine, temperature and salinity as most significant variables were optimized for enhanced melanin yield. The correlation between melanin yield and the four critical control factors (variables) were analyzed by Box-Behnken design, the following quadratic model polynomial equation was obtained to explain melanin yield in mg/ml (Y).

$$Y = +3.40 - 0.092 * A - 0.042 * B - 0.13 * C - 0.15 \\ * D + 0.12 * A * B + 0.000 * A * C + 0.20 \\ * A * D - 0.100 * B * C - 0.10 * B * D - 0.30 \\ * C * D - 0.67 * A2 - 0.74 * B2 - 0.75 \\ * C2 - 0.53 * D2 \tag{1}$$

The statistical significance of the equation 1 was checked by F- test and the results of ANOVA are shown in Table 1. The model F value of 251.68 implies the model is more significant (<0.0001). The coefficient determination ($R^2$) value was found to be 0.9960, which implies that the variation of 99.60% for the melanin yield was attributed to the independent variables and only 0.40% of the total variation could not be explained by the model. The $R^2$ value found in this study was closer to 1 show that the developed model could effectively increase the melanin production (3.4 mg/ml).

The 3D response surface plots showed the effect of medium components and fermentation conditions on the production of melanin (Figure 2). The response surface curve was plotted with two factors varied at a time when the other two factors as being remained at a fixed level. Higher melanin yield (3.4 mg/ml) was obtained

with 12.5 g/l of tyrosine and 25 g/l of sucrose in the medium and maintaining the other parameters such as salinity (2.5%), pH (7.5) and temperature (35°C) as constant (Figure 2). When the pH was below 7.5 (6.0 – 6.5) and the temperature above 35°C, the growth of *N. alba* MSA10 as well as melanin production has declined drastically. The pigment production consequently increased with increasing the temperature up to 35°C, but the growth of *N. alba* MSA10 was found optimum at the

**Table 1 ANOVA for response surface quadratic model of melanin production**

| Source | Sum of squares | df | Mean square | F value | p-value prob > F |
|---|---|---|---|---|---|
| Model | 9.02 | 14 | 0.64 | 251.68 | < 0.0001** |
| A-Sucrose | 0.10 | 1 | 0.10 | 39.40 | < 0.0001** |
| B-Tyrosine | 0.021 | 1 | 0.021 | 8.14 | 0.0128* |
| C-Temperature | 0.21 | 1 | 0.21 | 83.35 | < 0.0001** |
| D-Salinity | 0.27 | 1 | 0.27 | 105.49 | < 0.0001** |
| AB | 0.063 | 1 | 0.063 | 24.42 | 0.0002* |
| AC | 0.000 | 1 | 0.000 | 0.000 | 1.0000 |
| AD | 0.16 | 1 | 0.16 | 62.51 | < 0.0001** |
| BC | 0.040 | 1 | 0.040 | 15.63 | 0.0014* |
| BD | 0.040 | 1 | 0.040 | 15.63 | 0.0014* |
| CD | 0.36 | 1 | 0.36 | 140.65 | < 0.0001** |
| A² | 2.88 | 1 | 2.88 | 1126.34 | < 0.0001** |
| B² | 3.57 | 1 | 3.57 | 1394.02 | < 0.0001** |
| C² | 3.69 | 1 | 3.69 | 1441.40 | < 0.0001** |
| D² | 1.82 | 1 | 1.82 | 709.64 | < 0.0001** |

**More significant, *Significant; R-Squared 0.9960, Adj R-Squared 0.9921.

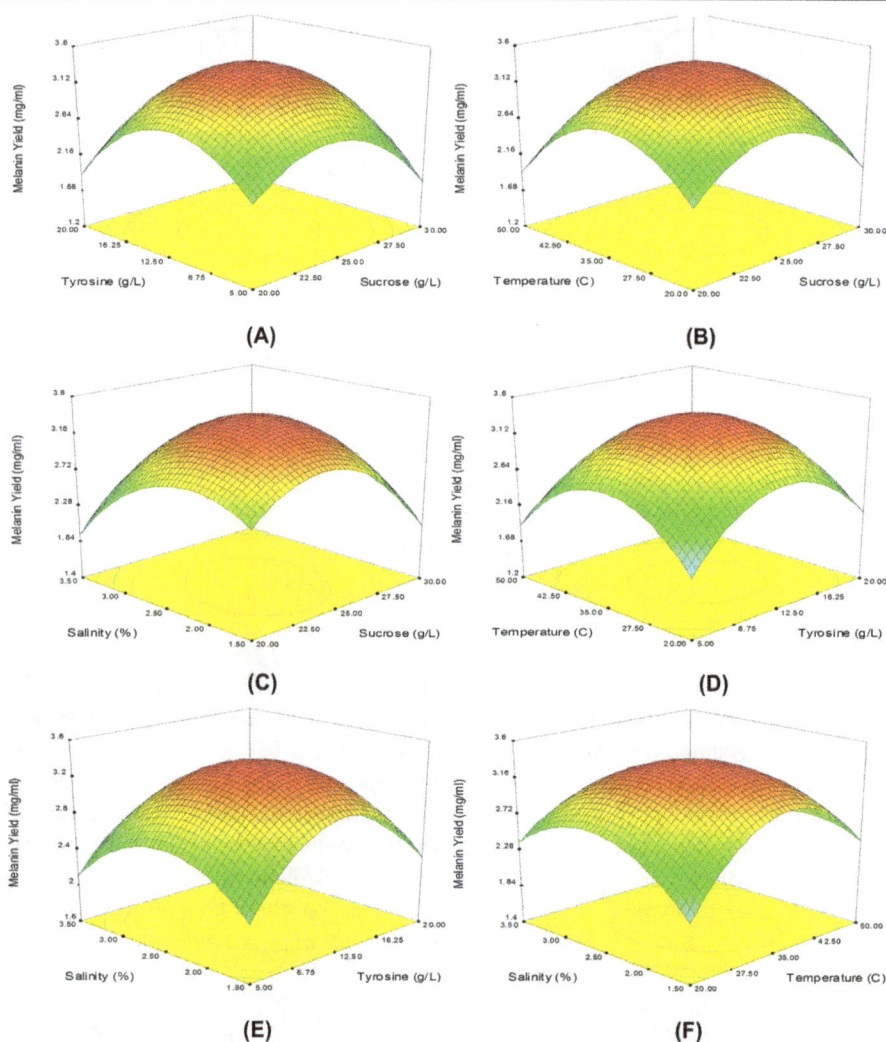

**Figure 2** The response surface plot shows the most significant variables interaction on melanin production. **(A)** tyrosine and sucrose, **(B)** temperature and sucrose, **(C)** salinity and sucrose, **(D)** temperature and tyrosine, **(E)** salinity and tyrosine, and **(F)** salinity and temperature.

temperature of 28 – 30°C. It was found that at pH 7.5, the growth of *N. alba* MSA10 and melanin production was found to be linear. This suggests that the near neutral pH was optimum for higher biomass and melanin production.

### Lights on melanin production

Light is considered as important environmental parameters for melanin production. Literature evidenced that pigments absorbed light at a particular wavelength and emits different colors. In this study, the various light sources such as green, red and yellow light on enhanced melanin production were investigated. It was found that the green light excitation had resulted in highest melanin production with the formation of dark brown color. Considerable pigment production was observed in red light and there is no pigment production in yellow light source in the culture plate, but slight production was

observed in the fermentation medium at 144 h of incubation (Figure 3).

### Characterization of melanin pigment

The chromatogram of violet color spot on TLC plate showed an $R_f$ value of 0.74 related to melanin pigment. A strong peak at 220 nm was obtained for UV- visible spectrum of *Nocardiopsis* melanin (data not shown). The colorimeter L* (lightness ranges from 0–100 (dark – light) a* (red- green) and b* (yellow-blue) values of melanin reflects the dark brown color. The L*, a* and b* values of melanin was found to be 2.74, 0.22 and 0.96 respectively. The low lightness value shows the dark color pigment. The values of a* (R-G) and b* (Y-B) represents the *Nocardiopsis* melanin pigment as a dark brown color. The FT-IR spectrum of column purified melanin showed the absorbance bands at 1118, 1385, 2077 cm$^{-1}$ and a strong band at 3397, 1638, 674 cm$^{-1}$. The intense broad band at

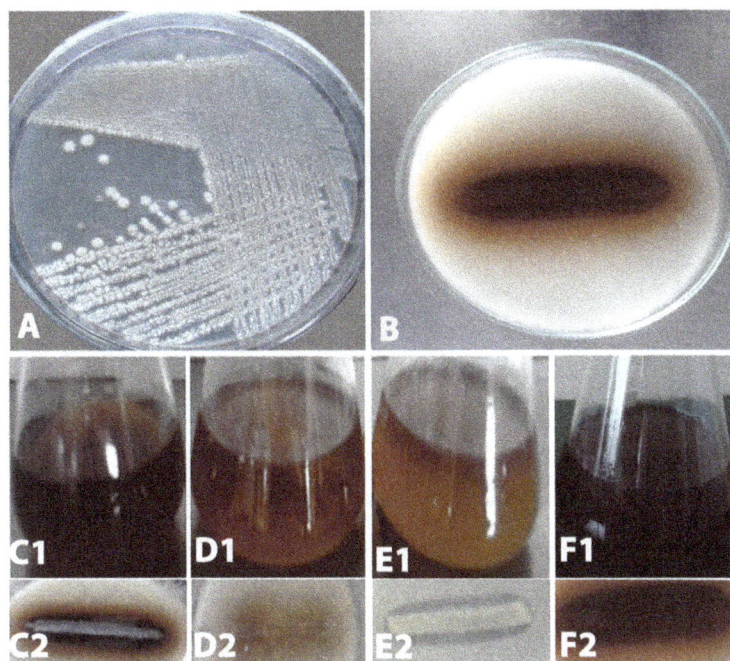

**Figure 3 Effects of light source on melanin production.** Growth of *N. alba* MSA 10 on Actinomycetes isolation agar medium **(A)**, Melanin production on tyrosine (1%) agar medium **(B)**, Green light source on melanin production in tyrosine broth **(C1)** and tyrosine agar medium **(C2)**, which produces dark brown pigment. Red light source produces brown pigment, tyrosine broth **(D1)** and tyrosine agar **(D2)**. Yellow light shows light brown pigment production on tyrosine broth **(E1)** at 144 h incubation, but no pigment production in tyrosine agar medium **(E2)**. The pigment production at normal light source on tyrosine broth **(F1)** and tyrosine agar medium **(F2)** produces dark brown pigment.

3397 cm$^{-1}$ corresponds to the OH groups of polymeric structure, the band at 1638 and 1118 associated with primary amine NH and primary amine CN stretch vibrations of melanin respectively. The band at 1385 cm$^{-1}$ is assigned to methylene scissoring of C-H groups and the band around 2077 arises from the carbonyl stretching vibrations. The TLC chromatogram and FT-IR spectrum analysis confirmed the melanin pigment produced by *N. alba* MSA10.

### Physico chemical properties of purified melanin

The *Nocardiopsis* melanin was found to be dissolving immediately in alkaline water and hexane when compared to water at room temperature. A precipitation was formed when the melanin was allowed to dissolve in ethanol, methanol and HCl. It remains insoluble in ether, chloroform and ethyl acetate. *Nocardiopsis* melanin was stable at the range of temperatures (20–100°C) even for 3 h (Figure 4) and light sources including UV, natural sun light and complete darkness. The stability of different pH (3–12) of melanin tested had showed slight variation of absorption spectrum scanned at 190–220 nm (data not shown). The strong peak at 215 was observed (peak value 3.9) in the alkaline pH (9, 10 and 12), which indicates the relative stability of melanin in alkaline conditions when compared to neutral and acidic conditions. Similar water solubility nature of melanin has been reported in a mutant strain of *Bacillus thuringiensis* [27]. The other physico chemical properties are similar to melanin obtained from *Osmanthus fragrans* seeds [28].

### Antioxidant activity and reducing power of melanin

The antioxidant assay is based on the reduction of Mo (VI) to Mo (V) by melanin with the formation of a green phospho molybdenum complex at different temperatures. Even though the green – complex formation takes place at room temperature, the formation of maximum phospho molybdenum complex increases with the increasing temperature of 90 and 180°C. Figure 5 shows the antioxidant property exhibited by *Nocardiopsis* melanin. Similar results were obtained with the melanin from berry of *Cinnamomum burmannii* and *Osmanthus fragrans* [29]. The reducing capabilities of *Nocardiopsis* melanin from $Fe^{3+}$ to $Fe^{2+}$ was clearly investigated with the standard BHT (Figure 6) and the results evidenced and validated the antioxidant property. Presence of antioxidant substances enhance the reduction of $Fe^{3+}$/Ferricyanide complex to the $Fe^{2+}$ form, which can be monitored at 700 nm [30].

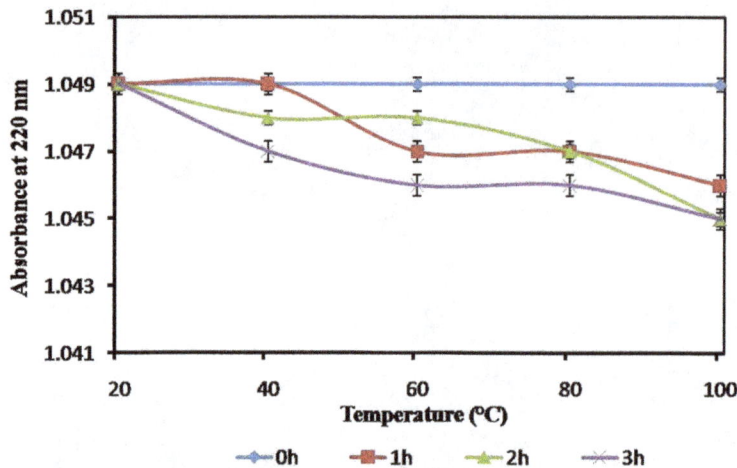

**Figure 4** Temperature stability of melanin produced by *Nocardiopsis alba* MSA 10 with different incubation times measured by UV absorbance at 220 nm (mean ± S.D., n = 3).

## Melanin mediated synthesis of silver nanostructures and antimicrobial assay

The synthesis of melanin mediated silver nanostructures was confirmed by the appearance of strong peaks in the UV- visible spectra at 420–460 nm. With the increase in temperature (100°C) stable and rapid synthesis of same sized particles takes place (Figures 7A and 8B). The synthesis pattern of UV- Visible spectrum at different temperature profile is depicted in Figure 7B. It is evident from the UV – absorbance spectrum that the temperature at 100°C shows effective synthesis. It is noticeable that the temperature stability of *Nocardiopsis* melanin tested before showed stability at 100°C over 3 h. The antioxidant and reductive capabilities of the melanin compound enhances the rapid synthesis of silver nanostructures without adding any capping agent. Thus, melanin acts as both reducing and capping agent of silver nano- sized structures

synthesis. The synthesis at various time interval shows that increasing incubation time at 30 min gives more stable particles when compared to 0, 10 and 20 min (Figure 7B). The FT-IR spectrum of melanin mediated silver nano-structures shows (Figure 9) characteristics absorbance bands of 3466, 3400, 2083, 1638, 1420, 1370, 1234, 1099 and 664 cm$^{-1}$ respectively. The shift in the bands at 1118 of pure *Nocardiopsis* melanin towards their lower frequency to 1099 is attributed to the binding of primary amine (N-H) to the silver ions. The shifting of symmetrical stretch of carboxylate group at 1385 to 1370 and 1234 clearly shows the reaction between silver particles to carboxylate group of melanin. The appearance of new band at 3466, shifting of band at 3397–3400 and 2077–2083 cm$^{-1}$ were suspected to cause heating of melanin with silver nitrate solution. The TEM results (Figure 8) showed that all synthesized particles were

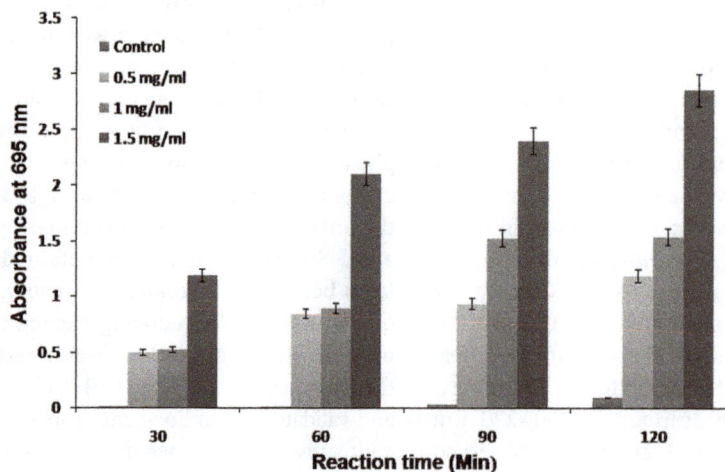

**Figure 5** Antioxidant activity in different concentration of *Nocardiopsis* melanin.

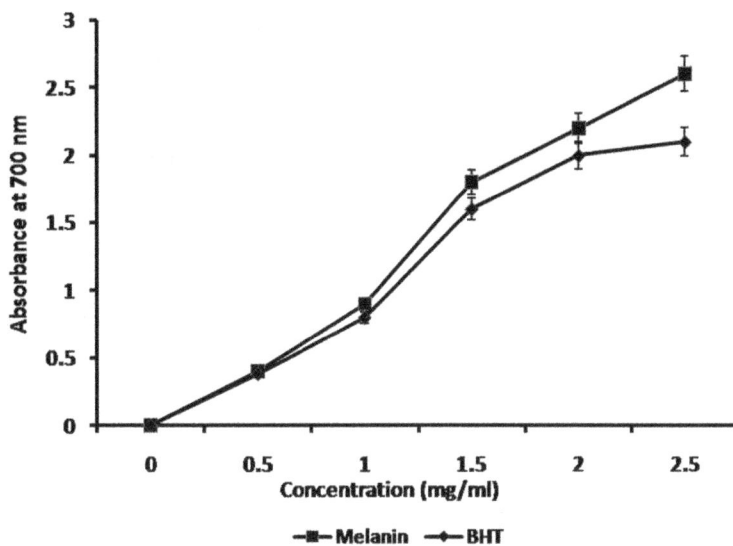

**Figure 6** Reductive capabilities in various concentrations of *Nocardiopsis* melanin and a standard BHT.

spherical in shape and found to be well dispersed in aqueous medium. The particle sizes ranging from 20 – 50 nm were formed. The melanin-silver nanoparticles showed (Figure 10) antimicrobial activity against all food pathogens tested but the highest activity was found against *B. cereus* (140 mm$^2$) and *P. fragi* and *E. coli* (120 mm$^2$ respectively).

## Discussion

Industrial production of colorants from microorganisms are more suitable due to factors such as ease of availability, culturing, higher production of pigments and the microbes' potential to be genetically manipulated. Isolation of new strain is still of particular interest because of the

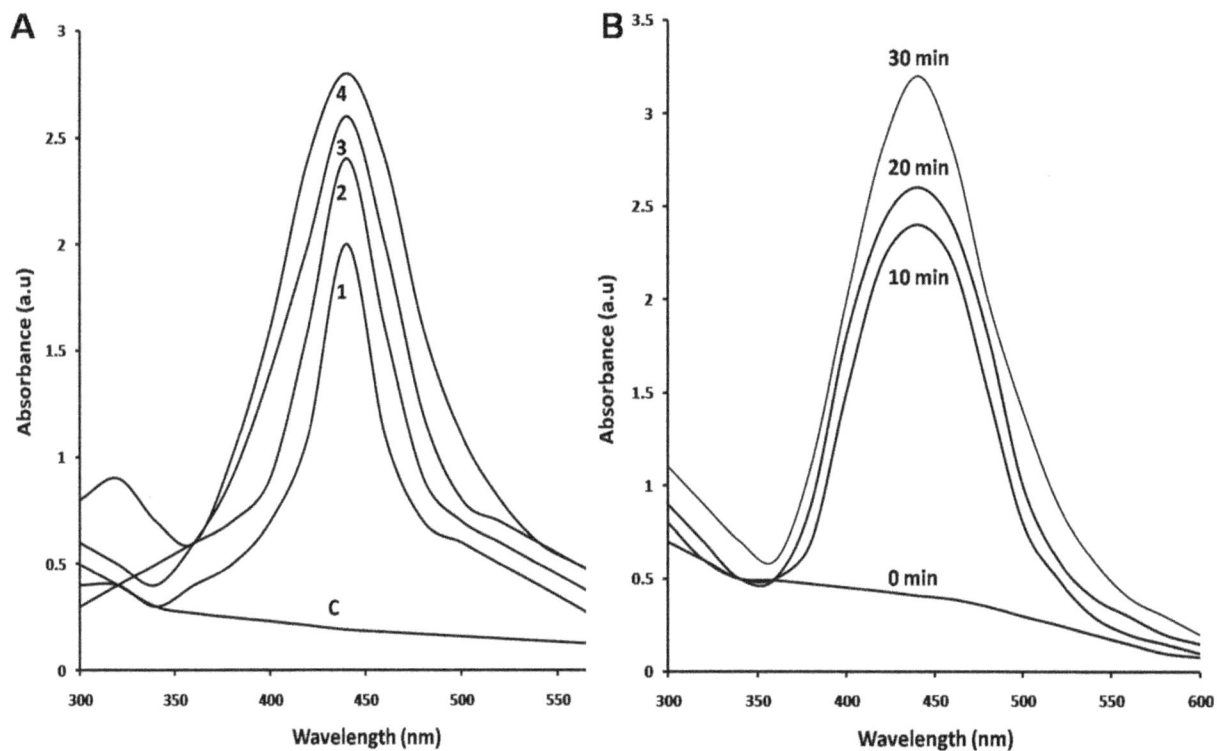

**Figure 7** UV absorption spectra of synthesized silver nanostructures. **(A)** At different temperatures. C - Control (AgNO$_3$ solution (1 mM), 1-40°C, 2-60°C, 3-80°C and 4-100°C **(B)** At different time intervals.

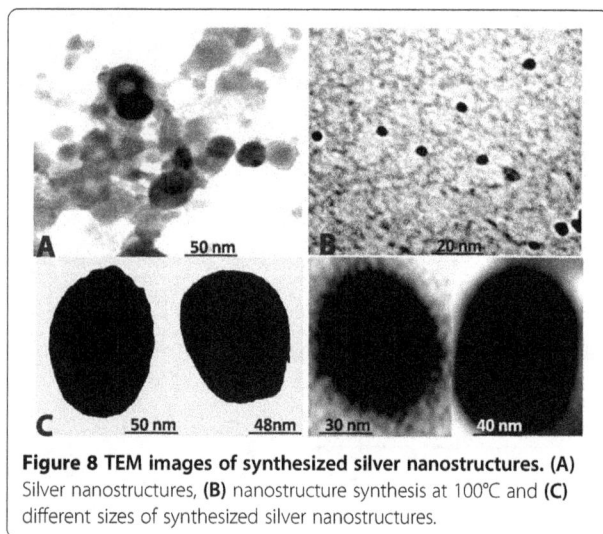

**Figure 8 TEM images of synthesized silver nanostructures. (A)** Silver nanostructures, **(B)** nanostructure synthesis at 100°C and **(C)** different sizes of synthesized silver nanostructures.

necessity to obtain microorganisms with suitable characteristics for submerged cultivation. Recently, sponge associated marine bacteria have been considered as a potential source of food- grade pigments [31].

The production of melanin by *N. alba* MSA10 was attributed to the supplement of tyrosine on the production medium via tyrosinase enzyme. The formation of dopachrome (red coloration) and the OD of 0.148 in tyrosinase assay were confirmed by tyrosinase activity of *N. alba* MSA10. The strain *N. alba* MSA10 utilized up to 12.5 g/l of tyrosine but on further addition up to 20 g/l, the melanin production rate gets declined. This shows that the strain *N. alba* MSA10 had produced melanin by the mediation of tyrosinase. According to Williams [32], about one third of the taxa of the genus *Streptomyces* produce

melanin. In strains including *Streptomyces antibioticus*, *S. glaucescens* and *S. lavendulae*, the tyrosinase gene for melanin production have been cloned, sequenced and recombinantly produced the protein which has sequence similarity to mammalian tyrosinase [33,34]. Melanin like pigments formed from L-tyrosine with different melanogenic pathway in *S. avermitilis* [35], *Xanthomonas campestris* [36], *Shewanella colwelliana* [37] and *Vibrio cholerae* [38] has been well deliberated.

Sucrose (25 g/l) as carbon source increased the melanin production up to 3.4 mg/ml significantly followed by glucose as alternative carbon source in *N. alba* MSA 10. Till date, there is no report on the production of melanin from sucrose as a sole carbon source. The red pigment produced by *Paecilomyces sinclarii* showed maximum mycelial growth in sucrose as carbon source, even though the highest pigment production has been attributed in soluble starch medium [39].

Stimulatory effects of various nitrogen sources including peptone, beef extract, yeast extract, urea and ammonium nitrate were tested by Placket-Burman experimental design, and the significant effect was found in case of beef extract and ammonium nitrate on melanin production. This shows that only trace amount of nitrogen source was utilized by *N. alba* MSA10 for melanin production, as the amino acid tyrosine mediates the melanin synthesis pathway. The strain MSA10 had utilized considerable level of nitrogen sources for their growth and mycelial development; however melanin production gets enhanced with the addition of tyrosine in the production medium. The strain grows optimum up to 3.5% of NaCl and the highest melanin production (3.4 mg/ml) has obtained at 2.5% of salinity. Further increasing salinity, the melanin

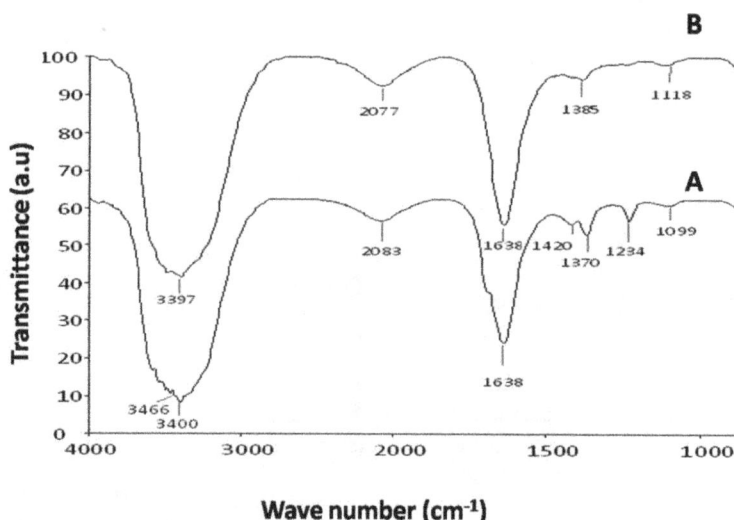

**Figure 9 FTIR spectra of synthesized silver nanostructures. (A)** Silver nanostructures, and **(B)** pure melanin.

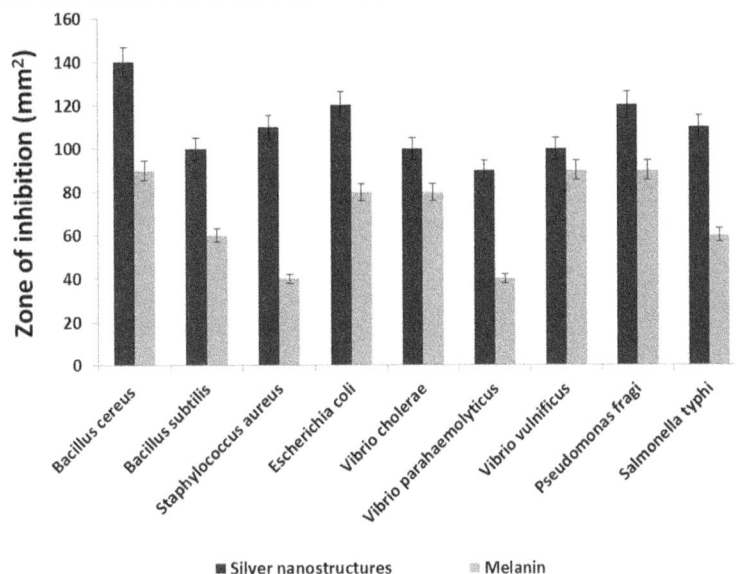

**Figure 10** Antimicrobial assay of melanin and melanin- mediated synthesis of silver nanostructures.

production was found to be decreased. Melanin production by *N. alba* MSA10 was highest at 35°C and pH 7.5. The highest yield of pigment from *Monascus* was reported at 30°C [40]. The initial pH at 6 and temperature of 32°C increased the pigment production by *Monascus* sp. [41].

The pigment production by *Monascus purpureus* with various light sources was well recognized by Babitha et al.[42] and this finding described that red light have little effect on growth and pigment production when compared to green and blue light sources which probably inhibits the pigment production, even though there is significant increase in biomass under green light. Despite the importance of influence of light on pigment production as investigated on *Monascus purpureus* [42], much has not yet been determined on actinomycetes melanin. Therefore, the strain *N. alba* MSA10 would be the first record among the actinomycetes produced melanin under illumination of the green light source.

The predicted melanin yield was found to be closer to actual melanin yield and the production rate was increased one fold over the wild strain *N. alba* MSA10. It reveals that the generated Box- Behnken design showed the interaction and actual relationships between the critical control factors. The RSM-based experiments showed that *N. alba* MSA10 has higher melanin (3.4 mg/ml) productivity potential.

The FT-IR absorbance band of *Nocardiopsis* melanin ranging from 3400 cm$^{-1}$ to 674 cm$^{-1}$ had high degree of similarity to the BC58 melanin, standard melanin sigma [43] and synthetic pyomelanin, pyomelanin extracted from *Aspergillus fumigatus* [44]. The shifting of band related to primary amine (N-H), carboxylate group and

C-O stretch vibration clearly evidenced that *Nocardiopsis* melanin reduces silver nitrate and at the same time it stabilizes the synthesized silver nanostructures. The stability of melanin mediated silver nanostructures were determined by synthesized particles which was allowed to stands for 3 months at room temperature. It was found that the color intensity of silver particles increased with aging and no aggregation was observed in duration of 3 months. The free amine or carboxylate group of proteins can bind with silver particles [45]. The interaction of melanin with metal ions, protein [46] and double stranded DNA [47] was extensively studied.

The melanin mediated silver nanostructures found to be most effective on food pathogens such as *B. cereus*, *P. fragi* and *E. coli*. Thereby, the incorporation of melanin mediated synthesized silver nanostructures in food packaging materials can effectively inhibit the growth of food pathogens and increase the shelf life of packed food products. Nanomaterials are being explored for their promising role in food industry such as providing longer shelf-life for foods, better barrier properties, improved heat resistance and temperature control, and antimicrobial and fungal protections [48]. Silver nanoparticles that act as antibacterial agents or nanoclay coatings are currently used in food packaging [49].

The future studies can be focused on the rapid formation of different shape of silver nanostructures under optimal conditions with melanin as reducing and capping agent. The size and shape based silver nanostructures has many positive attributes such as good conductivity, chemical stability, catalytic and antimicrobial activity that make them suitable for many practical food packaging applications. The melanin mediated silver nanostructures can be incorporated in food packaging materials. The efficacy of melanin-silver nano-conjugates on the shelf life of packed food products is needs to be investigated.

## Conclusion

The melanin pigment has been successfully purified and characterized from *N. alba* MSA10. The cultural conditions and environmental factors for enhanced yield of melanin were optimized through RSM- Box- Behnken design. The purified melanin has been used to synthesize and stabilizes the silver nanostructures *in vitro*. The antioxidant activity, reducing power and physicochemical properties of *Nocardiopsis* melanin was well characterized. The antioxidant, antimicrobial and natural coloring potential of *Nocardiopsis* melanin can be used as food additives, which significantly reduces the usage of artificial or synthetic colorants and antioxidants. The UV protective roles, withstanding higher temperatures, stability in alkaline conditions and water solubility nature of *Nocardiopsis* melanin increased their application in food, cosmetics and biomedical industries. Thus, the synthesis and stabilization of silver nanostructures by *Nocardiopsis* melanin demonstrates the metal interacting nature of pigment. Furthermore the antibacterial properties against food pathogens would facilitate its applicability in food processing and food packaging industries.

## Methods
### Isolation, screening and identification of melanin producing marine actinobacterium

The marine actinobacteria were isolated from marine sponge *Dendrilla nigra* as described by Selvin et al. [50]. The isolated actinobacteria were screened for melanin production on tyrosine agar medium (g/l): (peptone 5 g, sodium chloride 20 g, Beef extract 1.5 g, yeast extract 1.5 g, tyrosine 10 g, agar 20 g and pH 7.3) and were incubated at 30–35°C for 6–7 days. The pigment production was confirmed by the formation of a brownish color around the colonies. The melanin producer strains were identified based on morphological, biochemical and phylogenetic analysis [50].

### Fermentation by shake flask culture

The melanin production was carried out in five sets of 250 ml Erlenmeyer flasks under shake flask culture containing 100 ml of tyrosine medium. The culture flasks were incubated at 35°C for 7 days on a rotary shaker (Oasis) at 200 rpm. Samples were removed after initial color change periodically for biomass and melanin yield determination. The biomass yield was estimated by washing the cells with phosphate buffered saline (g/l) (NaCl 8 g, KCl 0.20 g, $Na_2HPO_4$ 1.44 g, $KH_2PO_4$ 0.24 g,

pH 7.4) and dried at 50–60°C for 2 h. The melanin supernatant was first adjusted to pH 9 with 10 N NaOH to ensure polymerization and then adjusted to pH 3 with 5 N HCl to precipitate melanin. The precipitated melanin was centrifuged at 10,000 rpm for 15 min (Eppendorf), washed thrice with deionized water and lyophilized for dry weight determination.

### Formulation of fermentation media for melanin production

To formulate the media with various concentrations of media constituents on melanin production by MSA10, different carbon, nitrogen sources, metal ions and organic solvents were used. The carbon sources used in this study include 20 g/L of glucose, dextrose, sucrose, mannitol and galactose. The organic nitrogen sources include 15 g/L of peptone, yeast extract, beef extract, and inorganic nitrogen sources of urea and ammonium nitrate are at the concentration of 100 mg/L. The pH of the melanin pigment production was studied using shake flask cultures at different initial values of pH (4–10). The effect of temperature on pigment production was determined with different incubation temperatures (25–60°C). The NaCl requirement for pigment production was optimized with 0.5 to 3.5% NaCl supplementation. Different metal ions such as $CuSO_4$, $FeSO_4$, $MgSO_4$, $MnCl_2$ and $MnSO_4$ were added in tyrosine broth at 100 mg/L concentration to determine the effect of metal ions.

### Experimental design and statistical analysis

For all experiments, fermentation was conducted in 500 ml of Erlenmeyer flasks containing different media constituents on melanin production. All experiments were carried out in triplicate and the final melanin yield was taken as the response (y). The Box-Behnken experimental design with four variables (A, B, C and D) such as sucrose, tyrosine, temperature and salinity respectively and three levels high (+), middle (0), and a low (−) was employed to optimize the fermentation conditions and thereby to obtain maximum melanin yield. The experimental design with four variables is summarized in Table 2. Based on Placket-Burman experimental design, the most significant variables sucrose, tyrosine, temperature and salinity were identified from the 11 variables analyzed such as glucose, sucrose, yeast extract, mannitol, tyrosine, ammonium nitrate, ferrous sulphate, pH, temperature, salinity and inoculums size (data not shown). The experimental data was analyzed using the software Design expert 8.0.4.1 trial version (Stat-Ease, Inc, USA).

### Light source on melanin production

The effect of light on melanin production by MSA 10 was studied by passing different wavelengths of light, red (620 – 750 nm), blue (450–475 nm), green (495–570 nm) on

**Table 2 Box - Behnken experimental design with four independent variables (coded values) and the melanin production rate**

| Run | A | B | C | D | Melanin yield (mg/ml) | |
|---|---|---|---|---|---|---|
| | | | | | Actual value | Predicted value |
| 1 | 0 | −1 | 1 | 0 | 1.90 | 1.94 |
| 2 | 1 | 0 | 0 | −1 | 2.00 | 1.98 |
| 3 | −1 | 0 | 0 | 1 | 2.00 | 2.06 |
| 4 | 0 | 0 | 1 | −1 | 2.40 | 2.34 |
| 5 | 0 | 1 | −1 | 0 | 2.10 | 2.12 |
| 6 | 1 | −1 | 0 | 0 | 1.80 | 1.82 |
| 7 | −1 | −1 | 0 | 0 | 2.20 | 2.25 |
| 8 | 0 | −1 | −1 | 0 | 2.00 | 1.98 |
| 9 | −1 | 1 | 0 | 0 | 1.90 | 1.91 |
| 10 | 0 | 1 | 0 | −1 | 2.40 | 2.43 |
| 11 | 0 | 1 | 0 | 1 | 1.80 | 2.40 |
| 12 | −1 | 0 | −1 | 0 | 2.20 | 2.16 |
| 13 | 1 | 1 | 0 | 0 | 2.00 | 1.96 |
| 14 | 0 | 0 | 0 | 0 | 3.40 | 3.40 |
| 15 | 0 | 0 | 1 | 1 | 1.50 | 1.53 |
| 16 | −1 | 0 | 1 | 0 | 2.00 | 1.95 |
| 17 | 0 | 0 | −1 | 1 | 2.40 | 2.43 |
| 18 | 0 | −1 | 0 | 1 | 2.10 | 2.14 |
| 19 | 0 | 1 | 1 | 0 | 1.60 | 1.63 |
| 20 | −1 | 0 | 0 | −1 | 2.60 | 2.65 |
| 21 | 0 | 0 | 0 | 0 | 3.40 | 3.40 |
| 22 | 0 | 0 | 0 | 0 | 3.40 | 3.40 |
| 23 | 0 | 0 | 0 | 0 | 3.40 | 3.40 |
| 24 | 1 | 0 | 0 | 1 | 2.20 | 2.16 |
| 25 | 0 | −1 | 0 | −1 | 2.30 | 2.22 |
| 26 | 0 | 0 | 0 | 0 | 3.40 | 3.40 |
| 27 | 0 | 0 | −1 | −1 | 2.10 | 2.10 |
| 28 | 1 | 0 | −1 | 0 | 2.00 | 2.02 |
| 29 | 1 | 0 | 1 | 0 | 1.80 | 1.75 |

A: Sucrose −1(20 g/l), 0 (25 g/l), +1(30 g/l); B: Tyrosine, -1(5 g/l), 0 (12.5 g/l), +1 (20 g/l); C: Temperature, -1(20°C), 0(35°C), +1(50°C); D: Salinity, -1(1.5%), 0(2.5%), +1(3.5%).

fermentation medium. The culture flasks were exposed to the light intensity of 32 W m$^{-2}$ for 7 days.

## Assay for tyrosinase activity

Tyrosinase activity was assessed by growing the MSA10 isolates in to glutamate medium [51] and 2 ml of culture supernatant mixed with 2 ml of 0.1 M phosphate buffer (pH 5.9), finally 1 ml DOPA was (10 mM) added. The reaction mixture was incubated at 37°C for 5 min. Red coloration resulting from dopachrome formation was observed and read spectrophotometrically at 475 nm (PG Instruments).

## Characterization of melanin pigment

The cell free supernatant was collected from fermented broth by centrifugation at 10,000 g for 15 min (Eppendorf 5804 R). The supernatant was filtered through Whatman No.1 filter paper to remove residue cell debris. The initial purification of melanin was performed according to Wan et al. [20]. Briefly, the melanin supernatant was first adjusted to pH 9 with 10 N NaOH to ensure polymerization and then adjusted to pH 3 with 5 N HCl to precipitate melanin. The precipitated melanin was centrifuged, washed thrice with deionized water and lyophilized for further use. The absorbance spectrum of melanin produced by MSA10 was measured with UV/VIS Spectrophotometer (PG instruments) over a range of wavelengths from 190 to 500 nm. The color intensity of melanin was measured by CR- 300 colorimeter with the HunterLab color system. The L* (lightness ranges from 0–100 (dark – light)), a* (red- green) and b* (yellow-blue) values were determined. The lyophilized melanin pigment was spotted on the TLC plate and the chromatogram was performed with the solvent system n-butanol: acetic acid: water (70:20:10). After drying, the pigment spot was sprayed with ninhydrin. The TLC purified pigment were applied to a column of DEAE-Cellulose (Bio-Rad, 1 × 30 cm) that had been equilibrated with 25 mM Tris–HCl buffer (pH 8.6) containing 50 mM sodium chloride. The column was eluted at a flow rate of 100 ml/h with 1:1 volume gradient from 0.1 M to 2 M NaCl in the same buffer.

## Physico-chemical properties of the melanin

The physico – chemical properties of *Nocardiopsis* melanin was analyzed according to Wang et al. [21]. The solubility of purified melanin was checked by adding 0.05 g of the melanin in 10 ml of water, aqueous acid, alkali (such as $Na_2CO_3$, NaOH solution), and organic solvents such as chloroform, ethyl acetate, ethanol, methanol, acetic acid, petroleum ether, hexane with stirring at 25°C for 1 h, then filtered and the absorption of the solutions were recorded spectrophotometrically at 220 nm. The temperature stability of melanin pigment was measured after treatment with various temperatures in a thermostatically controlled water bath at 20, 40, 60, 80 and 100°C for 3 h and subsequently the absorption of the solutions were recorded at 220 nm. Light stability of melanin was detected by holding the melanin solution (5 mg/ml) under natural light, at dark place and under the Ultraviolet-light far from 30 cm for two days and every 12 h interval the maximum absorbance was measured at 220 nm. The pH stability was assessed by adjusting the melanin solution (5 mg/ml) in to a varied pH range (3, 4, 6, 7,9,10 and 12) with 0.5 N NaOH and HCl. All the samples were held for 30 min at 25°C, and the absorption spectrum (190–220 nm) was scanned.

## Determination of antioxidant activity and reducing power of melanin

The antioxidant activity of *Nocardiopsis* melanin was determined by a standard spectroscopic method [52]. Briefly, Aliquots of 2 ml of different concentration of melanin solution (0.5, 1 and 1.5 mg/ml) prepared in phosphate buffer (0.2 M, pH 6.6) and mixed with 2 ml of reagent solution (0.6 M sulfuric acid, 28 mM sodium phosphate, and 4 mM ammonium molybdate). The tubes were capped and incubated in a thermal block at 95°C for 120 min. Every 30 min the absorbance of the mixture was measured at 695 nm against a blank.

The reducing power of the melanin pigment was determined by standard method [53]. Briefly, different concentrations of melanin were mixed with phosphate buffer (2.5 ml, 0.2 M, pH 6.6) and potassium ferricyanide [$K_3Fe(CN)_6$] (2.5 ml, 1%). The mixture was incubated at 50°C for 20 min. 2.5 ml of TCA (10%) was added to the mixture, which was then centrifuged at 3000 rpm for 10 min. The supernatant (1.0 ml) was mixed with distilled water (7.0 ml) and $FeCl_3$ (0.5 ml, 0.1%), and the absorbance was measured at 700 nm. The Butylated hydroxytoluene (BHT in ethanol solution) was used as the standard and the obtained value was used to compare and interpret the result with melanin.

## Synthesis of melanin mediated nanostructures by boiling method

Silver nanostructures were synthesized *in vitro* by adding 10 ml purified melanin solution (20 µg/ml) to 40 ml of 1 mM $AgNO_3$ (Sigma) and vigorously stirred for 5 minutes. The mixture was incubated at 60°C for 30 min. Both melanin and $AgNO_3$ was maintained separately as control. Silver nanostructures synthesis at different temperature range from 40–100°C and different time intervals (0- 30 min) were studied at 1 mM $AgNO_3$. Then the nanostructures were characterized by UV–vis spectrophotometer (PG instruments), FT-IR spectrum (Spectrum RX1) and TEM analysis. TEM measurements were performed on a TECHNAI 10 PHILIPS model instrument operating at an accelerating voltage of 80 kV.

## Antimicrobial assay of melanin and silver nanostructures against food pathogens

The silver nanostructures and the column purified melanin compound were tested for antimicrobial activity using well diffusion method and the area of the halo was measured [54]. The synthesized nanostructures were tested against common food pathogens such as *Bacillus subtilis* (MTCC 1305), *Bacillus cereus* (MTCC 1307), *Staphylococcus aureus* (MTCC 2940), *Escherichia coli* (MTCC 739), *Vibrio cholerae* (MTCC 3906), *Vibrio parahaemolyticus* (MTCC 451), *Vibrio vulnificus* (MTCC 1145), *Pseudomonas fragi* (MTCC 2458) and *Salmonella*

*typhi* (MTCC 734). These were cultured on Muller Hinton agar (Himedia). Well was made with a sterile steel cork borer (1 cm diameter) and 50 µl of purified melanin and silver nanostructures were added in the wells, incubated at 30°C for 24 h. After incubation the clear halo was measured and the area of inhibition in $mm^2$ was calculated.

**Abbreviations**
MSA: Marine sponge associated actinobacteria; UPGMA algorithm: Unweighted pair group method with arithmetic mean algorithm; RDP-II: Ribosomal database project; ANOVA: Analysis of variance; TLC: Thin layer chromatography; UV- visible spectrum: Ultra violet – visible spectrum; FT-IR: Fourier transform infrared; BHT: Butylated hydroxytoluene; TEM: Transmission electron microscopy; OD: Optical density; RSM: Response surface methodology.

**Competing interests**
The authors declare that they have no competing interests.

**Authors' contributions**
DA and ANL performed the experiments and GSK wrote the manuscript. MVA and NAA set the rationale of the experiments. JS designed the study and helped for preparing the manuscript. All authors read and approved the final manuscript.

**Acknowledgements**
ANL is thankful to DBT for Junior Research Fellowship. GSK thankful to DST for Young Scientist Fellowship. DA is thankful to DST for INSPIRE fellowship. JS is thankful to DBT for research grant. (No BT/PR14678/AAQ/03/538/2010). Thank Addiriyah Chair for Environmental Studies, Department of Botany and Microbiology, College of Science, King Saud University, Saudi Arabia for their support.

**Author details**
[1]Department of Food Science and Technology, Pondicherry University, Puducherry 605014, India. [2]Department of Microbiology, Bharathidasan University, Tiruchirappalli 620 024, India. [3]Department of Microbiology, Pondicherry University, Puducherry 605014, India. [4]Department of Botany and Microbiology, Addiriyah Chair for Environmental Studies, College of Science, King Saud University, P. O. Box 2455, Riyadh 11451, Saudi Arabia.

**References**
1. Kampman Y, de Clerck E, Kohn S, Patchala DK, Langerok R, Kreyenschmidt J: Study of the antimicrobial effect of silver-containing inner liners in refrigerators. *Appl Microbiol* 2008, **104**:1808–1814.
2. Quintavalla S, Vicini L: Antimicrobial food packaging in meat industry. *Meat Sci* 2002, **62**:373.
3. Appendini P, Hotchkiss JH: Review of antimicrobial food packaging. *Innov Food Sci Emerg Technol* 2002, **3**:113–126.
4. U.S. Food and Drug Administration: *FDA approved Food Contact Substances.* http://www.fda.gov.
5. Fernandez A, Picouet P, Lloret E: Cellulose-silver nanoparticle hybrid materials to control spoilage-related microflora in absorbent pads located in trays of fresh-cut melon. *Int J Food Microbiol* 2010, **142**:222–228.
6. Fayaz AM, Balaji K, Girilal M, Kalaichelvant PT, Venkatesan R: Mycobased synthesis of silver nanoparticles and their incorporation into sodium alginate films for vegetable and fruit preservation. *J Agric Food Chem* 2009, **57**:6246–6252.
7. Emamifar A, Kadivar M, Shahedi M, Soleimanian ZS: Evaluation of nanocomposite packaging containing Ag and ZnO on shelf life of fresh orange juice. *Innov Food Sci Emerg Technol* 2010, **11**:742–748.
8. Zhou L, Lv S, He G, He Q, Shi B: Effect of PE/Ag2O nano-packaging on the quality of apple slices. *J Food Qual* 2011, **34**:171–176.
9. de Azeredo HMC: Antimicrobial nanostructures in food packaging. *Trends Food Sci Tech* 2013, **30**:56–69.
10. Rozenberg BA, Tenne R: Polymer-assisted fabrication of nanoparticles and nanocomposites. *Prog Polym Sci* 2008, **33**:40–112.

11. Bajpai SK, Mohan YM, Bajpai M, Tankhiwale R, Thomas V: Synthesis of polymer stabilized silver and gold nanostructures. *J Nanosci Nanotechnol* 2007, **7**:2994–3010.

12. Zhang W, Qiao X, Chen J, Wang H: Preparation of silver nanoparticles in water-in-oil AOT reverse micelles. *J Colloid Interface Sci* 2006, **302**:370–373.

13. Kiran G, Sabu A, Selvin J: Synthesis of silver nanoparticles by glycolipid biosurfactant produced from marine *Brevibacterium casei* MSA19. *J Biotech* 2010, **148**:221–225.

14. Apte M, Girme G, Bankar A, RaviKumar A, Zinjarde S: 3, 4-dihydroxy-L-phenylalanine-derived melanin from *Yarrowia lipolytica* mediates the synthesis of silver and gold nanostructures. *J Nanobiotechnol* 2013, **11**:2.

15. Kim JK, Park SM, Lee SJ: Novel antimutagenic pigment produced by *Bacillus licheniformis* SSA3. *J Microbiol Biotechnol* 1995, **5**:48–50.

16. Unagul P, Wongsa P, Kittakoop P, Intamas S, Srikiti-Kulchai P, Tanticharoen M: Production of red pigments by the insect pathogenic fungus *Cordyceps unilateralis* BCC 1869. *J Ind Microbiol Biotechnol* 2005, **32**:135–140.

17. Riley PA: Melanin. *Int J Biochem Cell Biol* 1997, **29**:1235–1239.

18. Langfelder K, Streibel M, Jahn B, Haase G, Brakhage AA: Biosynthesis of fungal melanins and their importance for human pathogenic fungi. *Fungal Genet Biol* 2003, **38**:143–158.

19. Nappi A, Ottaviani E: Cytotoxicity and cytotoxic molecules in invertebrates. *Bio Essays* 2000, **22**:469–480.

20. Jacobson ES: Pathogenic roles for fungal melanins. *Clin Microbiol Rev* 2000, **13**:708–717.

21. Wan X, Liu HM, Liao Y, Su Y, Geng J, Yang MY, Chen XD, Shen P: Isolation of a novel strain of *Aeromonas media* producing high levels of dopa-melanin and assessment of the photoprotective role of the melanin in bio-insecticide applications. *J Appl Microbiol* 2007, **103**:2533–2541.

22. Wang Y, Casadevall A: Decreased susceptibility of melanized *Cryptococcus neoformans* to the fungicidal effects of ultraviolet light. *Appl Environ Microbiol* 1994, **60**:3864–3866.

23. Nappi AJ, Christensen MG: Melanogenesis and associated cytotoxic reactions: applications to insect innate immunity. *Insect Biochem Mol Biol* 2005, **35**:443–459.

24. Selvin J: Exploring the antagonistic producer *Streptomyces* MSI051: Implications of polyketide synthase gene type II and a ubiquitous defense enzyme phospholipase A2 in host sponge *Dendrilla nigra*. *Curr Microbiol* 2009, **58**:459–463.

25. Lam KS: Discovery of novel metabolites from marine Actinomycetes. *Curr Op Microbiol* 2006, **9**:245–251.

26. Gandhimathi R, Kiran GS, Hema TA, Selvin J, Rajeetha Raviji T, Shanmughapriya S: Production and characterization of lipopeptide biosurfactant by a sponge-associated marine actinomycetes *Nocardiopsis alba* MSA10. *Bioprocess Biosyst Eng* 2009, **32**:825–835.

27. Aghajanyan A, Hambardzumyan A, Hovsepyan A, Asaturian R, Vardanyan A, Saghiyan A: Isolation, purification and physicochemical characterization of water soluble *Bacillus thuringiensis* melanin. *Pigment Cell research* 2005, **18**:130–135.

28. Wang HS, Pan YM, Tang XJ, Huang ZQ: Isolation and characterization of melanin from Osmanthus fragrans' seeds. *LWT-Food Sci Technol* 2006, **39**:496–502.

29. Huang S, Pan Y, Gan D, Ouyang X, Tang S, Ekunwe SIN, Wang H: Antioxidant activities and UV-protective properties of melanin from the berry of *Cinnamomum burmannii* and *Osmanthus fragrans*. *Med Chem Res* 2011, **20**:475–481.

30. Chung YC, Chang CT, Chao WW, Lin CF, Chou ST: Antioxidative activity and safety of the 50% ethanolic extract from red bean fermented by *Bacillus subtilis*. IMR-NK1. *J Agric Food Chem* 2002, **50**:2454–2458.

31. Dharmaraj S, Ashokkumar B, Dhevendaran K: Food-grade pigments from *Streptomyces* sp. isolated from the marine sponge *Callyspongia diffusa*. *Food Research International* 2009, **42**:487–492.

32. Williams ST, Goodfellow M, Alderson G, Wellington EMH, Sneath PHA, Sackin MJ: Numerical classification of Streptomyces and related genera. *J Gen Microbiol* 1983, **129**:1743–1813.

33. Huber M, Hintermann G, Lerch K: Primary structure of tyrosinase from *Streptomyces glaucescens*. *Biochem* 1985, **24**:6038–6044.

34. Kawamoto S, Nakamura M, Yashima S: Cloning, sequence and expression of the tyrosinase gene from *Streptomyces lavendulae* MA406 A-1. *J Ferment Bioeng* 1993, **76**:345–355.

35. Denova CD, Skinner DD, Morgenstern MR: A *Streptomyces avermitilis* gene encoding 4-hydroxyphenylpyruvic acid dioxygenase-like protein that directs the production of homogentisic acid and orchronotic pigment in *Escherichia coli*. *J Bacteriol* 1994, **176**:5312–5319.

36. Goodwin PH, Sopher CR: Brown pigmentation of *Xanthomonas campestris* pv *phaseoli* associated with homogentisic acid. *J Microbiol* 1994, **40**:28–34.

37. Fuqua WC, Weiner RM: The melA gene is essential for melanin biosynthesis in the marine bacterium *Shewanella colwelliana*. *J Gen Microbiol* 1993, **139**:1105–1114.

38. Ruzafa C, Sanchezamat A, Solano F: Characterization of the melanogenic system in *Vibrio cholerae* ATCC 14035. *Pigment Cell Res* 1995, **8**:147–152.

39. Cho YJ, Park JP, Hwang HJ, Kim SW, Choi JW, Yun JW: Production of red pigment by submerged culture of *Paecilomyces sinclairii*. *Lett Appl Microbiol* 2002, **35**:195–202.

40. Pongrawee N, Saisamorn L: Improving solid- state fermentation of *Monascus purpureus* on agricultural products for pigment production. *Food Bioprocess Technol* 2009, **4**:1384–1390.

41. Lin CF, Suen SJT: Isolation of hyper pigment productive mutants of *Monascus* sp. F-2. *J Ferment Tech* 1973, **51**:757–759.

42. Babitha S, Julio C, Scowl CR, Pandey A: Effect of light on growth, pigment production and culture morphology of *Monascus purpureus* in solid - state fermentation. *Word J Microbiol Biotehnol* 2008, **24**:2671–2675.

43. Zhang J, Cai J, Deng Y, Chen Y, Ren G: Characterization of melanin produced by a wild-type strain of *Bacillus cereus*. *Front Biol* 2007, **2**:26–29.

44. Schmaler-Ripcke J, Sugareva V, Gebhardt P, Winkler R, Kniemeyer O, Heinekamp T, Brakhage AA: Production of pyomelanin, a second type of melanin, via the tyrosine degradation pathway in *Aspergillus fumigatus*. *Appl Environ Microbiol* 2009, **75**:493–503.

45. Huang J, Li Q, Sun D, Lu Y, Su Y, Yang X, Wanh H, Wang Y, Shao W, He N, Hong J, Chen C: Biosynthesis of silver and gold nanoparticles by novel sun dried *Cinnamomum canphora* leaf. *Nanotech* 2007, **18**:1–11.

46. Bowness JM, Morton RA: The association of zinc and other metals with melanin and a melanin-protein complex. *J Biochem* 1953, **53**:620–626.

47. Geng J, Yuan P, Shao C, Yu SB, Zhou B, Zhou P, Chen XD: Bacterial melanin interacts with double-stranded DNA with high affinity and may inhibit cell metabolism in vivo. *Arch Microbiol* 2010, **192**:321–329.

48. Karpilov A: *Nanomaterials in food packaging: promise and potential peril.* http://www.iopp.org/files/public/RITkarpilovIPTAsubmission.pdf.

49. *Nanomaterials in food packaging: keeping food fresh for longer, less waste.* European Commission's Joint Research Centre, Institute for Health and Consumer Protection. http://ihcp.jrc.ec.europa.eu/our_activities/nanotechnology/Nano_Food.

50. Selvin J, Thangavelu T, Kiran GS, Gandhimathi R, Shanmughapriya S: Culturable heterotrophic bacteria from the marine sponge Dendrilla nigra: isolation and phylogenetic diversity of actinobacteria. *Helgoland Mar Res* 2009, **63**:239–247.

51. Arai T, Mikami Y: Chromogenecity of streptomyces. *Appl Microbiol* 1972, **23**:402–406.

52. Prieto P, Pineda M, Miguel A: Spectrophotometric quantitation of antioxidant capacity through the formation of a phosphomolybdenum complex: specific application to the determination of vitamin E1. *Anal Biochem* 1999, **269**:337–341.

53. Oyaizu M: Studies on product of browning reaction prepared from glucose amine. *Jpn J Nutr* 1986, **44**:307–315.

54. Cappuccino JG, Sherman N: *Microbiology - a laboratory manual*. 4th edition. Harlow, England: Addison Wesley Longman, Inc; 1999:199–204.

# Investigation of antibacterial effect of Cadmium Oxide nanoparticles on *Staphylococcus Aureus* bacteria

Bahareh Salehi[1*], Sedigheh Mehrabian[2] and Mehdi Ahmadi[3]

## Abstract

**Background:** Inorganic antibacterial factors provide high bacterial resistance and thermal stability. Inorganic nanomaterial consists of modern formulation, biological, chemical, and physical properties produced on the basis of their function and influenced by their nano scales, the reason for which they have become very popular. The antibacterial effect of Cadmium Oxide Nanoparticles on *Staphylococcus Aureus* has been studied for the first time in this research because of their resistance to antibiotics.

**Materials and methods:** Different concentrations consist of 10 *µg/ml*, 15 *µg/ml*, and 20 *µg/ml* have been provided and their effects were studied in the agar and broth against the foregoing bacteria. Needless to say, the optimization of their non-microbial effect in variable times, pH, and temperatures of exposure was analyzed.

**Results:** The results represented that there is a direct association between the nanoparticles applied dosage and the restrain effect augmentation of applied dosage results in increase in restrain effect. In the study of environmental factors (pH and temperature), the results are in line with the inherent physiology of the bacteria; however, there was a significant decline in the number of analyzed bacteria cells due to the "Double Effect" of nanoparticle-pH variations as well as nanoparticle-temperature variables. In the very study, the promotion of Cadmium Oxide nanoparticles concentration leads to the elevation of antimicrobial feature and the reduction of bacteria growth rate is consistent with the other surveys about the nanoparticles effects on microorganisms to be more specific, one can come to this conclusion that the presence of nanoparticles prompts cellular destruction.

**Conclusion:** In the recent study, by elevation in Cadmium Oxide nanoparticles concentration, the antimicrobial property augments and the bacteria growth rate declines, that are in line with other researches about the nanoparticles effect on microorganisms.

**Keywords:** Antibacterial effect, Cadmium Oxide nanoparticles, *Staphylococcus Aureus*, Environmental factors

## Background

Nanoparticles high potentials caused their application in variable and precise processes, more specifically, in the biology and pharmacology which has been noticed by many biologists. Recently, along with the raising importance of healthcare, a large number of researches have been conducted to improve the antibacterial feature of nanoparticles. However, the application of certain antimicrobial materials has been restricted due to their lesion or toxicity. Inorganic antibacterial factors have a very high bacterial resistance and thermal stability. In recent years, researchers have highly noticed the Cadmium Oxide, its applications and properties in optoelectronic devices such as: solar cells [1], optical transistors, glassy electrodes, gas sensors, etc. [2]. These applications of Cadmium Oxide have been resulted from its individual electrical and optical features. Cadmium Oxide nanoparticles have been applied by many scholars up to now. In the same way, we have used Cadmium Oxide nanoparticles to confront bacteria which are pathogenic [3,4]. *Staphylococci* are Gram-positive sphere shaped cells that generally array in form of irregular groups like grape

* Correspondence: bahar.salehi007@gmail.com
[1]Young Researchers and Elites Club, North Tehran Branch of Islamic Azad University, Tehran, Iran
Full list of author information is available at the end of the article

clusters and grow in many mediums as well. *Staphylococcus Aureus* produces variable toxins and enzymes which are the major reason of bacteria survival; proteins, fats, and carbohydrates breakdown in order to provide necessitate materials, resistance against drugs and the ability of bacteria to cause disease. Some of these enzymes are Coagulase, Hemolysin, Leukocidin, Penicillinase, Lipase, Hyaluronidase, Catalase, and Protease. The enterotoxins of this microbe are dispersed by bacteria cells into the food or medium. The enterotoxin producing *staphylococci* are always able to produce Coagulase, but not all the positive Coagulase *Staphylococci* are usually capable of producing enterotoxin [5]. The synthesis of the nanomaterial effective on bacteria with high efficiency can be applied for disinfection and the elimination of environmental and industrial bacteria. It is expected that nanomaterial obtained in a variety of synthesis procedures enjoying different properties; hence, its antibacterial effect is essential. As it is difficult for most people to cope with the rising cost of combating pathogenic bacteria, finding a low price and prompt method to control its development and activity is a matter of the utmost importance. According to the fact that the bacteria are more resistant to prevalent drugs, the use of nanoparticles in hygiene and medicine is putative and they can be appropriate alternatives for traditional antibiotics; moreover, the production cost is lower and their storage is much easier compared to any other medicine. *Staphylococcus Aureus* is one of hospital's infectious resistant to traditional antibiotics, such as Beta-lactam, and is responsible for Gastroenteritis led by producing enterotoxin in food. Due to the importance of noted issues, in this research, we intend to study the effect of Cadmium oxide nanoparticles on *Staphylococcus Aureus*. We analyze the antibacterial effect of Cadmium Oxide nanoparticles on *Staphylococcus Aureus* bacteria in this study.

## Results and discussion
### Absorbance spectrums UV–Vis of Cadmium Oxide nanoparticles

This spectrometry is in regard to the transmissions between the electron scales. Generally, such transmissions are made between bonding orbital or non-bonding electron pairs and non-bonding orbital. Consequently, the link between the absorbance peaks wavelength and bonds emerged in the case study species seems to be feasible [6]. Visible-Ultraviolet spectrums of Cadmium Oxide nanoparticles are appeared in Figure 1. Although the wavelength of spectrum is limited by means of the light source, the absorbance band of nanoparticles represents a conversion in color location resulted from the amount of available limitation in the specimen comparing to the Cadmium Oxide nanoparticles. This optical phenomenon represents that these nanoparticles illustrate the level of

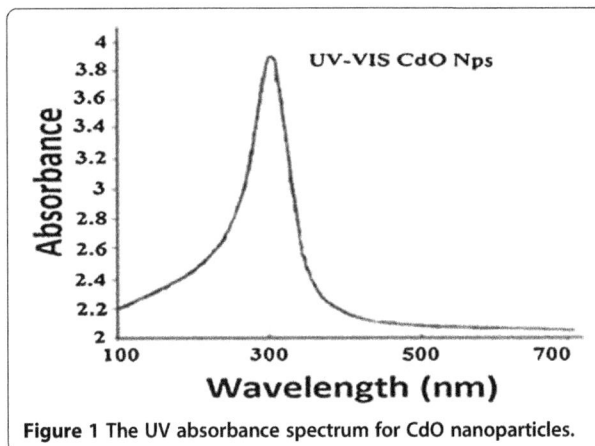

**Figure 1** The UV absorbance spectrum for CdO nanoparticles.

quantum effects [7]. At the very level, the development of nanoparticles depends on the surfactant and organic solvent, since the Cetyl Trimethyl Ammonium Bromide (CTAB) surfactant helps to the cohesion of synthesized nanoparticles' surface. Therefore, as a result of this interaction, stabilizing of particles and balancing the development or growth of the particles' cores are emerged to achieve a high level of uniformity [8]. The Acetic acid and Ethanol solvent assist the dispersion of particles identically, a deliberate growth of particles in limited sizes, and the prevention of the particles integration [9].

### Electron microscope analysis
The image of synthesized Cadmium Oxide nanoparticles is visible in the Figure 2. This image is taken with the composite electron microscope with the magnification of 13000 times representing an approximate 30 nm diameter of synthesized nanoparticles.

### The analysis of inhibitory effect of Cadmium Oxide nanoparticles on *staphylococcus aureus* bacteria in the agar
The growth phenomenon and the activity of bacteria against variable concentrations of 10 µg/ml, 15 µg/ml, and

**Figure 2** Image of electron microscope scanning of synthesized Cadmium Oxide nanoparticles with the magnification of 13000 times.

20 µg/ml of nanoparticles were analyzed. As it is obvious, the inhibition zone diameter which the *Staphylococcus Aureus* bacterium develops against the Cadmium Oxide nanoparticles indicates it could stop the growth of bacteria at a high level and there would be a direct relationship between the inhibitory effect and the nanoparticles applied dosage and it could be applied as a material including antibacterial properties (Table 1).

### The analysis of Cadmium Oxide nanoparticles inhibitory effect on *staphylococcus aureus* bacteria in the broth

To analyze the Cadmium Oxide nanoparticles inhibitory effect on *Staphylococcus Aureus* bacteria in the broth, different concentrations 10 µg/ml, 15 µg/ml, and 20 µg/ml of nanoparticles were applied. No nanoparticle was used for the control group. Therefore, the analysis carried out on the broth consists of four groups. The lids of the dishes containing treated broths and the control broth were shut tightly with the corks and they have been cultured to 37°C for 24 hours. Optical absorbance in the wavelength of 600 nm was utilized, the absorbance of the above solutions in Spectrophotometer was analyzed in order to accurately measure the bacteria concentration, and its diagram was drawn (Figure 3) for the purpose of analyzing the bacteria growth in every condition.

The 20 µg/ml concentration of Cadmium Oxide nanoparticles has provided the most inhibition effect on *Staphylococcus Aureus* bacteria. The effects of diverse Cadmium Oxide nanoparticles concentrations were studied on the number of *Staphylococcus Aureus* bacteria. The most inhibition effect has emerged on higher nanoparticles' concentrations and the value of *OD* is significantly declined by the statistical aspect ($P < 0.05$).

### The analysis of inhibitory effect of Cadmium Oxide nanoparticles on the number of *staphylococcus aureus* bacteria

The most inhibition effect ($P < 0.05$) with regard to 20 µg/ml concentration represents the antibacterial properties of Cadmium Oxide nanoparticles; the very property exceeds by an increase both in the concentration and in the maximum concentration made the bacteria to deteriorate to less than the 15% of original quantity (Table 2 and Figure 4).

Along with the enhancement in the concentration of Cadmium Oxide nanoparticles, the number of bacteria is deteriorated. The number of cells has a direct relationship with the applied concentration of nanoparticles in *Staphylococcus Aureus* bacteria, and it can be concluded from the regression ratio output that this relationship is negative, i.e., by the increase in concentration, the number of cells descends, but the P value in less than 0.05 and it represents the significance of this relationship.

### The analysis of temperature on the *Staphylococcus Aureus* bacteria in broth

According to the given data from Figure 5, it is crystal clear that, the minimum temperature required for the *Staphylococcus Aureus* bacteria growth is 8°C and the maximum temperature for the growth of this bacterium is 45°C; moreover, the optimum temperature for this bacterium is between 35°C and 37°C. The achieved results have conformity with the bacterium physiology. The results of thermal effect for this bacterium show that these results are in line with the inherent physiology of this bacterium against the temperature. However, as it is clear, due to the Double Effect phenomenon of both nanoparticle and the temperature conversions, a significant deterioration ($P < 0.05$) was emerged in the number of bacteria cells (Figure 6).

### The pH effect analysis on *Staphylococcus Aureus* bacteria in the broth

The pH effect on *Staphylococcus Aureus* bacteria in the absence of Cadmium Oxide nanoparticles and the presence of Cadmium Oxide nanoparticles was analyzed in this study and the results are demonstrated in Figures 7.

According to the data adopted from Figure 8, it is found that the minimum required pH for the growth of *Staphylococcus Aureus* bacteria is 4.5 and the maximum is 9.3, and the optimum pH is between 7 and 7.5. The obtained results ensure conformity with the physiology of the bacteria. These results are in line with the inherent physiology of the bacteria against pH. As it is obvious in Figure 8, as a result of Double Effect made by both of nanoparticle and the temperature conversions, a significant deterioration ($P < 0.05$) has been developed in the number of bacteria cells.

### Optimization of Cadmium Oxide nanoparticles antibacterial effect in variable times

In the final analysis, the living cells of *Staphylococcus Aureus* bacteria were exposed to maximum concentration (20 µg/ml) of Cadmium Oxide nanoparticles in 37°C water. The results illustrate that in the control group, a decline in *Staphylococcus Aureus* bacteria concentrations reaches from 6.3 log CFU/ml to non-evaluable concentrations after 14 days. Neverthless, by adding nanoparticles to the broths of these bacteria, the viability of bacteria deteriorates from 14 days to less than two (Figure 9).

**Table 1 Evaluation results of the inhibition zone diameter in variable concentrations of Cadmium Oxide nanoparticles on case study bacteria**

| 20 mg/ml | 15 mg/ml | 10 mg/ml | control | Concentration |
|---|---|---|---|---|
| 19 ± 2 mm | 14 ± 2 mm | 10 ± 2 mm | 0.0 mm | *Staphylococcus Aureus* |

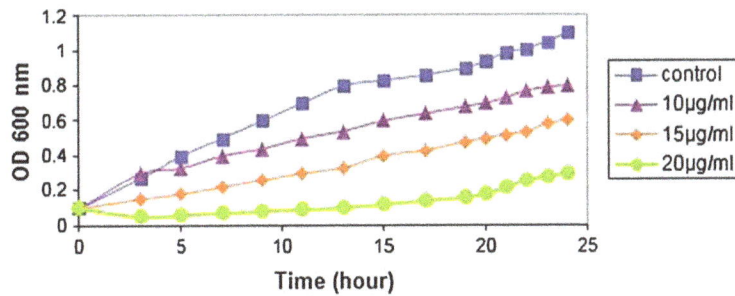

**Figure 3** The effect of variable Cadmium Oxide nanoparticles concentrations on *Staphylococcus Aureus* bacteria.

Through an elevation in concentration and time *OD* declines and there is a significant relationship between *OD* and time and concentration of applied nanoparticles in Dunnett test. Furthermore, in two-way *ANOVA* test, the identical assumption of *OD* value in different times and concentrations is rejected, because: ($P < 0.05$).

Nanomaterials have a huge number of functions in medicine; as antibiotics eradicate a little number of pathogenic factors, those are capable of about 650 types of pathogenic factors [10-12]. Cadmium oxide nanoparticles in 20 µg/ml concentration showed antibacterial effect on the resistant bacterium against antibiotics that was applied for this research. In research held by Buzby et al., they studied the antibacterial effects of Ag nanoparticles by employing them in 10-15 nm size and reported that an increase in nanoparticles effect depends on consumed values. [13]. It was found that by an increase in Cadmium oxide nanoparticles number, the antibacterial effect increases and the bacterium growth rate decreases. In another research by Sundrerajan, the antibacterial effects of MgO nanoparticles on gram positive *Staphylococcus Aureus* and gram negative *E.Coli* were studied and it was found that the inhibition zone for gram positive bacterium had been greater than gram negative one [14]. In the other research, the researchers studied antimicrobial effects of $CeO_2$ nanoparticles on *Staphylococcus Aureus*. Their applied nanoparticles size was 37.6 nm and their results showed that the antibacterial effects of $CeO_2$ depend on applied doses and the inhibition property developed by nanoparticles attested their antibacterial effect [15]. In 2009, Ayala et al. could inhibit *Staphylococcus Aureus* by Ag nanoparticles. They could also verify the antibacterial effect of Ag nanoparticles by well diffusion method protocol, subsequently,

they could define minimum inhibitory concentration (MIC) of which by macro dilution method [16]. In research by Rafie et al. [17] in Egypt, they could control *Staphylococcus Aureus* and *E.Coli* by using Ag nanoparticles inside cotton crop. In a research, the researchers studied the effect of Titanium Oxide nanoparticles on *Staphylococcus* and they concluded that those have descent antibacterial effect on this gram positive bacteria [18] of which the same results of present study have achieved in antibacterial effect. Experiments were done to study the CrO nanoparticles toxicity on gram positive bacteria of human immune system and gram negative bacteria such as *Shigella* that their results imply the toxic nature of CrO nanoparticle for variable microbial systems and human T-lymphocytes [19]. In 2011, researchers studied CrO and $CoFe_2O_4$ nanoparticles impacts on *Staphylococcus* and the results showed that CrO has higher bacterial killing power against *Staphylococcus Aureus* in comparison with $CoFe_2O_4$ and in general, both nanoparticles have antibacterial effect; however, CrO had a better function [20] showing what we approved in the antibacterial effect of Cadmium oxide nanoparticles in this research. As for the high toxicity of Cadmium oxide nanoparticles, they are used to eliminate *Staphylococcus Aureus* in vitro and to eradicate the environment bacteria as well as cleaning medical supplies and equipment contaminated by that bacteria.

## Conclusion

In the recent study, along with the elevation in Cadmium Oxide nanoparticles concentration, the antimicrobial property augments and the bacteria growth rate declines being in line with the other researches about the nanoparticles effect on microorganisms. It can be concluded that at the presence of nanoparticles the cell destruction prompts. It is proposed to use cadmium oxide nanoparticles in elimination of environmental bacteria resistant to traditional antibiotics.

## Methods

Two different methods were used to analyze the sensibility of *Staphylococcus Aureus* to Cadmium Oxide nanoparticles

**Table 2 Analysis results relevant to the effect of Cadmium Oxide nanoparticles concentration on the number of case study bacteria**

| 20 mg/ml | 15 mg/ml | 10 mg/ml | control | Concentration |
|---|---|---|---|---|
| 100,000 ± 100 | 600,000 ± 100 | 1,200,000 ± 100 | 2,000,000 ± 100 | *Staphylococcus Aureus* |

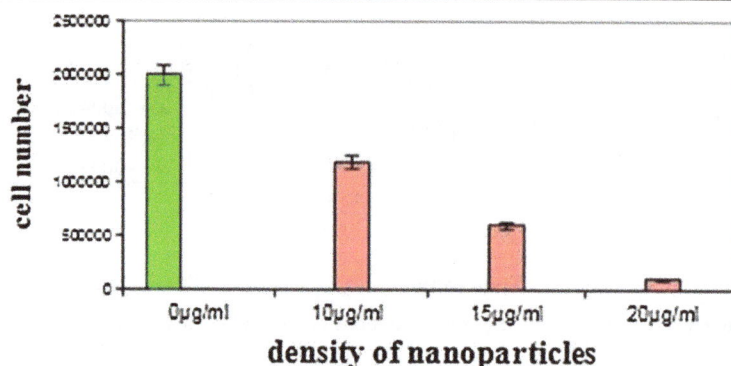

**Figure 4** The inhibitory effect of Cadmium Oxide nanoparticles on *Staphylococcus Aureus* Bacteria.

in order to confirm the results. Different concentrations of each nanoparticle were separately cultured on *Staphylococcus Aureus* (according to half McFarland) in the Mueller Hinton Agar bearing different Cadmium Oxide nanoparticles. No nanoparticles medium was used for control group. The bacteria grew in 37°C for 24 hours. Then, the number of colonies was compared with the specimen, and to enumerate a hemocytometer, a scaled lam with definite volume was used. Subsequently, the number of bacteria in the spaces of lam was enumerated, next, 9 ml of the medium suspension was mixed with 1 ml of methylene blue and it was found that the bacteria took the color and converted to blue. In the second method, the bacteria in Trypticase™ Soy Broth were placed separately. The number of bacteria was measured every one hour. Optical density (*OD*) in the wavelength of 600 nm was applied by the absorbance of foregoing solutions in the spectrophotometric device to measure the concentration of bacteria. The number of bacteria was compared with the control group in variable times and turbidimetric analysis was carried out for enumerating the bacteria in the medium. In this method, some of the bacteria were posed in cultured suspension in a test tube with definite diameter and the tube in was placed the optical beam of spectrophotometer with 600 nm wavelength in a way that, initially, 100 ml of nutrient broth sterile

was added to any of 11 sterile tubes and then 100 µl of nanomaterial suspension was added to the tube no. 1. After shaking the content of the tube, 1 ml of the material was transferred to the tube no. 2. In the same way, we provided the dilution series up to the tube no. 9 and shook the content of each tube subsequently. The tubes no. 10 and 11 were considered as the evidence. The tube no. 10 was considered as the evidence for nanoparticles suspension (containing bacteria and no nanoparticles) and the tube no. 11 as the evidence for the medium (with no bacteria cells and nanoparticles). By addition of 100 µl of the bacteria suspension to serial dilution tubes and heating the tubes to reach a concentration of half McFarland in 37°C, the minimum of growth inhibitory concentration was determined accurately for each bacterium.

### Preparation and the method of Cadmium Oxide nanoparticles analysis

To produce Cadmium Oxide nanoparticles in this study, in one experiment, the first solution sample with 0.06M Acetic acid and 0.03M Cadmium Sulfate was provided with 40 mg Cetyl Trimethyl Ammonium Bromide (CTAB) as the surfactant in 1 liter of double distilled water. Subsequently, the first solution was added to the second one and the obtained sediment was filtered by

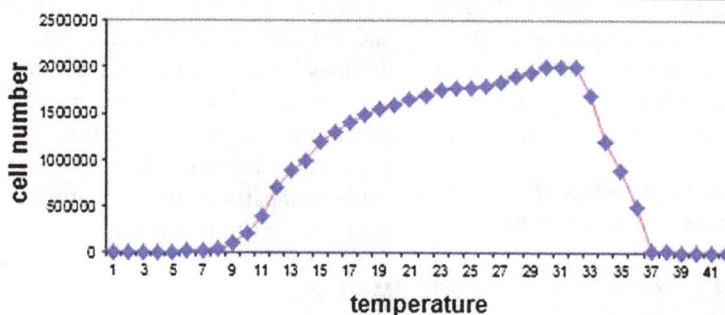

**Figure 5** The temperature effect on *Staphylococcus Aureus* bacteria in the presence of Cadmium Oxide nanoparticles.

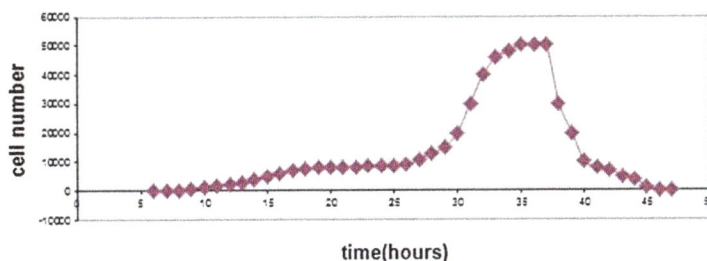

**Figure 6** The temperature effect on *Staphylococcus Aureus* bacteria in the broth.

the Whatman filter paper, then, it was dried out in hot air stove of 80°C for 1 hour approximately. In the next level, it was transmitted to Silica crucible (41°C) and burned in 400°C about 2 hours, then, the resulted powder was cleansed with Ethanol to eliminate the impurities available in the particles for 3-4 times. The characteristics of obtained nanoparticles were studied by the use of X-ray diffraction, visible Spectroscopy absorbance - Ultraviolet, and it was utilized in order to study its antibacterial characteristics. Morphological analysis and the observation of synthesized Cadmium Oxide nanoparticles were carried out with the visible Spectroscopy-Ultraviolet device, double beam TU-1901, X-ray diffraction device D/Max-RA employing CuKα radiation and composite Electron microscope JEM-200CX [21].

### Selection, separation and cultivation of clinical specimens

To provide clinical specimens, the *Staphylococcus Aureus* bacteria were separated from urine culture of the hospital remedial ward.

### Provision of bacteria and cultures

*Staphylococcus Aureus* bacteria were provided from Shiraz University of Medical Sciences and approved by means of microbiological common methods. Broths and agars were purchased from the Merck Company of Germany.

### Preparation of Mueller Hinton agar

The Agar was provided in compliance with the instruction written on the package; 11.4 g of the powdered material for agar were solved in the distilled water under 25°C heat to reach the pH equal to $7.3 \pm 0.2$. Initially, the Agar was heated and corked firmly, and held on flame to make the culture identical, then, it was placed in autoclave to be sterilized in 121°C for 15 min. and put in refrigerator afterward. Subsequently, they were put in Petri dish; their lids were shut and kept reversely (to prevent vapor infiltration).

### Preparation of nutrient broth

The medium was prepared in compliance with the instruction written on the package. 0.65 g of the powder was taken and poured into the beaker with 50 ml of distilled water being stirred until it acquired the pH of $7.3 \pm 0.2$. This is a nutrient broth, poured into a flask and placed in autoclave. To remove the microbe from the broth and place it in the agar, a swab sterilized by the autoclave in 120°C for 15 min was used, then, they are kept in the refrigerator up to their time-of-use.

### The culture of bacteria and study of Cadmium Oxide nanoparticles effect in agar

Initially, four holes were emerged in specific points of the agar to hold three of Cadmium Oxide nanoparticles

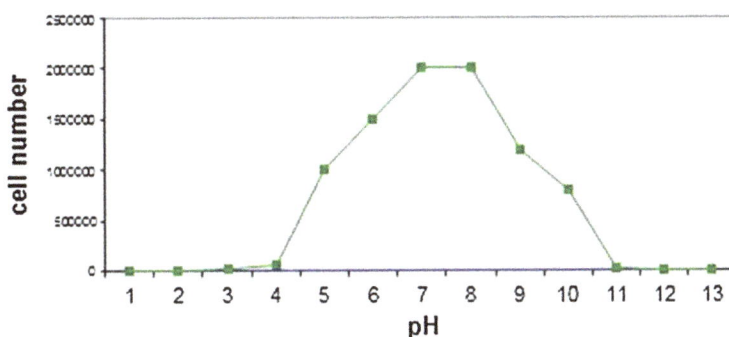

**Figure 7** pH effect on *Staphylococcus Aureus* bacteria in the broth.

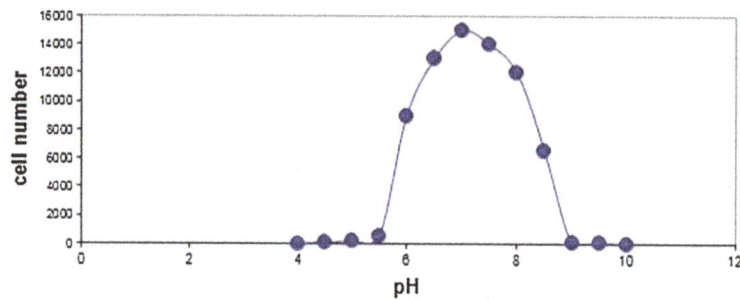

**Figure 8** The pH effect on *Staphylococcus Aureus* bacteria in the presence of Cadmium Oxide Nanoparticles.

and the fourth for the control group (at the middle of agar). Subsequently, the sterile swab was entered in nutrient broths which *Staphylococcus Aureus* bacteria were grown in and was previously provided, and smeared it with the bacteria, then it was spread in the agar plates in three dimensions to cover the plate surface wholly. Later on, we selected the foregoing nanoparticles' solutions provided with the mean and steady concentration of 15 μg/ml amongst the three concentrations of 10 μg/ml, 15 μg/ml, and 20 μg/ml for the purpose of standardizing, and infused into the determined shaft, and in the mid shaft, we infused the control solution which is distilled water in this study. Afterwards, the plates were placed into the incubator in 37°C for 24 hours and observed the effect of nanoparticles on the growth of bacteria, and finally, the colonies of bacteria were enumerated.

### Preparation of TSB (Trypticase™ Soy Broth), culture of bacteria and analysis of Cadmium Oxide nanoparticles effect in broth

The broth was prepared to comply with the instruction written on the package. 3 g of the powder was taken and infused into the beaker with 100 ml distilled water of which a pH of 7.3 ± 0.2 was acquired, and finally the solution poured and divided into eight test tubes. Then, we sterilized them in the Autoclave. We cultured the

*Staphylococcus Aureus* bacteria in four separate tubes and considered one of these four as the control group. Consequently, we added every three different concentrations of Cadmium Oxide nanoparticles for each bacterium in three reminder tubes complying with half McFarland and shut the lids of containers holding treated broths (bacteria + Cadmium Oxide nanoparticles); the control broth was shut tightly with a cork, shook aerobically, and placed in the incubator in 37°C for 24 hours. Afterwards, the optical density with the wavelength of 600 nm was used applying the foregoing solutions absorbance in Spectrophotometer to measure the concentration of the bacteria and its diagram was drawn so that, we could study the growth of bacteria in any situation. The antimicrobial effect of Cadmium Oxide nanoparticles in broth was analyzed, as their control group was distilled water. Their absorbance was read with the Spectrophotometer every one hour and their absorbance diagram in time unit were drawn for the specimens. All of obtained conclusions in tests were compared with the control group.

### Statistical analysis

Statistically, all the results were shown as the average and deviation forms. After determining the data distribution, to compare the results of every value in each of

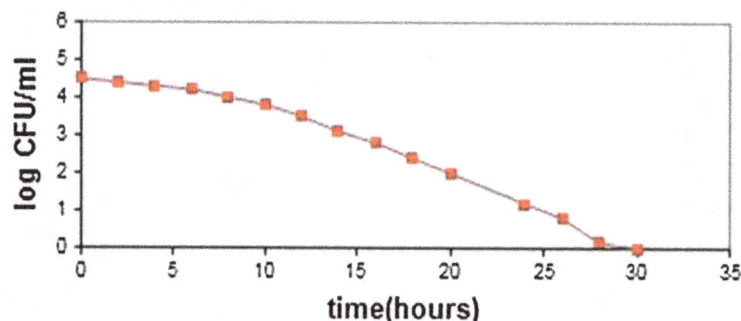

**Figure 9** The effect of maximum concentration (20 μg/ml) Cadmium Oxide nanoparticles on *Staphylococcus Aureus* bacteria viable cells.

the groups before and after the study, the ANOVA test with continuous measurement and to compare the groups the one-way ANOVA, *Dunnett* test, and regression were used. Moreover, a significant level less than 0.05 was assigned for all the analyses.

## Abbreviations
TEM: Transmission electron microscopy; UV-Vis: Visible-Ultraviolet spectrums; OD: Optical density; CTAB: Cetyl Trimethyl Ammonium Bromide surfactant.

## Competing interests
The authors declare that they have no competing interests.

## Authors' contributions
BS conceived the project, and revised the manuscript and the biological experiment and performed the experiments of nanoparticle assembly and *in vitro* test. SM designed the experiment. MA has done statistical analysis. All authors read and approved the final manuscript.

## Acknowledgments
We would like to thank all everyone who for their support in this study. This article was extracted from the thesis of Bahareh Salehi and the expenses of this research were discharged by Corresponding Author.

## Author details
[1]Young Researchers and Elites Club, North Tehran Branch of Islamic Azad University, Tehran, Iran. [2]Microbiology Group, Biological Sciences Faculty, North Tehran Branch of Islamic Azad University, Tehran, Iran. [3]Institute of Biochemistry and Biophysics, University of Tehran, Tehran, Iran.

## References
1. Zaien M, Ahmed NM, Hassan Z: Fabrication and characterization of an n-CdO/p-Si solar cell by thermal evaporation in a vacuum. *Int J Electrochem Sci* 2013, **8**:6988–6996.
2. Leary SP, Liu CY, Apuzzo ML: Toward the emergence of nanoneurosurgery: part III-nanomedicine: targeted nanotherapy, nanosurgery, and progress toward the realization of nanoneurosurgery. *Neurosurgery* 2006, **58**(6):1009.
3. Gao PX, Ding Y, Mai WJ, William LH, Lao CS, Wang ZL: Conversion of zinc oxide nanobelts into superlattice-structured nanohelices. *Sci* 2005, **309**(5741):1700.
4. Guo T, Ma YL, Guo P, Xu ZR: Antibacterial effects of the Cu (II)-exchanged montmorillonite on Escherichia coli K88 and Salmonella choleraesuis. *Vet Microbiol* 2005, **105**(2):113.
5. Jones N, Ray B, Koodali TR, Adhar CM: Antibacterial activity of ZnO nanoparticle suspensions on a broad spectrum of microorganisms. *FEMS Microbiol Lett* 2008, **279**(1):71.
6. Axelsson-Olsson D, Waldenstrom J, Broman T, Olsen B, Holmberg M: Protozoan acanthamoebapolyphaga as a potential reservoir for campylobacter jejuni. *Appl Environ Microbiol* 2005, **71**(2):987.
7. Barth WF, Segal K: Reactive arthritis (Reiters syndrome). *Am Fam Physician* 1999, **60**(2):499.
8. Barbarino A: Helicobacter pylori-related iron deficiency anaemia: a review. *Helicobacter* 2002, **7**(2):71.
9. Bell BP, Goldoft M, Griffin PM, Davis MA, Gordon DC, Tarr PI, Bartleson CA, Lewis JH, Barrett TJ, Wells JG, Roy B, John K: A multi-state outbreak of Escherichia coli O157:H7-associated bloody diarrhoea and haemolyticuraemic syndrome from hamburgers. *JAMA* 1994, **272**(17):1349.
10. Oberdörster G, Oberdörster E, Oberdörster J: An emerging discipline evolving from studies of ultrafine particles Environ. *J Nanotoxicol* 2005, **113**(7):823.
11. Hernandez-Sierra J, Ruiz F, Pena D, Martinez-Gutierrez F, Martinez AE, Guillen AJ, Tapia-Perez H, Castanon GM: The antimicrobial sensitivity of Streptococcus mutans to nanoparticles of silver, zinc oxide, and gold. *Nanomed Nanotechnol* 2008, **4**(3):237.
12. Rai M, Yadav A, Gade A: Silver nanoparticles as a new generation of antimicrobials. *Biotechnol Adv* 2009, **27**(1):76.
13. Buzby JC, Roberts T: Economic and trade impacts of microbial foodborne illness. *World Health Stat Q* 1997, **50**(1&2):57.
14. Sundrarajan M, Suresh J, Rajiv GR: A comparative study on antibacterial properties of MgO nanoparticles prepared under different calcination temperature. *Digest J Nanomaterials Biostructures* 2012, **7**(3):983–989.
15. Negahdary M, Mohseni G, Fazilati M, Parsania S, Rahimi G, Rad S, RezaeiZarchi S: The Antibacterial effect of cerium oxide nanoparticles on Staphylococcus aureus bacteria. *Ann Biol Res* 2012, **3**(7):3671–3678.
16. Ayala-Núñez NV, Lara Villegas HH, del Carmen ITL, Rodríguez PC: Silver nanoparticles toxicity and bactericidal effect against methicillin-resistant staphylococcus aureus: Nanoscale does matter. *Nanobiotechnology* 2009, **5**(1):2–9.
17. Rafie M, Mohamed AA, Shaheen TI, Hebeish A: Antimicrobial effects of silver nanoparticles produced by fungal process on cotton fabrics. *Carbohydr Polymeris* 2010, **80**:779–782.
18. Nataraj N, Anjusree GS, Madhavan AA, Priyanka P, Sankar D, Nisha N, Lakshmi SV, Jayakumar R, Balakrishnan A, Biswas R: Synthesis and anti-staphylococcal activity of TiO2 nanoparticles and nanowires in ex vivo porcine skin model. *J Biomed Nanotechnol* 2014, **10**(5):864–870.
19. Ghosh SK, Pal T: Interparticle coupling effect on the sur- face plasmon resonance of cobalt nanoparticles: from theory to applications. *Chem Rev* 2007, **107**:4797–4862.
20. Saeed R-Z, Saber I, Zand A m, MojtabaSaadatiand Zahra Z: Study of bactericidal properties ofcarbohydrate-stabilized platinum oxidenanoparticles. *Int Nano Lett* 2012, **2**:21.
21. Singh AK, Nakate UT: Microwave synthesis, characterization, and photoluminescence properties of nanocrystalline zirconia. *Sci World J* 2014, 349457(2014):7.

# Engineering of near infrared fluorescent proteinoid-poly(L-lactic acid) particles for *in vivo* colon cancer detection

Michal Kolitz-Domb, Igor Grinberg, Enav Corem-Salkmon and Shlomo Margel*

## Abstract

**Background:** The use of near-infrared (NIR) fluorescence imaging techniques has gained great interest for early detection of cancer owing to the negligible absorption and autofluorescence of water and other intrinsic biomolecules in this region. The main aim of the present study is to synthesize and characterize novel NIR fluorescent nanoparticles based on proteinoid and PLLA for early detection of colon tumors.

**Methods:** The present study describes the synthesis of new proteinoid-PLLA copolymer and the preparation of NIR fluorescent nanoparticles for use in diagnostic detection of colon cancer. These fluorescent nanoparticles were prepared by a self-assembly process in the presence of the NIR dye indocyanine green (ICG), a FDA-approved NIR fluorescent dye. Anti-carcinoembryonic antigen antibody (anti-CEA), a specific tumor targeting ligand, was covalently conjugated to the P(EF-PLLA) nanoparticles through the surface carboxylate groups using the carbodiimide activation method.

**Results and discussion:** The P(EF-PLLA) nanoparticles are stable in different conditions, no leakage of the encapsulated dye into PBS containing 4% HSA was detected. The encapsulation of the NIR fluorescent dye within the P(EF-PLLA) nanoparticles improves significantly the photostability of the dye. The fluorescent nanoparticles are non-toxic, and the biodistribution study in a mouse model showed they evacuate from the body over 24 h. Specific colon tumor detection in a chicken embryo model and a mouse model was demonstrated for anti-CEA-conjugated NIR fluorescent P(EF-PLLA) nanoparticles.

**Conclusions:** The results of this study suggest a significant advantage of NIR fluorescence imaging using NIR fluorescent P(EF-PLLA) nanoparticles over colonoscopy. In future work we plan to broaden this study by encapsulating cancer drugs such as paclitaxel and/or doxorubicin, within these biodegradable NIR fluorescent P(EF-PLLA) nanoparticles, for both detection and therapy of colon cancer.

**Keywords:** Proteinoid nanoparticles, Fluorescent nanoparticles, NIR fluorescence, Optical imaging, Colon cancer

## Background

In recent years, several types of nanoparticles have been introduced to the field of cancerous tissue detection. When referring to colon cancer specifically, the early detection of adenomatous colonic polyps is a major concern, as the early detection is the key to survival [1,2]. Up until now, colorectal cancer screening includes either stool-based tests or endoscopic and radiological examination of the colon [3,4]. These techniques are considered to be invasive and insensitive, causing poor patient compliance. Thus, colon cancer continues to be a major cause of death. As recently studied and published by several research groups, novel near-infrared (NIR) fluorescent nanoparticles may serve as a valuable tool in the field of colon tumor detection [5,6]. The nanoparticles introduce the use of fluorescence in the NIR region (700–1000 nm), where autofluoresence, light scattering and absorption of the light by normal tissues is not a concern. This way, the imaging has an improved signal-to-noise ratio, as the background is non-fluorescent in the NIR region and the detection of the fluorophore is optimal [7-9]. Nanoparticles

* Correspondence: shlomo.margel@mail.biu.ac.il
Department of Chemistry, The Institute of Nanotechnology and Advanced Materials, Bar-Ilan University, Ramat-Gan 52900, Israel

containing NIR dyes have already been developed and proved to have significant advantages over free NIR dyes, including biocompatibility, improved fluorescence signal, enhanced photostability and the presence of functional groups on the nanoparticle surface allowing easy conjugation to bioactive molecules [6,10-12]. Among various NIR fluorescent dyes, cyanine dyes are already approved and used in a wide range of biological applications, since they are known as water-soluble, stable, sensitive and have sharp fluorescence bands [13].

Recently, proteinoid particles were studied by several groups as new drug delivery systems [14-16]. The thermal condensation of amino acids into proteinoids was first described and characterized by Fox and Harada [17-20]. When proteinoids are incubated in an aqueous solution they form hollow particles that range in size according to the environment conditions [21]. Since proteinoid particles are considered biodegradable, non-immunogenic and non-toxic, they can be used as a delivery vehicle in the body [16]. In this work, a new version of proteinoid is synthesized and studied. In order to synthesize suitable particles for this application, proteinoids made of natural amino acids along with low molecular weight poly (L-lactic acid) (PLLA) were synthesized. The natural amino acids L-glutamic acid (E) and L-phenylalanine (F) are thermally polymerized with and without 2000 Da PLLA. All of the monomers used are safe, without exception, as PLLA is widely used in many biomedical applications as medical implants, such as screws, pins, sutures, rods [22,23], etc.

After preparation, the crude proteinoids can go through a self-assembly process to form micro- and nano-sized particles [24,25]. If a molecule of a dye or a drug is introduced during the self-assembly, the process may include the encapsulation of the molecule within the proteinoid particle [16]. Here, indocyanine green (ICG), a well-known fluorescent cyanine dye, was encapsulated by the proteinoid-PLLA to form new NIR fluorescent proteinoid-PLLA nanoparticles. Leakage of the entrapped NIR dye into PBS in the absence and the presence of 4% albumin was not detected.

The NIR fluorescent P(EF-PLLA) nanoparticles, containing ICG, were tested for their in vivo biodistribution in a mouse model. Additionally, the NIR fluorescent P (EF-PLLA) nanoparticles were conjugated to a bioactive targeting molecule: anti-cacinoembryonic antigen antibodies (anti-CEA) [12]. The bioactive-conjugated NIR fluorescent P(EF-PLLA) nanoparticles were found to specifically detect colon cancer tumors, as demonstrated in tumor implants in a chicken embryo model and in a mouse model.

## Results and discussion

In the first stage of this study P(EF) and P(EF-PLLA) polymers and particles have been prepared and characterized as described in the Materials and methods section. Table 1 compares the physical and chemical properties of the formed polymers and particles.

The incorporation of low molecular weight PLLA segments (2000 Da) in the copolymer backbone contributes to the overall polymer biodegradability and the smaller nanometric-scale size of the particles made of it. Furthermore, it does not affect significantly the molecular weights, improves slightly the polydispersity and decreases significantly the size and size distribution of the obtained nanoparticles from $196 \pm 24$ to $103 \pm 11$ nm. Additionally, the presence of the PLLA segments provides an additional safe way for biodegradation through ester hydrolysis [22]. Hence, P(EF-PLLA) was selected as the better polymer for this specific application.

Several P(EF-PLLA) copolymers were synthesized, changing the PLLA percentage in the total monomer weight, as specified in the Materials and methods section. Figure 1 exhibits the effect of changing the PLLA content on the different P(EF-PLLA) particle sizes.

As shown, the optimal particles, judging by their size and size distribution, are the particles made of P(EF-PLLA) where PLLA is 10% of the total monomer. In this case, the particles formed are nanoparticles of the smallest nanometric size with a narrow size distribution, $103 \pm 11$ nm. When PLLA is of lower percentage in the total monomer weight, the copolymers self-assemble into larger particles in size, 232–502 nm. Overall, as the PLLA fraction in the copolymer rises, the particles formed are smaller in size, indicating the importance of incorporating PLLA into the proteinoid particles, where these hydrophobic moieties are well-packed to form the interior of the nanoparticle. However, P(EF-PLLA) with 20% PLLA does not self-assemble into particles at the specified conditions in the Materials and methods section.

The P(EF-PLLA) nanoparticle density measured as described in the method section was 0.005 g/mL, indicating that the particles formed have a relatively high volume and a very low mass, probably hollow particles, as already

**Table 1 Characterization of the P(EF) and P(EF-PLLA) polymers and particles**

|  | P(EF) | P(EF-PLLA)[f] |
|---|---|---|
| **Mw (kDa)[a]** | 165 | 168 |
| **Mn (kDa)[a]** | 138 | 156 |
| **PDI[a]** | 1.19 | 1.07 |
| **Optical activity $[\alpha]_D^{25°C}$ (°)[b]** | −9.0 | −4.5 |
| **Particle diameter (nm)[c]** | $196 \pm 24$ | $103 \pm 11$ |
| **Particle density (g/mL)[d]** | 0.001 | 0.005 |

[a]Molecular masses were measured by GPC, PDI is the polydispersity index, given by Mw/Mn; [b]specific optical rotation (c = 1, in $H_2O$, at 25°C); [c]hydrodynamic particle diameter, measured by DLS, as described in the Materials and methods section; [d]particle density was measured as described in the Materials and methods section; [f]P(EF-PLLA), PLLA 10% of the total monomer weight.

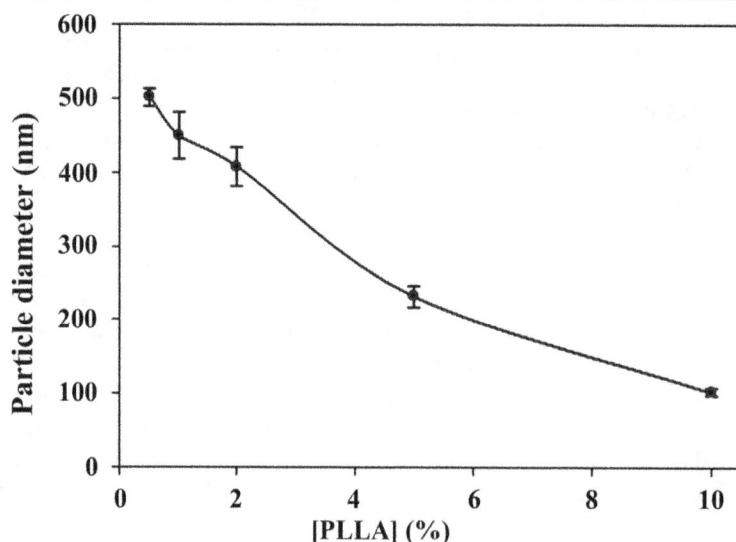

**Figure 1** Effect of the PLLA weight% of the total monomer weight in synthesized P(EF-PLLA) on the hydrodynamic particle diameter.

known for the proteinoid particles reported in the literature [26]. This is significantly important, since hollow particles may be used for the encapsulation of drugs and dyes, etc.

The optimal P(EF-PLLA) nanoparticles were used to encapsulate ICG. Figures 2A and 2B show that the dry (A) and hydrodynamic (B) diameters of the NIR fluorescent P(EF-PLLA) nanoparticles are $70 \pm 15$ nm and $145 \pm 20$ nm, respectively. The difference in diameters is due to the fact that the hydrodynamic diameter takes into account the water molecules around and within the hydrophilic nanoparticles. Figures 2C and 2D exhibit the fluorescence and absorbance spectra of the NIR fluorescent P(EF-PLLA) nanoparticles compared to those of the free dye in solution. The absorbance spectra shows no shift in the absorbance, but a change in the maximal absorbance peak from 779 nm in free ICG to 718 nm in ICG-containing nanoparticles, probably since the ICG molecules get close to each other inside the nanoparticle interior and aggregation may occur and cause this change in absorbance [27,28]. Moreover, a blue-shift of 12 nm in the emission spectrum of the NIR fluorescent nanoparticles compared to the free ICG in solution was also observed, probably due to the dye molecule aggregation inside the particle.

Leakage of the encapsulated ICG into PBS not-containing and containing 4% albumin at room temperature was not observed, indicating that the dye is strongly associated within the P(EF-PLLA) nanoparticles, probably due to physical interactions between the dye and the polymer.

### Optimization of the ICG concentration encapsulated within the P(EF-PLLA) nanoparticles

In order to optimize the ICG-containing P(EF-PLLA) nanoparticles fluorescence intensity, different concentration

(0.5-5% w/w relative to P(EF-PLLA)) of ICG were encapsulated in the P(EF-PLLA) nanoparticles, as described in the Materials and methods section. The concentration of encapsulated ICG that provided the maximum fluorescence intensity of the resultant NIR fluorescent P(EF-PLLA) nanoparticles was 1% w/w relative to P(EF-PLLA). At higher dye concentrations, fluorescence quenching was observed, as the distance between the dye molecules encapsulated within the nanoparticle is shorter, resulting in non-emissive energy transfer between them.

### Photobleaching of the NIR fluorescent P(EF-PLLA) nanoparticles

Photostability experiments of the free and the encapsulated dye within the P(EF-PLLA) nanoparticles were performed. Samples were illuminated at 800 nm and the fluorescence intensities over 20 minutes of illumination were recorded. Figure 3 shows that the fluorescence intensity of free ICG in solution is decreased significantly with time, as opposed to the fluorescence intensity of the encapsulated ICG in the nanparticles, which remains intact. The encapsulation of the dye protects the dye from reactive oxygen species, other oxidizing or reducing agents, temperature, exposure time and illumination levels, which may reduce the fluorescence intensity irreversibly [5,10].

### Nanoparticle stability

Long-term refrigeration of particle dispersions and freeze-drying of particles are common ways to store nanoparticles for long-time periods. P(EF-PLLA) nanoparticle were kept in these conditions as mentioned in the Materials and methods section, in order to determine their stability. The nanoparticles were kept in refrigeration for 6 months

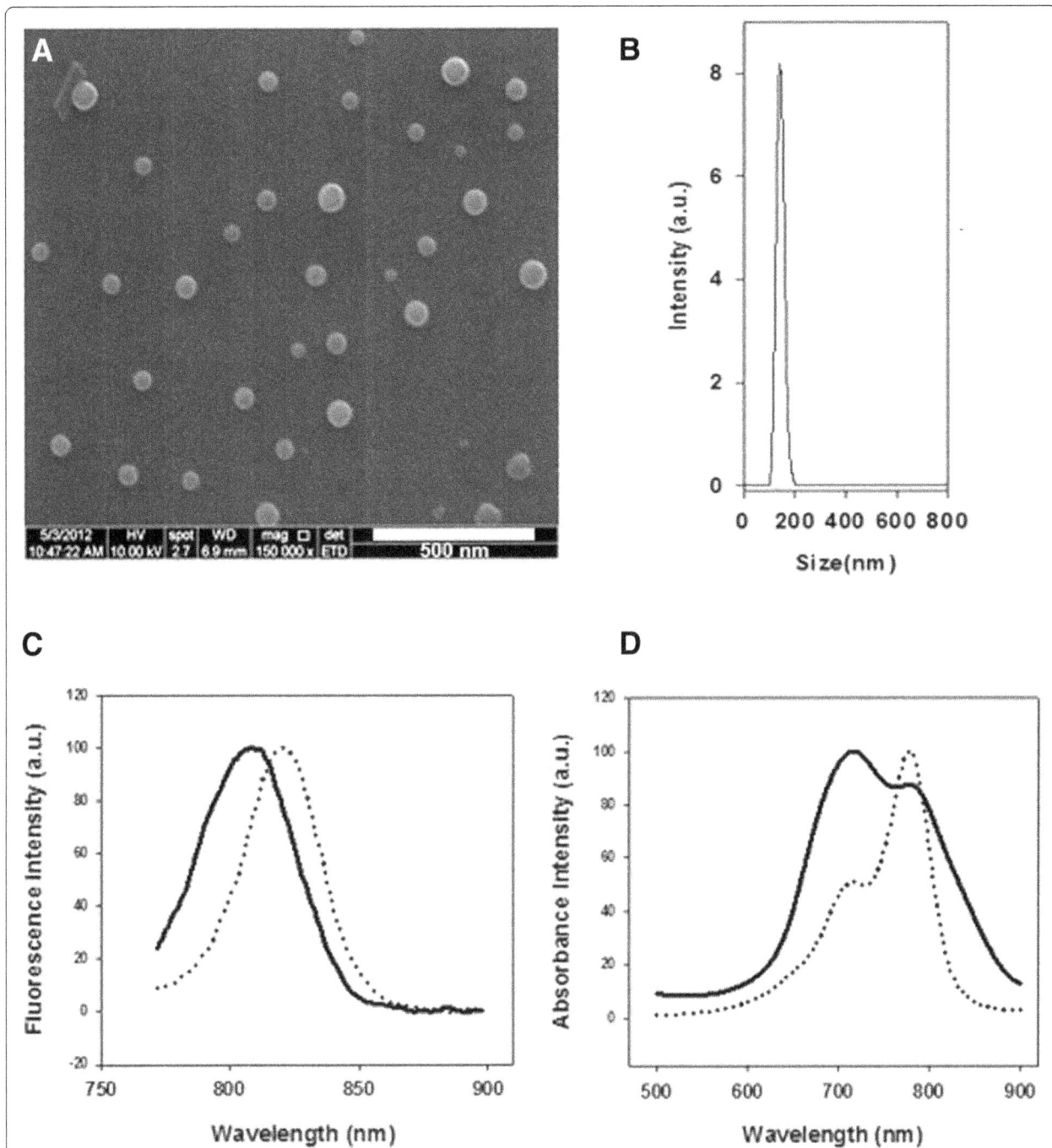

**Figure 2 Characterization of the NIR fluorescent P(EF-PLLA) nanoparticles. (A, B)** SEM image and hydrodynamic size histogram of the NIR fluorescent P(EF-PLLA) nanoparticles; **(C, D)** emission and absorbance spectra of free ICG (dotted lines) and nanoparticles containing ICG (solid lines), respectively.

as particle dispersions in PBS, and were found to remain their size and size distribution over this period of time. Additionally, no free ICG or soluble P(EF-PLLA) were detected in the aqueous phase, and the fluorescence intensity of the dispersion remained unaltered. These results indicate that the NIR fluorescent nanoparticles may be kept for long periods of time under refrigeration. The NIR fluorescent nanoparticles were also lyophilized to dryness and then redispersed in an aqueous phase to their original concentration. Yet again, the particles size, size distribution and their fluorescence were not affected. This indicates that the nanoparticles may be stored and handled as a freeze-dried powder and redispersed upon use without the addition of cryoprotectants prior to drying, as common to

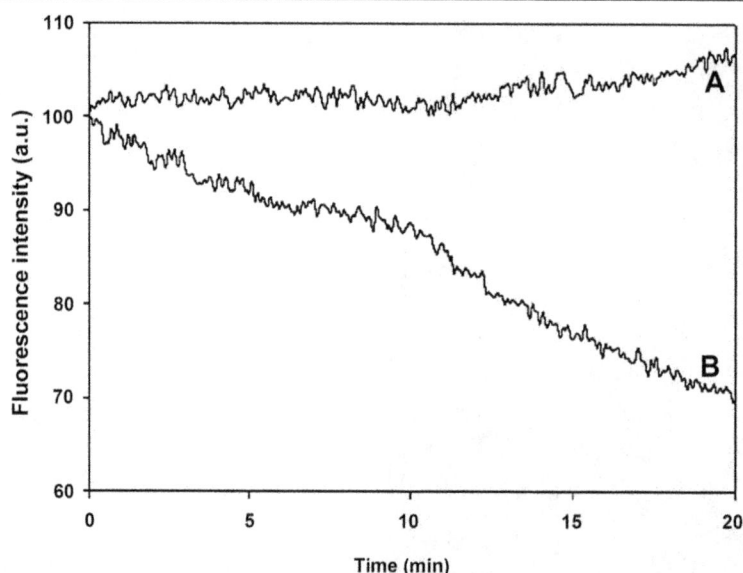

**Figure 3** Photostability of the ICG-containing P(EF-PLLA) particles (A) and free ICG (B) as function of time. Samples of ICG-containing P (EF-PLLA) particles and free ICG were illuminated with a Xenon flash lamp for 20 min as described in the Materials and methods section.

be done in order to prevent significant agglomeration of nanoparticles [29].

The stability of the fluorescent P(EF-PLLA) nanoparticles as function of pH was tudied by ζ-potential measurements. Figure 4 illustrates the ζ-potential curve of the nanoparticles aqueous dispersion as function of pH. As presented in the figure, there is a decrease in the ζ-potential of the particles as the pH of the particle aqueous dispersions increases: when pH was increased from 1.3 to 10.1, the ζ-potential decreased from 3.0 to −34.7 mV. The isoelectric point of the P(EF-PLLA) nanoparticles is around pH 2, probably due to the large amount of negative charge

carboxylate groups that self-assemble on the surface of the nanoparticles.

### In vitro cytotoxicity of the P(EF-PLLA) nanoparticles

In order to revoke cell toxicity of the NIR P(EF-PLLA) nanoparticles, *in vitro* cytotoxicity of the particles was tested by using human colorectal adenocarcinoma LS174t, SW480 and HT29 cell lines. Cell cytotoxicity was assessed by measuring the release of cytoplasmic lactate dehydrogenase (LDH) into cell culture supernatants. LDH is an intracellular enzyme which catalyzes the reversible oxidation of lactate to pyruvate. Since LDH is predominantly in

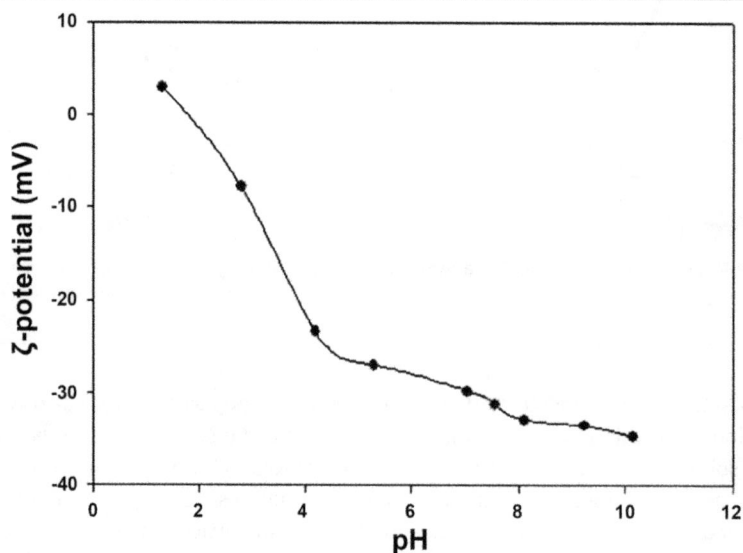

**Figure 4** ζ-potential of the NIR fluorescent P(EF-PLLA) nanoparticles as function of pH.

the cytosol, the enzyme is released into the supernatant only upon cell damage or lysis [30]. Figure 5 exhibits the cytotoxicity levels of the P(EF-PLLA) particles at two different concentrations (1.25 and 2.5 mg/mL). It can be seen that at both concentrations, the P(EF-PLLA) particles have no significant cytotoxic effect on all three cell lines, compared to untreated cells, meaning that the nanoparticles may be used for biomedical applications as suggested, including drug delivery.

### In vivo biodistribution in a mouse model

NIR fluorescent P(EF-PLLA) nanoparticles (2 mg/mL, 0.01 mg/kg body weight per mouse) were injected i.v. into mice through the tail vein and checked at several time intervals over 24 h. Figure 6 shows whole body images of mice injected with the nanoparticles over time: at 5 min, 20 min, 1 h and 24 h from injection. 5 min post injection, there is an initial burst of fluorescence which subsided quickly, while the majority of the fluorescent nanoparticles concentrated in the liver, at 20 min. 24 h post injection, the fluorescence is almost nonexistent, signifying the nanoparticle clearance from the body over 24 h. Biodistribution was tested for free ICG as well, and no significant differences in distribution and kinetics were found between nanoparticles containing ICG and free ICG up to 24 h post injection. These findings were in complete agreement with previous reports of ICG and ICG-containing nanoparticles pharmacokinetics and biodistribution, as the free dye in solution, derivatives of the free dye and ICG-containing nanoparticles are all evacuated from the body after 1 h and completely vanished 24 h after i.v. injection [31,32].

*Ex vivo* fluorescence images of specific organs and blood were also obtained. Organs from mice were harvested and blood was drawn 5 min, 20 min, 1 h and 24 h post injection of the nanoparticles into the tail vein. Figure 7 shows the calculated fluorescence intensities of the lungs, bones, brain, colon, duodenum, heart, liver, kidney, spleen and blood screening. Evidently, this analysis shows that the nanoparticles penetrated and were found in all checked organs. It is shown clearly that by 20 min most of the inserted quantity of the fluorescent nanoparticles is cleared from the blood. The nanoparticles concentrate mostly at the liver and are probably evacuated from the body. Interestingly, it is also apparent that the nanoparticles pass the blood–brain barrier (BBB), since they are found in the brain at 20 min post injection. This may open up a scope of drug targeting to the brain for drug molecules which are usually blocked. Overall, it was demonstrated that following a single i.v. injection of the nanoparticles, fluorescence intensity at all organs decreased over time, and only traces of fluorescence could be seen after 24 h.

### In vivo optical detection of human colon tumors on a CAM model

To demonstrate the feasibility of using the NIR fluorescent P(EF-PLLA) nanoparticles for the detection of colon tumors, monocloneal antibodies against CEA (anti-CEA) and anti-rabbit IgG were conjugated to the nanoparticles. Carcinoembryonic antigen, CEA, is a highly glycosylated glycoprotein expressed in most human carcinomas, and therefore is used as an effective biomarker in several modalities of human carcinoma. As

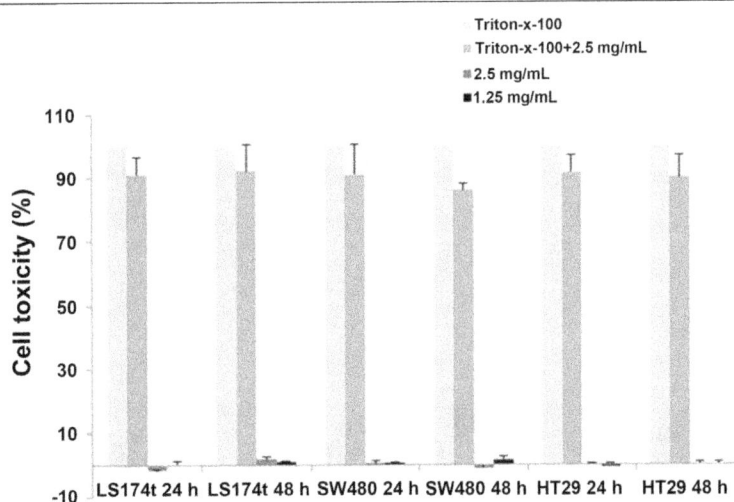

**Figure 5 Cytotoxic effect of the NIR fluorescent P(EF-PLLA) nanoparticles on human colorectal adenocarcinoma LS174t, SW480 and HT29 cell lines measured by the LDH assay.** Cells ($3 \times 10^5$) were incubated for 24 and 48 h with the P(EF-PLLA) nanoparticles (1.25 and 2.5 mg/mL in PBS). Cells were incubated with 1% Triton-x-100 as positive control (100% toxicity). In addition, cells were incubated with Triton-x-100 1% and the P(EF-PLLA) nanoparticles (2.5 mg/mL) to revoke any interaction of the nanoparticles with the LDH kit components. Untreated cells (negative control) were similarly incubated. Each bar represents mean ± standard deviations of 4 separate samples.

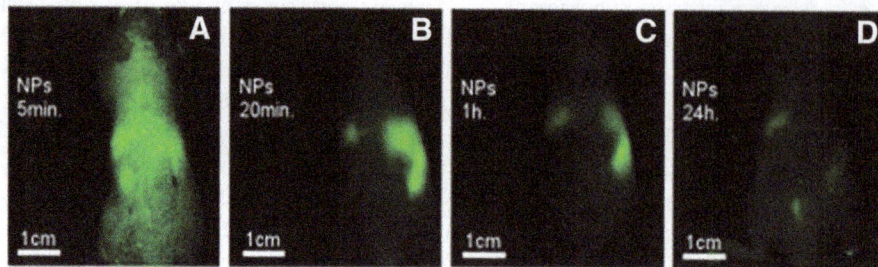

**Figure 6 Typical whole body fluorescence images of the NIR fluorescent P(EF-PLLA) nanoparticles at 5 min (A), 20 min (B), 1 h (C) and 24 h (D) after i.v injection.** 12 mice (each experiment group contained 3 mice) were anesthetized and treated with NIR fluorescent P(EF-PLLA) nanparticles (2 mg/mL, 0.01 mg/kg body weight per mouse). Blood was drawn and organs were harvested at each time point. 2 uninjected mice served as negative control. 12 mice were injected correspondingly with free ICG solution, giving similar results (not shown). The experiment was repeated twice with similar results.

stated in the literature, CEA is upregulated on the mucosal side of the LS174t colorectal cancer cell line, as opposed to SW480 (at least x$10^3$ less) [33]. Anti-rabbit IgG was conjugated to nanoparticles as a non-specific binding agent, with the intention of inactivating the conjugated particles in terms of tumor detection. As clearly illustrated in Figure 8, LS174t tumors treated with anti-CEA-conjugated nanoparticles (B) gained greater fluorescence compared to those treated with non-conjugated nanoparticles (A) or anti-rabbit IgG-conjugated nanoparticles (C). This can be explained by the effective ligand-receptor interaction. Furthermore, the SW480 tumors treated with the anti-CEA-conjugated nanoparticles gained less fluorescence (about 3.5 times) compared to the LS174t tumors treated the same way. The fluorescent signal of LS174t tumors labeled by anti-CEA-conjugated nanoparticles was 4 times higher than

that of the the tumors labeled by the anti-rabbit IgG-conjugated nanoparticles. Anti-rabbit IgG "blocks" the particle from interacting with the tumor receptors by the conjugation to the surface active moieties, thus serving as a negative control in colon tumor labeling. The results discussed and calculated are an average of 3 different experiments, wherein each experiment contained 6 eggs in a group, altogether 18 eggs in each group.

### *In vivo* optical detection of human colon tumors in a mouse model

Labeling of human colorectal tumors was performed using orthotopic mouse model (22 mice) with colonic tumors originated from LS174t cells injected to the colon wall 2 weeks before the experiment. Mice were anesthetized and treated with 0.1% bioconjugated NIR fluorescent P (EF-PLLA) nanoparticles dispersion in PBS through the

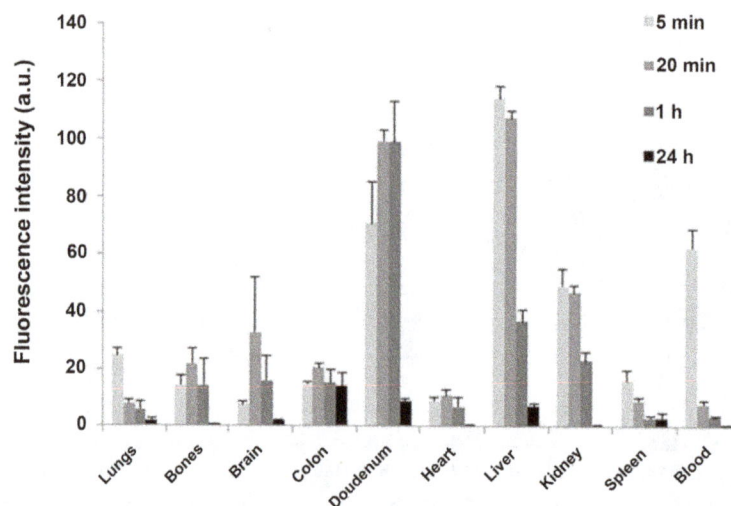

**Figure 7 Fluorescence intensities of different organs taken at 5 min, 20 min, 1 h and 24 h post i.v. injection into mice tail veins.** 12 mice (each experiment group contained 3 mice) were anesthetized and treated with NIR fluorescent P(EF-PLLA) nanoparticles (2 mg/mL, 0.01 mg/kg body weight per mouse). Blood was drawn and organs were harvested at each time point. 2 uninjected mice served as negative control. The experiment was repeated twice with similar results.

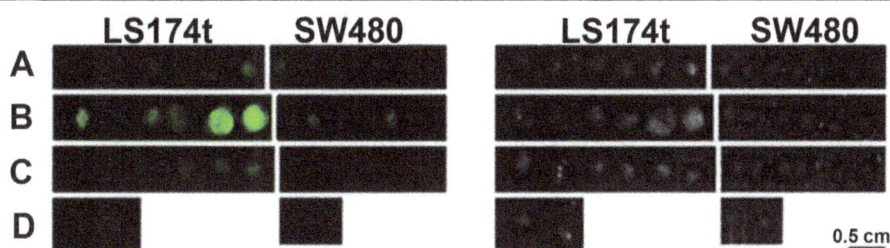

**Figure 8 Fluorescent and grayscale images from a typical experiment of LS174t and SW480 human tumor cell lines implanted on chicken embryo CAM treated with the non-conjugated (A), anti-CEA-conjugated (B) and anti-rabbit IgG-conjugated (C) NIR fluorescent P(EF-PLLA) nanoparticles.** Images of untreated tumors are shown in **(D)**. The experiment was repeated 3 times with similar results.

anus. After 20 min, the colons were extensively washed with PBS and were left to recover for 4 h. The colons were then removed and prepared for the imaging as described in the Materials and methods section. Figure 9 shows typical (8 out of 10 mice) fluorescent and grayscale images of the mice colons after treatment with anti-CEA (A) and anti-rabbit IgG (B) conjugated nanoparticles. As illustrated in Figure 9A, the anti-CEA-conjugated nanoparticles detected the tumors specifically and selectively with good signal to background ratio (SBR), the background refers to the surrounding non-pathological tissue. Moreover, as illustrated in Figure 9B, the "inactive" anti-rabbit IgG-conjugated nanoparticles did not produce a significant signal of the tumors.

## Conclusions

Proteinoid and proteinoid-PLLA copolymer were made from (L) glutamic acid, (L) phenylalanine and poly(L-lactic acid) (2 kDa). The optimal copolymer P(EF-PLLA) (10% PLLA), was used to encapsulate ICG to yield NIR fluorescent P(EF-PLLA) nanoparticles. The new NIR fluorescent nanoparticles discussed in the work have the potential to assist in early diagnosis of colonic neoplasms. They are stable, avoiding leakage and photobleaching of the dye over time and non-toxic. The nanoparticles penetrate a variety of organs, including the brain and bones, and are evacuated almost completely over 24 h. The anti-CEA-conjugated NIR fluorescent nanoparticles may be very useful for tumor diagnosis *in vivo*, as they specifically label LS174t colon tumors in the chicken embryo model and in mice.

In future work, we plan to extend the study to include other *in vivo* tumor detection devices, such as whole body imaging or a fluorescent endoscopy camera. Additionally, other tumor-targeting ligands, including antibodies, peptides and proteins (TRAIL and EGF) [34,35] may be conjugated to the nanoparticles. We also plan to encapsulate anti-cancer drugs such as doxorubicin and/or paclitaxel within the NIR fluorescent P(EF-PLLA) nanoparticles, providing a strategy for both diagnosis and therapy of colon cancer.

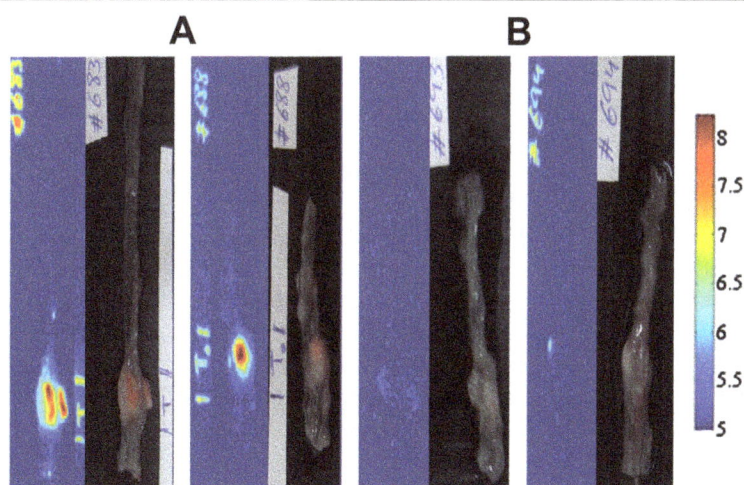

**Figure 9 Fluorescent and grayscale images of typical LS174t colon tumors treated with anti-CEA (A) and anti-rabbit IgG (B) -conjugated NIR fluorescent P(EF-PLLA) nanoparticles.** 20 mice (10 in each experiment group) were anesthetized and treated with 0.1% particle dispersion in PBS, as described in the Materials and methods section. 2 untreated mice served as a control group.

## Materials and methods

### Materials

The following analytical-grade chemicals were purchased from commercial sources and were used without further purification: L-glutamic acid (E), L-phenylalanine (F), indio-cyaninegreen (ICG), human serum albumin (HSA), bovine plasma fibrinogen, 1-ethyl-3-(3-dimethylaminopropyl) carbodiimide (EDC), Matrigel, Triton-x-100 and monoclonal anti-CEA antibodies (T86-66) from Sigma (Rehovot, Israel); Poly(L-lactic acid) MW 2,000 Da from Polysciences (Warrington, PA, USA); N-hydroxysulfosuccinimide (Sulfo-NHS) and 2-morpholino ethanesulfonic acid (MES, pH 6) from Thermo Fisher Scientific (Rockford, IL, USA); Phosphate Buffered Saline (PBS), Minimum Essential Medium (MEM) eagle, McCoy's 5A medium and Dulbecco's modification of eagle's medium (DMEM), Fetal Bovine Serum (FBS), glutamine, penicillin, streptomycin and mycoplasma detection kit from Biological Industries (Bet Haemek, Israel); cell cytotoxicity assay kit (LDH) from Roche (Switzerland); LS174t, SW480 and HT29 cell lines from American Type Culture Collection (ATCC); donkey anti-rabbit IgG from Jackson Immuno-Research Laboratories (West Grove, PA, USA); water was purified by passing deionized water through an Elga-stat Spectrum reverse osmosis system (Elga Ltd., High Wycombe, UK).

### Synthesis and characterization of the P(EF) and P(EF-PLLA) proteinoids

L-glutamic acid (E) was heated at 180°C in an oil bath, under nitrogen atmosphere until melting. The liquefied mass was stirred at 180°C for 30 min. To this, L- phenylalanine (F) and poly(L-lactic acid) (PLLA, Mw = 2000 Da) were added, and kept at 180°C under nitrogen. The total monomer weight was 5 g, PLLA was added at different percentages (1-20% of the total monomer weight) or not added at all. The mixture was mechanically stirred at 150 rpm for 3 h. The product is a highly viscous orange-brown paste, which hardens to give a glassy mass when cooled to room temperature. Then, water (10 mL) was added to the crude product, and the mixture was stirred for 20 min. The solution was then intensively dialyzed through a cellulose membrane (3500 Da MWCO) against distilled water. The content of the dialysis tube was then lyophilized to obtain a yellow-white powder.

The molecular weights and polydispersity indices of the dried P(EF) and P(EF-PLLA) were determined using Gel Permeation Chromatography (GPC) consisting of a Waters Spectra Series P100 isocratic HPLC pump with an ERMA ERC-7510 refractive index detector and a Rheodyne (Coatati, CA) injection valve with a 20 μL loop (Waters, MA). The samples were eluted with super-pure HPLC water through a linear BioSep SEC-s3000 column (Phenomenex) at a flow rate of 1 mL/min.

The molecular weight was determined relative to poly (ethylene glycol) standards (Polymer Standards Service-USA, Silver Spring, MD) with a molecular weight range of 100–450000 Da, Human Serum Albumin (HSA, 67 kDA) and bovine plasma fibrinogen (340 kDa), using Clarity chromatography software. The optical activities of the P (EF-PLLA) and P(EF) were determined using a PE 343 polarimeter (PerkinElmer). The measurements were done in water, at 589 nm at 25°C.

### Synthesis of the non-fluorescent and NIR fluorescent P(EF) and P(EF-PLLA) nanoparticles

The P(EF) and P(EF-PLLA) nanoparticles were prepared by a self-assembly process [21]. Briefly, 100 mg of the dried fabricated P(EF) or P(EF-PLLA) were resuspended in 10 mL of $10^{-5}$ N NaCl solution. The mixture was then heated to 80°C while stirring for 15 min. The mixture was removed from the hot plate and was allowed to return to room temperature. During the cooling process, particles formed and precipitated from the aqueous solution. The formed nanoparticles were then dialyzed versus 4 L of $10^{-5}$ NaCl solution overnight at room temperature.

NIR fluorescent particles P(EF-PLLA) nanoparticles were prepared by the same procedure, with the addition of ICG (1 mg, 1% of the copolymer) to the hot solution, prior to particle formation.

### Characterization of the non-fluorescent and NIR fluorescent P(EF) and P(EF-PLLA) nanoparticles

Particle size and size distribution were determined using DLS with photon cross-correlation spectroscopy (Nano-phox particle analyzer, Sympatec GmbH, Germany) and by Scanning Electron Microscopy (SEM, JOEL, JSM840 Model, Japan). In SEM samples, the diameters of more than 200 particles were measured with AnalySIS Auto image analysis software (Soft Imaging System GmbH, Germany). The density of the particles was determined by pycnometry. Briefly, dry pre-weighed particles were put in a calibrated pycnometer, which was then filled with water. The density of the particles was calculated from the known density of the water, the weight of the pycnometer filled only with water, the weight of the pyc-nometer containing both the sample and water, and the weight of the sample, as described in the literature [36]. Absorbance spectra were obtained using a Cary 100 UV-Visible spectrophotometer (Agilent Technologies Inc.). Excitation and emission spectra were recorded using a Cary Eclipse spectrofluorometer (Agilent Technologies Inc.). ζ-potential measurements were performed by gradual titration of the nanoparticles' aqueous dispersion at pH range from 11 to 2.5 using 1 M HCl (Zetasizer zeta potential analyzer 3000 Has Model, Malvern Instruments, England).

## Determination and optimization of the encapsulated ICG concentration in the NIR fluorescent P(EF-PLLA) nanoparticles

The encapsulated ICG concentration was determined for 1 mg/mL nanoparticles, using a calibration curve of the integrals of absorbance peaks of standard free ICG solutions in PBS at 630–900 nm.

Different ICG quantities were encapsulated within the P(EF-PLLA) nanoparticles at weight% ratios of 0.5, 1, 2, and 5% relative to the P(EF-PLLA). The nanoparticle dispersions were diluted to 1 mg/mL in PBS and their fluorescence intensities at 809 nm were measured.

## Leakage extent of the encapsulated ICG from the nanoparticles dispersed in PBS containing 4% HSA

NIR fluorescent P(EF-PLLA) nanoparticles dispersions (1 mg/mL in PBS containing 4% HSA) was shaken at 37°C for 24 h and then filtered via a 300-kDa filtration tube (VS0241 VIVA SPIN) at 4000 rpm (Centrifuge CN-2200 MRC). The fluorescence intensity of the supernatant was then measured at 809 nm.

## Photostability of the NIR fluorescent P(EF-PLLA) nanoparticles

The photostability of the NIR fluorescenet P(EF-PLLA) nanoparticles was examined by recording the fluorescence intensity over a period of 20 min of nanoparticle dispersion in PBS and a PBS solution of free ICG (0.05 M). The nanoparticle dispersion was diluted to give comparable fluorescence intensity to the ICG solution, with λex set at 780 nm and λem set at 800 nm. The excitation and emission slits were opened to 20 nm and 5 nm, respectively. Each of the samples was illuminated continuously with a xenon lamp, and the fluorescence intensity was recorded using a Cary Eclipse fluorescence spectrophotometer (Agilent Technologies Inc.). Intensity values were normalized for comparison.

## Particle stability in storage conditions

NIR fluorescent P(EF-PLLA) nanoparticles aqueous dispersions (1 mg/mL) were put in a refrigerator at 4°C for 6 months. Samples were taken at different time periods, filtered through a centrifugation tube (Vivaspin 3000 Da MWCO) and the filtrate was checked by UV at 200–210 nm, to find soluble P(EF-PLLA) and at 630–900 nm to find free ICG. The nanoparticle size, size distribution and fluorescence were checked as well by the same procedures mentioned above.

Additionally, the nanoparticles were freeze-dried in order to check their stability after drying. Following lyophilization, the nanoparticles were resuspended in PBS to their original concentration and the dispersions were retested for particle size, size distribution and fluorescence intensity.

## In vitro cytotoxicity of the NIR fluorescent P(EF-PLLA) nanoparticles

In vitro cytotoxicity of the P(EF-PLLA) nanoparticles was tested by using human colorectal adenocarcinoma LS174t, SW480 and HT29 cell lines. The cell lines are adherent to the used culture dishes. LS174t cells were grown in Minimum Essential Medium (MEM) eagle supplemented with heat-inactivated fetal bovine serum (FBS, 10%), penicillin (100 IU/mL), streptomycin (100 μg/mL) and L-glutamine (2 mM). SW480 cells were maintained in Dulbecco's MEM supplemented with heat-inactivated fetal bovine serum (FBS, 10%), penicillin (100 IU/mL), streptomycin (100 μg/mL) and L-glutamine (2 mM). HT29 cells were maintained in McCoy's 5A medium supplemented with FBS (10%), penicillin (100 IU/mL), streptomycin (100 μg/mL) and L-glutamine (2 mM). Cells were screened to ensure they remained mycoplasma-free using Mycoplasma Detection Kit [37].

Cell cytotoxicity was assessed by measuring the release of cytoplasmic lactate dehydrogenase (LDH) into cell culture supernatants. LDH activity was assayed using the Cytotoxicity Detection Kit according to the manufacturer's instructions [30]. Cells ($3 \times 10^5$ cells per well) were seeded and grown to 90–95% confluency in 24 well plates before treatment with the P(EF-PLLA) nanoparticles. Cell cultures that were not exposed to the nanoparticles were included in all assays as negative controls. Cell cultures that were treated with 1% Triton-x-100 were used as positive controls. To test if the nanoparticles can interact with LDH kit compounds, cell cultures were exposed to a mixture containing maximal nanoparticles concentration (2.5 mg/mL) dispersed in PBS and 1% Triton-x-100.

The P(EF-PLLA) nanoparticles were freshly dispersed in PBS (1.25 and 2.5 mg/mL) and then added to the 95% confluent cell culture in culture medium. The cell cultures were further incubated at 37°C in a humidified 5% CO₂ incubator and then checked for cellular cytotoxicity at intervals of 24 h. The percentage of cell cytotoxicity was calculated using the formula shown in the manufacturer's protocol [30]. All samples were tested in tetraplicates.

## Biodistribution in a mouse model

Male BALB/C mice (Harlan Laboratories, Israel) were utilized in this study under a protocol approved by the Institutional Animal Care and Use Committee at Bar-Ilan University. The biodistribution of the NIR fluorescent P(EF-PLLA) nanoparticles was studied in normal 8-weeks-old mice, weighing 20–25 g at the time of experiment. Prior to the experiment, mice were anesthetized by intraperitoneal injection of Ketamine (40–80 mg/kg body weight) and Xylazine (5–10 mg/kg body weight), and the mice's skin was shaved with an electric animal clipper.

100 μL of either nanoparticle dispersion or free ICG solution (0.01 mg/kg body weight, dissolved in PBS)

were administered to the mice through tail vein injection at a concentration of 2 mg/mL. During image acquisition, mice remained anesthetized by the intraperitoneal injection of Ketamine/Xylazine. Image cubes were obtained from the mice at several time points up to 24 h after injection. Each treatment group includes 3 mice for each time point (5 min, 20 min, 1 h and 24 h); 2 uninjected mice served as negative control. The experiment was repeated twice, testing a total of 52 mice. At the end of the experiment, the mice were euthanized by cervical dislocation, and organs were taken for imaging (liver, spleen, kidney, duodenum, colon, brain, heart, tibia bone and blood).

Whole body fluorescence images were acquired using a Maestro II *in vivo* fluorescence imaging system (Cambridge Research &Instrumentation, Inc., Woburn, MA). The system is equipped with a fiber-delivered 300 W xenon excitation lamp, and images can be acquired from $\lambda = 500$-950 nm by a 1.3 megapixel CCD camera (Sony ICX285 CCD chip). Each pixel within the image cube therefore has an associated fluorescence spectrum. The software for the Maestro system (Maestro 2.10.0) contains several algorithms to process the spectral data cubes to remove undesired auto-fluorescence signal and generate overlaid images for multiple fluorophores. A deep red excitation/emission filter set was used for our experiments ($\lambda$ex: 700–770 nm, $\lambda$em > 780 nm). The liquid crystal tunable filter (LCTF) was programmed to acquire image cubes from $\lambda = 780$ nm-860 nm with an increment of 10 nm per image. The camera was set to 150 ms (whole body image), 15 ms (liver), 500 ms (spleen), 7000 ms (kidney), 10 ms (duodenum), 500 ms (colon), 1000 ms (brain), 1000 ms (tibia bones), 200 ms (heart) and 1000 ms (blood) exposure times. Fluorescence intensity measurements were performed using ImageJ NIH (National Institutes of Health) software.

### Conjugation of tumor-targeting ligands to the NIR fluorescent P(EF-PLLA) nanoparticles

Anti-CEA was covalently conjugated to the nanoparticles through carbodiimide activation of the carboxylate groups on the particle surface [6]. Briefly, EDC (1 mg) and Sulfo-NHS (1 mg) were dissolves separately in 1 mL 0.1 M MES containing 0.5 M NaCl. The EDC solution (10 μL, 1 mg/mL) was then added to an aqueous anti-CEA solution (62.5 μL, 0.25 mg) followed by the addition of the Sulfo-NHS solution (25 μL, 1 mg/mL). The mixture was shaken at room temperature for 15 min, and then the NIR fluorescent nanoparticle dispersion was added (2.5 mg in 1 mL PBS). The mixture was shaken for an additional 90 min. The obtained anti-CEA-conjugated fluorescent nanoparticles were then washed from excess reagents by dilution and filtration through a 30-kDa filtration tube (VS2021 VIVA SPIN) at 1000 rpm (Centrifuge

CN-2200 MRC) for 2 min, repeated three times. Anti-rabbit IgG was conjugated to the NIR fluorescent nanoparticles through a similar procedure. The concentration of bound anti-CEA and anti-rabbit IgG ($1.9 \pm 0.2$ μg/mg nanoparticles) was determined using a mouse IgG ELISA kit (Biotest, Israel).

### Optical detection of human colon tumor with the non-conjugated and bio-conjugated NIR fluorescent P(EF-PLLA) in a chicken embryo chorioallantoic membrane (CAM)

Human colorectal adenocarcinoma LS174t and SW480 cell lines were used for each of the experiments and maintained as mentioned above. Tumor cells were grafted on CAM according to the literature [38]. Briefly, fertile chicken eggs obtained from a commercial supplier were incubated at 37°C at 60–70% humidity in a forced-draft incubator. On day 3 of incubation, an artificial air sac was formed, allowing the CAM to drop. After 8 days of incubation, a window was opened in the shell and the CAM was exposed. Tumor cells were collected by trypsinization, washed with culture medium and pelleted by gentle centrifugation. Following removal of the medium, $5 \times 10^6$ cells were resuspended in 30 μL ice-cold Matrigel and inoculated on the CAM at the site of the blood vessels. Eggs were then sealed and returned to incubation. On day 6 post-grafting, day 14 of incubation, the tumor diameter ranged from 3 to 5 mm with visible neoangiogenesis.

Chicken embryos with 6-days-old human adenocarcinoma tumors (LS174t and SW480 cancer cell lines) implanted on the CAM were treated with the non-conjugated, anti-CEA-conjugated and anti-rabbit IgG-conjugated NIR fluorescent P(EF-PLLA) nanoparticles (40 μL, 2 mg/mL). Additionally, non-pathological CAM treated with nanoparticles and untreated tumors served as control groups. After 40 minutes, the nanoparticle dispersions were removed and the tumors were washed with PBS. Then, the tumors and the non-pathological CAM were removed from the eggs, washed again with PBS and spread on a mat black background for observation using a Maestro II *in vivo* imaging system (Cambridge Research & Instrumentation, Inc., Woburn, MA). A NIR excitation/emission filter set was used for the experiments ($\lambda$ex: 710–760 nm, $\lambda$em > 750 nm). The Liquid Crystal Tunable Filter (LCTF) was programmed to acquire image cubes from $\lambda = 790$ nm to 860 nm with an increment of 10 nm per image. Fluorescence intensity measurements were calculated as average intensity over the tumor surface area, using ImageJ software.

### Optical detection of human colon tumor with the non-conjugated and bio-conjugated NIR fluorescent P(EF-PLLA) in a mouse model

Experiments were performed according to the protocols of the Israeli National Council for Animal Experiments by Harlan Biotech, Israel. Cancerous cells (30 μL containing $2 \times 10^6$ LS174t cells) were injected into the mouse intestinal wall. 2 weeks later the nude mice were anaesthetized and treated with the bio-conjugated NIR fluorescent P (EF-PLLA) nanoparticles (0.1%, 200 μL), through the anus, using the guidance of a mini-colonoscope. 20 min later each colon was washed with PBS (5 × 1 mL) and mice were allowed to recover for 4 h. The mice were sacrificed and the colons were removed. Each colon was spread on a solid surface and imaging was performed using the Odyssey Infrared Imaging System (Li-Cor Biosciences, Lincoln, NE, USA) with excitation wavelength of 780 nm and emission wavelength of 800 nm.

## Competing interests
The authors declare that they have no competing interests.

## Authors' contributions
MKD carried out the synthesis and characterization of the nanoparticles. IG and ECS carried out the biological studies, including toxicity assays and the in vivo assays. SM supervised the study, and participated in it's the design and coordination. All authors read and approved the final manuscript.

## Acknowledgments
The authors thank Harlan Biotech Israel for their assistance in performing the in vivo mice experiments. Thanks also Dr. Ronen Yehuda of the Life Sciences Faculty, Bar-Ilan University for his assistance in performing the fluorescence imaging.

## References
1. Burt RW, Barthel JS, Dunn KB, David DS, Drelichman E, Ford JM, Giardiello FM, Gruber SB, Halverson AL, Hamilton SR: Colorectal cancer screening. J Natl Compr Canc Netw 2010, 8:8–61.
2. Nelson RS, Thorson AG: Colorectal cancer screening. Curr Oncol Rep 2009, 11:482–489.
3. Mandel JS, Church TR, Bond JH, Ederer F, Geisser MS, Mongin SJ, Snover DC, Schuman LM: The effect of fecal occult-blood screening on the incidence of colorectal cancer. N Eng J Med 2000, 343:1603–1607.
4. Labianca R, Beretta GD, Mosconi S, Milesi L, Pessi MA: Colorectal cancer: screening. Ann Oncol 2005, 16:127–132.
5. Altınoğlu Eİ, Adair JH: Near infrared imaging with nanoparticles. Wiley Interdiscip Rev Nanomed Nanobiotechnol 2010, 2:461–477.
6. Cohen S, Pellach M, Kam Y, Grinberg I, Corem-Salkmon E, Rubinstein A, Margel S: Synthesis and characterization of near IR fluorescent albumin nanoparticles for optical detection of colon cancer. Mat Sci Eng C 2012, 33:923–931.
7. Jiang S, Gnanasammandhan MK, Zhang Y: Optical imaging-guided cancer therapy with fluorescent nanoparticles. J R Soc Interface 2009, 7:3–18.
8. Santra S, Dutta D, Walter GA, Moudgil BM: Fluorescent nanoparticle probes for cancer imaging. Technol Cancer Res Treat 2005, 4:593–602.
9. Gluz E, Rudnick Glick S, Mizrahi DM, Chen R, Margel S: New biodegradable bisphosphonate vinylic monomers and near infrared fluorescent nanoparticles for biomedical applications. Polym Advan Technol 2014, 25:499–506.
10. Sharrna P, Brown S, Walter G, Santra S, Moudgil B: Nanoparticles for bioimaging. Adv Colloid Interfac 2006, 123:471–485.
11. Gluz E, Grinberg I, Corem-Salkmon E, Mizrahi D, Margel S: Engineering of new crosslinked near-infrared fluorescent polyethylene glycol bisphosphonate nanoparticles for bone targeting. J Polym Sci Pol Chem 2013, 51:4282–4291.
12. Metildi CA, Kaushal S, Luiken GA, Talamini MA, Hoffman RM, Bouvet M: Fluorescently labeled chimeric anti-CEA antibody improves detection and resection of human colon cancer in a patient derived orthotopic xenograft (PDOX) nude mouse model. J Surg Oncol 2014, 109:451–458.
13. Saxena V, Sadoqi M, Shao J: Enhanced photo-stability, thermal-stability and aqueous-stability of indocyanine green in polymeric nanoparticulate systems. J Photoch Photobio B 2004, 74:29–38.
14. Quirk S: Triggered release of small molecules from proteinoid microspheres. J Biomed Mater Res, Part A 2009, 91A:391–399.
15. Quirk S: Triggered release from peptide-proteinoid microspheres. J Biomed Mater Res, Part A 2010, 92A:877–886.
16. Kumar ABM, Rao KP: Preparation and characterization of pH-sensitive proteinoid microspheres for the oral delivery of methotrexate. Biomaterials 1998, 19:725–732.
17. Fox SW, Waehneld T: Thermal synthesis of neutral and basic proteinoids. Biochim Biophys Acta 1968, 160:246–249.
18. Fox SW, Nakashima T, Przybylski A, Syren RM: The updated experimental proteinoid model. Int J Quantum Chem 1982, 22:195–204.
19. Fox SW, Harada K: The thermal copolymerization of amino acids common to protein. J Am Chem Soc 1959, 82:3745–3751.
20. Harada K, Fox SW: The thermal condensation of glutamic acid and glycine to linear peptides. J Am Chem Soc 1957, 80:2694–2697.
21. Kokufuta E, Sakai H, Harada K: Factors controlling the size of Proteinoid microspheres. BioSystems 1983, 16:175–181.
22. Auras R: Poly (lactic acid). In Encyclopedia Of Polymer Science and Technology. Hoboken, NJ: Wiley Interscience; 2010.
23. Tsuji H: Poly (lactic acid). In Bio-Based Plastics: Materials and Applications. Edited by Kabasci S. Chichester, UK: John Wiley & Sons; 2013:171–239.
24. Bahn PR, Pappelis A, Bozzola J: Protocell-like microspheres from thermal Polyaspartic acid. Orig Life Evol Biospheres 2006, 36:617–619.
25. Syren RM, Sanjur A, Fox SW: Proteinoid microspheres more stable in hot than in cold water. BioSystems 1985, 17:275–280.
26. Steiner S, Rosen R: Delivery systems for pharmacological agents encapsulated with proteinoids, vol. U.S. Patent No. 4,925,673. Washington, USA: 1990.
27. Haritoglou C, Freyer W, Priglinger SG, Kampik A: Light absorbing properties of indocyanine green (ICG) in solution and after adsorption to the retinal surface—an ex-vivo approach. Graef Arch Clin Exp 2006, 244:1196–1202.
28. Zweck J, Penzkofer A: Microstructure of indocyanine green J-aggregates in aqueous solution. Chem Phys 2001, 269:399–409.
29. Ma XH, Santiago N, Chen YS, Chaudhary K, Milstein SJ, Baughman RA: Stability study of drug-loaded proteinoid microsphere formulations during freeze-drying. J Drug Target 1994, 2:9–21.
30. Decker T, Lohmannmatthes ML: A quick and simple method for the quantitation of lactate-dehydrogenase release in measurements of cellular cyto-toxicity and Tumor Necrosis Factor (Tnf) activity. J Immunol Methods 1988, 115:61–69.
31. Yaseen MA, Yu J, Jung B, Wong MS, Anvari B: Biodistribution of encapsulated indocyanine green in healthy mice. Mol Pharm 2009, 6:1321–1332.
32. Mizrahi DM, Ziv-Polat O, Perlstein B, Gluz E, Margel S: Synthesis, fluorescence and biodistribution of a bone-targeted near-infrared conjugate. Eur J Med Chem 2011, 46:5175–5183.
33. Kaushal S, McElroy MK, Luiken GA, Talamini MA, Moossa A, Hoffman RM, Bouvet M: Fluorophore-conjugated anti-CEA antibody for the intraoperative imaging of pancreatic and colorectal cancer. J Gastrointest Surg 2008, 12:1938–1950.
34. Bae S, Ma K, Kim TH, Lee ES, Oh KT, Park ES, Lee KC, Youn YS: Doxorubicin-loaded human serum albumin nanoparticles surface-modified with TNF-related apoptosis-inducing ligand and transferrin for targeting multiple tumor types. Biomaterials 2012, 33:1536–1546.
35. Capdevila J, Elez E, Macarulla T, Ramos FJ, Ruiz-Echarri M, Tabernero J: Anti-epidermal growth factor receptor monoclonal antibodies in cancer treatment. Cancer Treat Rev 2009, 35:354–363.
36. Heiskanen J: Comparison of three methods for determining the particle density of soil with liquid pycnometers. Commun Soil Sci Plant Anal 1992, 23:841–846.

37.  Epsztejn S, Glickstein H, Picard V, Slotki IN, Breuer W, Beaumont C, Cabantchik ZI: **H-ferritin subunit overexpression in erythroid cells reduces the oxidative stress response and induces multidrug resistance properties.** *Blood* 1999, **94:**3593–3603.

38.  Noiman T, Buzhor E, Metsuyanim S, Harari-Steinberg O, Morgenshtern C, Dekel B, Goldstein RS: **A rapid in vivo assay system for analyzing the organogenetic capacity of human kidney cells.** *Organogenesis* 2011, **7:**140–144.

# Development of an insecticidal nanoemulsion with *Manilkara subsericea* (Sapotaceae) extract

Caio Pinho Fernandes[1,2]*, Fernanda Borges de Almeida[1], Amanda Nunes Silveira[3], Marcelo Salabert Gonzalez[4], Cicero Brasileiro Mello[4], Denise Feder[4], Raul Apolinário[4], Marcelo Guerra Santos[5], José Carlos Tavares Carvalho[6], Luis Armando Cândido Tietbohl[7], Leandro Rocha[7] and Deborah Quintanilha Falcão[3]

## Abstract

**Background:** Plants have been recognized as a good source of insecticidal agents, since they are able to produce their own defensives to insect attack. Moreover, there is a growing concern worldwide to develop pesticides with low impact to environment and non-target organisms. Hexane-soluble fraction from ethanolic crude extract from fruits of *Manilkara subsericea* and its triterpenes were considered active against a cotton pest (*Dysdercus peruvianus*). Several natural products with insecticidal activity have poor water solubility, including triterpenes, and nanotechnology has emerged as a good alternative to solve this main problem. On this context, the aim of the present study was to develop an insecticidal nanoemulsion containing apolar fraction from fruits of *Manilkara subsericea*.

**Results:** It was obtained a formulation constituted by 5% of oil (octyldodecyl myristate), 5% of surfactants (sorbitan monooleate/polysorbate 80), 5% of apolar fraction from *M. subsericea* and 85% of water. Analysis of mean droplet diameter (155.2 ± 3.8 nm) confirmed this formulation as a nanoemulsion. It was able to induce mortality in *D. peruvianus*. It was observed no effect against acetylcholinesterase or mortality in mice induced by the formulation, suggesting the safety of this nanoemulsion for non-target organisms.

**Conclusions:** The present study suggests that the obtained O/A nanoemulsion may be useful to enhance water solubility of poor water soluble natural products with insecticidal activity, including the hexane-soluble fraction from ethanolic crude extract from fruits of *Manilkara subsericea*.

**Keywords:** *Dysdercus peruvianus*, *Manilkara subsericea*, Nanoemulsions

## Background

Chemical pesticides have been used to control pest insects, however, they are usually toxic to environment. There is a growing concern worldwide regarding indiscriminate use of these substances, which are associated to environmental pollution and toxicity risk to non-targeted organisms [1]. Plant species are well recognized by their ability to produce defensive substances, in order to protect themselves from insect attack [2]. These natural products appear as potential sources of new biodegradable insecticides with wide range of mechanisms of action, being an important alternative for insect pest management in agriculture [3]. One of the most promising and recognized group of substances with insecticidal activity are the triterpenes [4].

*Manilkara subscericea* (Mart.) Dubard (Sapotaceae) is an endemic species of Brazilian Atlantic Forest [5] and widely distributed at Restinga de Jurubatiba National Park (Rio de Janeiro State, Brazil) [6]. Several non-polar pentacyclic triterpenes have been described as major constituents of *M. subsericea*, mainly alpha- and beta-amyrin esters [7,8]. Hexane-soluble fraction from ethanolic crude extract from fruits of *M. subsericea* and its major substances (alpha- and beta-amyrin acetate) was able to induce mortality, delayed development and inhibition of moulting in *Dysdercus peruvianus* [9], a

* Correspondence: caio_pfernandes@yahoo.com.br
[1]Programa de Pós, Graduação em Biotecnologia Vegetal, Centro de Ciências da Saúde, Universidade Federal do Rio de Janeiro – UFRJ, Bloco K, 2º andar – sala 032, Av. Brigadeiro Trompowski s/n, CEP: 21941-590 Ilha do Fundão, RJ, Brazil
[2]Laboratório de Farmacotécnica, Colegiado de Ciências Farmacêuticas, Universidade Federal do Amapá, Campus Universitário Marco Zero do Equador, Rodovia Juscelino Kubitschek – KM – 02-Jardim Marco Zero, CEP: 68903-419 Macapá, AP, Brazil
Full list of author information is available at the end of the article

hemiptera species which causes serious loss of cotton crops [10]. This apolar fraction and its triterpenes have poor water solubility and are soluble in toxic organic solvents, such as chloroform and dichloromethane, being this intrinsic characteristic a technological challenge if development if a viable product is desired.

Nanotechnology has emerged as a promising area for development of products in a wide range of applications, including pesticide agents. Considering that many of the insecticides known today are organic compounds with poor water solubility, development of nanoproducts appear to solve this main problem, enhancing water solubility, bioavailability and resulting in stable formulations without utilization of organic toxic solvents [11]. Nanoemulsions are one of the most important formulations to enhance solubility and dissolution properties of poorly water soluble substances [12]. They are also referred as miniemulsions or ultrafine emulsions and have small droplet size (20-200 nm). They are transparent or translucent, often presenting a bluish reflect and have high kinetic stability [13]. Low energy methods have been used to achieve nanoemulsions, including reverse-phase composition (RPC) and temperature of inversion phase (TIF) [14]. Formulation screening stage is crucial if development of a stable nanoformulation is desired, especially if a low energy method is employed, being determination of required HLB value of an oil [15] and construction of pseudo-ternary phase diagrams [16] very useful, especially to achieve nanoemulsions.

On this context, the aim of the present study was to develop an insecticidal nanoemulsion containing apolar fraction from fruits of *Manilkara subsericea* and verify its effects against *Dysdercus peruvianus* and non-target organisms.

## Results and discussion

Preliminary solubility studies were performed regarding choice of oil phase and surfactants. Octyldodecyl myristate (MOD®) was the best oil, being able to solubilize equal amount (1:1, w/w) of hexane-soluble fraction from fruits of *M. subsericea* (HF). It is frequently necessary to use blends, such as a pair of hydrophilic and lipophilic non-ionic surfactants, to achieve droplets with small diameter [17]. Sorbitan oleate and polysorbate 80 were considered the best pair (Data not shown). These surfactants have been used in low energy methods, being able to produce nanoemulsions with smaller mean droplet size, when compared to other surfactants. This could be explained by the ability of this couple to induce formation of a looser film, which is associated to generation of nanoemulsions [12]. Addition of water to a surfactant in oil solution was employed in the present study, since it provided better results, when compared to addition of oil to an aqueous surfactant solution (Data not shown).

This could be attributed to phase transitions and changes in the curvature of the surfactant from W/O to O/W during emulsification process [18].

In order to predict the best ratio of surfactants to be used, several emulsions were prepared varying the relative amounts of sorbitan oleate and polysorbate 80. Most of them presented instable behavior, including critical macroscopical changes, such as creaming and phase separation. Surfactants can be classified according to their Hydrophile-Lipophile Balance (HLB), a semi-empirical scale [19] and several HLB values can be obtained using different amounts of each component of a couple of surfactants [20]. Emulsions with HLB values of 10 (sorbitan oleate/polysorbate ratio, 1.0/1.1) and 11 (sorbitan oleate/polysorbate ratio, 1.0/1.7) were considered more stable. A second set of emulsions within this HLB range was prepared and the obtained formulations presented translucent aspect and bluish reflect, which is characteristic for nanoemulsions [13]. Mean droplet size analysis indicated that nanoemulsion with HLB value of 10.75 (sorbitan oleate/polysorbate ratio, 1.0/1.5) presented the smallest mean diameter (50.6 ± 0.4 nm) and low polydispersity (0.164 ± 0.021). Stable formulations with low mean droplet size can be obtained when HLB value of the surfactant couple coincides with required HLB value of the oil [12,20]. Thus, required HLB of oil can be determined by calculating the HLB value of emulsifier or emulsifier mixture which was able to induce formation of the most stable formulation, among a set of emulsions prepared with different blends of a couple of emulsifiers in a wide range of HLB value [21]. Our results indicate that 10.75 should be the required HLB value of MOD® used in the present study.

It is observed that not every combination of components produces nanoemulsions over the whole range of possible compositions [22]. Thus, a total of 28 emulsions were prepared using different percentages of water, MOD® and surfactants (sorbitan monoleate/polysorbate 80, HLB 10.75) and mean droplet size of each formulation was analyzed. Mean droplet size ranged from 45.9 ± 0.4 nm (oil 15%, surfactants 15%, water 70%) to 421.5 ± 50.4 nm (oil 2.5%, surfactants 7.5%, water 90%) (Table 1). Composition of each emulsion obtained can be expressed as a pseudo-ternary phase diagram, which is represented by equilateral triangle in which four or more constituents are investigated [22] and is very useful to determine relation between phase behavior of a mixture and its composition [23]. In all, 23 emulsion presented mean droplet size bellow 200 nm, ranging 45.9 ± 0.4 nm to 196.4 ± 12.5 and were used to delineate the nanoemulsion region (Figure 1). Small mean droplet sizes, such as 48.7 ± 0.2 nm (7.5% of oil, 10% of surfactants, 82.5% of water), 51.7 ± 0.2 nm (10% of oil, 12.5% of surfactants, 77.5% of water) and 45.9 ± 0.4 nm (15% of oil, 15% of surfactants, 70% of water) were

**Table 1 Composition, mean droplet size and polydispersity of each formulation prepared during construction of pseudo-ternary phase diagram for delimitation of nanoemulsion region**

| | % of oil | % of surfactants | % of water | Mean diameter (nm) | Polydispersity |
|---|---|---|---|---|---|
| 1[a] | 5 | 5 | 90 | 50.6 ± 0.4 | 0.164 ± 0.021 |
| 2 | 2.5 | 5 | 92.5 | 234.2 ± 12.5 | 0.025 ± 0.012 |
| 3 | 2.5 | 7.5 | 90 | 421.5 ± 50.4 | 0.005 ± 0.000 |
| 4[a] | 5 | 7.5 | 87.5 | 196.4 ± 12.5 | 0.178 ± 0.044 |
| 5[a] | 7.5 | 5 | 87.5 | 145.7 ± 8.6 | 0.132 ± 0.038 |
| 6 | 7.5 | 2.5 | 90 | 256.9 ± 8.6 | 0.016 ± 0.011 |
| 7[a] | 5 | 2.5 | 92.5 | 151.7 ± 6.0 | 0.155 ± 0.018 |
| 8[a] | 10 | 5 | 85 | 139.4 ± 9.6 | 0.078 ± 0.049 |
| 9[a] | 10 | 7.5 | 82.5 | 133.5 ± 0.5 | 0.247 ± 0.011 |
| 10[a] | 7.5 | 7.5 | 85 | 139.5 ± 4.7 | 0.073 ± 0.039 |
| 11[a] | 10 | 10 | 80 | 86.8 ± 0.9 | 0.294 ± 0.006 |
| 12[a] | 7.5 | 10 | 82.5 | 48.7 ± 0.2 | 0.313 ± 0.002 |
| 13 | 5 | 10 | 85 | 234.4 ± 2.3 | 0.270 ± 0.016 |
| 14 | 5 | 12.5 | 82.5 | 298.5 ± 31.8 | 0.005 ± 0.000 |
| 15[a] | 7.5 | 12.5 | 80 | 85.4 ± 1.2 | 0.350 ± 0.005 |
| 16[a] | 10 | 12.5 | 77.5 | 51.7 ± 0.2 | 0.331 ± 0.005 |
| 17[a] | 7.5 | 15 | 77.5 | 162.4 ± 1.3 | 0.335 ± 0.011 |
| 18[a] | 7.5 | 17.5 | 75 | 159.2 ± 2.3 | 0.334 ± 0.007 |
| 19[a] | 10 | 15 | 75 | 68.7 ± 0.8 | 0.360 ± 0.004 |
| 20[a] | 10 | 17.5 | 72.5 | 87.9 ± 1.5 | 0.365 ± 0.003 |
| 21[a] | 12.5 | 15 | 72.5 | 170.6 ± 6.2 | 0.144 ± 0.027 |
| 22[a] | 12.5 | 12.5 | 75 | 66.6 ± 1.1 | 0.304 ± 0.005 |
| 23[a] | 12.5 | 10 | 77.5 | 120.1 ± 1.0 | 0.225 ± 0.003 |
| 24[a] | 15.0 | 12.5 | 72.5 | 95.3 ± 0.6 | 0.255 ± 0.006 |
| 25[a] | 15 | 15 | 70 | 45.9 ± 0.4 | 0.271 ± 0.005 |
| 26[a] | 12.5 | 7.5 | 80 | 75.4 ± 2.3 | 0.340 ± 0.005 |
| 27[a] | 17.5 | 15.0 | 67.5 | 161.6 ± 1.9 | 0.266 ± 0.007 |
| 28[a] | 12.5 | 17.5 | 70 | 97.2 ± 1.0 | 0.256 ± 0.002 |

Oil – MOD®.
Surfactants – sorbitan monooleate/polysorbate 80 at HLB of 10.75.
[a]Formulations in the nanoemulsion region.

obtained (Table 1). This data may be important for further studies or development of nanoformulations using MOD®, sorbitan oleate, polysorbate 80 and water.

Special attention was given to formulation comprised by 5% of MOD, 5% of surfactants and 90% of water, which also presented stable behavior, small mean droplet size (50.6 ± 0.4 nm) and low polidispersity (0.164 ± 0.021). Low surfactant percentage could be considered an advantage, since further preparation of this formulation would reduce toxicity and costs with raw materials, when compared to other nanoemulsions with higher concentrations of surfactants. Thus, this formulation

was chosen to prepare a nanoemulsion with hexane-soluble fraction from fruits of *Manilkara subsericea* dispersed through internal phase (HFNE).

Concentration of extract corresponded to equal percentage of MOD®, based on its intrinsic solubility. This amount was discounted from water percentage, being HFNE constituted by 5% of MOD, 5% of surfactants, 5% of hexane-soluble fraction from fruits of *M. subsericea* and 85% of water. A blank nanoemulsion without hexane-soluble fraction of *M. subsericea* extract (HF) was prepared for negative control. Both nanoemulsions presented a characteristic bluish reflect, associated to Tyndall effect [13] (Figure 2). It was observed an increase in the nanoemulsion mean droplet size when HF was dispersed through oil phase (Figure 3). This fact could be explained due to deposition of substances, which may reduce the flexibility of the surfactant film and result in more compact films instead of looser films and smaller mean droplets [12]. Previous gas chromatography analysis of HF indicated a high relative percentage of pentacyclic triterpenes, including beta-amyrin acetate (10.27%), alpha-amyrin acetate (42.34%), beta-amyrin caproate (5.46%), alpha-amyrin caproate (7.26%), beta-amyrin caprylate (2.44%) and alpha-amyrin caprylate (5.04%) [8]. These substances may be contributing to the result described above.

HFNE and blank nanoemulsion presented zeta potential values of − 47.4 ± 3.2 and − 59.6 ± 4.1, respectively. Zeta potential is a special parameter that should be analyzed, in order to determine stability of nanoemulsions and is associated to surface potential of the droplets [24]. Maximum stability is observed when zeta potential value is above ± 30 mV [25]. The high stability of formulations with great zeta potential values is associated to repulsive forces that exceed attracting Van der Waals forces, resulting in dispersed particles and a deflocculated system [23]. Macroscopical analysis of the nanoemulsion with HF and blank nanoemulsion indicated that these formulations maintained their original fine appearance and bluish reflection. It was observed no phase separation, creaming and sedimentation under room temperature (25 ± 2°C) and accelerated stability evaluation. Long term physical stability of a nanoemulsion related to its small droplets, making this type of formulation being also referred as "approaching thermodynamic stability" [26,13].

Insecticidal assay was performed in order to verify if HFNE is able to induce mortality in *D. peruvianus*. During the whole experimental period, it was observed that HFNE (treated group) did not interfere in body weight, when compared to untreated group, indicating the absence of antifeedant effect. This effect was also not detected in negative control group. It was not observed overaged, extranumerary nymphs or insects with body

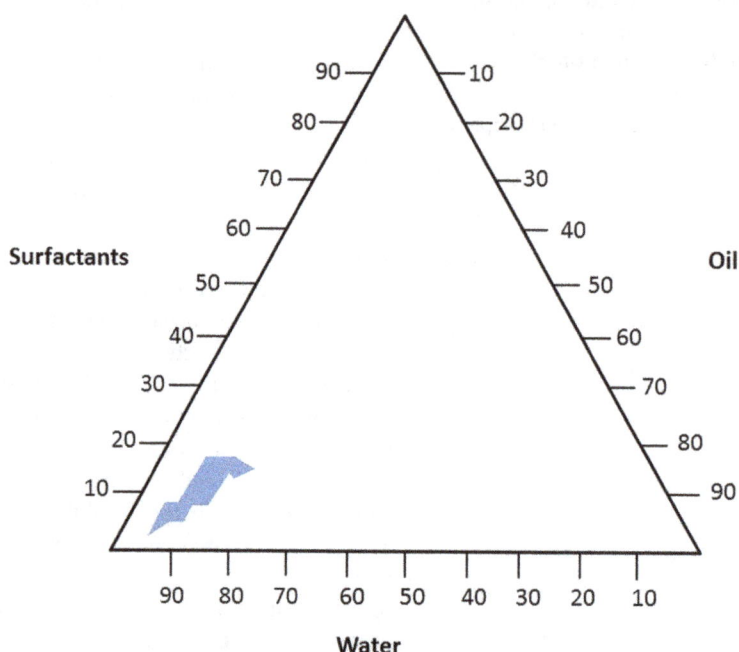

**Figure 1 Pseudo-ternary phase diagram constructed with water, MOD® and surfactants (sorbitan monoleate/polysorbate 80, HLB =10.75) at different compositions.** Nanoemulsion region is delimited in blue.

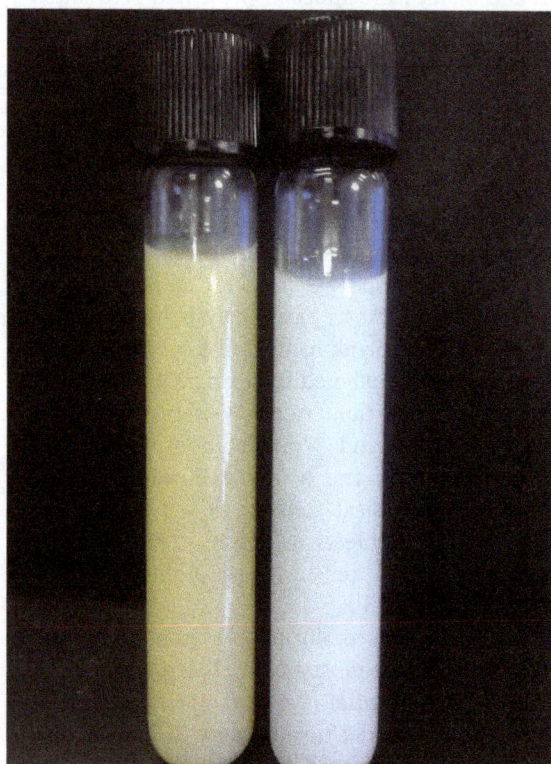

**Figure 2 Nanoemulsions obtained by low energy method.** HFNE shown in left side and blank nanoemulsion shown in right side of the picture.

deformations. Figure 4 indicates that mortality in the untreated group ranged from $3.3 \pm 1.15\%$, between 5° and 14° days of observation, to $10 \pm 1.53\%$, between 15° and 30° of observation. Negative control group (treated with blank nanoemulsion) presented higher levels of mortality throughout the experimental period, reaching $(6.6 \pm 1.15\%)$ $(p < 0.001)$ after 4 days, $13.3 \pm 2.52\%$ $(p < 0.001)$ after 14 days and $21.10 \pm 3.06\%$ $(p < 0.001)$ after 30 days of treatment. Treatment of insects with HFNE exhibited significantly higher levels of mortality. It was observed that mortality began on the first day after treatment $(12.23 \pm 0.58\%)$ $(p < 0.001)$, reached $22.23 \pm 1.73\%$ $(p < 0.001)$ after 14 days and $44.43 \pm 6.66\%$ $(p < 0.0001)$ after the end of the experiment. Significant differences between the group of insects treated with hexanic nanoemulsion containing extract of *M. subsericea* and the control group were detected in almost all days of observation until the end of the experiment (ANOVA, $p < 0.005$). However, it is worth to note that there was no statistical difference between HFNE-treated and blank nanoemulsion-treated insects among days 21-23 after treatment. Perhaps, differences in the speed of absorption between HFNE and blank nanoemulsion by insect metabolic systems may explain this not expectable result. Physiological mechanisms of metabolization of these compounds by invertebrates remains unknown [9] and is now under investigation by our research group.

Changes in the time period in which occur the processes of molt and metamorphosis were observed in

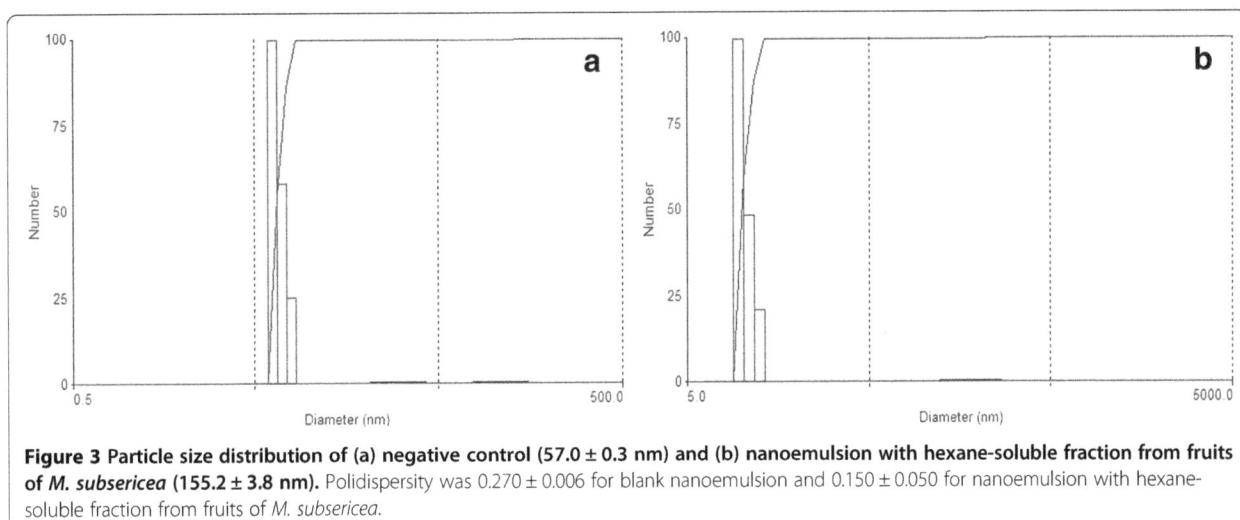

**Figure 3 Particle size distribution of (a) negative control (57.0 ± 0.3 nm) and (b) nanoemulsion with hexane-soluble fraction from fruits of *M. subsericea* (155.2 ± 3.8 nm).** Polidispersity was 0.270 ± 0.006 for blank nanoemulsion and 0.150 ± 0.050 for nanoemulsion with hexane-soluble fraction from fruits of *M. subsericea*.

group treated with HFNE and negative group (Data not shown). Moreover, an associated high mortality rate were displayed continuously and gradually increasing throughout insects lifecycle regardless whether the insects were in the nymphal or adult stage. This observation point out to a physiological connection between the neuroendocrine control of the insect development and the reduced longevity obtained after treatments. These results suggest that HFNE may able to release insecticidal components from HF, while formulation used as blank nanoemulsion may be used to disperse other insecticidal agents.

In order to evaluate if the formulation interfere with acetylcholinesterase, HFNE was tested using a colorimetric assay. Positive control was performed by preparing a nanoemulsion with eserine, a recognized lipophilic anticholinesterase agent dispersed through oil phase (MOD®). Mean droplet analysis confirmed this formulation, constituted by 5% of MOD®, 5% of surfactants

(HLB of 10.75), 0.05% of eserine and 89.95% of water, as a nanoemulsion (67.3 ± 0.3 nm). IC50 of this substance could not be determined, since lowest eserine concentration (0.6 ppm) was able to inhibit 90% of enzyme activity (Figure 5). Results suggest that eserine may be able to displace from disperse phase of the nanoemulsion to the aqueous external phase and induced a dose-dependent inhibition of acetylcholinesterase, since it should be in external phase to bind to the enzyme. Pesticides are used as an important tool to protect crops worldwide, however, residues of these substances can be found in many environments, including rivers, estuaries and oceans [27,28]. Most insecticides provide harmful impacts on non-target species, especially aquatic organisms, such as fishes. These animals are especially susceptive to acetylcholinesterase inhibitors, probably due to lacking of detoxification systems and sharing same neurological and respiratory mechanisms [27,29]. Acetylcholinesterase inhibitory assay indicated that no significant inhibition of

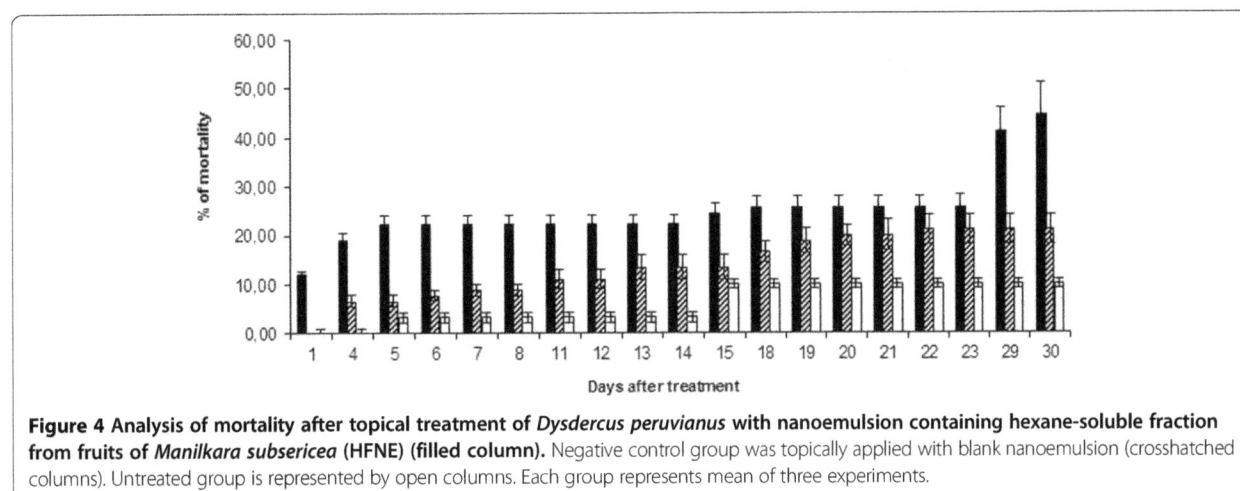

**Figure 4 Analysis of mortality after topical treatment of *Dysdercus peruvianus* with nanoemulsion containing hexane-soluble fraction from fruits of *Manilkara subsericea* (HFNE) (filled column).** Negative control group was topically applied with blank nanoemulsion (crosshatched columns). Untreated group is represented by open columns. Each group represents mean of three experiments.

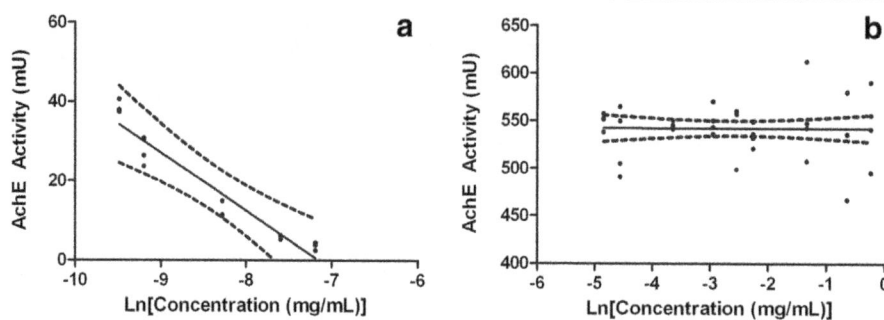

**Figure 5** Linear regression between AchE activity (mU) x natural logarithm of (a) effective concentration of eserine ($p < 0.05$) and (b) effective concentration of hexane-soluble fraction from fruits of *Manilkara subsericea* ($p > 0.05$).

acetylcholinesterase modulated by HFNE (Figure 5). Considering that acetylcholinesterase used in this assay is from fish origin, our results suggest that HFNE may not induce harmful effects over aquatic non-target animals, indicating the potential of this nanoemulsion as an insecticidal agent. Nanoemulsion without *M. subsericea* extract also did not interfere with the enzyme (Data not shown).

Acute toxicity evaluation was performed in order to verify effects of HFNE in mice. It was not observed any behavioral change during all tested period and mortality in all groups. Analysis of body weight also indicated absence of significant difference between HFNE and negative control group (Table 2). It was also not observed significant difference in food and water consumption, macroscopical aspects and weight of organs between groups treated with HFNE and negative control group (Data not shown). Treatment with HFNE was performed at a single high dose corresponding to 3 g/kg of extract per animal. Since no death or toxic signals were observed, LD50 could not be estimated and HFNE, suggesting that HFNE may be considered non-toxic [30].

## Conclusions

Previous study performed by our research group indicated that hexane-soluble fraction from ethanolic crude extract from fruits of *Manilkara subsericea* presented

insecticidal activity against *Dysdercus peruvianus*. This activity may be partially attributed to beta-and alpha amyrin acetates, which may be used as chemical markers for quality control of products with *M. subsericea* extracts. However, these substances, as well as the active fraction are poorly water soluble. As part of our ongoing studies with this species, we decided to develop an insecticidal nanoemulsion. This formulation was able to induce mortality in insects and our results suggest that it may be safe for non-target organisms and environment. The present study suggests the obtained O/A nanoemulsion may be useful to enhance water solubility of poor water soluble natural products with insecticidal activity, including the hexane-soluble fraction from ethanolic crude extract from fruits of *Manilkara subsericea*. The absence of organic toxic solvents and stability makes this nanoemulsion a potential insecticidal product.

## Materials and methods
### Chemicals
Sorbitan oleate (HLB: 4.3) and Polysorbate 80 (HLB: 15) were purchased from La Belle Ativos Ltda (Paraná, Brazil). Octyldodecyl myristate (MOD®) was purchase from Brasquim Ltda (São Paulo, Brazil). Acetylthiocholine iodide (ATCI), 5,5-dithiobis-2-nitrobenzoic acid (DTNB), physostigmine (eserine) and acetylcholinesterase from electric eel (type VI-S, C3389-2UK, lyophilized powder) were purchased from Sigma (Sigma-Aldrich Corporation, St Louis, MO). Hexane-soluble fraction from fruits of *M. subsericea* was previously obtained [9] and stored at 4°C for further utilization.

### Emulsification method
Emulsions were prepared by temperature of inversion phase method [31]. The required amounts of both emulsifiers were dissolved in the oil phase and heated at 75 ± 5°C, while the aqueous phase was separately heated at same temperature. When both phases reached the same temperature, aqueous phase was gently added and mixed with the oil phase, using a mechanic agitator model

**Table 2 Weight variation in adult female and male Swiss albino mice (*Mus musculus*) treated with HFNE (5% of MOD®, 5% of surfactants (HLB of 10.75), 5% of hexane-soluble fraction from fruits of *M. subsericea* and 85% of water) by oral route, corresponding to 3 g/kg of extract**

|  | Body weight (Male) | | Body weight (Female) | |
|---|---|---|---|---|
|  | Initial (g) | Final (g) | Initial (g) | Final (g) |
| HFNE | 49.06 ± 0.43 | 50.39 ± 1.37 | 52.72 ± 1.62 | 50.64 ± 0.63 |
| Control | 50.65 ± 1.50 | 52.05 ± 2.71 | 50.64 ± 0.63 | 51.76 ± 1.59 |

Control groups received same volume of blank nanoemulsion (5% of MOD, 5% of surfactants and 90% of water).
$p > 0.05$.

Fisatom 713D at 400 rpm for 10 min and additional 5 min of agitation under cooling. Aditional constituents was weight an placed together with oil and surfactants mixture, being its mass discounted from water mass.

## Required HLB determination

Each emulsion was prepared at a final mass of 25 g, containing 90% (w/w) of distilled water, 5% (w/w) of MOD® and 5% of a mixture of emulsifiers [32]. Series of emulsions were prepared using sorbitan oleate (HLB = 4.3) and polysorbate 80 (HLB = 15), allowing a wide range of HLB values from 4.3 (5% w/w of sorbitan oleate) to 15 (5% w/w of polysorbate 80) by blending together the emulsifiers in different ratios.

## Pseudo-ternary phase diagram

Nanoemulsion region was determined using pseudo-ternary phase diagram. Each corner corresponded to 100% of water, surfactants and MOD®. Surfactants blend was kept constant and corresponded to ratio which results on required HLB value of oil phase. Composition (w/w) which allowed required HLB value determination was used as starting point (90% of distilled water, 5% of oil and 5% of surfactants blend) and mean droplet size of each prepared composition was performed in order to determine nanoemulsion region.

## Macroscopical analysis

Stability of all emulsions was evaluated immediately and after 1, 15 and 30 days of manipulation by macroscopic analysis, such as color, visual aspect, phase separation, creaming and sedimentation. During this period all emulsions were maintained under room temperature ($25 \pm 2°C$) in screw-capped glass test tubes [32]. Acelerated stability evaluation was performed keeping emulsion under controlled temperature ($40 \pm 5°C$).

## Droplet size and zeta potential analysis

The droplet size, polydispersity **and zeta potential** were determined by photon correlation spectroscopy using a ZetaPlus (Brookhaven Inst. Corp., USA). Each emulsion was diluted using ultra-pure Milli-Q water (1:25). Measurements were performed in quintuplicate and average droplet size was expressed as the mean diameter.

## Insect bioassay

*Dysdercus peruvianus* were obtained from the colony maintained in the Laboratory of Insect Biology of the Universidade Federal Fluminense (GBG-UFF), being kept at 24-25°C, relative humidity of 70-75% and a 16:8 h light:dark cycle [9].

Fourth-instar insects were randomly chosen and separated in two treated groups, being one group topically applied with a nanoemulsion containing hexane-soluble fraction from fruits of *M. subsericea* (HFNE) (5% of MOD®, 5% of surfactants (HLB of 10.75), 5% of HF and 85% of water), corresponding to 50 µg of extract per insect, while negative control group was treated with blank nanoemulsion (5% of MOD®, 5% of surfactants and 90% of water). Untreated insects received no treatment, being only fed. Biological evaluation was performed in order to determine mortality levels during the entire time required for development from the fourth instar to the adult stage [9,33,34]. All experiments were repeated at least three times with samples from 30 insects ($n = 30$ in each triplicate). Significance of the results was analysed using ANOVA and Tukey's test21 according to Stats Direct Statistical Software, v.2.2.7 for Windows 98. Differences between treated group and control.

## Anticholinesterase assay

Anticholinesterase activity was performed according to method described by Ellman et al. (1961) [35] with some modifications [36], using a 96-well microplate. A total volume of 200 µL of test media was composed by 65 µL of Phosphate buffered saline (PBS), 60 µL of 5,5′-dithiobis-(2-nitrobenzoic acid) (DTNB) 1,5 mM, 25 µL of electric eel acetylcholinesterase (Sigma) (AchE) 550 mU/mL, 25 µL of nanoemulsion and 25 µL of acetylthiocholine iodide (ASCh). Different concentrations of HF and eserine (positive control) were obtained by dilution of each nanoemulsion with PBS. Negative control was performed using a blank nanoemulsion, without inhibitor or extract. The spontaneous hydrolysis of substrate was calculated replacing the enzyme solution by PBS. Absorbance was measured at 412 nm. The statistical analysis of the anticholinesterase assay was performed on GraphPad Prism 5.04 program using Pearson's correlation coefficient with 95% confidence interval.

## Acute toxicity
### Animals

This study was approved by the Ethics Committee of the Universidade Federal do Amapá (CEP – UNIFAP – 005AP/2013). All procedures were performed according to the International Committee for animal care in accordance with established national regulations for animal experimentation. The experiments were performed using adult female and male Swiss albino mice (*Mus musculus*), 12 weeks age, provided by the Central Laboratory of the State of Amapá – Macapá (LACEN/AP). Each experimental group was composed of 5 animals. They were kept in polyethylene cages on a temperature-controlled rack ($25°C \pm 2°C$) under a 12-hour light-dark cycle. They had free access to food and water, except for the 24 hours before the experiments, when they had access only to water.

## Experimental protocol

Acute toxicity studies were performed using both sexes of mice according to Pina et al. (2012) [30], with some modifications. Treated groups received a single dose of HFNE (5% of MOD®, 5% of surfactants (HLB of 10.75), 5% of hexane-soluble fraction from fruits of *M. subsericea* and 85% of water) by oral route, corresponding to 3 g/kg of extract. Negative control groups received a blank nanoemulsion (5% of MOD®, 5% of surfactants and 90% of water).

Observations were performed at 30, 60, 120, 240, 360 and 720 min after the oral treatment and daily for fourteen days. Behavioral changes (agitation, convulsions, vocal fremitus, irritation, stereotyped movements, touch response, salivation, tremors, writhing, body distension, ptose, sleepiness, defecation, diarrhea, piloerection), weight, food and water intake, clinical signs of toxicity and mortality were recorded daily. At the end of fourteen days, they were sacrificed by cervical dislocation and taken to autopsy for macroscopic observation of the organs (heart, lung, liver, kidney and spleen). Statistical analysis was performed by Student t test with 95% confidence intererval, using GraphPad Prism 5.04. Differences between organs, body weight and food and water intake were considered significant when $p < 0.05$.

### Competing interests

All authors declare no conflict of interests.

### Authors' contributions

CPF contributed in collecting plant sample, running the laboratory work, analysis of the data and drafted the paper. FBA and ANS contributed in preparation of extracts, HLB determination and nanoemulsions preparation. MSG, CBM and DF contributed in insect bioassay. MGS contributed in plant identification and herbarium confection. LACT contributed in AChE bioassay. JCTC contributed to critical reading of the manuscript and acute toxicity assay. LR and DQF designed the study, supervised the laboratory work and contributed to critical reading of the manuscript. All the authors have read the final manuscript and approved the submission.

### Authors' information

Caio Pinho Fernandes is a professor at Universidade Federal do Amapá and has been working with natural products, including phytochemistry, nanotechnology and biological activities of these compounds.
Fernanda Borges de Almeida is an undergraduate student at Universidade Federal do Amapá and participated in this project as part of her scientific initiaion program.
Amanda Nunes Silveira is an undergraduate student at Universidade Federal Fluminense and participated in this Project as part of her scientific initiaion program.
Marcelo Salabert Gonzalez is professor at Universidade Federal Fluminense and has been working with complementary strategies to control insects with secondary metabolites from plant species.
Cicero Brasileiro Mello is professor at Universidade Federal Fluminense and has been working with complementary strategies to control insects with secondary metabolites from plant species.
Denise Feder is professor at Universidade Federal Fluminense and has been working with complementary strategies to control insects with secondary metabolites from plant species.
Raul Apolinário is undergraduate student at Universidade Federal Fluminense and participated in this project as part of her scientific initiation program and did al experiments with insects.

Marcelo Guerra Santos is professor at Universidade Estadual do Rio de Janeiro. He is a botanist and has been working with species from sandbanks of Parque Nacional da Restinga de Jurubatiba (RJ) Brazil.
José Carlos Tavares Carvalho is professor and President of the Universidade Federal do Amapá (Brazil) and has been working with natural products pharmacology.
Luis Armando Cândido Tietbohl is a Master's student at Universidade Federal Fluminense and has been working with acetylcholinesterase inhibition.
Leandro Rocha is professor at Universidade Federal Fluminense and has been working with natural products and its biological activities.
Deborah Quintanilha Falcão is professor at Universidade Federal Fluminense and has been working with nanotechnology of natural products.

### Acknowledgement

Authors would like to thank CAPES (no 3292/2013 AUXPE), CNPQ and FAPERJ for the finantial support and "Centro Brasileiro de Pesquisas Físicas" for the use of Zeta Potential Analyzer.

### Author details
[1]Programa de Pós, Graduação em Biotecnologia Vegetal, Centro de Ciências da Saúde, Universidade Federal do Rio de Janeiro – UFRJ, Bloco K, 2º andar – sala 032, Av. Brigadeiro Trompowski s/n, CEP: 21941-590 Ilha do Fundão, RJ, Brazil. [2]Laboratório de Farmacotécnica, Colegiado de Ciências Farmacêuticas, Universidade Federal do Amapá, Campus Universitário Marco Zero do Equador, Rodovia Juscelino Kubitschek – KM – 02-Jardim Marco Zero, CEP: 68903-419 Macapá, AP, Brazil. [3]Laboratório de Tecnologia Farmacêutica I, Faculdade de Farmácia, Universidade Federal Fluminense, Rua: Mario Viana, 523, Santa Rosa, CEP: 24241-000 Niterói RJ, Brazil. [4]Laboratório de Biologia de Insetos – LABI, Departamento de Biologia Geral (GBG), Universidade Federal Fluminense, Morro do Valonguinho S/No, CEP 24001-970 Niterói, RJ, Brazil. [5]Faculdade de Formação de Professores, UERJ, Rua: Dr. Francisco Portela, 1470 – Patronato, CEP: 24435-005 São Gonçalo, Rio de Janeiro, Brazil. [6]Laboratório de Pesquisa em Fármacos, Colegiado de Ciências Farmacêuticas, Universidade Federal do Amapá, Rodovia Juscelino Kubitschek – KM – 02 – Jardim Marco Zero, CEP: 68903-419 Macapá, AP, Brazil. [7]Laboratório de Tecnologia de Produtos Naturais – LTPN, Departamento e Tecnologia Farmacêutica, Faculdade de Farmácia, Universidade Federal Fluminense – UFF Rua, Mario Viana, 523, CEP: 24241-000, Santa Rosa, Niterói, RJ, Brazil.

### References

1. Rao JV, Shilpanjali D, Kavitha P, Madhavendra SS: **Toxic effects of profenofos on tissue acetylcholinesterase and gill morphology in a euryhaline fish, Oreochromis mossambicus.** *Arch Toxicol* 2003, **77**:227–232.
2. Gobbo-Neto L, Lopes NP: **Plantas medicinais: fatores de influência no conteúdo de metabólitos secundários.** *Quim Nova* 2007, **30**:374–381.
3. Rattan RS: **Mechanism of action of insecticidal secondary metabolites of plant origin.** *Crop Prot* 2010, **29**:913–920.
4. Viegas C Jr: **Terpenos com atividade inseticida: uma alternativa para o controle químico de insetos.** *Quim Nova* 2003, **26**:390–400.
5. Almeida EB Jr: *Manilkara in Lista de Espécies da Flora do Brasil. Jardim Botânico do Rio de Janeiro.* Available at: <http://floradobrasil.jbrj.gov.br/jabot/floradobrasil/FB14473> Accessed in November 2013.
6. Santos MG, Fevereiro PCA, Reis GL, Barcelos JI: **Recursos vegetais da Restinga de Carapebus.** *Rev Biol Neotrop* 2009, **6**:35–54.
7. Fernandes CP, Corrêa AL, Cruz RAS, Botas GS, Silva-Filho MV, Santos MG, de Brito MA, Rocha L: **Anticholinesterasic activity of Manilkara subsericea (Mart.) Dubard triterpenes.** *Lat Am J Pharm* 2011, **30**:1631–1634.
8. Fernandes CP, Corrêa AL, Lobo JFR, Caramel OP, Almeida FB, Castro ES, Souza KFCS, Burth P, Amorim LMF, Santos MG, Ferreira JLP, Falcão DQ, Carvalho JCT, Rocha L: **Triterpene esters and biological activities from edible fruits of Manilkara subsericea (Mart.) Dubard, Sapotaceae.** *Bio Med Res Int* 2013, **Article ID 280810**:7 p.
9. Fernandes CP, Xavier A, Pacheco JPF, Santos MG, Mexas R, Raticliffe NA, Gonzalez MS, Mello CB, Rocha L, Feder D: **Laboratory evaluation of the effects of Manilkara subsericea (Mart.) Dubard extracts and triterpenes on the development of Dysdercus peruvianus and Oncopeltus fasciatus.** *Pest Manag Sci* 2013, **69**:292–301.

10. Stanisçuaski F, Ferreira-da-Silva CT, Mulinari F, Pires-Alves M, Carlini CR: Effects of canatoxin on the cotton stainer bug *Dysdercus peruvianus* (Hemiptera: Pyrrhocoridae). *Toxicon* 2005, **45**:753–760.

11. Margulis-Goshen K, Magdassi S: Nanotechnology: An Advanced Approach to the Development of Potent Insecticides. In *Advanced Technologies for Managing Insect Pests*. Dordrecht: Springer; 2013:295–314.

12. Wang L, Dong J, Chen J, Eastoe J, Li X: Design and optimization of a new self-nanoemulsifying drug delivery system. *J Colloid Interface Sci* 2009, **330**:443–448.

13. Solans C, Izquierdo P, Nolla J, Azemar N, Garcia-Celma MJ: Nano-emulsions. *Curr Opin Colloid Interface Sci* 2005, **10**:102–110.

14. Quintão FJO, Tavares RSN, Vieira-Filho SA, Souza GHB, Santos ODH: Hydroalcoholic extracts of *Vellozia squamata*: study of its nanoemulsions for pharmaceutical or cosmetic applications. *Braz J Pharmacog* 2013, **23**:101–107.

15. Schmidts T, Dobler D, Guldan AC, Paulus N, Runkel F: Multiple W/O/W emulsions—Using the required HLB for emulsifier evaluation. *Colloids Surf A Physicochem Eng Asp* 2010, **372**:48–54.

16. Hadzir NM, Basri M, Rahman MBA, Salleh AB, Rahman RNZRA, Basri H: Phase behavior and formation of fatty acid esters nanoemulsions containing piroxicam. *AAPS PharmSciTech* 2013, **14**:456–463.

17. Gulapalli RP, Sheth BB: Influence of an optimized non-ionic emulsifier blend on properties of oil-in-water emulsions. *Eur J Pharm Biopharm* 1999, **48**:233–238.

18. Forgiarini A, Esquena J, González C, Solans C: Studies of the relation between phase behavior and emulsification methods with nanoemulsion formation. *Progr Colloid Polym Sci* 2000, **115**:36–39.

19. Griffin WC: Classification of surface-active agents by HLB. *J Soc Cosmet Chem* 1949, **1**:311–326.

20. Orafidiya LO, Oladimeji FA: Determination of the required HLB values of some essential oils. *Int J Pharm* 2002, **237**:241–249.

21. Rodríguez-Rojo S, Varona S, Núnez M, Cocero MJ: Characterization of rosemary essential oil for biodegradable emulsions. *Ind Crop Prod* 2012, **37**:137–140.

22. Shakeel F, Ramadan W, Faisal MS, Rizwan M, Faiyazuddin M, Mustafa G, Shafiq S: Transdermal and topical delivery of anti-inflammatory agents using nanoemulsion/ microemulsion: an updated review. *Curr Nanosci* 2010, **6**:184–198.

23. Mahdi ES, Noor AM, Sakeena MH, Abdullah GZ, Abdulkarim MF, Sattar MA: Formulation and in vitro release evaluation of newly synthetized palm kernel oil esters-based nanoemulsion delivery system for 30% ethanolic dried extract derived from local *Phyllanthus urinaria* for skin antiaging. *Int J Nanomedicine* 2011, **6**:2499–2512.

24. Bruxel F, Laux M, Wild LB, Fraga M, Koester LS, Teixeira HF: Nanoemulsões como sistemas de liberação parenteral de fármacos. *Quim Nova* 2012, **35**:1827–1840.

25. Araújo FA, Kelmann RG, Araujo BV, Finatto RB, Teixeira HF, Koester LS: Development and characterization of parenteral nanoemulsions containing thalidomide. *Eur J Pharm Sci* 2011, **42**:238–245.

26. Izquierdo P, Esquena J, Tadros TF, Dederen C, Garcia MJ, Azemar N, Solans C: Formation and stability of nano-emulsions prepared using the phase inversion temperature method. *Langmuir* 2002, **18**:26–30.

27. Sánchez-Bayo F: Insecticides mode of action in relation to their toxicity to non-target organisms. *J Environ Anal Toxicol* 2011, **S**:4.

28. Brooks ML, Fleishman E, Brown LR, Lehman PW, Werner I, Scholz N, Mitchelmore C, Lovvorn JR, Johnson ML, Schlenk D, van Drunick S, Drever JI, Stoms DM, Parker AE, Dugdale R: Life histories, salinity zones, and sublethal contributions of contaminants to Pelagic Fish declines illustrated with a case study of San Francisco Estuary, California, USA. *Estuar Coast* 2012, **35**:603–621.

29. Hedayati A, Tarkhani R, Shadi A: Investigation of acute toxicity of two pesticides Diazinon and Deltamethrin, on Blue Gourami. *Trichogaster trichopterus (Pallus) Global Veterinaria* 2012, **8**:440–444.

30. Pina EML, Araújo FWC, Souza IA, Bastos IVGA, Silva TG, Nascimento SC, Militão GCG, Soares LAL, Xavier HS, Melo SJ: Pharmacological screening and acute toxicity of bark roots of *Guettarda platypoda*. *Braz J Pharmacog* 2012, **22**:1315–1322.

31. Aulton ME: *Delineamento de Formas Farmacêuticas*. 2ath edition. São Paulo: Artmed; 2005.

32. Fernandes CP, Mascarenhas MP, Zibetti FM, Lima BG, Oliveira RPRF, Rocha L, Falcão DQ: HLB value, an important parameter for the development of essential oil phytopharmaceuticals. *Braz J Pharmacog* 2013, **23**:108–114.

33. Mello CB, Uzeda CD, Bernardino MV, Mendonça-Lopes C, Kelecom A, Fevereira PCA, Guerra MS, Oliveira AP, Rocha LM, Gonzalez MS: Effects of the essential oil obtained from *Pilocarpus spicatus* on the development of *Rhodnius prolixus* nymphae. *Rev Bras Farmacogn* 2007, **17**:514–520.

34. Mello CB, Mendonça-Lopes D, Feder D, Uzeda CD, Carneiro RM, Rocha MA, Gonzales MS: Laboratory evaluation of the effects of triflumuron on the development of *Rhodnius prolixus* nymph. *Mem Inst Oswaldo Cruz* 2008, **103**:839–842.

35. Ellman GL, Courtney KD, Andres V, Featherstone RM: A new and rapid colorimetric determination of acetylcholinesterase activity. *Biochem Pharmacol* 1961, **7**:88–95.

36. Rhee IK, Van De Meent M, Ingkaninan K, Veerporte R: Screening for acetylcholinesterase inhibitors from Amaryllidacea using silica gel thin-layer chromatography in combination with bioactivity staing. *J Chromatogr A* 2001, **915**:217–223.

# Effects in cigarette smoke stimulated bronchial epithelial cells of a corticosteroid entrapped into nanostructured lipid carriers

Maria Luisa Bondì[1*†], Maria Ferraro[2†], Serena Di Vincenzo[2], Stefania Gerbino[2], Gennara Cavallaro[3], Gaetano Giammona[3], Chiara Botto[3], Mark Gjomarkaj[2] and Elisabetta Pace[2]

## Abstract

**Background:** Nanomedicine studies have showed a great potential for drug delivery into the lung. In this manuscript nanostructured lipid carriers (NLC) containing Fluticasone propionate (FP) were prepared and their biocompatibility and effects in a human bronchial epithelial cell line (16-HBE) stimulated with cigarette smoke extracts (CSE) were tested.

**Results:** Biocompatibility studies showed that the NLC did not induce cell necrosis or apoptosis. Moreover, it was confirmed that CSE increased intracellular ROS production and TLR4 expression in bronchial epithelial cells and that FP-loaded NLC were more effective than free drug in modulating these processes. Finally, the nanoparticles increased GSH levels improving cell protection against oxidative stress.

**Conclusions:** The present study shows that NLC may be considered a promising strategy to improve corticosteroid mediated effects in cellular models associated to corticosteroid resistance. The NLC containing FP can be considered good systems for dosage forms useful for increasing the effectiveness of fluticasone decreasing its side effects.

**Keywords:** Nanostructured lipid carriers, Corticosteroid, Fluticasone propionate, Cigarette smoke, Airway epithelial cell, Chronic obstructive pulmonary disease, Asthma

## Background

Pulmonary drug delivery is an important research area with a potential high impact in the treatment of various obstructive pulmonary diseases including asthma and chronic obstructive pulmonary disease. It can provide rapid responses and can minimize the required drug dose being the drug delivered directly into the lungs and specifically at the site of activity [1].

Nanomedicine is used for modified and targeted drug delivery. It is based on nanostructured materials at colloidal size (1–500 nm) and is able to release biologically active agents, chemically or physically incorporated, into specific sites and within very well defined time frames. These systems are characterized by: 1) nanoscaled dimensions, able to allow their direct interaction at molecular levels with cell components of the damaged tissue; 2) the ability to incorporate elevated amounts of active molecules with subsequent increase of the efficiency of the drug delivery systems; 3) the ability to deliver the drugs by increasing their bioavailability and decreasing administered doses; 4) the ability to obtain an efficient localization of the drug in the target site [2]. Nanomedicine provides new solutions to clinical problems, particularly in pulmonary diseases, promising better delivery of therapeutics to disease sites [3,4]. These advantages can be properly exploited for the administration of inhaled corticosteroids, especially during long-term therapies like in patients with chronic obstructive pulmonary disease (COPD); potentially, it might be possible to utilise these nanosystems in inhalatory therapies in order to maximize local effects into the lung and to reduce systemic effects as well as the frequency of administration. Moreover, long-term use of high-dose

* Correspondence: marialuisa.bondi@ismn.cnr.it
†Equal contributors
[1]Istituto per lo Studio dei Materiali Nanostrutturati- U.O.S. di Palermo-Consiglio Nazionale delle Ricerche-via Ugo La Malfa, 153 90146 Palermo, Italy
Full list of author information is available at the end of the article

inhaled corticosteroids (ICS) has the potential to cause undesirable side effects. Conversely, a modified delivery system provides constant levels of drug at the prime site of action for a prolonged time and it would enable better control of the disease [5,6].

To obtain these results it is important the choice of the material forming the nanodevices. In particular the use of pegylated lipid for the production of Nanostructured Lipid Carrier (NLC) combines the advantages of the safety of lipids and the possibility of large-scale production, with the mucoadhesive properties useful for improving residence time of nanodevices on airways surface and to contrast the effect of the abnormal production of mucus, occurring in COPD, with the consequent dramatic reduction of corticosteroids absorption [7]. In this context, colloidal lipid nanoparticles such as NLC could give great benefit in designing new drug delivery systems with great potential advantages.

The aim of the present work was to realize a novel drug delivery system to improve the drug bioavailability, making the drug able to achieve an increase of permeability through the membrane cell and consequently to reduce the administered dose. In this paper we report the preparation and characterization of NLC by using a pegylated lipid such as Compritol HD5 ATO, for the delivery through inhalator route of Fluticasone propionate (FP). Pegylated NLC containing Fluticasone (FP-loaded NLC) as well as empty NLC, as control, were prepared and characterized in terms of size, polydispersity index (PDI), surface charge, stability, and in vitro drug release. Moreover, the biological efficiency of this new drug delivery system was evaluated in vitro by using the human bronchial epithelial cell line (16-HBE) considering the effects of the drug, in the loaded-NLC or free form on Reactive Oxygen Species (ROS) production, GSH levels, and TRL4 expression in cigarette smoke extracts (CSE) stimulated cells.

**Results and discussion**

In this paper, in order to improve the FP efficacy in the treatment of respiratory diseases such as COPD, FP-loaded NLC were developed. Due to the lypophilic characteristics of FP, FP-loaded NLC were prepared by the precipitation method [8-10]. In particular, Compritol HD5 ATO was chosen as lipid matrix for obtaining NLC with or without FP because of its good biocompatibility and the presence of PEG in its structure. In this regards, it has been shown that the PEG chains can play an important role in transport across mucosae of nanoparticles since their presence can improve their transport across the nasal epithelia [7,11-13].

Since some physicochemical and technological properties are quite critical for biopharmaceutical behaviour of NLC, either empty and drug-loaded samples, after preparation and purification, were characterized in terms of particle size, PDI and $\zeta$ potential in three different dispersing aqueous media by light scattering measurements, and analytical data are reported in Table 1.

Empty and drug loaded-NLC have size of about 116 and 130 nm in bi-distilled water respectively, and greater in the all other investigated media; these differences could be attributed to the different ionic strength of the media. Moreover, all these systems possessed quite low PDI values, which indicated a good dimensional homogeneity of particles that, together with small size, make them suitable for inhalatory administration.

The $\zeta$ potential values of these structures, also reported in Table 1, were rather high (absolute value) in bi-distilled water and decreased when they were determined in PBS and NaCl 0.9% aqueous solution. The presence of electrolytes causes a diminution of surface charge for the potential screening effect of solution ions. The surface charge of nanoparticles is important because it makes the nanosystems more stable when dispersed into an aqueous solution, reducing the occurrence of the aggregation phenomenon. Several systems were prepared with different size and surface characteristics (data not shown) but for the in vitro tests the system with better physical-chemical characteristics was chosen.

In order to confirm the nanometer size and to investigate the morphology of empty or FP-loaded NLC, SEM was used and the obtained images are reported in Figure 1.

These images were consistent with the findings obtained from dimensional analysis and also revealed a spherical shape of investigated samples.

Moreover, an important aspect to be taken into account in the formulations of NLC as possible carrier to be aerosolized for the pulmonary delivery of drugs is their capability to give colloidal dispersions stable during storage. The occurrence of aggregation phenomena can lead to a significant worsening of the biopharmaceutical features of colloidal suspensions, above all in terms of ability to be uptaken into the cells.

Therefore in order to evaluate the stability of these systems during storage empty and FP-loaded NLC were kept for 4 months and 10 months at 4°C and subsequently characterized in terms of size, PDI, and $\zeta$ potential. The results (Table 1) showed that either empty or FP-loaded NLC were stable during storage under tested conditions.

*DL %* and *EE %* of FP loaded in NLC were equal to about as 4.8% and 76.8% respectively. In order to evaluate the ability of these NLC of retaining the encapsulated drug under sink conditions and to release it slowly in physiological media, a release study was carried out in PBS at pH 7.4/ethanol mixture 80:20 (v/v) by evaluating the amount of released drug from NLC at prefixed time

**Table 1 Mean size (nm), polydispersity index (PDI) and ζ-potential (mV) in bi-distilled water, phosphate buffer solution (PBS) and NaCl 0.9 wt% of empty and Fluticasone propionate (FP)-loaded nanoparticles**

| Sample | Dispersing medium | Mean size (nm) | PDI | Zeta potential (mV)(± S.D.) | EE% (w/w) |
|---|---|---|---|---|---|
| Empty | PBS pH 7.4 | 133.7 | 0.243 | −15.1 ± 3.78 | ——— |
|  | NaCl 0.9% | 132.5 | 0.215 | −12.3 ± 4.16 | ——— |
|  | $H_2O$ | 115.9 | 0.285 | −27.8 ± 3.21 | ——— |
| FP-loaded | PBS pH 7.4 | 178.7 | 0.266 | −14.3 ± 2.44 | 76.8 ± 0.04 |
|  | NaCl 0.9% | 189.6 | 0.244 | −13.3 ± 4.56 | 76.8 ± 0.05 |
|  | $H_2O$ | 129.9 | 0.330 | −31.3 ± 4.50 | 76.8 ± 0.04 |
| Empty 4 months) | $H_2O$ | 137.9 | 0.247 | −26.6 ± 3.13 | ——— |
| Empty (10 months) | $H_2O$ | 149.9 | 0.235 | −25.4 ± 2.15 | ——— |
| FP-loaded (4 months) | $H_2O$ | 143.2 | 0.323 | −28.3 ± 2.72 | 74.7 ± 0.06 |
| FP-loaded (10 months) | $H_2O$ | 145.6 | 0.334 | −27.5 ± 4.50 | 72.5 ± 0.03 |

Efficiency entrapment % (EE%) of FP-loaded NLC.

intervals across a dialysis tube (Spectra/Por®, MWCO 12,000-14,000 Da), in accordance to the European Pharmacopoeia [14,15]. In Figure 2, the drug release profile from FP-loaded NLC was reported until 72 hrs incubation.

As shown, after 1 hr, the amount of FP released from NLC was equal to 15%. An initial burst effect in the drug-release profile of FP-NLC is evident and it can be probably ascribed to the presence of the drug absorbed on the nanoparticle surface. Moreover, these studies revealed that in a physiological-mimicking medium, FP was not completely released from nanoparticles until 72 hrs, supporting the hypothesis that these systems at the contact with the airways mucosae, could efficiently enter in the colloidal form improving then the drug internalization into the cells and its accumulation into human bronchial epithelial cells. The drug burst release shown in this study could be exploited to deliver a high initial dose when desired. The gradual release after the initial

burst would also be important in order to maintain an effective drug concentration in the target organ.

An innovative drug delivery system has to be tested for its safety. This aspect is much more relevant in the case of pulmonary delivery, since several side effects may result from an unsafe material [16]. Taking also into account the possibility to incorporate into aerosol droplets the FP-loaded pegylated nanoparticles and to administer them by inhalation, safety of empty NLC and FP-loaded NLC was evaluated in vitro by using 16-HBE cells as a model of epithelial cells.

Cytotoxicity of FP-loaded pegylated nanoparticles in 16-HBE cells (Figures 3 and 4) was evaluated by using the PI/Annexin V binding method [17].

Neither FP-loaded NLC nor empty NLC at both tested concentrations and time points induced relevant numbers of necrotic (PI positive) or of apoptotic (Annexin V positive) 16-HBE cells (Figures 3 and 4), evidencing the high biocompatibility of obtained nanoparticles.

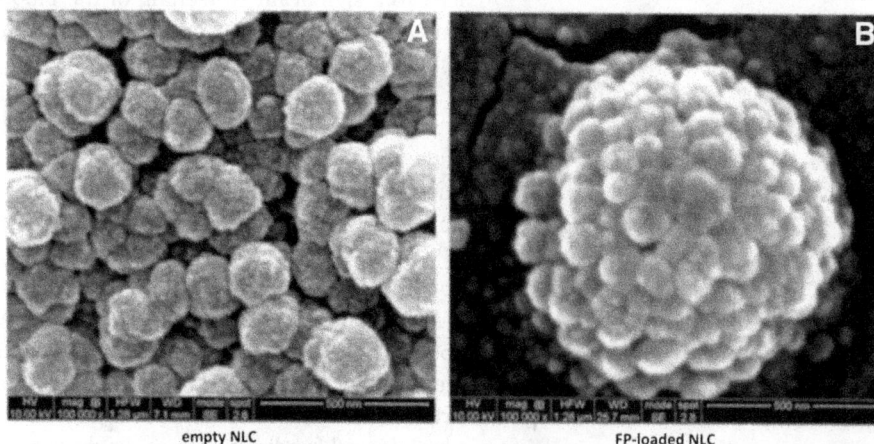

**Figure 1 Scanning electron microscopy.** Representative SEM images of NLC, empty (A) and loaded with FP (B), respectively. The bars represent 500 nm.

Effects in cigarette smoke stimulated bronchial epithelial cells of a corticosteroid entrapped...

47

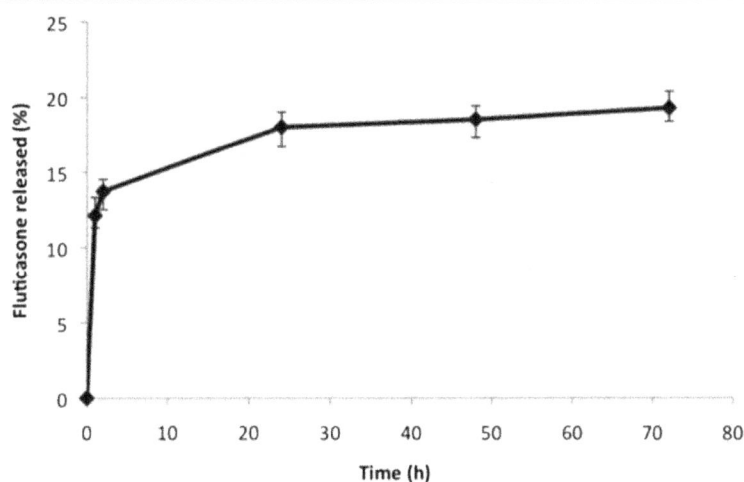

**Figure 2** Drug release profile from FP- loaded NLC.

Our results indicate the potential of the obtained NLC as carriers for FP delivered by intra-bronchial route.

The increased oxidative stress present in COPD patients is related to the increased burden of inhaled oxidants such as cigarette smoke and to the increase in ROS generated by several inflammatory, immune, and structural airways cells [18]. We initially tested the effect of CSE in ROS production by bronchial epithelial cells. When the cells were exposed to CSE, an increased ROS expression occurred. The presence of free FP tended to increase ROS expression in CSE stimulated cells but this increase was not statistically different. FP-loaded NLC significantly reduced the CSE induced ROS expression and the effect was significantly greater than that exerted by free FP (Figure 5A and B).

Glutathione (GSH) is one of the most important defensive mechanisms against oxidative stress [19,20]. The effects of unloaded FP and FP-loaded NLC as GSH expression in CSE stimulated bronchial epithelial cells were explored. CSE did not significantly increase GSH expression in bronchial epithelial cells. Free FP and empty NLC did not significantly induce GSH expression (data not shown). On the contrary FP-loaded NLC significantly increased GSH expression in CSE stimulated bronchial epithelial cells (Figure 6).

Therefore the increase of GSH suggests a protective effect against oxidative stress into the cells induced by FP when it is administered by NLC.

A key component of the innate immunity and of the innate defence mechanisms against infections is represented by the toll like receptor (TLR4) family. After stimulation of these receptors the cell is triggered to produce inflammatory mediators. Since CSE increased TLR4 expression in bronchial epithelial cells [21,22], the effect of free FP and FP-loaded NLC in CSE induced TLR4 expression was assessed. FP-loaded NLC at $10^{-8}$M concentration was more effective in reducing TLR4 expression in CSE stimulated cells in comparison to the other two tested concentrations (Figure 7). According to the results of these experiments, 24 hours of incubation was selected as the best time point (Figure 8). Free FP as well as empty NLC did not significantly affect the CSE induced TLR4 expression while FP-loaded NLC significantly reduced the CSE induced TLR4 expression (Figure 9A and B).

To further investigate why the use of FP-loaded NLC was more effective than the free drug in reducing CSE-mediated effects, the intracellular and extracellular contents of FP in 16-HBE cells treated with unloaded FP or with FP-loaded NLC were assessed by UV analysis. The content of FP was higher within the cells treated with FP-loaded NLC than within the cells treated with free FP at all time points. These results further supported the obtained biological data. Figure 10 shows that the intracellular concentrations of FP loaded into the NLC were always higher than those found in cells treated with free FP. Furthermore, the intracellular and extracellular concentrations were lower than the ones used for the experiments because an aliquot of FP was probably degraded by enzymes present in the cells, thus confirming the data reported in the literature [23].

Cigarette smoking is the major cause of chronic obstructive pulmonary disease, which is associated with increased oxidative stress and altered innate and adaptive immunity [24].

Cigarette smoke-mediated oxidative stress other than producing protein denaturation [25], lipid peroxidation, and DNA damage, contributes to reducing corticosteroid activity [26]. In the presence of mucus hypersecretion, a phenomenon frequently present in the airways of COPD patients, lipophilic substances, such as corticosteroids, can be remarkably impeded in reaching their receptors,

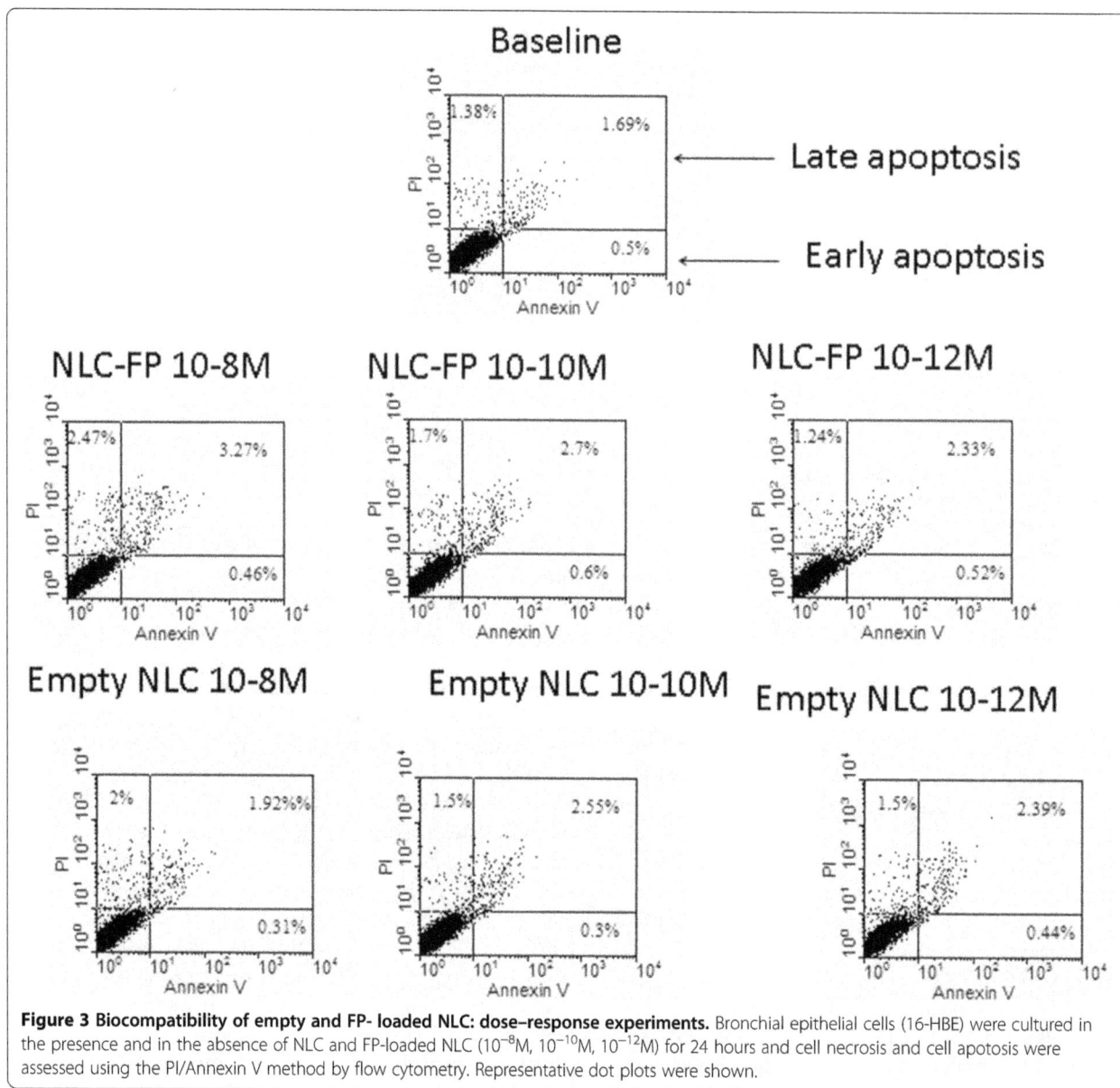

**Figure 3 Biocompatibility of empty and FP- loaded NLC: dose–response experiments.** Bronchial epithelial cells (16-HBE) were cultured in the presence and in the absence of NLC and FP-loaded NLC ($10^{-8}$M, $10^{-10}$M, $10^{-12}$M) for 24 hours and cell necrosis and cell apotosis were assessed using the PI/Annexin V method by flow cytometry. Representative dot plots were shown.

which are localized within the cytoplasm of bronchial epithelial cells. A modified delivery system that provides constant levels of drug at the prime site of action for a prolonged time can contribute to better control this disease [27]. Lipid nanoparticles such as NLC may increase cell uptake of the drug and may improve drug stability and these events may contribute to increased efficacy of the drug. By varying the composition of lipids, structure and size, they could offer a controlled and prolonged duration of the effect of the encapsulated drugs as well as a regional and cell-specific drug targeting within the airways. This kind of drug carriers presents many advantages [28,29].

On the basis of the well-known capability of the NLC to solubilize adequate amounts of hydrophobic drugs, to

pass through the mucus layer associated with bronchial inflammatory diseases escaping from pulmonary phagocytosis due to their bulky hydrophilic outer shell [11], the potential of NLC containing PEG chains as inhalatory delivery systems for FP, was investigated.

The present study describes the preparation of NLC loaded with FP and demonstrates their biocompatibility and their efficacy in controlling oxidative stress and innate immune responses in bronchial epithelial cells exposed to cigarette smoke extracts. The method used for preparing cigarette smoke extracts was previously validated [30,31] and samples of CSE were filtered to remove bacteria and other macromolecules. These large particles in vivo do not reach the deeper airways because they are deposited in the oral cavity or in the upper

Effects in cigarette smoke stimulated bronchial epithelial cells of a corticosteroid entrapped...

49

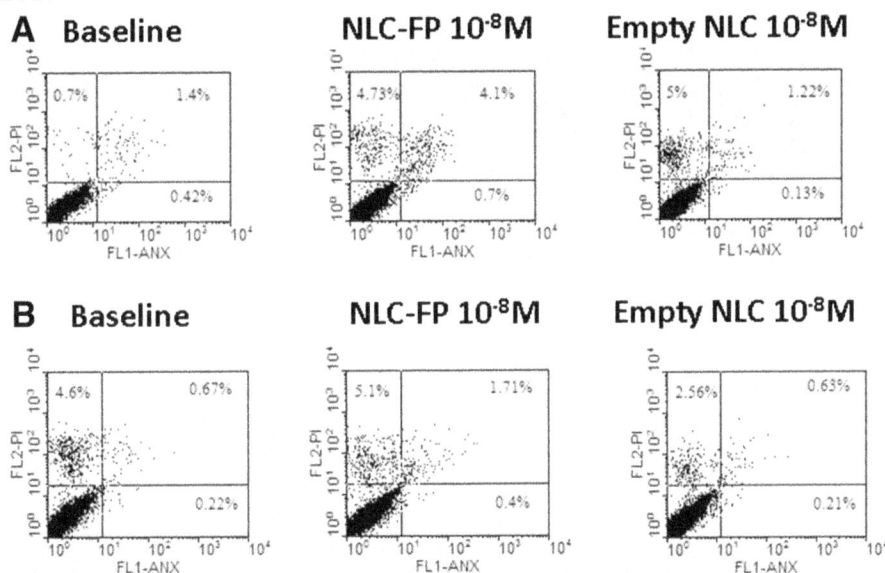

**Figure 4 Biocompatibility of empty and FP- loaded NLC: time-dependent experiments.** Bronchial epithelial cells (16-HBE) were cultured in the presence and in the absence of FP, NLC and FP-loaded NLC ($10^{-8}$M) for 48 **(A)** and 72 **(B)** hours and cell necrosis and cell apoptosis were assessed using the PI/Annexin V method by flow cytometry. Representative dot plots were shown.

respiratory tract. Furthermore, CSE 10% was selected because we showed that this concentration was not toxic for the cells and was able to significantly increase both ROS production [17] and TLR4 expression [21] in bronchial epithelial cells. In addition, this concentration as mentioned in a previous study [32] corresponds to exposures associated with smoking equivalent to two packs per day of cigarettes.

Lungs are unique because they have a large epithelial surface area that is at risk for oxidant-mediated attack. The tracheobronchial tree and the alveolar space are exposed to reactive oxidizing species in the form of inhaled airborne pollutants, tobacco smoke, and products of inflammation. The ROS play an integral role in the modulation of several physiological functions but can also be destructive if produced in excessive amounts. Cigarette smoke results in an imbalance between oxidants and antioxidants in favor of oxidants thus promoting increased oxidative stress [33]. Oxidative stress leads to cause oxidative lung damage including apoptosis [34], senescence and inflammation, all of which have been described in the airways of smokers with COPD [23]. Signal transducers and activators of transcription (STAT), nuclear factor-κB, and transcription factor activator protein-1 (AP-1) are activated in epithelial cells and inflammatory cells during oxidative stress [35]. Increased oxidative stress leads to reduced histone deacetylase (HDAC) activity contributing to the low response to corticosteroids [36]. Upon the exposure of cigarette smoke extracts in airway epithelial cells, the ubiquitin proteasome system is unable to cope with severely damaged proteins that

accumulate in the cell in the form of insoluble polyubiquitinated aggregates [37]. In the present study we confirm that CSE increase ROS and here we provide data supporting the efficacy of FP when it is entrapped into NLC containing PEG reducing TLR4 expression any more than compared to NLC without PEG (see below) and thus suggesting an important role of the PEG nanoparticles in reducing the effects of CSE in the modulation of innate immunity responses. Although the majority of COPD cases can be directly related to smoking, only a quarter of smokers actually develop the disease. A potential reason for the disparity between smoking and COPD may involve an individual's ability to mount a protective adaptive response to cigarette smoke. The GSH system belongs to enzymatic anti-oxidant systems. It is highly concentrated in the lung epithelial lining fluid and protects against many inhaled oxidants. The exposure to cigarette smoke in airway epithelial cells leads to the exhaustion of the pool of reduced GSH thus promoting a lack of antioxidant protection [38]. Under our experimental conditions in vitro CSE do not increase GSH levels as well as no effect on GSH levels were recorded by using free FP. On the contrary we provide data supporting the efficacy of FP-loaded NLC in increasing GSH levels in the presence of CSE.

Furthermore, the surface of the airway epithelium represents a battleground in which the host intercepts signals from pathogens and external insults and activates epithelial defences mainly represented by the innate host immune system. Innate immunity relies on pattern recognition receptors that recognize molecular structures

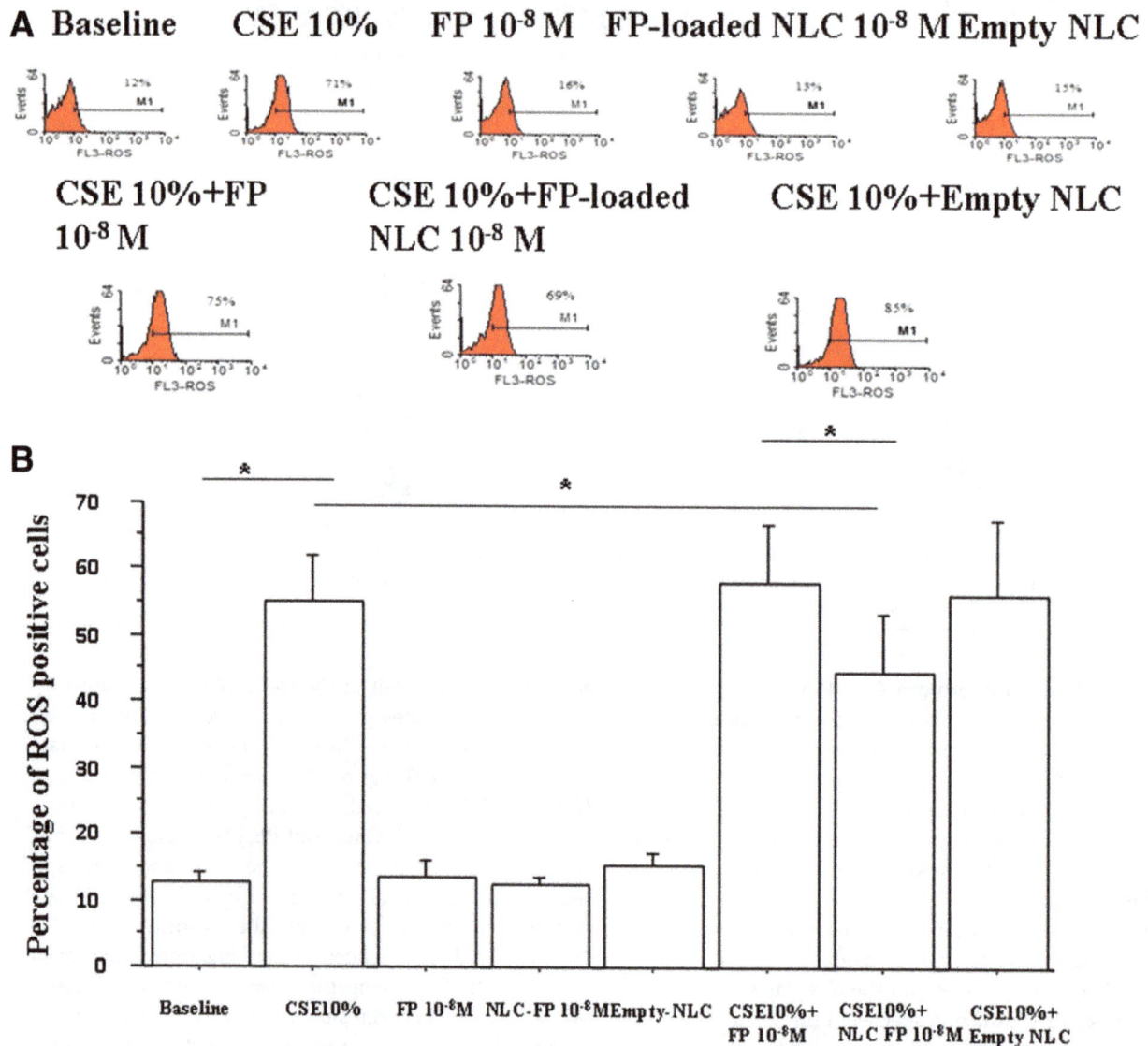

**Figure 5 Effects of FP-loaded NLC on ROS production.** Bronchial epithelial cells (16-HBE) were cultured in the presence and in the absence of CSE (10%), FP, NLC and FP-loaded NLC for 24 hours and then were used for assessing ROS production by flow cytometry (see Materials and methods for details). **(A)** Data are expressed as percentage of ROS positive cells ± SD. *p < 0.05. **(B)** Representative histogram plots are shown.

common to many micro-organisms, such as lipopolysaccharides (LPS), and endogenous ligands such as heat shock proteins. TLR4 is a transmembrane protein that participates in the recognition of LPS and plays a crucial role in the activation of innate host immune system. LPS via TLR4 activation increases inflammatory responses and stimulates MUC5AC expression thus contributing to airway mucus hypersecretion [39]. TLR4 expression is increased in the bronchial epithelium of smokers [40] and cigarette smoke increases TLR4 expression in bronchial epithelial cells [21] and, via activation of the TLR4 signaling cascade, mediates MMP-1 expression [41] and increases IL-8 release [42]. All together these findings suggest that the over-expression and over-activation of

TLR4 can contribute to many phenomena associated to COPD pathogenesis. In the present study we confirm that CSE increase TLR4 expression and demonstrate that FP-loaded NLC, but not unloaded FP, significantly reduce the effects of CSE in increasing TLR4 expression. Moreover, when FP is entrapped into the pegylated NLC is more effective than FP entrapped into the un-pegylated NLC (data not shown).

## Conclusions

In the present study NLC based on a pegylated lipid have been prepared and tested as carrier of FP. Either empty and drug-loaded NLC showed negative ζ potential values and a mean size in the nanometer scale with low

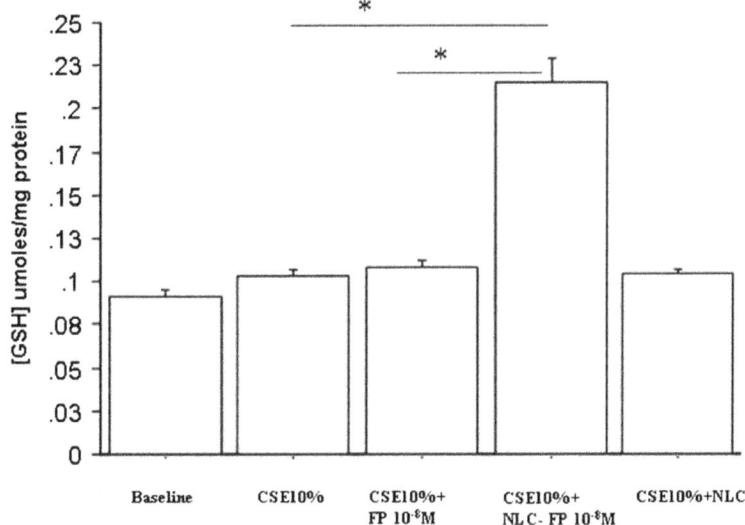

**Figure 6 Effects of FP-loaded NLC on GSH expression.** Bronchial epithelial cells (16-HBE) were cultured in the presence and in the absence of CSE (10%), FP, NLC and FP-loaded NLC for 24 hours and then were used for assessing GSH content (see Materials and methods for details). Data are expressed as GSH μmoles/mg proteins ± SD. *p < 0.05.

PDI values, which indicated a good dimensional homogeneity of particles such as make them suitable for inhalatory administration.

Excellent stability was also showed by this system during storing in the dried form at 4°C, being unchanged their size, PDI and ζ potential.

The release study showed that the investigated carrier had a great stability, being able to retain about 80% of initially entrapped corticosteroid even after 72 h. This result is in agreement with the hypothesis that this system, able to keep inside the drug, at the contact with airways mucosae, could improve drug cell uptake because FP-loaded NLC could enter the cells by endocytosis.

On the other hand a greater amount of FP was found into bronchial epithelial cells treated with FP-loaded NLC in comparison with that treated with free FP.

*In vitro* studies on 16-HBE cells revealed that neither unloaded FP nor FP-loaded NLC induced relevant numbers of necrotic or of apoptotic cells. In 16-HBE cells exposed to CSE, FP-loaded NLC were able to control oxidative stress increasing oxidant/anti-oxidant balance in favour of anti-oxidant responses and to limit innate immune responses, and were similar or superior to unloaded FP in these effects. These observations suggest the use of this system for the FP administration in inhalation therapies because of the ability of NLC to solubilize an adequate amount of the drug and to penetrate into the airway epithelial cells. These findings suggest a potential role of these nanocarriers in the therapy of chronic obstructive pulmonary diseases such as COPD.

## Experimental
### Materials and methods

Fluticasone propionate, sodium taurocholate, and acetonitrile for HPLC were purchased from Sigma Aldrich (Milan, Italy). Dichlorometane for HPLC was obtained from Merck (Germany). Compritol HD5 ATO (behenoyl polyoxyl-8 glycerides) was a gift sample from Gattefossè (France). Epikuron 200 (soybean lecithin) was a gift sample from Lucas Meyer Company (Germany). HPLC (UFLC-Prominence system, Shimadzu Instrument, Japan) was equipped with two pumps LC-20 AD, an UV-visible detector SPD-20 AV, an autosample SIL-20A HT and a column Gemini® $C_{18}$ Phenomenex (250 mm, 5 μm particle size, 110 Å pores size).

### Preparation of NLC

Pegylated NLC, empty or FP-loaded were prepared by the precipitation method [8-10]. Briefly, Compritol HD5 ATO (180 mg) was heated at 5–10°C above its melting point (m.p. 60°-67°C). For obtaining drug-loaded NLC, FP (10 mg; m.p. 272°-273°C) was added, under mechanical stirring, to the melted lipid phase. An ethanolic solution (2 ml) of Epikuron 200 (78.5 mg) was then added to the melted lipid phase containing FP and the resulting organic dispersion was dispersed into bidistilled water (100 ml) containing sodium taurocholate (177.4 mg) at 2-3°C and stirred by using an Ultraturrax T125 (IKA Labortechnik, Germany) at 13,500 rpm for 10 minutes. Finally, the colloidal aqueous dispersion of NLC was purified by exhaustive dialysis (dialysis tube with 12,000/

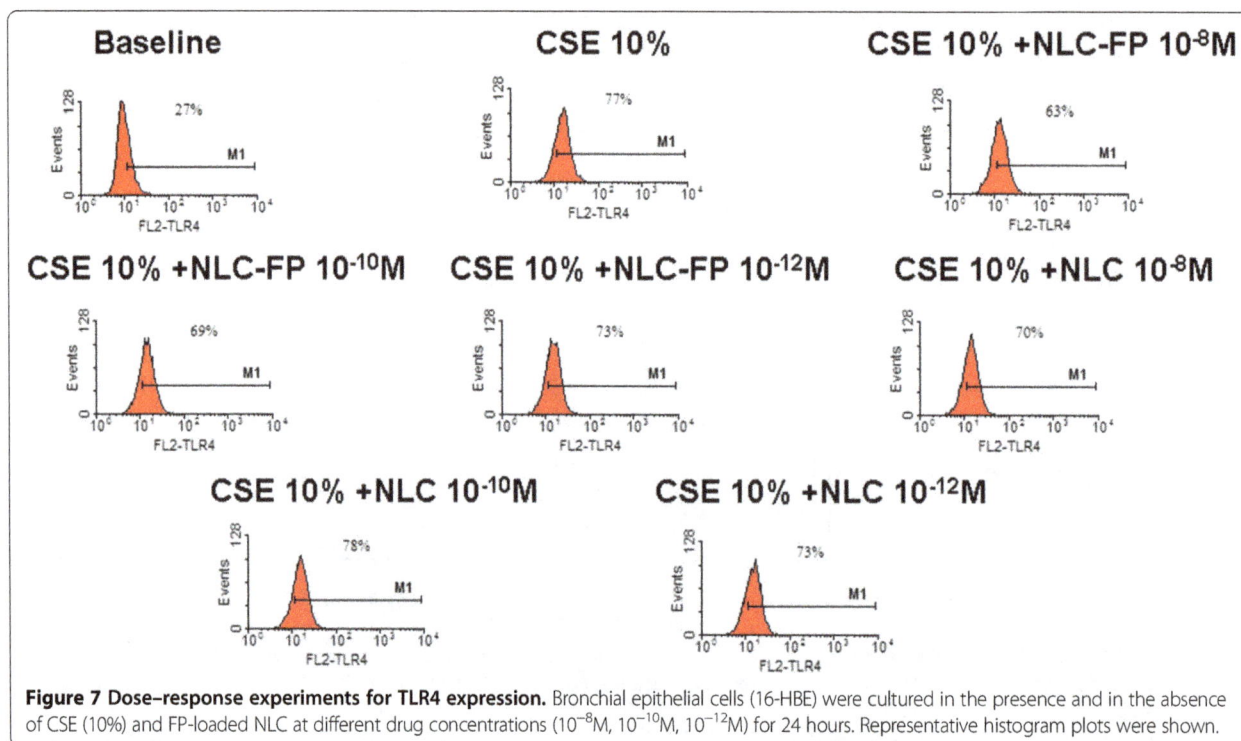

**Figure 7 Dose–response experiments for TLR4 expression.** Bronchial epithelial cells (16-HBE) were cultured in the presence and in the absence of CSE (10%) and FP-loaded NLC at different drug concentrations ($10^{-8}$M, $10^{-10}$M, $10^{-12}$M) for 24 hours. Representative histogram plots were shown.

14,000 Dalton cut-off (Spectra/Por®, USA) and freeze-dried. NLC samples (m.p. about 70°C) were stored at $4 \pm 1$°C for successive characterization.

### Scanning Electron Microscopy (SEM) analysis

For morphological studies, freeze-dried samples were observed by using an ESEM FEI Quanta 200F scanning electron microscope. Samples were dusted on a double-sided adhesive tape, previously applied on a stainless steel stub. All samples were then sputter-coated with gold prior to microscopy examination.

### Particle size analysis

The hydrodynamic diameter (z-average) and the width of distribution (polydispersity index, PDI) of the nanosuspensions were investigated by Photon Correlation Spectroscopy (PCS) by using a Zetasizer Nano ZS (Malvern Instrument Ltd, UK). The nanoparticles were diluted until the appropriate concentration and then the measurements performed at a temperature of 25°C, at a fixed angle of 173° (NIBS = non-invasive backscattering detection) in respect to the incident beam. Bidistilled water, isotonic aqueous solution (NaCl 0.9% w/w), and phosphate buffered saline solution (PBS) at pH 7.4 as suspending media were used. When the measurement was carried out in NaCl 0.9 wt%, the instrument setting conditions were: $\mu = 0.902$, RI = 1.331; in PBS at pH 7.4, the setting conditions were: $\mu = 0.980$, RI = 1.334. Results of light scattering experiments are given as the average values obtained using samples from three different batches. Each sample was measured in triplicate.

### ζ potential measurements

The surface charge or ζ potential is considered as one of the benchmark of stability of a colloidal system. It indicates the degree of repulsion between similarly charged particles into a dispersion. For the nanoparticles, a high value of ζ potential will confer stability and the nanosuspensions will resist aggregation phenomena. When the ζ potential is low, attraction exceeds repulsion and the dispersions will flocculate.

The analysis was performed at a temperature of $25 \pm 1$°C using appropriately diluted samples in the same media used for size measurements. Instrument setting conditions were equal to those described above for size measurements.

Results of these experiments are given as the average values obtained using samples from three different batches. Each sample was measured in triplicate.

### HPLC analysis

An adequate HPLC method was developed to reveal FP and to study its stability in PBS at pH 7.4, as well as Loading Capacity (LC%) and drug release profiles from drug-loaded NLC. The HPLC analysis was performed at room temperature by using the instrument described above. A column Gemini® $C_{18}$ Phenomenex (above described) was used as stationary phase and a mixture of $CH_3CN/H_2O$ 80/20 (v/v), with a flow rate of 0.8 ml/min, was used as mobile phase with an isocratic method. The drug peak was measured at wavelength of 239 nm and quantitatively determined by comparison with a standard curve obtained using FP organic solutions in a mixture of $CH_2Cl_2{:}CH_3CN$

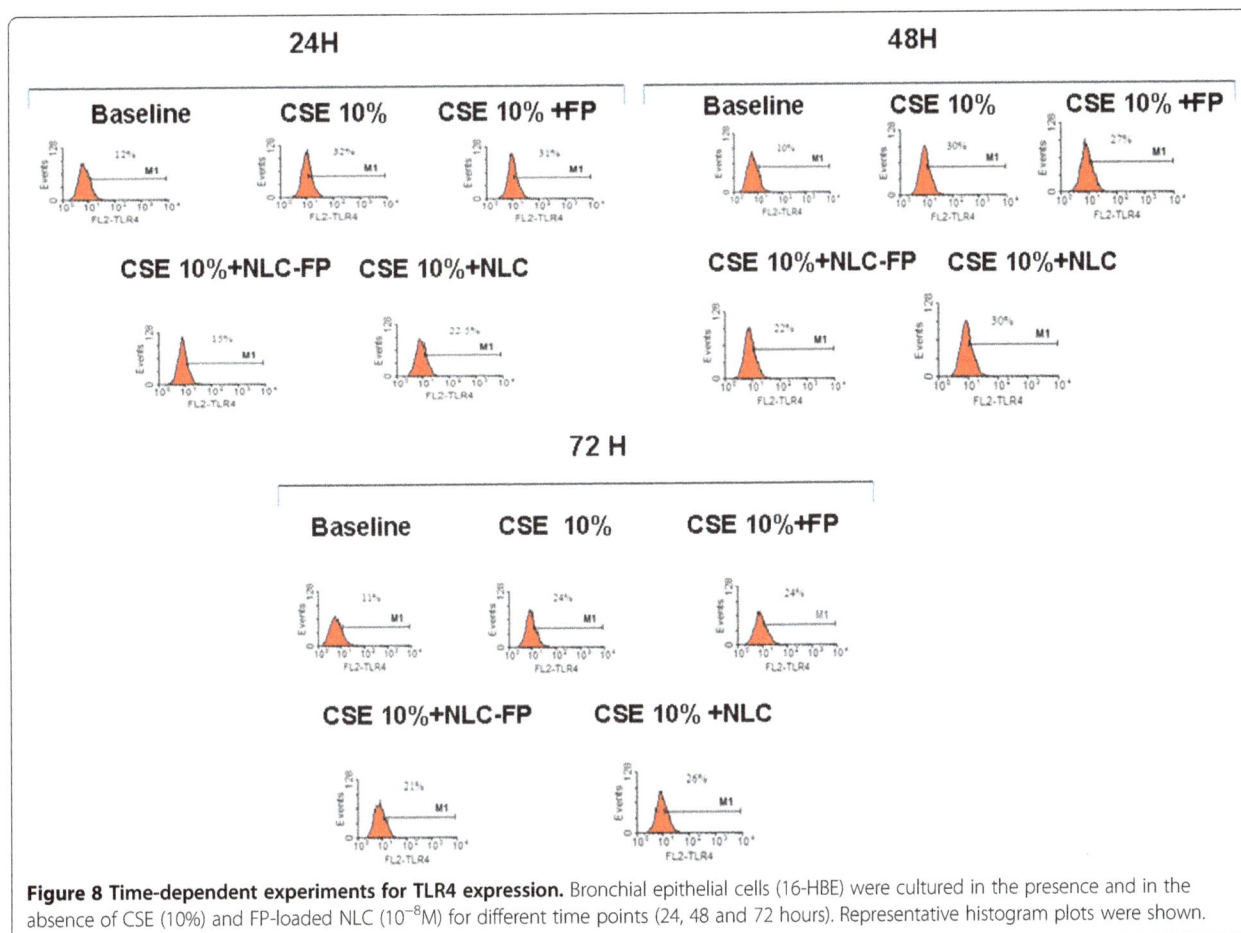

**Figure 8 Time-dependent experiments for TLR4 expression.** Bronchial epithelial cells (16-HBE) were cultured in the presence and in the absence of CSE (10%) and FP-loaded NLC ($10^{-8}$M) for different time points (24, 48 and 72 hours). Representative histogram plots were shown.

3:2 (v/v) at known concentrations ($t_r$ =7.03 min). The straight-line equation was: $y = 4 \cdot 10^5 \, x$ and the linear regression value was: $r^2 = 0.9993$. The linearity of the method was studied in the range 0.30-1.20 μg/ml.

### Drug loading and entrapment efficiency determination

Loading capacity (LC%) was determined by solubilizing the nanoparticles into an organic mixture ($CH_2Cl_2:CH_3CN$ 3:2 (v/v)), filtered with 0.45 μm PTFE syringe filters (Puradisc Whatman) and analyzed by the HPLC method described above. Drug loading capacity (DL%) was calculated as drug analyzed in the nanoparticles versus the total amount of the drug and the lipid added during preparation, according to the following equation, where $W_{drug}$ is the amount drug found inside nanoparticles and $W_{NPS}$ is the weight of drug-loaded nanoparticle:

$$DL \% = \frac{W_{drug}}{W_{NPs}} \times 100$$

Results are given as the average values obtained using samples from three different batches and were expressed as the percentage of the FP amount contained in 100 mg of dried material (LC%). Moreover, entrapment efficiency

(EE%) was determined using the HPLC method above described on purified NLC following their disruption with a mixture of $CH_2Cl_2$ and $CH_3CN$ (3:2 v/v). The encapsulated amount of FP was expressed dividing the found amount of FP and the total amount used to prepare the nanoparticles. The following equation was used to calculate the EE%, where $W_f$ is the amount drug found and $W_i$ is the initial amount of drug for the preparation:

$$EE \% = \frac{W_f}{W_i} \times 100$$

### Stability studies in PBS/ethanol

In order to obtain a release profile of FP from NLC under sink conditions, a mixture of PBS at pH 7.4 and ethanol 80:20 (v/v) was used. The term "sink conditions" refers to release conditions in which the volume of the buffer used is sufficient to dissolve all drug present into NLC. Such conditions are used to assure that the amount of drug released is not limited by the degree of solubility in the buffer or solvent used. In particular, lyophilized FP-loaded nanoparticles (5 mg) were suspended into the mixture release medium above described (5 ml) and transferred inside of a Spectra/Por® dialysis membrane that was

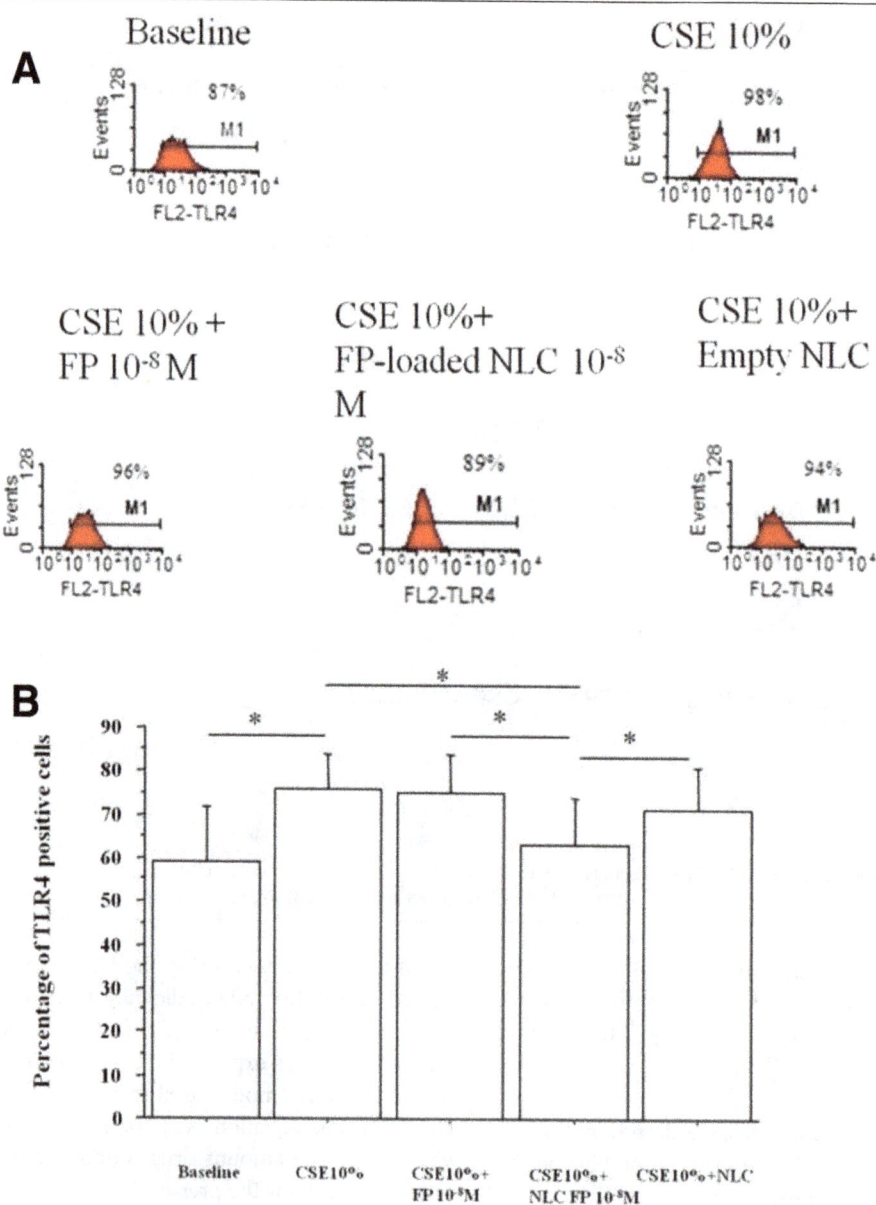

**Figure 9 Effects of FP-loaded NLC on TLR4 expression.** Bronchial epithelial cells (16-HBE) were cultured in the presence and in the absence of CSE (10%), FP, NLC and FP-loaded NLC ($10^{-8}$M) for 24 hours and then were used for assessing TLR4 expression by flow cytometry. **(A)** Representative histogram plots are shown. **(B)** Data are expressed as percentage of TLR4 positive cells $\pm$ SD. *$p < 0.05$.

immersed into the same pre-heated medium (25 ml) and incubated at $37 \pm 0.1°C$, under continuous stirring, in a Benchtop Incubator Orbital Shaker model 420 (Thermo-Scientific Instruments, CA).

At scheduled time, solution aliquots were taken out from the outside of the dialysis membrane and replaced with equal volumes of the fresh PBS/ethanol mixture. In order to determine the released FP amount, the drawn samples were filtered by 0.2 μm cellulose syringe filters (Millipore) and analyzed by HPLC, following the method above described. Profile releases were determined by comparing the amount of released drug as a function of

incubation time with the total amount of drug loaded into the nanoparticles.

Moreover, in order to determine the amount of FP entrapped into residual NLC samples, the PBS suspension containing FP-loaded NLC was freeze-dried (FreeZone®-Freeze Dry System, Labconco Corporation, Missouri, USA). Successively, an organic mixture ($CH_2Cl_2:CH_3CN$ 3:2 (v/v)), was added to lyophilized product, which was filtered through 0.2 μm (PTFE membrane) filters and analysed by HPLC, as reported above.

Finally, in order to determine the diffusion behaviour of the unloaded drug a control experiment was also

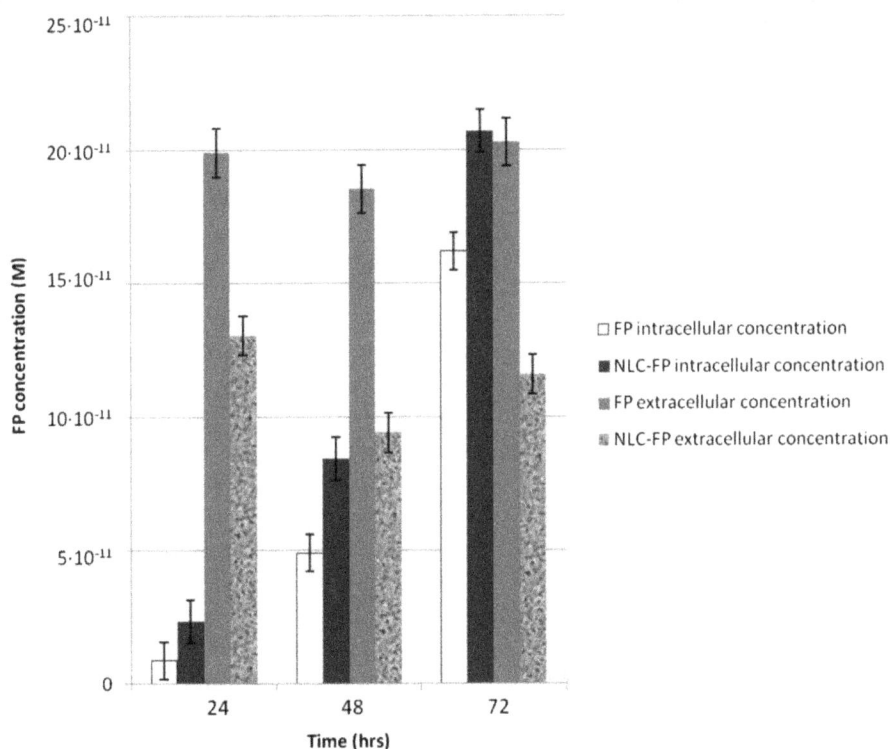

**Figure 10 Intracellular and extracellular concentrations of FP at 24, 48 and 72 hrs.** Data are expressed as Molarity.

performed. At this purpose, an appropriate amount of FP (equal to whom of FP-loaded nanoparticles) was dispersed in the mixture release medium (5 ml), placed inside a dialysis tube (MWCO 12,000-14,000 Da) and immersed into the same medium (25 ml). The amount of FP was detected by HPLC, as reported above.

### Storage and colloidal stability evaluation

Both lyophilised empty and FP-loaded NLC were stored at 4°C for 4 and 10 months in the dark. The stability test was carried only at 4°C and not at room temperature because the NLC are prepared with lipids that must be stored at a temperature not exceeding 10°C so as reported in the data sheet of the supplier of Gattefosse Compritol HD5 ATO. After this period of storage, samples were dispersed in bidistilled water and characterized in terms of mean size, PDI, ζ potential and drug stability.

### Preparation of cigarette smoke extracts (CSE)

Commercial cigarettes (Marlboro) were used in this study. Cigarette smoke solution was prepared as described previously [21]. Each cigarette was smoked for 5 min and two cigarettes were used per 20 ml of PBS to generate a CSE-PBS solution. The CSE solution was filtered through a 0.22 μm-pore filter to remove bacteria and large particles as previously described and standardised [30,31]. The smoke solution was then adjusted to pH 7.4 and used within 30

minutes of preparation. This solution was considered to be 100% CSE and diluted to obtain the desired concentration in each experiments. The concentration of CSE was calculated spectrophotometrically measuring the optical density (OD) as previously described at the wavelength of 320 nm [21]. The presence of contaminating LPS on undiluted CSE was assessed by a commercially available kit (Cambrex Corporation, East Rutherfort, New Jersey, USA) and was below the detection limit of 0.1 EU/ml.

### Stimulation of bronchial epithelial cell lines

The SV40 large T antigen-transformed 16-HBE cell line (16-HBE) was used for these studies [21]. 16-HBE is a cell line that retains the differentiated morphology and function of normal airway epithelial cells. 16-HBE was maintained in Eagle's minimum essential medium (MEM) supplemented with 10% heat-inactivated (56°C, 30 min) fetal bovine serum (FBS), 1% MEM (non-essential amino acids, Euroclone), 2 mM L-glutamine and gentamicin 250 μg/ml. Cell cultures were maintained in a humidified atmosphere of 5% $CO_2$ in air at 37°C. 16HBE were plated in 12-well plates. 70.000 cells in 1ml MEM 10% FBS were seeded for each well. At confluence 16-HBE cells were treated in 1 ml MEM 1% FBS in presence of CSE (10%) and with or without FP ($10^{-8}$M), empty-NLC or FP-loaded NLC ($10^{-8}$M) for 24 hrs. 1% FBS was used during cell stimulation to limit the basal activation of the cells due to serum proteins.

Preliminary experiments aimed to identify the best time point (24, 48 and 72 hrs) as well as the best drug concentration (FP-loaded NLC $10^{-8}$M, $10^{-9}$M, $10^{-10}$M) were performed. At the end of stimulation, cells were collected for further evaluations.

## Cell apoptosis by annexin V binding method

Cell apoptosis in the presence of free FP ($10^{-8}$M), empty NLC and FP-loaded NLC ($10^{-8}$M) was evaluated by staining with annexin V-fluorescein isothiocyanate and propidium iodide (PI) using a commercial kit (Bender Med-System, Vienna, Austria) following the manufacturer's directions. Cells were analyzed using a FACS Calibur (Becton Dickinson, Mountain View, CA) analyzer equipped with an Argon ion Laser (Innova 70 Coherent) and Consort 32 computer support.

## Analysis of intracellular reactive oxygen species (ROS)

Intracellular ROS were measured by the conversion of the non-fluorescent dichlorodihydrofluorescein diacetate (DCFH-DA; Sigma Aldrich, Milan, Italy) in a highly fluorescent compound, dichlorofluorescein (DCF), by monitoring the cellular esterase activity in the presence of peroxides. The ROS generation was assessed by uptake of 1 µM DCFH-DA, incubation for 10 min at room temperature in the dark, followed by flow cytometric analysis.

## Measurement of cellular glutathione (GSH) content

Intracellular total GSH content was assessed in cell extracts as previously reported [43]. Briefly, cell extracts were prepared in 0.1 M potassium phosphate extraction buffer containing 0.6% (w/v) sulfosalicylic acid, 0.1% (v/v) Triton X-100, 5 mM EDTA. After harvesting and resuspension in extraction buffer, cells were sonicated in ice-cold water and underwent two cycles of freezing and thawing. Supernatants/extracts were collected by centrifugation and used for the following colorimetric assay: 10 µl of extracts were incubated in presence of 60 µl 0.6 mg/ml 5,5′-Dithiobis(2-nitrobenzoic acid) (DTNB) and 60 µl of 250 U/ml glutathione reductase for 30 seconds at room temperature; 50 µl of 0.6 mg/ml β-NADPH were added and formation of 2-nitro-5-thiobenzoic acid was immediately evaluated by measuring the absorbance at 412 nm in a microplate reader. Concentration of GSH in cell extracts was calculated using a standard curve, normalized by the total protein content and expressed as nmol/mg protein.

## Expression of TLR4 in 16-HBE

The total TLR4 protein expression (inside the cells and on their surface) was assessed in permeabilized cells. For cell permeabilization, a commercial fix-perm cell permeabilization kit (Caltag Laboratories, Burlingame, CA, USA) was used. Cells were incubated in the dark (30 min, 4°C) with PE anti-human TLR4 monoclonal mouse antibody

(eBioscence San Diego CA) and then evaluated by flow-cytometry (FACS Calibur).

Negative controls were performed using mouse immunoglobulins negative control (Dako). Data are expressed as percentage of positive cells.

## Intracellular and extracellular concentrations of FP and NLC-FP

Cell cultures were maintained in a humidified atmosphere of 5% $CO_2$ in air at $37 \pm 1°C$. Cell lines were cultured in the presence and in the absence of free FP ($10^{-8}$M) and FP-loaded NLC ($10^{-8}$M) for 24, 48 and 72 hrs. At the end of stimulation, cells and supernatants were collected for assessing the intracellular and extracellular content of FP by UV analysis. In particular, the supernatants were sucked from the wells and then collected by centrifugation at 1300 rmp for 10 min. Cells were detached from the wells by tripsin, washed with PBS and stored as dry pellet at −20°C. After several cycles of freezing and thawing, the cells as well as the previously recovered culture supernatants, were used for testing their FP content. FP was extracted both from the cells from the supernatants with an organic solution (4 ml) of $CH_2Cl_2:CH_3CN$ (3:2 v/v), filtered with 0.45 µm PTFE syringe filters (Puradisc Whatman) and the absorbance was determined by ultraviolet–visible (UV–vis) Spectrophotometer (UV-1800 Shimadzu, Kyoto, Japan) at 239 nm.

## Statistics

Data are expressed as mean counts ± standard deviation. Comparison between different experimental conditions was evaluated by paired $t$ test. $P < 0.05$ was accepted as statistically significant.

### Abbreviations
AP-1: Activator protein-1; COPD: Chronic obstructive pulmonary disease; CSE: Cigarette smoke extracts; DCF: Dichlorofluorescein; DCFH-DA: Dichlorodihydrofluorescein diacetate; DL: Drug loading capacity; DTNB: 5,5′-Dithiobis(2-nitrobenzoic acid); EDTA: Ethylenedyaminetetraacetic acid; EE: Entrapment efficiency; FP: Fluticasone propionate; FBS: Fetal bovine serum; GSH: Glutathione; 16-HBE: Human bronchial epithelial cell line; HDAC: Histone deacetylase; HPLC: High pressure liquid chromatography; ICS: Inhaled corticosteroids; LC: Loading capacity; LPS: Lipopolysaccharides; MEM: Eagle's minimum essential medium (MEM); NADPH: Nicotinamide adenine dinucleotide phosphate; NLC: Nanostructured lipid carriers; OD: Optical density; PBS: Phosphate buffered saline solution; PCD: Photon correlation spectroscopy; PDI: Polydispersity index; PI: Propidium iodide; ROS: Reactive oxygen species; SEM: Scanning electron microscopy; STAT: Signal transducers and activators of transcription; TLR4: Toll-like receptor 4.

### Competing interests
The authors declare that they have no competing interests.

### Authors' contributions
CB, SDV and SG are the PhD students who carried out the laboratory work. MF carried out the biological work in laboratory. EP was the supervisor of the biological study and helping to develop the study parameters and design. MLB was the principal, scientific supervisor of the study. She conceived the study, supervised the students in the laboratory, directed the analysis and wrote the manuscript. All authors read and approved the final

draft of the manuscript. MG, GC and GG have revised the final version of manuscript.

## Acknowledgements
Maria Luisa Bondì and Maria Ferraro contributed equally to this manuscript. This work was supported by the Italian National Research Council. Authors thanks also FFR 2012 of University of Palermo for funding.

## Author details
[1]Istituto per lo Studio dei Materiali Nanostrutturati- U.O.S. di Palermo-Consiglio Nazionale delle Ricerche-via Ugo La Malfa, 153 90146 Palermo, Italy. [2]Istituto di Biomedicina e Immunologia Molecolare-Consiglio Nazionale delle Ricerche – via Ugo La Malfa, 153 90146 Palermo, Italy. [3]Laboratory of Biocompatible Polymers-Dipartimento di Scienze e Tecnologie, Biologiche, Chimiche e Farmaceutiche (STEBICEF), Università di Palermo -via Archirafi, 32-90123 Palermo, Italy.

## References
1. Chaturvedi NP, Solanki H: **Pulmonary drug delivery system: review.** *Nit J Appl Pharm* 2013, **5**:7–10.
2. Torchilin VP: **Nanocarriers.** *Pharm Res* 2007, **24**:2333–2334.
3. Terzano C, Allegra L, Alhaique F, Marianecci C, Carafa M: **Non-phospholipid vesicles for pulmonary glucocorticoid delivery.** *Eur J Pharm Biopharm* 2005, **59**:57–62.
4. Buxton DB: **Nanomedicine for the management of lung and blood diseases.** *Nanomedicine (Lond)* 2009, **4**:331–339.
5. Nassimi M, Schleh C, Lauenstein HD, Hussein R, Hoymann HG, Koch W, Pohlmann G, Krug N, Sewald K, Rittinghausen S, Braun A, Müller-Goymannet C: **A toxicological evaluation of inhaled solid lipid nanoparticles used as a potential drug delivery system for the lung.** *Eur J Pharm Biopharm* 2010, **75**:107–116.
6. Doktorovová S, Araújo J, Garcia ML, Rakovský E, Souto EB: **Formulating fluticasone propionate in novel PEG-containing nanostructured lipid carriers (PEG-NLC).** *Colloids Surf B Biointerfaces* 2010, **75**:538–542.
7. Üner M, Yener G: **Importance of solid lipid nanoparticles (SLN) in various administration routes and future perspectives.** *Int J Nanomedicine* 2007, **2**:289–300.
8. Bondì ML, Craparo EF, Giammona G, Cervello M, Azzolina A, Diana P, Martorana A, Cirrincione G: **Nanostructured lipid carriers-containing anticancer compounds: preparation, characterization, and cytotoxicity studies.** *Drug Deliv* 2007, **14**:61–67.
9. Bondì ML, Azzolina A, Craparo EF, Capuano G, Lampiasi N, Giammona G, Cervello M: **Solid lipid nanoparticles (SLNs) containing nimesulide: preparation, characterization and in cytotoxicity studies.** *Curr Nanosci* 2009, **5**:39–44.
10. Bondì ML, Craparo EF, Picone P, Di Carlo M, Di Gesù R, Capuano G, Giammona G: **Curcumin entrapped into lipid nanosystems inhibits neuroblastoma cancer cell growth and activate Hsp70 protein.** *Curr Nanosci* 2010, **6**:439–445.
11. Alonso MJ: **Nanomedicine for overcoming biological barriers.** *Biomed Pharmacother* 2004, **58**:168–172.
12. Tobío M, Gref R, Sánchez A, Langer R, Alonso MJ: **Stealth PLA-PEG nanoparticles as protein carriers for nasal administration.** *Pharm Res* 1998, **15**:270–275.
13. Vila A, Sánchez A, Tobío M, Calvo P, Alonso MJ: **Design of biodegradable particles for protein delivery.** *J Control Release* 2002, **78**:15–24.
14. Craparo EF, Teresi G, Bondì ML, Licciardi M, Cavallaro G: **Phospholipid-polyaspartamide micelles for pulmonary delivery of corticosteroids.** *Int J Pharm* 2011, **406**:135–144.
15. Pitarresi G, Casadei MA, Mandracchia D, Paolicelli P, Palumbo FS, Giammona G: **Photocrosslinking of dextran and polyaspartamide derivatives: a combination suitable for colon-specific drug delivery.** *J Control Release* 2007, **119**:328–338.
16. Marianecci C, Paolino D, Celia C, Fresta M, Carafa M, Alhaique F: **Non-ionic surfactant vesicles in pulmonary glucocorticoid delivery: characterization and interaction with human lung fibroblast.** *J Control Release* 2010, **147**:127–135.
17. Pace E, Ferraro M, Di Vincenzo S, Cipollina C, Gerbino S, Cigna D, Caputo V, Balsamo R, Lanata L, Gjomarkaj M: **Comparative cytoprotective effects of carbocysteine and fluticasone propionate in cigarette smoke extract-stimulated bronchial epithelial cells.** *Cell Stress Chaperones* 2013, **18**:733–743.
18. Faux SP, Tai T, Thorne D, Xu Y, Breheny D, Gaca M: **The role of oxidative stress in the biological responses of lung epithelial cells to cigarette smoke.** *Biomarkers* 2009, **14**:90–96.
19. Biswas SK, Rahman I: **Environmental toxicity, redox signaling and lung inflammation: the role of glutathione.** *Mol Aspects Med* 2009, **30**:60–76.
20. Ghezzi P: **Role of glutathione in immunity and inflammation in the lung.** *Int J Gen Med* 2011, **4**:105–113.
21. Pace E, Ferraro M, Siena L, Melis M, Montalbano A, Johnson M, Bonsignore MR, Bonsignore G, Gjomarkaj M: **Cigarette smoke increases TLR4 and modifies LPS mediated responses in airway epithelial cells.** *Immunology* 2008, **124**:401–411.
22. Pace E, Ferraro M, Uasuf CG, Giarratano A, La Grutta S, Liotta G, Johnson M, Gjomarkaj M: **Cilomilast counteracts the effects of cigarette smoke in airway epithelial cells.** *Cell Immunol* 2011, **268**:47–53.
23. Olsson B, Bondesson E, Borgström L, Edsbäcker S, Eirefelt S, Ekelund K, Gustavsson L, Hegelund-Myrbäck T: **Pulmonary Drug Metabolism, Clearance, and Absorption.** In *Controlled Pulmonary Drug Delivery.* Edited by Smith HDC, Hickey AJ. Germany: Springer; 2011:21–50.
24. Brusselle GG, Joos GF, Bracke KR: **New insights into the immunology of chronic obstructive pulmonary disease.** *Lancet* 2011, **378**:1015–1026.
25. Kelsen SG, Duan X, Ji R, Perez O, Liu C, Merali S: **Cigarette smoke induces an unfolded protein response in the human lung: a proteomic approach.** *Am J Respir Cell Mol Biol* 2008, **38**:541–550.
26. Smola M, Vandamme T, Sokolowski A: **Nanocarriers as pulmonary drug delivery systems to treat and to diagnose respiratory and non respiratory diseases.** *Int J Nanomedicine* 2008, **3**:1–19.
27. Rouse JJ, Whateley TL, Thomas M, Eccleston GM: **Controlled drug delivery to the lung: influence of hyaluronic acid solution conformation on its adsorption to hydrophobic drug particles.** *Int J Pharm* 2007, **330**:175–182.
28. Jaspart S, Bertholet P, Piel G, Dogné JM, Delattre L, Evrard B: **Solid lipid microparticles as a sustained release system for pulmonary drug delivery.** *Eur J Pharm Biopharm* 2007, **5**:47–56.
29. Bondì ML, Montana G, Craparo EF, Di Gesù R, Giammona G, Bonura A, Colombo P: **Lipid nanoparticles as delivery vehicles for the Parietaria judaica major allergen Par j 2.** *Int J Nanomedicine* 2011, **6**:2953–2962.
30. Liu X, Conner H, Kobayashi T, Kim H, Wen F, Abe S, Fang Q, Wang X, Hashimoto M, Bitterman P, Rennardt SI: **Cigarette smoke extract induces DNA damage but not apoptosis in human bronchial epithelial cells.** *Am J Respir Cell Mol Biol* 2005, **33**:121–129.
31. Moretto N, Facchinetti F, Southworth T, Civelli M, Singh D: **Patacchini R: α, β-Unsaturated aldehydes contained in cigarette smoke elicit IL-8 release in pulmonary cells through mitogen-activated protein kinases.** *Am J Physiol Lung Cell Mol Physiol* 2009, **296**:839–848.
32. Su Y, Han W, Giraldo C, De Li Y, Block ER: **Effect of cigarette smoke extract on nitric oxide synthase in pulmonary artery endothelial cells.** *Am J Respir Cell Mol Biol* 1998, **19**:819–825.
33. MacNee W: **Oxidants/antioxidants and COPD.** *Chest* 2000, **117**:303–317.
34. Baglole CJ, Bushinsky SM, Garcia TM, Kode A, Rahman I, Sime PJ, Phipps RP: **Differential induction of apoptosis by cigarette smoke extract in primary human lung fibroblast strains: implications for emphysema.** *Am J Physiol Lung Cell Mol Physiol* 2006, **291**:19–29.
35. Rahman I, MacNee W: **Oxidative stress and regulation of glutathione in lung inflammation.** *Eur Respir J* 2000, **16**:534–544.
36. Barnes PJ: **Theophylline for COPD.** *Thorax* 2006, **61**:742–744.
37. Luppi F, Aarbiou J, van Wetering S, Rahman I, de Boer WI, Rabe KF, Hiemstra PS: **Effects of cigarette smoke condensate on proliferation and wound closure of bronchial epithelial cells in vitro: role of glutathione.** *Respir Res* 2005, **6**:140–151.
38. Van der Toorn M, Smit-de Vries MP, Slebos DJ, de Bruin HG, Abello N, Van Oosterhout AJ, Bischoff R, Kauffman HF: **Cigarette smoke irreversibly modifies glutathione in airway epithelial cells.** *Am J Physiol Lung Cell Mol Physiol* 2007, **293**:1156–1162.
39. Chen L, Wang T, Zhang JY, Zhang SF, Liu DS, Xu D, Wang X, Chen YJ, Wen FQ: **Toll-like receptor 4 relates to lipopolysaccharide-induced mucus hypersecretion in rat airway.** *Arch Med Res* 2009, **40**:10–17.
40. Pace E, Ferraro M, Minervini MI, Vitulo P, Pipitone L, Chiappara G, Siena L, Montalbano AM, Johnson M, Gjomarkaj M: **Beta defensin-2 is reduced in central but not in distal airways of smoker COPD patients.** *PLoS One* 2012, **7**:e33601.

41. Geraghty P, Dabo AJ, D'Armiento J: **TLR4 protein contributes to cigarette smoke-induced matrix metalloproteinase-1 (MMP-1) expression in chronic obstructive pulmonary disease.** *J Biol Chem* 2011, **286**:30211–30218.

42. Mortaz E, Henricks PA, Kraneveld AD, Givi ME, Garssen J, Folkerts G: **Cigarette smoke induces the release of CXCL-8 from human bronchial epithelial cells via TLRs and induction of the inflammasome.** *Biochim Biophys Acta* 1812, **2011**:1104–1110.

43. Rahman I, Kode A, Biswas SK: **Assay for quantitative determination of glutathione and glutathione disulfide levels using enzymatic recycling method.** *Nat Protoc* 2006, **1**:3159–3165.

# Investigation of magnetically controlled water intake behavior of Iron Oxide Impregnated Superparamagnetic Casein Nanoparticles (IOICNPs)

Anamika Singh, Jaya Bajpai and Anil Kumar Bajpai[*]

## Abstract

Iron oxide impregnated casein nanoparticles (IOICNPs) were prepared by in-situ precipitation of iron oxide within the casein matrix. The resulting iron oxide impregnated casein nanoparticles (IOICNPs) were characterized by Scanning electron microscopy (SEM), Transmission electron microscopy (TEM), X-ray photoelectron spectroscopy (XPS), Fourier transform infrared (FTIR), Vibrating sample magnetometer (VSM) and Raman spectroscopy. The FTIR analysis confirmed the impregnation of iron oxide into the casein matrix whereas XPS analysis indicated for complete oxidation of iron (II) to iron(III) as evident from the presence of the observed representative peaks of iron oxide. The nanoparticles were allowed to swell in phosphate buffer saline (PBS) and the influence of factors such as chemical composition of nanoparticles, pH and temperature of the swelling bath, and applied magnetic field was investigated on the water intake capacity of the nanoparticles. The prepared nanoparticles showed potential to function as a nanocarrier for possible applications in magnetically targeted delivery of anticancer drugs.

**Keywords:** Casein, IOICNPs, Swelling behaviour, pH sensitive, Magnetic drug targeting

## Introduction

The concept of magnetic drug targeting (MDT) simply entails retaining specially designed magnetic drug nanocarriers at a specific site in the body using an externally applied magnetic field. The two key properties for an effective nanocarriers are (a) efficient targeting to specific tissue and cells, and (b) avoiding rapid clearance (i.e. remaining in circulation) for a significant amount of time to increase particle uptake in target tissue. Circulation time, targeting, and the ability to overcome biological barriers depend on the shape (e. g, aspect ratio), chemical coating and size of the nanoparticles [1]. Thus a precise control over the shape and size of nanoparticles is a challenging task and must be addressed in order to achieve high performance drug delivery systems.

In order to achieve an efficient MDT, various polymeric/inorganic hybrid materials have been suggested that offer unique properties because of their small size,

limited toxicity [2], low production cost, ease of separation and detection [3]. The magnetic nanoparticles often tend to form large aggregates owing to the strong magnetic dipole–dipole attractions among particles. To improve their chemical stability and biocompatibility, the surface of magnetic nanoparticles have been modified with various surfactants [4] or biopolymer compounds having multiple functional groups capable of binding to the particle surfaces (multidentate ligands). The rational for using magnetic nanoparticles to tumor targeting is based on the fact that nanoparticles will be able to deliver a concentrate dose of drug in the vicinity of the tumor targets via the enhanced permeability and retention effect or active targeting by ligands on the surface of nanoparticles [5]. Moreover, the extent of drug loading onto the nanoparticles greatly depends on the hydrophilic nature of the biopolymer also and, therefore, proteins could be an excellent option to design such magnetic nanocarriers.

Use of milk proteins, like caseins, in drug delivery applications is relatively a new trend and this is mainly

* Correspondence: akbmrl@yahoo.co.in
BMRL, Department of Chemistry, Government Model Science College, Jabalpur, India

due to its amphiphilic nature that allows them to interact with both the drug and solvent. In fact, caseins may be regarded as block copolymers with high level of hydrophilic and hydrophobic amino acid residues and thus they exhibit a strong tendency to self assemble into spherical micelles. Moreover, thier biodegradability, not-toxicity, metabolizablity and feasibility to surface modification enable them to interact with the targeting ligand. In an study, the complexation of curcumin with $\beta$-casein micelles increased the solubility of curcumin at least 2500-fold with enhanced curcumin cytotoxicity to a human leukemia cell line [6]. Shapira et al. [7] showed that $\beta$-CAS micelles could entrap and deliver hydrophobic chemotherapeutics such as mitoxantrone and paclitaxel, allowing them to be thermodynamically stable in aqueous solutions for oral delivery applications. Thus, the hydrophobic and hydrophilic domains of casein are responsible not only for their water sorption capacity, but also for the nature and type of drug to be encapsulted in the casein nanoparticles.

Thus, motivated by the pharmaceutical specialities of the casein protein the authors were pused to undertake a systemic investigation of synthesis and characteriztion of magnetic casein nanoparticles and their water sorption behavior to judge their suitability in designing swelling controlled and magnetic mediated drug delivery system. The present study aims at designing controllable size iron oxide impregnated casein nanoparticles (IOICNPs) by co-precipitation of iron salts within the casein nanoparticles matrix. As the It is also proposed to characterized the so prepared IOICNPs by various analytical techniques and investigate their water sorption potential in the presence of applied magnetic field.

## Experimental

### Materials
Casein was purchased from Merck, Mumbai, India and used without any pretreatment. $FeCl_2.H_2O$, $FeCl_3.6H_2O$, glutaraldehyde (used as a crosslinker) were obtained from Loba Chemie, Mumbai, India. Toluene was obtained from Sigma Aldrich Co., USA, and used for preparing oil phase. Other chemicals like acetone, NaOH etc. were of analytical reagent (AR) grade and double distilled water was used throughout the experiments.

### Preparation of Iron oxide impregnated casein nanoparticles (IOICNPs)
Preparation of magnetic casein nanoparticles consists of a two steps process. In the first step the casein nanoparticles are prepared by emulsion crosslinking method while in the second one iron oxide nanoparticles are impregnated within casein nanoparticles matrix by in situ precipitation.

### Preparation of casein nanoparticles (CNPs)
In order to prepare CNPs the microemulsion crosslinking method was adopted as described in literature [8]. In brief, an aqueous phase was prepared by dissolving known amount of casein in 1% NaOH whereas toluene was used to prepare the oil phase. The above two solutions were mixed with vigorous shaking (shaking speed 1000 RPM, 5 L capacity, Remi, India) for 30 min and to this emulsion 1 mL glutaraldehyde was added as crosslinker with constant stirring. The crosslinking reaction was allowed to take place for 30 min at room temperature (30°C) and $H_2SO_4$ was added to the solution for the solidification of particles. The nanoparticles were cleaned by washing them thrice with acetone and stored in air-tight polyethylene bags.

### Impregnation of Iron oxide in to the casein nanoparticles
The dried CNPs were placed in an aqueous mixture of $Fe^{2+}$ and $Fe^{3+}$ chloride salts at 1:2 molar ratio and allowed to swell for 24 h so that both $Fe^{2+}$ and $Fe^{3+}$ ions were entrapped in to the biopolymer matrix. Prior to putting them in salt solution, a dry stream of $N_2$ was flushed for at least 15 min to control the reaction kinetics, which is strongly related to the oxidation speed of iron species. Bubbling nitrogen gas through the solution not only protects critical oxidation of the magnetite but also reduces the particle size [9]. The ferrous and ferric ions loaded casein nanoparticles were added to NaOH solution for a definite time period so that magnetite is precipitated within the biopolymer matrix according to the following chemical reactions [10].

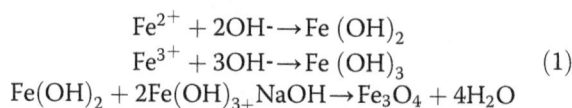

$$Fe^{2+} + 2OH\text{-} \rightarrow Fe\,(OH)_2$$
$$Fe^{3+} + 3OH\text{-} \rightarrow Fe\,(OH)_3 \qquad (1)$$
$$Fe(OH)_2 + 2Fe(OH)_{3+} NaOH \rightarrow Fe_3O_4 + 4H_2O$$

Furthermore, for synthesis of nanoparticles with either the ferrous or ferric ion alone, the chemical reactions pass through different mechanisms as shown in Equations (2) and (3), respectively.

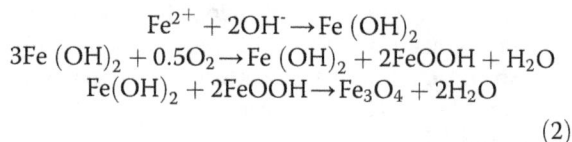

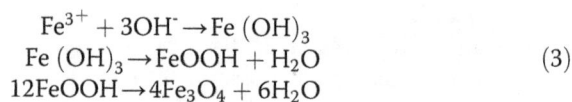

$$Fe^{2+} + 2OH^- \rightarrow Fe\,(OH)_2$$
$$3Fe\,(OH)_2 + 0.5O_2 \rightarrow Fe\,(OH)_2 + 2FeOOH + H_2O$$
$$Fe(OH)_2 + 2FeOOH \rightarrow Fe_3O_4 + 2H_2O$$
$$(2)$$

$$Fe^{3+} + 3OH^- \rightarrow Fe\,(OH)_3$$
$$Fe\,(OH)_3 \rightarrow FeOOH + H_2O \qquad (3)$$
$$12FeOOH \rightarrow 4Fe_3O_4 + 6H_2O$$

These two mechanisms provide a larger particle size as compared to the case of the mixture of ferrous and ferric ions because the particle size of magnetite also depends on the nature of the intermediate form. Since the methods adopted require longer reaction times for the transformations of ferric ions and ferrous ions so that the intermediates can continuously grow. However, care must be taken while adding NaOH because according to the thermodynamics of this reaction, a complete precipitation of $Fe_3O_4$ occurs in the range of 9 to 14 pH and in molar ratio of 1:2 for $Fe^{2+}$: $Fe^{3+}$ under a non oxidizing oxygen free environment. Otherwise, $Fe_3O_4$ might get oxidized as,

$$Fe_3O_4 + 0.25O_2 + 4.5H_2O \rightarrow 3Fe\,(OH)_3 \qquad (4)$$

The change in color of the casein nanoparticles from orange to dark brown also confirms the formation of oxides of iron. The prepared nanoparticles were washed, dried at room temperature and stored in airtight polyethylene bags. The chemical reaction is shown in Figure 1. The percentage impregnation of iron oxide was calculated using following equation.

$$\mathbf{Impregnation}(\%) = \frac{\mathbf{Wimpregnated\text{-}Wdry}}{\mathbf{Wdry}} \times \mathbf{100}$$

$$(5)$$

In fact, the process of formation of iron oxide involves the diffusion of ferrous/ferric ions into the polymer matrix and their subsequent in situ precipitation in an alkaline medium. It is therefore expected that the higher is the water uptake greater will be, the iron oxide formation [11].

## Characterization

### FTIR Spectral analysis

The FTIR spectra of casein and IOICNPs were recorded on a FTIR- 8400, Shimadzu Spectrophotometer. Samples for the spectral analysis were prepared by mixing nanoparticles and KBr in 1:10 proportion and the spectra were obtained in the range of 4000 to 400 $cm^{-1}$ with a resolution of 2 $cm^{-1}$.

### SEM analysis

Morphological studies of cross-linked CNPs and IOICNPs were performed using SEM, Philips 515, fine coater (Philips, Eindhoven, The Netherlands). Drops of the polymeric nanoparticles suspension were placed on a graphite surface and freeze-dried. The sample was then coated with gold by ion sputter at 20 mA for 4 minutes, and observations were made at 10 kV.

### TEM analysis

The size and morphology of the nanoparticles were determined by conducting TEM analysis of casein and IOICNPs on Morgagni 268-D Transmission Electron Microscope with an accelerating voltage of 80.0 kV. The samples for TEM measurements were prepared by dispersing a drop of the sample solution on Formvar-coated C grids.

### Raman spectral analysis

In order to investigate the impregnation of iron oxide nanoparticles in to the matrix of casein nanoparticles,

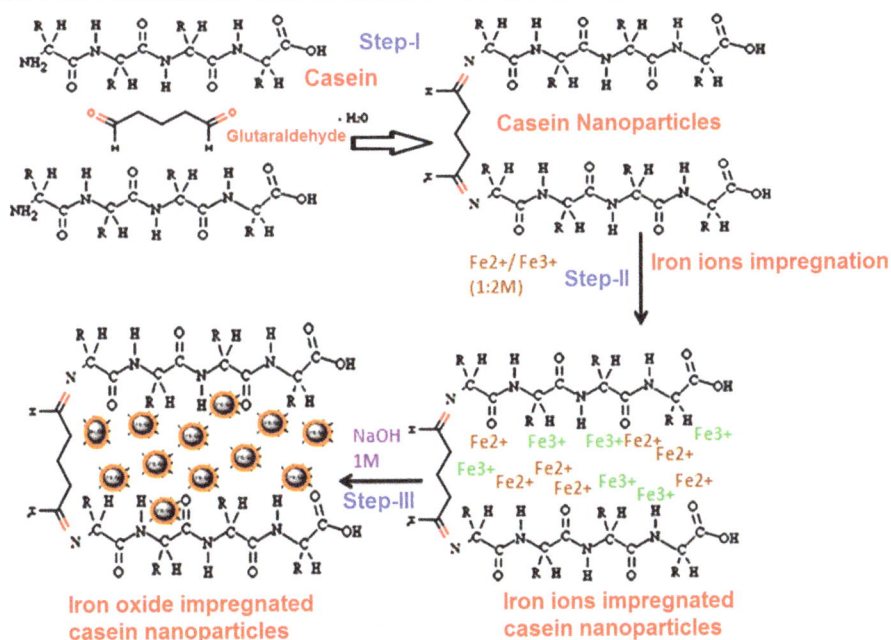

**Figure 1** Schematic formation of IOICNPs (Step-I) Formation of casein nanoparticles by crosslinking of casein macromolecules by reaction with glutaraldehyde. (Step-II) Impregnation of iron ions within the casein nanoparticls network by swelling in iron salts solution. (Step-III) In situ precipitation of iron oxide within the casein nanoparticles matrix to yield IOICNPs.

Raman spectroscopy was used and the spectra were obtained in the range of 200–1800 $cm^{-1}$. The characteristic peak position of magnetite ($Fe_3O_4$) and its possible oxidation product maghemite ($\gamma$-$Fe_2O_3$) and hematite ($\alpha$- $Fe_2O_3$) were determined in the Raman region of 100–1200 $cm^{-1}$. For correct assignment of the band positions and phase identification present in the samples, combined Raman data were used for key iron oxides bands. The Raman spectra of casein and IOICNPs were recorded on a Micro Raman Spectrometer, Jobin Yvon Horibra LABRAM-HR.

## VSM analysis

The magnetization versus magnetic field measurements (M–H first magnetization curve and hysteresis loop) at 300 K, for the IOICNPs (powder sample) were done on 14 T PPMS- vibrating sample magnetometer.

## XPS analysis

The samples were also analyzed by X-ray photoelectron spectroscopy (XPS) on a modified laser ablation system, Riber LDM-32, using a Cameca Mac3 analyzer. Photoelectron spectra were collected by acquiring data for every 1.0 eV with an energy resolution of 3 eV. Narrow-scan photoelectron spectra were recorded for C 1 s, N 1 s, O 1 s, and Fe 2p by acquiring data for every 0.2 eV and the energy resolution was 0.8 eV.

## In vitro cytotoxicity test

In order to determine in vitro cytotoxicity of the prepared materials test on extract method (ISO10993-5,2009) was applied. In brief, a test sample of the nanoparticles, negative control and positive control in triplicate were placed with subconfluent monolayer of L-929 mouse fibroblast cells. After incubation of cells with test samples at 37 ± 1°C, for 24 h, cell culture was examined microscopically for cellular response around and under the test samples.

## Water sorption capacity

The extent of swelling of CNPs and IOICNPs in both presence and absence of magnetic field were determined by a conventional gravimetric procedure as reported in the literature [12]. The swelling ratio was determined by the following equation:

$$\text{Swelling Ratio} = \frac{\text{Weight of swollen nanoparticles}(\mathbf{Ws})}{\text{Weight of drynanoparticles}(\mathbf{Wd})}$$

(6)

The amount of water imbibed by the sample provides information about the hydrophilic nature of the material, which is one of the criterions for biocompatibility.

## Effect of pH

The effect of pH on swelling of the nanoparticles was studied by preparing solutions over the pH range 1.8 to 9.0, and the desired pH was adjusted with the help of 0.1 M HCl and 0.1 M NaOH solutions. The pH was determined on a digital pH meter (Systronics, No. 362, Ahmadabad, India).

## Effect of temperature

The effect of temperature on swelling of the nanoparticles was studied by varying temperature of the swelling medium in the range of 10° to 40°C.

## Swelling studies in physiological fluid

In order to study the swelling of nanoparticles in simulated biological media, the following aqueous fluids (100 mL) were prepared: Saline water (0.9 g NaCl), synthetic urine (0.8 g NaCl, 0.10 g $MgSO_4$, 2.0 g urea, 0.6 g $CaCl_2$), urea 5.0 g, and D-glucose 5.0 g.

## Statistical analysis

All experiments were done at least thrice and Figures and data have been expressed along with their respective error bars and standard deviations, respectively.

## Results and discussion
### Effect of composition on impregnation

Impregnation of iron oxide into the polymer matrix is a result of inclusion of ferrous/ ferric ions into the polymer matrix and their subsequent *in situ* precipitation in alkaline medium. The impregnation process basically depends on the swelling capacity of the biopolymeric network which, in turn, varies as a function of chemical composition of the CNPs. Among various structural factors influencing water sorption capacity of a CNPs, the ratio of hydrophilicity to hydrophobicity plays a key role in determining swelling characteristic of the matrix. In the present study, the prepared matrix is composed of casein and glutaraldehyde which are hydrophilic biopolymer and crosslinker, respectively and their relative amounts in the CNPS are expected to affect extent of swelling and, consequently, the impregnation of iron oxide also.

### FTIR spectral analysis

The FT-IR spectra of native casein, CNPs and IOICNPs are shown in (Figure 2a, b and c), respectively. Figure 2a shows absorption bands at 3455, 3100, 1661, 1530 and 1235 $cm^{-1}$ which can be explained as follows: In the case of native casein, the amide A band at 3455 $cm^{-1}$ and amide B at 3100 $cm^{-1}$ are observed, which originate as a result of Fermi resonance between the first overtone of amide II and the N-H stretching vibration. Amide I and amide II bands are two major bands of the infrared

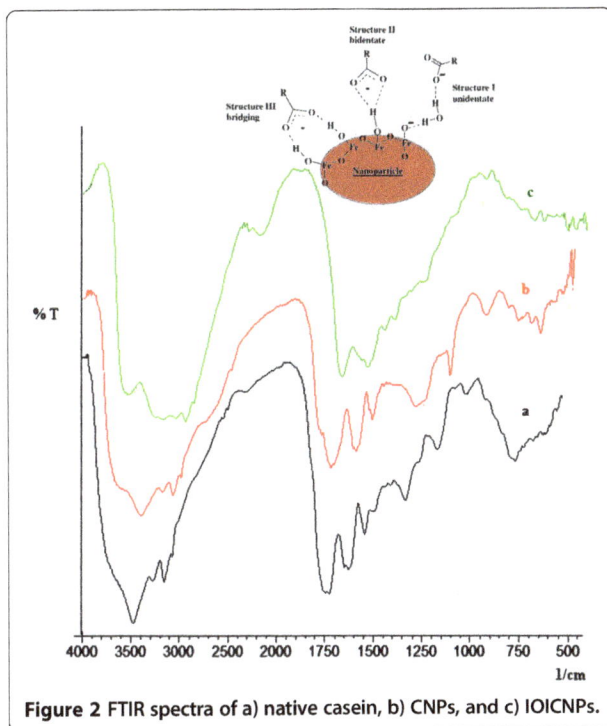

**Figure 2** FTIR spectra of a) native casein, b) CNPs, and c) IOICNPs.

spectrum of casein. The observed intense band for amide I appears at 1661 cm$^{-1}$ and is mainly associated with the C = O stretching vibration and depends on the backbone conformation and hydrogen bonding. The amide II bands obtained in the 1510 and 1580 cm$^{-1}$ region result from the N-H bending and the C-N stretching vibrations. The obtained bands at 1661 cm$^{-1}$ and 1531 cm$^{-1}$ for the amide I and amide II, respectively also confirm the alpha helical structure of the casein protein.

Casein also exhibits another characteristic band at 1415 cm$^{-1}$ which may be attributed to the carboxylate group (O-C-O). As shown in (Figure 2b), a band appears at 1683 cm$^{-1}$ and may be assigned to C = N stretching which confirms the presence of crosslinking between casein and glutaraldehyde. In (Figure 2c) the appearance of peaks around 450 and 480 cm$^{-1}$ may be assigned to Fe–O bonds of magnetite, which are characteristic peaks of iron oxide (e.g., polyhedral Fe$^{3+}$–O$^{2-}$ )stretching vibrations of iron oxide, and thus confirm the impregnation of iron oxide into the matrix of casein nanoparticles [13,14].

According to Deacon and Phillips [15], the carboxylate ion may be coordinated to a metal atom in one of the following structures:

- **structure I:** unidendate complex where one metal ion binds with one carboxylic oxygen atom
- **structure II:** bidendate complex where one metal ion binds with two carboxylate oxygens

- **structure III:** bridging complex where two metal ions bind with two carboxylate oxygen's. The FTIR spectra indicated the presence of two bands, 1415 cm$^{-1}$ (V$_s$: COO$^-$) and 1538 cm$^{-1}$ (V$_{as}$: COO$^-$), which may be attributed to the carboxylate ion of casein immobilized on the magnetite surface.

## SEM analysis
SEM images of CNPs and IOICNPs are shown in (Figure 3a and b), respectively which illustrate non-smooth morphology of CNPs and formation of iron oxide in the casein networks. The coating of iron oxide nanoparticles by the casein produces larger size particles due to the formation of the casein layers on the surfaces of iron oxide. During in-situ precipitation it may be inferred that iron oxides are assembled or attached inside the biopolymeric networks and on the casein surface as well. Loading of iron oxide inside the network affects its morphology and structural integrity. It is likely that the presence of intermolecular forces between casein macromolecular units facilitates formation of an extensive physical network of hydrogen bonds and other van der waal forces, which provide 'nano' domains for growth of the iron oxide nanoparticles as well as ensure their protection within the casein network. These biopolymeric networks may be considered as nanoreactors to construct or assemble iron oxide. The results may be attributable to contributions of a Fe-O$^-$ coordination bond on the surface, steric effect and a compartment effect of the network structures of casein, which limit the growth of iron oxide, and thus play an important role in the process of the formation of iron oxide aggregates [16].

## TEM analysis
In order to ascertain more precise morphology and size of the iron oxide particles at the nanoscale levels, TEM studies were performed. The size distribution of magnetic nanoparticles is an important parameter related to their biological applications and performance. Different shaped IOICNPs may be prepared by the facile co-precipitation method by adjusting the amounts of polymer, crosslinker and Fe$^{2+}$/Fe$^{3+}$ ratio, pH so as to investigate the influence on the shape and particle size of IOICNPs. It was observed that the synthesized IOICNPs displayed a relatively spherical distribution, good dispersion and a uniform morphology with distinct crystalline structure. From the TEM image of IOICNPs, it can be clearly demonstrated that magnetic nanoparticles comprise of core shell structure with homogenous incorporation of magnetite as a core of IOICNPs. The magnetic nanoparticles are homogeneously covered by the casein shells [17]. The particle size of IOICNPs may be controlled by the amounts of casein and glutaraldehyde. The TEM images of IOICNPs with different amounts of casein and glutaraldehyde are shown in

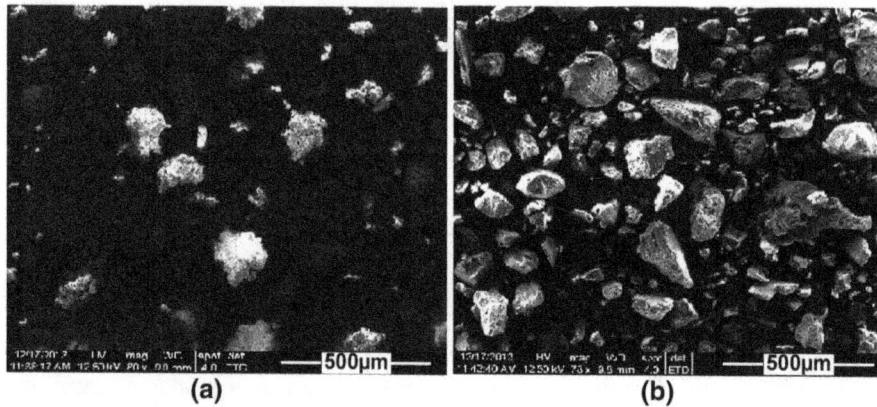

**Figure 3** SEM images of a) CNPs and b) IOICNPs.

Figure 4A and B, respectively, which indicate that an average size of IOICNPs falls in the range of 80–90 nm. As the amount of casein increases from 0.5 to 2.5 g, the size of the as-prepared IOICNPs increases from 15 nm to 50 nm as shown in (Figure 4A). The results also show that increasing amount of casein tends to produces largerer particle size because it can produce a bigger micelle [10,17]. Thus, on increasing the amount of casein, the size of IOICNPs also increases. The average particle size of the nanoparticles, as a function of the oil/water ratio in the emulsions, decreases from 20 to 8 nm with increasing oil/water ratio from 1.3 to 7.0. When the oil/water ratio is high, the molar ratio of casein to water is also increased, resulting in a high surface tension at the oil/water interface. This produces small water droplets and determines the size of iron nanoparticle [18]. The effect of glutaraldehyde concentration on the size of IOICNPs is shown in

(Figure 4B), which reveal that as the concentration of crosslinker increases, the number of more crosslink points increases thus resulting in an increase the crosslinking density. As a result, the network voids are minimized and the particle size decreases.

The Figures also show that the nanoparticles formed are present as aggregates due to the reason that they have a natural tendency to undergo clustering due to variety of charges and functional groups present on their surfaces.

### Raman spectral analysis

Raman spectra of IOICNPs are shown in (Figure 5a), which show characteristic Raman bands for casein due to amide I (CONH) at ~1666 cm$^{-1}$ and amide III band at ~1245 cm$^{-1}$. Between these Raman bands an intense peak is observed at 1450 cm$^{-1}$, which is attributed to

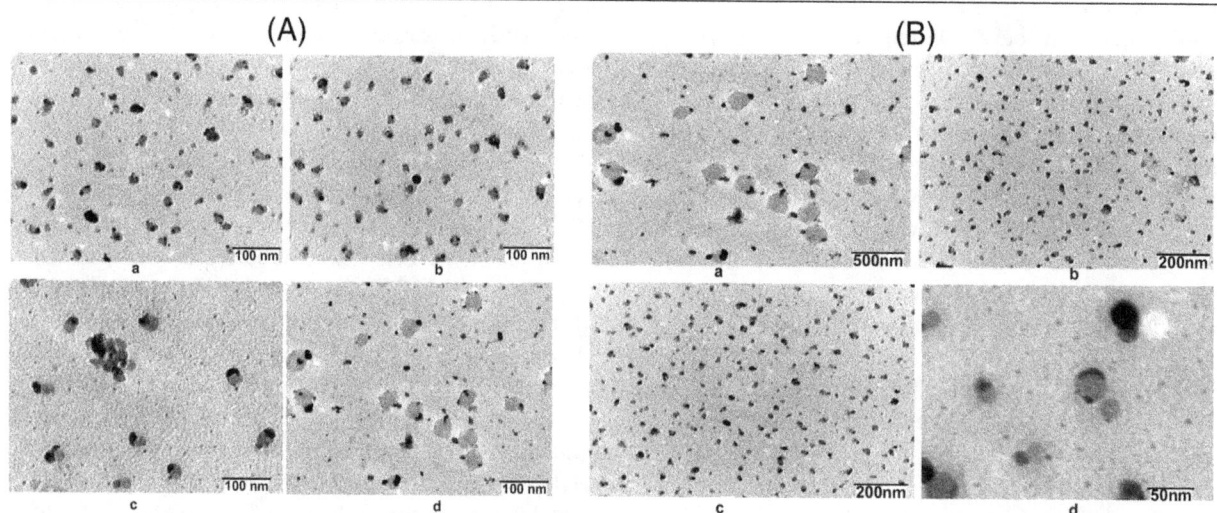

**Figure 4 TEM images of IOICNPs containing varying amount of casein 0.5 g, 1.0 g, 1.5 g, 2.0 g, and glutaraldehyde 5 mM, 10 mM, 15 mM and 20 mM. (A)** TEM images of IOICNPs containing varying amounts of casein a) 0.5 g, b) 1.0 g, c) 1.5 g, d) 2.0 g. **(B)** TEM images of IOICNPs containing varying amounts of glutaralehyde a) 5 mM, b) 10 mM, c) 15 mM, d) 20 mM.

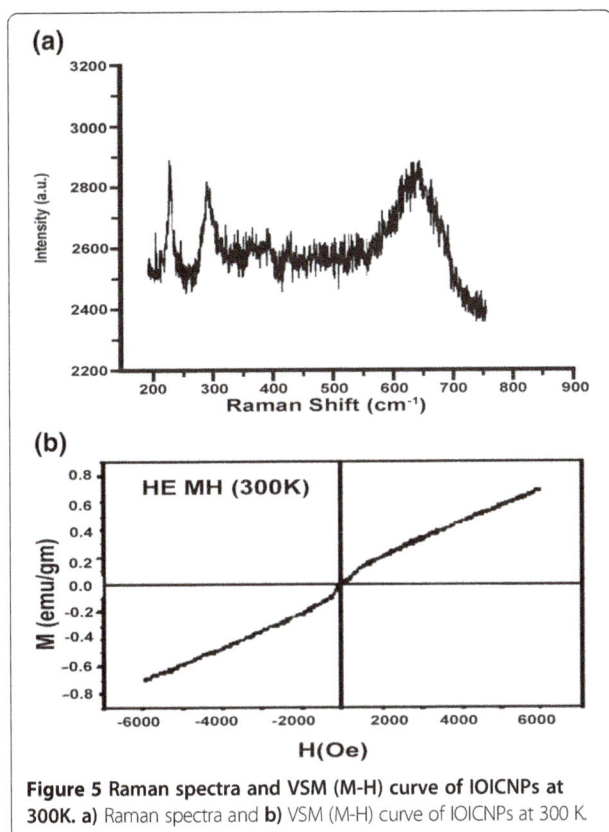

**Figure 5** Raman spectra and VSM (M-H) curve of IOICNPs at **300K. a)** Raman spectra and **b)** VSM (M-H) curve of IOICNPs at 300 K.

the $CH_2$ scissoring mode. In the Raman spectra weaker peaks observed at 193 $cm^{-1}$, 306 $cm^{-1}$ and 538 $cm^{-1}$ confirm the presence of iron oxide in the form of magnetite. Moreover, an additional strong peak is also observed at 668 $cm^{-1}$. For correct assignment of sample combined Raman data key can be used as follows [19,20].

(i)   $Fe_3O_4$: 193 (weak), 306 (weak), 538 (weak), 668 (strong),

(ii)  $gFe_2O_3$: 350 (strong), 500 (strong), 700 (strong); and

(iii) $aFe_2O_3$: 225 (strong), 247 (weak), 299 (strong), 412 (strong), 497 (weak), 613 (medium).

## VSM analysis

The magnetization versus magnetic field plot (M-H magnetization curve and hysteresis loop) at 300 K, for the impregnated casein nanoparticles was measured over the range of applied field between –6000 to +6000 Oe with a sensitivity of 0.1 emu/g, using vibrating sample magnetometer. The results are shown in (Figure 5b) which show that the saturation magnetization value is around 64 emu $g^{-1}$, and the hysteresis is very weak. The value obtained is lower than the reported value of 92–100 emu $g^{-1}$ for magnetite nanoparticles and may be attributed to the fact that below a critical size, nanocrystalline

magnetic particles may be of single domain and show unique phenomenon of superparamagnetism [21,22]. The reduction in saturation magnetization of $Fe_3O_4$ particles may be attributed to the presence of non-magnetic layer on the surface of the particles, charge distribution, superparamagnetic relaxation and spin effect because of ultrafine nature of the particles.

## XPS analysis

In the present study XPS analyses were done to monitor the iron oxide deposition in the cross-linked IOICNPs. The XPS spectra of the cross-liked IOICNPs is shown in the (Figure 6) Which reveal the presence of the C 1 s, O 1 s, and N 1 s core-level peak. However, after impregnation of iron oxide, the spectrum exhibits two more peaks associated with $Fe^{2+}$ and $Fe^{3+}$, due to the iron oxide deposition. It is worth to mention that the peak assignment is based on characteristic binding energies reported in the literature [23,24]. Furthermore, the O 1 s core-level spectra of the cross-linked IOICNPs were fitted using two peaks at 532.3 eV and 534 eV. The first one is associated with the binding energy of the [C = O] in the imide group and carboxylic acid group while the second one at the binding energy of the OH in the carboxylic acid group. The absence of the peak at 287.2 eV, associated with the binding energy of carboxylic acid groups, is accompanied by an increase in the intensity of the peak at 285.9 eV, due to the contribution of the carboxylate species in the crosslinked biopolymer [25]. These groups were also observed in the FTIR analysis. The O 1 s core-level spectrum from the resulting composite was fitted to peaks at 530.2 eV (g-$Fe_2O_3$), 531.4 eV (a-FeOOH). Furthermore, the spectrum displays two peaks associated with the iron oxide in the composite which are in good agreement with the magnetite, at 715.3 eV and 725.4 eV for $Fe^{2+}$ and $Fe^{3+}$ ions.

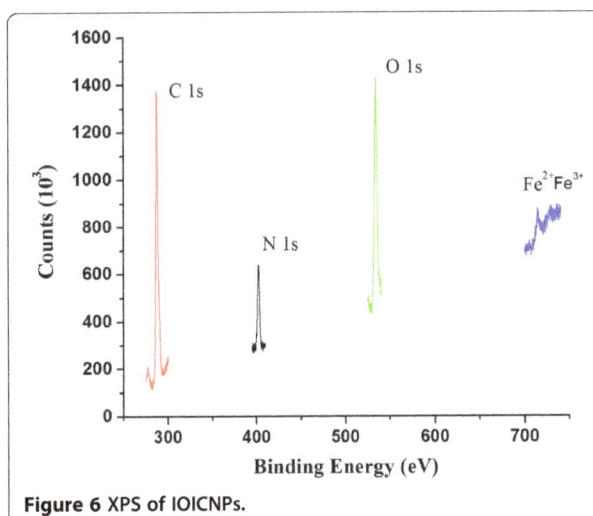

**Figure 6** XPS of IOICNPs.

The XPS was applied to provide elemental information of surface composition of IOICNPs after Fe$_3$O$_4$ loading. There were C, N, O and Fe elements in the magnetic IOICNPs, which further proved that Fe$_3$O$_4$ nanoparticles were in situ synthesized in the IOICNPs [26]. The different oxidation states of the iron in these nanoparticles can also be detected and distinguished from each other by XPS.

### In vitro cytotoxicity test

In order to determine in vitro cytotoxicity of the prepared materials Test on extract method was performed (ISO10993-5, 2009). In this method powdered (0.2 g) material was soaked in culture medium (1 mL) with serum and then the extract was prepared by incubating the presoaked test material with serum for 24 h. After incubation, the extract was filtered using 0.22 µm millex gp filter. 100% extract were diluted with culture medium to get 50% and 25% concentrations. Different dilutions of test sample extracts, positive control and 100% extracts of negative control in triplicate were placed on subconfluent monolayer of L-929 cells. After incubation of cells with extracts of the test sample and controls at 37 ± 1°C for 24 to 26 h, culture was examined microscopically for cellular response. For negative control the sample was prepared by incubating 1.25 cm$^2$ polyethylene disc with 1 mL culture medium with serum at 37 ± 1°C and positive control was prepared by diluting phenol stock solution (13 mg/mL) with culture medium with serum. The cytotoxicity reactivity were graded based on zone of lysis, vacuolization, detachment and membrane disintegration as 0, 1, 2, 3 and 4 representing none, slight mild, moderate and severe, respectively. The quantitative evaluation of reactivity for negative and positive controls and test sample are summarized in Table 1, while microscopic observation are depicted in (Figure 7a, b and c), respectively.

It was found that the test sample showed none reactivity to fibroblast cells after 24 h of contact. The achievement of numerical grade more than 2 is considered as cytotoxic effect. Since the polymer network material in the present work achieved a reactivity grade less than 2, the material is considered as not cytotoxic.

### Effect of Magnetic Field (MF) on the swelling

In magnetic drug targeting swelling-controlled system must have satisfactory swelling properties and high

### Table 1 Quantitative evaluation of in vitro cytotoxic reactivity of various samples

| S. no. | Sample | Grade | Reactivity |
|---|---|---|---|
| 1. | Negative control | 0 | None |
| 2. | Positive control | 4 | Severe |
| 3. | IOICNPs | 0 | None |

capacity of drug loading. The degree and time of swelling are important characteristics and have significant effect on the release kinetic of loaded drugs from swelling-controlled systems. To evaluate the effect of MF the on the swelling of IOICNPs, the MF was varied in the range of 1000 to 3000 G. The results are depicted in (Figure 8a) which reveal that the swelling ratio increases with increasing strength of magnetic field. The observed results may be attributed to the fact that the magnetic moment of a material **M** is proportional to the applied field H.

$$\mathbf{M} = \chi m \mathbf{H}, \tag{7}$$

where χm is magnetic susceptibility of the material. The possible reason for the observed increased swelling upon application of external magnetic field may be that under the applied field the magnetic nanoparticles get aligned with the applied field. Since the particles are in constant motion, they will experience fluctuating magnetic field which may cause agitation of the nanoparticles. This will produce a motion in the nanoparticles matrix and result in relaxation of polymer chains of the nanocomposite, thus, leading to a greater swelling of IOICNPs. This may also enlarge the nanostructure of the polymeric matrix to produce porous channels that cause enhanced diffusion process enabling easy swelling. The mechanical deformation generates compressive and tensile stresses, enhancing the penetration of water molecules. Similar type of result has also been reported elsewhere [27].

### Effect of chemical composition in CNPs and IOICNPs

In the present work, the influence of chemical composition of casein nanoparticles on their swelling ratio has been investigated by varying the amounts of casein and crosslinker (glutaraldehyde), in the feed mixture, respectively. The observed results may be discussed as below:

### Variation of casein in CNPs and IOICNPs

The effect of increasing biopolymer content on the swelling characteristics of CNPs and IOICNPs has been investigated by varying the amount of casein in the range 0·5–2.5 g while keeping the concentration of glutaraldehyde as constant. The results are shown in (Figure 8b), which clearly indicate that the swelling of IOICNPs is higher than CNPs and initially increases up to 0.5 g of casein content and thereafter decreases with further increase in the amount of casein. The results may be attributed to the fact that till 1.5 g of casein, highly compact nanoparticles are formed which restricts the inward movement of water molecules thus resulting in a decrease in swelling ratio. However, beyond 1.5 g of casein, swelling ratio is found to increase with increase in the amount of casein up to 2.5 g. The observed results are due to the fact that casein is a hydrophilic biopolymer and its increasing

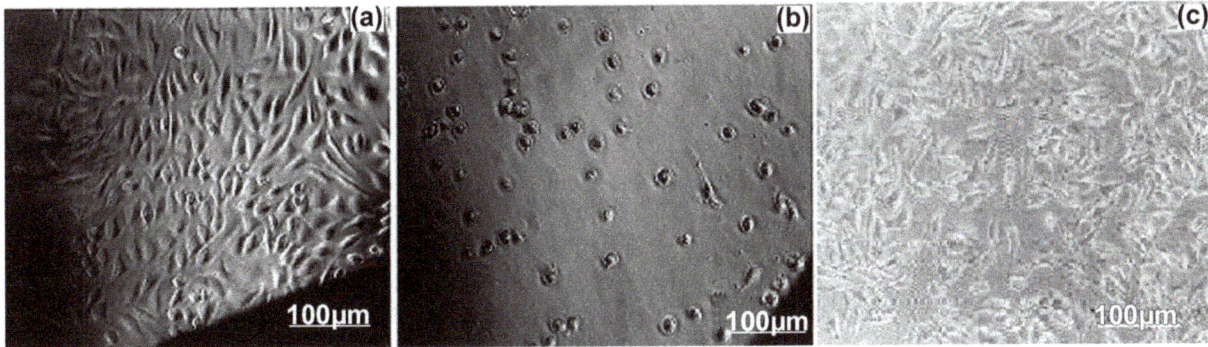

**Figure 7 Microscopic images showing L-929 cells. a)** negative control, **b)** positive control, and **c)** IOICNPs, respectively.

amount in the particles will obviously enhance the hydrophilicity of the nanoparticles and, thus, an increase in the swelling ratio is obtained.

In IOICNPs, the decrease in the total water content , might be attributed to effective interactions between the iron oxide and the polymer matrix. These interactions may arise from the carboxylic groups of casein that act as iron-binding sites [28,29]. Taking into account that the number of carboxylic groups increases on increasing the amount of casein, and the interactions between iron oxide and polymer matrix a decrease in the swelling behavior of IOICNPs may be justified.

The results also indicate that swelling of nanoparticles in magnetic field is higher than that in the absence of magnetic field. The results can be explained by the fact that due to the applied magnetic field, the magnetic moments of impregnated iron oxide nanoparticles tend to get aligned with the external magnetic field and while doing so, they produce a motion of macromolecular chains in the casein matrix. Thus due to the mobility of iron oxide nanoparticles, the macromolecule chains of casein get relaxed and facilitate greater inclusion of water molecules into the biopolymer matrix [30]. This clearly results in an enhanced swelling of nanoparticles.

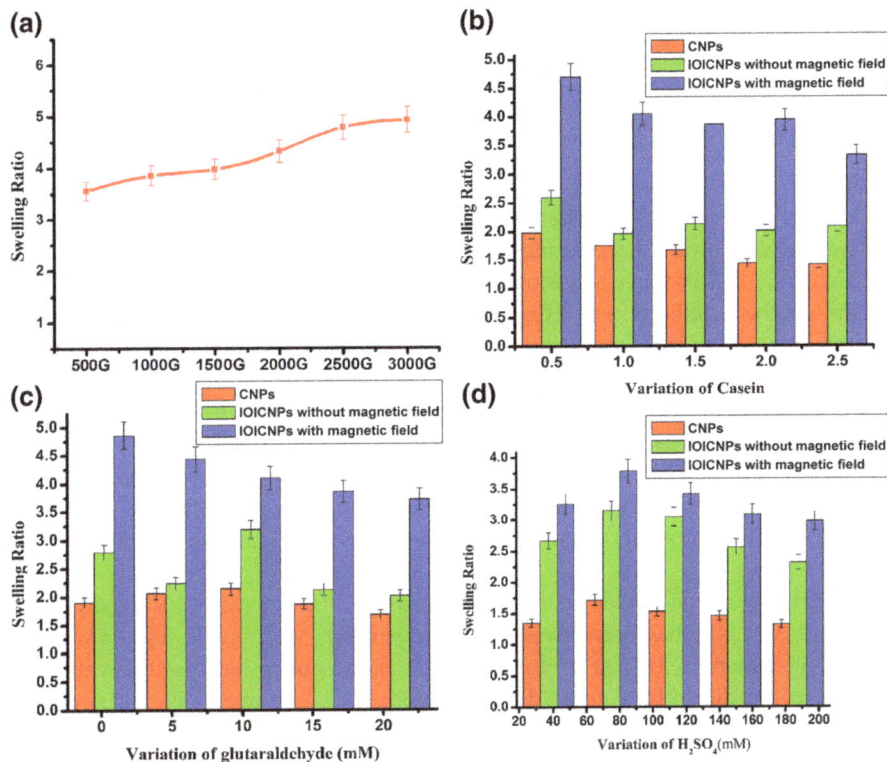

**Figure 8 Effect of magnetic field, casein, glutaraldehyde, and H$_2$SO$_4$ on swelling of CNPs and IOICNPs. a) magnetic field, b)** casein, **c)** glutaraldehyde, and **d)** H2SO4 on swelling of CNPs and IOICNPs.

## Variation of glutaraldehyde in CNPs and IOICNPs

The effect of crosslinker on the swelling profiles of CNPs and IOICNPs have been investigated by varying the concentration of glutaraldehyde in the range 0.02 to 20 mM. The results are shown in (Figure 8c), which clearly reveal that the water sorption by nanoparticles constantly increases while beyond 10 mM concentration a drop in swelling ratio is observed.

The observed increase may be attributed to the fact that glutaraldehyde is a low molecular weight crosslinking agent, and it crosslinks by reacting with the $NH_2$ groups of casein at its two terminals. Thus, a crosslinked network is developed as a result of crosslinking reaction between casein and glutaraldehyde thus creating a wide space in its structure and, therefore, possessing high capacity of accommodating water molecules into the network [31]. In this way, the capacity of nanoparticles to accommodate large number of water molecules results in an increased swelling ratio. The decrease observed beyond 10 mM of glutaraldehyde may be explained by the reason that much higher crosslinker content in the nanoparticles matrix reduces the free space in the nanoparticles network accessible to the penetrating water molecules and consequently results in a fall in the swelling capacity.

Some authors have also reported that introduction of crosslinker in to the polymer matrix enhances its glass transition ($T_g$), which because of glassy behavior of polymers restrains the mobility of network chains and, thus decreases the swelling [32]. The swelling results also indicate that the applied magnetic field also enhances the swelling of nanoparticles which has already been explained earlier.

## Variation of Sulphuric acid

Ionically crosslinked CNPs were successfully prepared by a two-step process. The first step involved the formation of oil droplets (emulsion) by an oil-in-water emulsion formation. The second step was the solidification by using sulphuric acid of the formed droplets by ionic crosslinking of casein (pH 2) enveloping the oil droplets with glutaraldehyde . By decreasing the pH of casein solution below its isoelectric point (4.6–4.8), the amino groups of casein become positively charged by protonation and can strongly attract the negatively charged aldehydic groups of glutaraldehyde, leading to the formation of ionically crosslinked nanoparticles [33].

The effect of solidifying agent (sulphuric acid) on the swelling profile of CNPs and IOICNPs has been investigated by varying the concentration of $H_2SO_4$ in the range 20–200 mM. The results are shown in (Figure 8d), which clearly reveal that the swelling of nanoparticles constantly increases while beyond 80 mM concentration a drop in swelling behavior is observed. The observed decrease could be attributed to the fact that as the amount of $H_2SO_4$ increases in the feed mixture, the number of positively charged amino groups increases which enhances the interaction between aldehydic group ( $^-O^{-+}C$-$H$ ) increases thus resulting high crosslinking density between casein and glutaraldehyde.

## Variation of $Fe^{2+}/Fe^{3+}$

The CNPs treated with $Fe^{2+}/Fe^{3+}$ solution contain iron ions dispersed throughout the casein network. The electrostatic forces between the iron ions and amide groups of casein are responsible for the increase in the swelling of IOICNPs, as compared to CNPs without ions. When the iron ions (**$Fe^{2+}0.5/Fe^{3+}$ 1 M**) loaded particles are treated with alkali, magnetite nanoparticles are formed inside the CNPs. This results in an increased swelling capacity due to increased electrostatic forces between the amide groups and iron oxide. The results are shown in (Figure 9a).

The swelling capacity follows the order CNPs < IOICNPs which has same trend as observed in PAM hydrogel–silver nanocomposites [34]. When the iron ions loaded nanoparticles are treated with NaOH, magnetite nanoparticles are produced inside the matrix. The particles treated with iron salts had iron ions dispersed throughout the polymeric network. The electrostatic forces between the iron ions and amide groups of casein matrix are responsible for the increase in the swelling of iron ion loaded nanoparticles, as compared to native casein nanoparticles without ions.

$Fe_3O_4$ magnetic nanoparticles were prepared with different concentrations of $Fe^{2+}$, $Fe^{3+}$ in aqueous phase, while other preparation conditions were kept same. The average size of $Fe_3O_4$ magnetic nanoparticles increases as concentration of iron salt solution increases. The particle size was found to increase with the increase in concentration of iron salts [35].

## Effect of pH

pH sensitive macromolecular devices have been most frequently used to design swelling controlled release drug formulations for oral administration which is the most clinically acceptable way of drug delivery. In the present study, the effect of pH on the swelling ratio of CNPs and IOICNPs has been studied by adjusting pH of swelling medium in the range of $1 \cdot 0$ to 9.0. The results are depicted in (Figure 9b), which clearly indicate that the swelling ratio of particles constantly increases with increase pH. The increase observed in the swelling ratio of the particles with increasing pH may be explained as follows. It is known that casein is cationic in nature and therefore a change in pH of the swelling medium also affects the charge profiles of casein as well as IOICNPs. The swelling results indicate a significant dependence of swelling behavior of the nanoparticles on different pH values. In both the cases of CNPs and IOICNPs, the

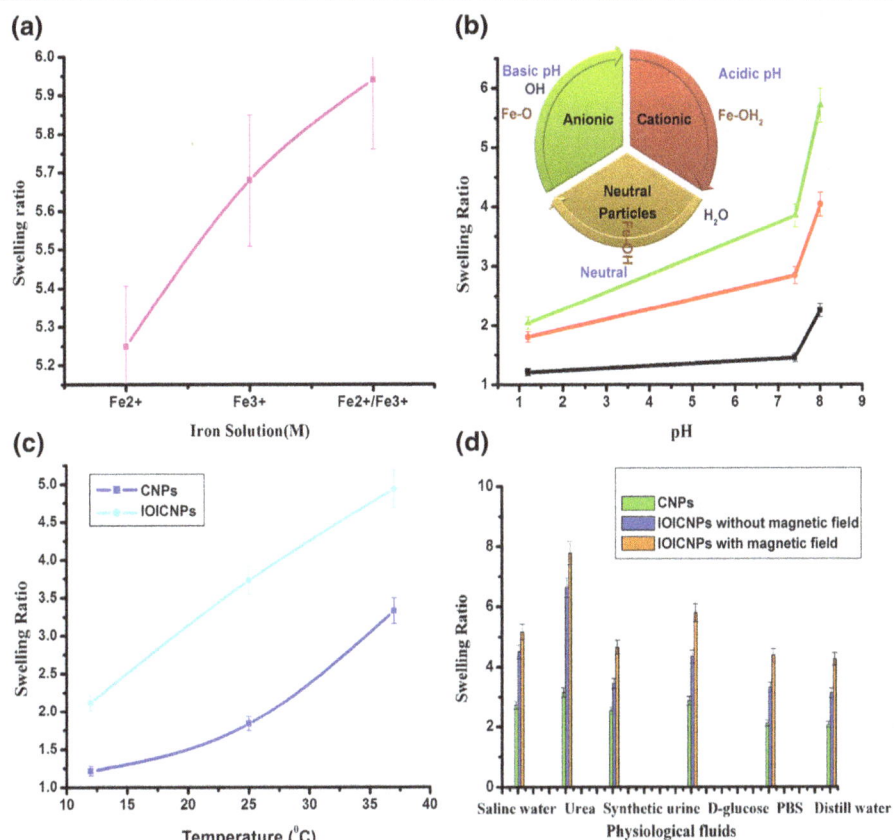

**Figure 9** Effect of iron salts, pH, temperature and simulated biofluids on swelling of CNPs and IOICNPs. **a)** iron salts, **b)** pH, **c)** temperature, and **d)** simulated biofluids on swelling of CNPs and IOICNPs.

results reveal that at pH 1.8, a lower swelling ratio was observed, because $PK_a$ of casein is about 4.2 and most of the carboxyl groups in the casein exist in the form of COOH at low pH medium (pH 1.7). In the macromolecular nanoparticle network of the hydrogen bonding produced by –COOH groups of casein led to the stronger interaction between polymer chains. Accordingly, the swelling ratio in pH 1.7 is relatively lower. At higher pH, the carboxylic groups get ionized and acquire $-COO^-$ form. Thus, weak hydrogen bonding between biopolymer chains and electrostatic repulsion between $-COO^-$ groups result in the higher swelling ratio [36].

Since pKa of casein is 4.2, the species involved in the interactions are $NH_3^+$ and COOH at pH 1–3, $NH_2$ and COO – at pH 7–13. In acidic conditions, the swelling is controlled mainly by the amino group ($NH_2$) which is a weak base with an intrinsic pKa value of about 6.2. So, it gets protonated and the increased charge density on the biopolymer enhances the osmotic pressure inside the network particles because of the $NH_3^+$-$NH_3^+$ electrostatic repulsion. This osmotic pressure difference between the internal and external solutions of the network is balanced by the swelling of the network. However, under very highly acidic conditions (pH < 3), a screening effect of the

counter ions, i.e. $Cl^-$, shield the charges of the ammonium cations and prevent an efficient repulsion. As a result, a remarkable decrease in equilibrium swelling is observed. Similarly, the screening effect of the counter ions ($Na^+$) limits the swelling at pH > 8.5 [37].

The swelling of IOICNPs is governed by negative charge possessed by iron oxide nanoparticles. When pH of the swelling medium is 9.0, the number of negatively charged groups ($Fe-O^-$) is large, so the swelling is maximum because of the strong electrostatic repulsion between negatively charged groups. When pH of the swelling medium is 7.0, the protons from the swelling medium neutralize most of the negatively-charged groups and, therefore, the swelling ratio decreases due to the reduced electrostatic repulsion. At 1.2 pH, all negatively-charged groups are neutralized; instead, there would be some positively charged amine groups and iron species ($Fe-OH_2^+$), because the amine groups in the IOICNPs are fewer than the carboxyl groups and the net charge is quite low, thus that the swelling is also very low [38].

When pH is less than isoelectric point (pI) of casein, the extent of increase in swelling ratio at acidic medium is small, because there are very few amine groups existing along protein chains so that the positive charges are very limited. The swelling becomes minimum when pH of the medium is close to the pI of the casein (4.6-4.8). This is because the net charge on casein molecules is close to zero at pI, which means almost no electrostatic repulsion exist between the casein chains. On the other hand, when pH > pI, there are lots of negatively-charged groups on the casein molecule, which results in an increased swelling ratio. The higher the pH, greater is the surface charge and consequently the higher electrostatic repulsive forces result in higher swelling ratio [39].

### Effect of temperature

In the present study, the effect of temperature on swelling was studied by varying the temperature in the range 12.5–37.5°C. The results are shown in (Figure 9c), which indicate that with increase in temperature, the swelling of nanoparticles increases from 2.5 to 4.5 in the whole studied range of temperature. The observed increase in the swelling of IOICNPs may be explained by the fact that a rise in temperature enhances the rate of diffusion of water molecules and segmental mobility of biopolymer chains which results in a greater degree of swelling [40]. However, beyond 30°C the swelling ratio decreases, which may be due to the reason that at higher temperature the hydrogen bonds holding water molecules and the polymer chains get broken and, therefore, the swelling ratio decreases.

### Effect of physiological fluids

The effect of nature of physiological fluids on the swelling of CNPs and IOICNPs in absence and presence of magnetic field has been investigated by performing swelling experiments in various simulated physiological fluids like urea, D- glucose, PBS, saline water, distilled water and synthetic urine. The results are presented in (Figure 9d), which indicate that swelling ratio is quite high in urea in comparison to other fluids and lower degree of water sorption is noticed. The possible reason for the higher swelling ratio in urea may be that the presence of urea works as hydrogen bond breaker in casein macromolecule and this tends to result in greater relaxation of biopolymer chains which eventually leads to higher swelling. In the case of other fluids the salt ions in the medium lowers the osmotic pressure difference between the casein matrix and solvent medium which causes a decrease in swelling ration of nanoparticles.

The change in swelling due to the presence of electrolyte concentration has also been predicted theoretically. Fernández-Nieves et al. [41] demonstrated that the addition of salts modifies the ionic distribution inside the casein matrix, altering the number of dissociated groups in the network, and changing the net charge. The change in the network charge state makes particles swell or

shrink, depending on whether the electrical repulsions increase or decrease, respectively.

### Conclusions

The controlled size magnetite nanoparticles have been successfully prepared via a convenient co-precipitation method and characterized by various analytical techniques, such as FTIR spectroscopy, TEM, SEM, XPS, VSM and Raman analyses which confirm the *in situ* impregnation of nanosized iron oxide within the matrix of casein nanoparticles. The biopolymeric nanoparticles clearly show the presence of characteristic functional groups of casein and iron oxide as confirmed by their FTIR spectra. SEM and TEM of the nanoparticles provide information about their size and morphology.

The IOICNPs show an optimum swelling when the casein content is 0.5 g while on increasing casein content further the degree of swelling decreases. Likewise, when the concentration of crosslinker increases from 0.02-20 mM, the quantity of water imbibed by the nanoparticles increases while beyond 10 mM of crosslinker concentration, the extent of swelling decreases. It is also found that in alkaline pH the nanoparticles show an enhanced swelling which thereafter decreases with further increase in pH. In the case of rising temperature the swelling ratio constantly increases. The prepared IOICNPs exhibit property of superparamagnetism which is a significant for biomedical applications. It is also observed that the swelling of nanoparticles is enhanced by the application of external magnetic field. Thus the present swelling system may be helpful in designing targeted drug delivery carriers using external magnetic field.

#### Competing interests
The authors declare that they have no competing interests.

#### Authors' contributions
AKB supervised the study, and contributed to the drafting of this article, selection of methodology, analysis, discussion of the results and finalized the manuscript. JB conceived the study, participated in its design and coordination. AS carried out all the experimental studies, contributed in the design of the study, analyzed data, and drafted the manuscript. All authors read and approved the final manuscript.

#### References
1.  Indira TK, Lakhshmi PK: **Magnetic nanoparticles.** *Int J Pharma Nanotechnol* 2010, 3:1035–1042.
2.  Figuerola A, Corato RD, Manna L, Pellegrino T: **From iron oxidenanoparticles towards advanced iron-based inorganic materials designed forbiomedical applications.** *Pharma Res* 2010, 62:126–143.
3.  Monson TC, Venturini EL, Petkov V, Ren Y, Lavin JM, Huber DL: **Large enhancements of magnetic anisotropy in oxide-free iron nanoparticles.** *J Magn Magn Mat* 2013, 331:156–161.
4.  Jiang F, Fu Y, Zhu Y, Tang Z, Sheng P: **Fabrication of iron oxide/silicacore–shell nanoparticles and their magnetic characteristics.** *J Alloys Compd* 2010, 543:43–48.
5.  Mohanraj VJ, Chen Y: **Nanoparticles - a review.** *J Pharm Res* 2006, 5:561–573.

6. Esmaili M, Ghaffari SM, Moosavi-Movahedi Z: **Beta casein–micelle as a nano vehicle for solubility enhancement of curcumin; food industry application.** *LWT Food Sci Technol* 2011, **44**:2166–2172.

7. Shapira A, Assaraf YG, Epstein D, Livney YD: **Beta-casein nanoparticles as an oral delivery system for chemotherapeutic drugs: impact of drug structure and properties on co-assembly.** *Pharm Res* 2010, **27**(10):2175–2186.

8. Bouchemal K, Briançon S, Perrier E, Fessi H: **Nano-emulsion formulation using spontaneous emulsification: solvent, oil and surfactant optimisation.** *Int J Pharm* 2004, **280**:241–251.

9. Lu AH, Salabas EL, Schüth F: **Magnetic nanoparticles: synthesis, protection, functionalization, and application.** *Angew Chem Int Ed Engl* 2007, **46**:1222–1244.

10. Petcharoena K, Sirivat A: **Synthesis and characterization of magnetite nanoparticles via the chemical co-precipitation method.** *Mater Sci Eng B* 2012, **177**:421–427.

11. Jiang W, Yang HC, Yang SY: **Preparation and properties of superparamagnetic nanoparticles with narrow size distribution and biocompatible.** *J Magn Magn Mat* 2004, **283**:210–214.

12. Choubey J, Bajpai AK: **Investigation on magnetically controlled delivery of doxorubicin from superparamagnetic nanocarriers of gelatin crosslinked with genipin.** *J Mater Sci-Mater Med* 2010, **21**:1573–1586.

13. Lien YH, Wu TM: **Preparation and characterization of thermosensitive polymers grafted onto silica-coated, iron oxide nanoparticles.** *J Colloid Interface Sci* 2008, **326**:517–521.

14. Sepulveda-guzman S, Lara L, Perez-Camacho O, Rodriguez-Fernandez O, Olivas A, Escudero R: **Synthesis and characterization of an iron oxixe poly (styrene- co- carboxybutylmaliemide) ferromagnetic composites.** *Polymer* 2007, **48**:720–727.

15. Deacon GB, Phillips RJ: **Relationship between the carbon–oxygen stretching frequencies of carboxylato complexes and the type of carboxylate coordination.** *Coord Chem Rev* 1980, **33**:227–250.

16. Murthy PSK, Mohan YM, Varaprasad K, Sreedhar B, Raju KM: **First successful design of semi-IPN hydrogel-silver nanocomposites: a facile approach for antibacterial application.** *J Colloid Interface Sci* 2008, **318**:217–224.

17. Wei S, Wang Q, Zhu J, Sun L, Hongfei Line H, Guo Z: **Multifunctional composite core–shell nanoparticles.** *Nanoscale* 2011, **3**:4474.

18. Zhang GI, Liao Y, Baker I: **Surface engineering of core/shell iron/iron oxide nanoparticles from microemulsions for hyperthermia.** *Mater Sci Eng C* 2010, **30**:92–97.

19. Chourpa I, Douziech-Eyrolles L, Ngaboni-Okassa L, Fouquenet JF, Cohen-Jonathan S, Souce M, Marchais H, Dubois P: **Molecular composition of iron oxide nanoparticles, precursors for magnetic drug targeting, as characterized by confocal raman microspectroscopy.** *Analyst* 2005, **130**:1395–1403.

20. De Faria DLA, Venâncio Silva S, De Oliveira MT: **Raman microspectroscopy of some iron oxides and oxyhydroxides.** *J Raman DR2013228Spectro* 1997, **28**:873–878.

21. Oh JK, Park JM: **Iron oxide-based superparamagnetic polymeric nanomaterials: design, preparation, and biomedical application.** *Prog Poly Sci* 2011, **36**:168–189.

22. Grüttner C, Teller J, Schütt W, Westphal F, Schümichen C, Paulke BR, Häfeli UO, Schütt W, Teller J, Zborowski M: *Scientific and Clinical Applications of Magnetic Carriers editors.* New York: Plenum Press; 1997:53.

23. Shen G, Anand MFG, Levicky R: **X-ray photoelectron spectroscopy and infrared spectroscopy study of maleimide-activated supports for immobilization of oligodeoxyribonucleotides.** *Nucleic Acids Res* 2004, **32**:5973–5980.

24. Beamson G, Briggs D: *High Resolution XPS of Organic Polymers.* New York, NY: John Wiley & Sons; 1992:293.

25. Alexander MR, Beamson G, Blomfield CJ, Leggett G, Duc TM: **Interaction of carboxylic acids with the oxyhydroxide surface of aluminium: poly (acrylic acid), acetic acid, and propionic acid on pseudobo-hemite.** *J Electron Spectrosc Relat Phenom* 2001, **121**:19–32.

26. Xuan SH, Fang QL, Hao LY, Jiang WQ, Gong XL, Hu Y, Chen ZY: **Fabrication of spindle Fe2O3@polypyrrole core/shell particles by surface-modified hematite templating and conversion to spindle polypyrrole capsules and carbon capsules.** *J Colloid Interface Sci* 2007, **314**:502–509.

27. Hu SH, Liu TY, Huang HY, Liu DM, Chen SY: **Magnetic-sensitive silica nanospheres for controlled drug release.** *Langmuir* 2008, **24**:239–244.

28. Hernandez R, Sacristan J, Nogales A, Fernandez M, Ezquerrab TA, Mijangos C: **Structure and viscoelastic properties of hybrid ferrogels with**

iron oxide nanoparticles synthesized in situ. *Soft Matter* 2010, **6**:3910–3917.

29. Choubey J, Bajpai AK: **Investigation on magnetic controlled delivery of doxorubincin from superparamagnetic nanocarriers of gelatin crosslinked with genipin.** *J Mater Sci Mater Med* 2010, **21**:1573–1586.

30. Likhitkar S, Bajpai AK: **Investigation of magnetically enhanced swelling behavior of superparamagnetic starch nanoparticles.** *Bulle Mater Sci* 2013, **36**(1):15–24.

31. Chouhan R, Bajpai AK: **A swellable in vitro release study of 5-Flurouracil (5-FU) from poly-(2-hydroxyethyl methacrylate) (PHEMA) nanoparticles.** *J Mater Sci Mater Med* 2009, **20**:1103–1114.

32. Chairam S, Somsook E: **Starch vermicelli template for synthesis of magnetic iron oxide nanoclusters.** *J Magn Magn Mater* 2008, **320**:2039–2043.

33. Elzoghby AO, Helmy MW, Samy WM, Elingdy NA: **Novel ionically crosslinked casein nanoparticles for flutamide delivery: formation, characterization and in-vivo pharmacokinetics.** *Int J Nanomedicine* 2013, **8**:1721–1732.

34. Vimala K, Sivudu KS, Mohan YM, Sreedhar B, Raju KM: **Controlled silver nanoparticles synthesis in semi-hydrogel networks of poly(acrylamide) and carbohydrates: rationalmethodology for antibacterial application.** *Carbohydr Polym* 2009, **75**:463–471.

35. Sivudu KS, Rhee KY: **Preparation and characterization of pH-responsive hydrogel magnetite nanocomposite.** *Colloids and Surfaces A: Physicochem Eng Aspects* 2009, **349**:29–34.

36. Zhao Y, Qiu Z, Huang J: **Preparation and analysis of Fe3O4 magnetic nanoparticles used as targeted-drug carriers.** *Chinese J Chem Eng* 2008, **16**:451–455.

37. Choi CY, Chae SY, Nah JW: **Theromosensititve poly (N- isopropylacrylamide)-b-poly (E-caprolactone) nanoparticles for efficient drug delivery system.** *J Polymer* 2006, **47**:4571.

38. Pourjavadi A, Mahdavinia GR: **Superabsorbency, pH-Sensitivity and swelling kinetics of partially hydrolyzed chitosan-g-poly (Acrylamide) Hydrogels.** *J Turk Chem* 2006, **30**:595.

39. Gunasekaran S, Ko S, Xiao L: **Use of whey proteins for encapsulation and controlled delivery applications.** *J Food Eng* 2007, **83**:31–40.

40. Likhitkar S, Bajpai AK: **Magnetically controlled release of cisplatin from superparamagnetic starch nanoparticles.** *Carbohydr Polym* 2012, **87**:300–308.

41. Fernandez-Nieves A, Fernandez-Barbero A, Nieves FJ D I, Vincent B: **Motion of microgel particles under an external electric field.** *J Phys Condens Matter* 2000, **12**:3605–3614.

# Activation of caspase-dependent apoptosis by intracellular delivery of cytochrome c-based nanoparticles

Moraima Morales-Cruz[2], Cindy M Figueroa[1], Tania González-Robles[1], Yamixa Delgado[2], Anna Molina[1], Jessica Méndez[1], Myraida Morales[3] and Kai Griebenow[1*]

## Abstract

**Background:** Cytochrome c is an essential mediator of apoptosis when it is released from the mitochondria to the cytoplasm. This process normally takes place in response to DNA damage, but in many cancer cells (i.e., cancer stem cells) it is disabled due to various mechanisms. However, it has been demonstrated that the targeted delivery of Cytochrome c directly to the cytoplasm of cancer cells selective initiates apoptosis in many cancer cells. In this work we designed a novel nano-sized smart Cytochrome c drug delivery system to induce apoptosis in cancer cells upon delivery.

**Results:** Cytochrome c was precipitated with a solvent-displacement method to obtain protein nanoparticles. The size of the Cytochrome c nanoparticles obtained was 100-300 nm in diameter depending on the conditions used, indicating good potential to passively target tumors by the Enhanced Permeability and Retention effect. The surface of Cytochrome c nanoparticles was decorated with poly (lactic-co-glycolic) acid-SH via the linker succinimidyl 3-(2-pyridyldithio) propionate to prevent premature dissolution during delivery. The linker connecting the polymer to the protein nanoparticle contained a disulfide bond thus allowing polymer shedding and subsequent Cytochrome c release under intracellular reducing conditions. A cell-free caspase-3 assay revealed more than 80% of relative caspase activation by Cytochrome c after nanoprecipitation and polymer modification when compared to native Cytochrome c. Incubation of HeLa cells with the Cytochrome c based-nanoparticles showed significant reduction in cell viability after 6 hours while native Cytochrome c showed none. Confocal microscopy confirmed the induction of apoptosis in HeLa cells when they were stained with 4',6-diamidino-2-phenylindole and propidium iodide after incubation with the Cytochrome c-based nanoparticles.

**Conclusions:** Our results demonstrate that the coating with a hydrophobic polymer stabilizes Cytochrome c nanoparticles allowing for their delivery to the cytoplasm of target cells. After smart release of Cytochrome c into the cytoplasm, it induced programmed cell death.

**Keywords:** Drug delivery, Protein nanoparticles, PLGA, Passive targeting, Triggered release

## Background

Cytotoxic drugs (such as, cis- and carboplatin, doxorubicin, 5-fluorouracil, gemcitabine, paclitaxel, out of a list of over 50) are still commonly used in chemotherapy to stop cancer cells from multiplying and dispersing. These drugs affect all growing cells, but since most normal cells do not divide as often as cancer cells they are proportionally less affected. Unfortunately, the lack of tumor specificity of these cytotoxins produces unwanted and often severe and dangerous side effects. One approach to overcome this is the development of nano-sized drug delivery systems (DDS), which have been shown to increase the drug accumulation in tumors *via* passive targeting, i.e., through the enhanced permeation and retention (EPR) effect [1-3]. Two types of nano-sized DDS, liposomes and albumin nanoparticles (NPs), i.e. Doxil® and Abraxane®, respectively, have caused notable improvements in the therapeutic efficacy of anticancer agents [4]. However, also these

* Correspondence: kai.griebenow@gmail.com
[1]Departments of Chemistry, University of Puerto Rico, Río Piedras Campus, San Juan, PR 00931, USA
Full list of author information is available at the end of the article

therapeutics still display significant side effects and only prolong life marginally. The lack of success in improving overall survival in many solid tumors and the still significant side effects of even the second generation drugs make it imperative to further develop and refine such DDS.

All currently US FDA approved chemotherapeutic DDS and most that are in current clinical trials or *in vivo* studies still employ traditional cytotoxic drugs as their therapeutic agents. Many of these chemotherapeutic agents (e.g., alkylating agents and antimetabolites) produce DNA damage which leads to p53-dependent apoptosis [5]. Indeed, the loss of the p53 tumor suppressor pathway contributes to the development of most human cancers [6]. Thus, inactivation of the apoptotic response provides an attractive explanation for the poor responsiveness of p53 mutant tumors to many traditional anticancer agents. Such limitations have spurred efforts to identify new and more effective chemotherapeutic agents that act independently of the p53 pathway. The high selectivity and low toxicity of many proteins make them attractive substitutes of cytotoxic drug. For example, Cytochrome c (Cyt c) is an important mediator of apoptosis when it is released from the mitochondria to the cytoplasm [7,8]. This process normally takes place in response to DNA damage, but in many cancer cells it is inhibited (most likely due to inactivation of the upstream components of the signaling pathway(s) that activates the Cyt c release, such as the p53 pathway). The targeted delivery of Cyt c directly to the cytoplasm could selectively initiate apoptosis in most cancer cell by circumventing inactivation of apoptosis due to damage to upstream events.

However, protein drugs also do have significant drawbacks, primarily related to their limited physical and chemical stability during storage and after administration [9-11]. Also, most proteins including Cyt c cannot cross lipid bilayer membranes [12]. This makes it necessary to develop methods allowing for the intracellular delivery of sufficient amounts of Cyt c to induce apoptosis in the target cells. The development of nanosized carriers for protein therapeutics allows maintaining protein stability and controlling release properties but having a sufficiently high protein loading remains challenging [11,13]. Recently our research group has been developing strategies for the intracellular delivery of Cyt c using nanosized DDS [14,15]. We demonstrated that Cyt c delivered from a smart nanosized system could potentially be an effective chemotherapeutic agent to treat cancer, but also recognized that the delivery system used needed improvement to achieve a better efficiency. Specifically, we demonstrated that mesoporous silica nanoparticles (MSN) could be utilized as efficient carrier for the delivery of apoptosis-inducing Cyt c [15]. However, Cyt c-MSN conjugates were not able to induce cell death in HeLa cells during the first 48 h. They showed 45% cell viability after 72 h

at a Cyt c concentration of 37.5 µg/ml. It was not possible to increment the Cyt c concentration since that resulted in a toxic concentration of MSN (>100 µg/ml) and the particles were already loaded to the maximum with the protein. This demonstrates fundamental limits of such drug-loaded systems in practical applications. Ideally, to avoid such limitations, the drug should itself form the DDS. Consequently, we evolved a delivery system in which the core consisted of Cyt c NPs and thus the drug itself.

Production of Cyt c NPs by a solvent displacement method is a straight-forward process (i.e., by solvent-induced nanoprecipitation) [16]. However, such NPs would simply dissolve when injected into the blood stream and thus need to be stabilized for application, e.g., by applying high temperature and/or chemical cross-linkers [17]. NPs, which were stabilized via a chemical cross-linker agent with thiol-cleavability, have been successfully used for triggered drug release mediated by the reducing environment inside the cell [18]. In this work we tested a new method for stabilizing Cyt c NPs by covalently coating them with the hydrophobic polymer poly (lactic-co-glycolic) acid (PLGA). PLGA, a biocompatible, biodegradable, and non-toxic polymer, has been intensely studied in the field of DDS and has received FDA approval for various applications including drug delivery [19,20]. Our idea was to coat the Cyt c NPs with PLGA using a hetero-bifunctional linker, such as succinimidyl 3-(2-pyridyldithio) propionate (SPDP), that includes a disulfide bond to be able to shed the polymer shell in the reducing environment of the cell (Figure 1A). Afterwards the Cyt c nanoparticle should dissolve in the cell thus inducing apoptosis (Figure 1B). In this work we demonstrate the potential of the designed anti-tumor Cyt c nanoparticle in cancer treatment.

## Results and discussion
### Nanoprecipitation of Cyt c
Cyt c, an apoptosis-initiating protein, could potentially be used to target and specifically destroy cancer cells if delivered to their cytoplasm. Protein-based nanomaterials have been studied as carriers of anti-cancer drugs because of their biodegradability, low toxicity, and multiple modification capacity [21]. It would be ideal to use the protein as both, nanosized delivery device and drug, simultaneously. To test this concept, in this study we generated and tested Cyt c-based NPs to kill cancer cells by inducing the apoptosis.

Cyt c nanoparticles were obtained by a solvent displacement method shown to induce a reversible nanoprecipitation of proteins without compromising their frequently fragile structure and function [14]. In a previous study we found that acetonitrile was a useful organic solvent for obtaining Cyt c nanoprecipitates when it was added in at least 4-fold excess to the aqueous protein solution [14]. In this work we optimized this procedure for obtaining

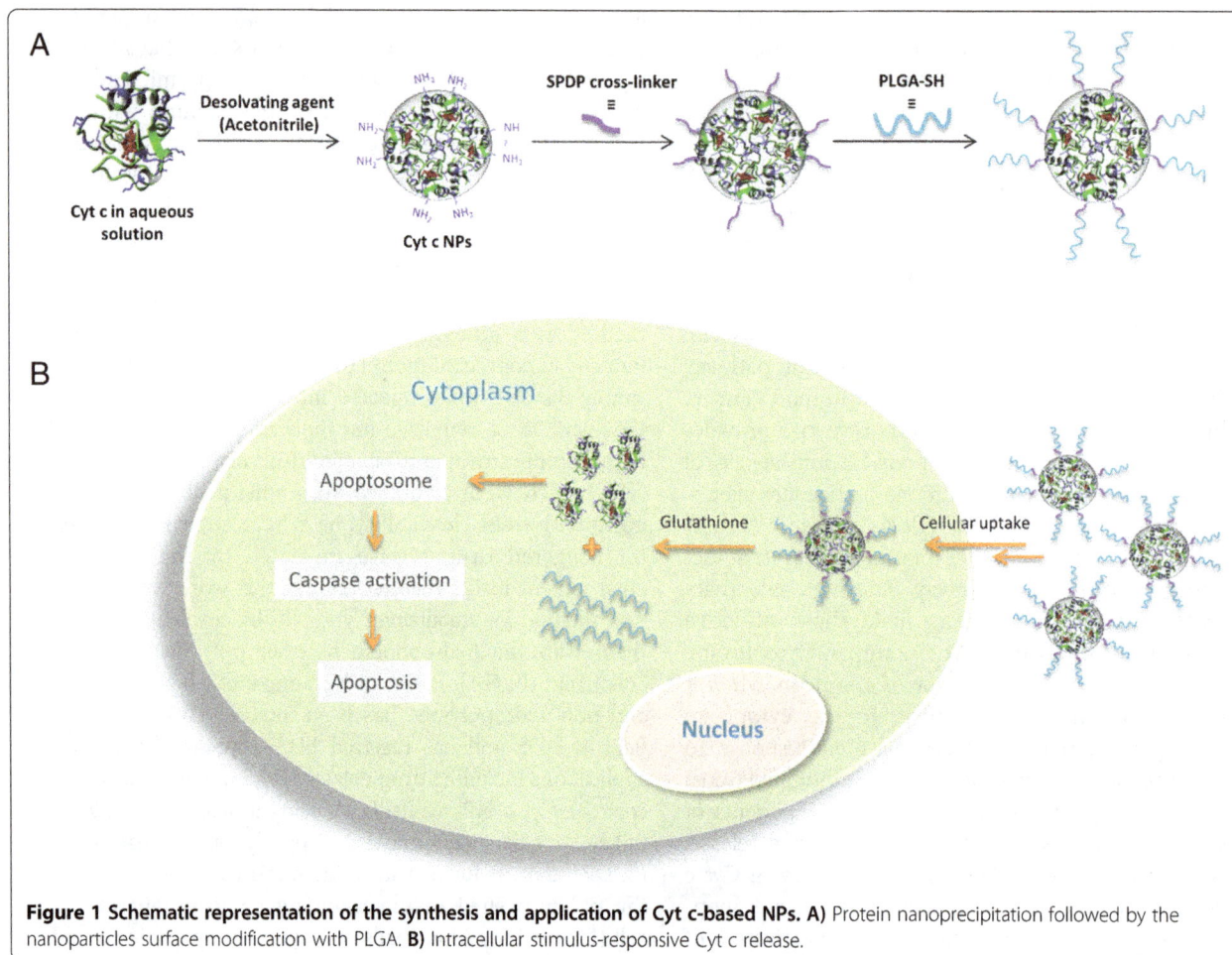

**Figure 1 Schematic representation of the synthesis and application of Cyt c-based NPs. A)** Protein nanoprecipitation followed by the nanoparticles surface modification with PLGA. **B)** Intracellular stimulus-responsive Cyt c release.

Cyt c nanoparticles and tested the effect of protein concentration on Cyt c precipitation and stability. While no buffer-insoluble aggregates were formed regardless of the protein concentration (data not shown), the precipitation yield decreased at higher protein concentrations (Figure 2). In agreement with our previous work for other proteins [14], we also found that at increasing protein concentration under otherwise constant conditions the particle size increased (Figure 2). We selected a Cyt c concentration of 5 mg/ml for further experiments since it had the highest precipitation efficiency (100%) and smallest particle diameter (around 150 nm). Note that a particle diameter of less than 400 nm should result in particles with good passive delivery properties [2,22,23].

The capability of Cyt c to still interact with Apoptotic Protease Activating Factor 1 (Apaf-1) and induce apoptosis after nanoprecipitation was verified next. Since Cyt c is a cell membrane impermeable protein the experiment was conducted in a cell-free system. The addition of Cyt c to fresh cytosol produces caspase activation [15]. Thus, the integrity of the soluble protein after the nanoprecipitation procedure was compared with native Cyt c and was found

to be around 90%. We tested the use of the excipient methyl-β-cyclodextrin (mβCD) to further increase the Cyt c integrity during the nanoprecipitation procedure (Figure 3). The excipient mβCD has been used before to improve enzyme activity in organic solvents [24]. The excipient was co-dissolved with the protein previous to the solvent precipitation step with acetonitrile. Because mβCD is soluble in acetonitrile, the excipient was removed by repeated centrifugation/washing cycles. Our results show that the excipient further improved bioactivity to 96% and decreased the particle diameter to 90 nm.

The impact of the nanoprecipitation procedure on Cyt c tertiary structure was investigated by circular dichroism (CD) measurements after dissolving the NPs in buffer (Figure 3B). The near-UV CD spectra show that the nanoprecipitation procedure did not cause irreversible changes in the protein conformation. The two minima at 286 and 293 nm, characteristic of native Cyt c [25], were present in all NP formulations. However, these two minima were somewhat reduced when mβCD was used at the very high 1:20 mass ratio during nanoprecipitation. Since these two minima correspond to Trp-59 [15] and cyclodextrins have

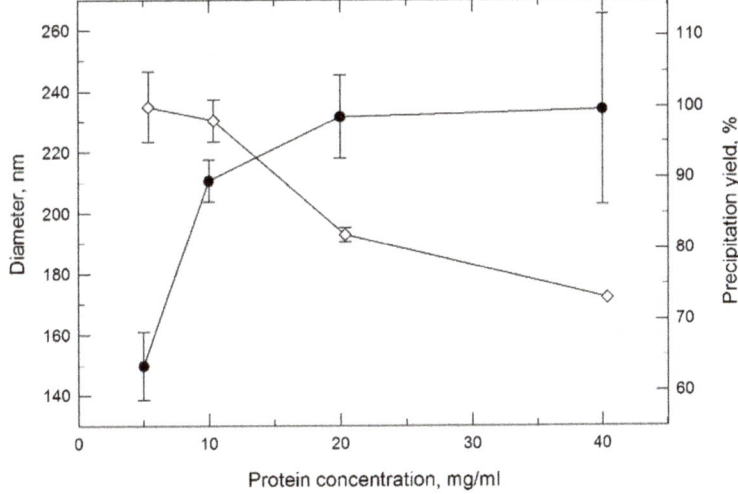

**Figure 2 Effect of the protein concentration during Cyt c nanoprecipitation.** The particle diameter (—●—) and precipitation yield (—◇—) were determined for different protein concentration.

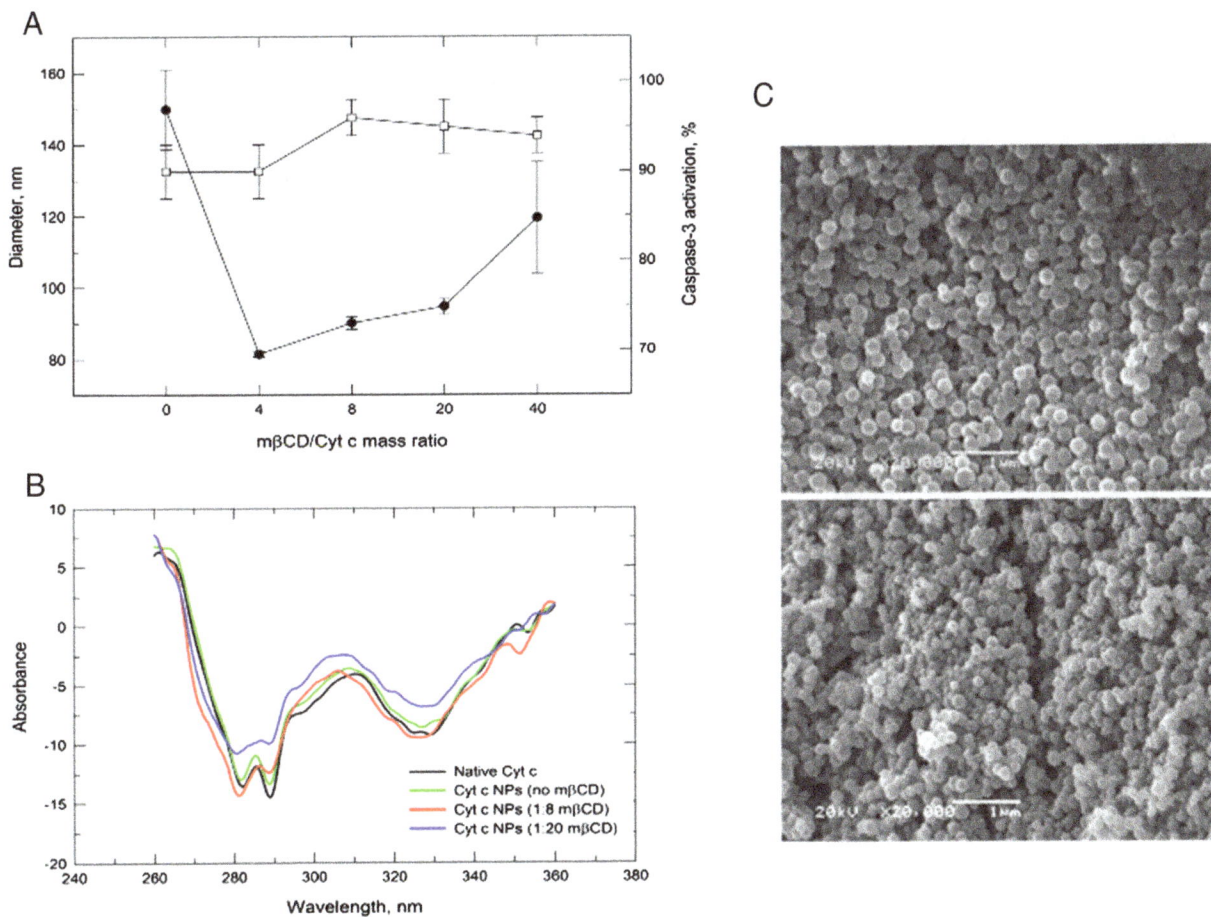

**Figure 3 The effect of methyl-β-cyclodextrin (mβCD) during the protein precipitation. (A)** Effect of mβCD on the particle diameter (—●—) and caspase-3 activation (—□—) after Cyt c nanoprecipitation. **(B)** Near-UV CD spectrum of Cyt c nanoprecipitated in absence or presence of mβCD. **(C)** SEM image of Cyt c NPs made without mβCD (upper picture), and at a 1:8 Cyt c-to-mβCD mass ratio (lower image).

the ability to sequester hydrophobic moieties on protein surfaces [26] it is possible that Cyt c was somewhat destabilized by the additive under these conditions. Such destabilization has been reported for some proteins under aqueous and non-aqueous conditions [24,27].

Scanning electron microscopy (SEM) was performed to investigate the shape of the NPs (Figure 3C). The SEM images of lyophilized Cyt c NP formulations show that the powder particles obtained had a spherical shape and confirm that the particle size was in the nanometer range.

Above findings demonstrate that the nanoprecipitation method is useful in obtaining nano-scale Cyt c without introducing irreversible functional loss. We also demonstrated that the protein was still able to interact with Apaf-1 to induce apoptosis upon rehydration.

## Surface modification of Cyt c NPs

To make the nanoparticles useful for delivery purposes we proceeded to modify their surface using a two-step procedure (Figure 4). First, the NPs' surface was modified with the linker SPDP to allow for their coating with PLGA-SH. The level of SPDP modification was determined by measuring the release of pyridine-2-thione at 343 nm after addition of dithiothreitol (DTT) to the reaction products. An increasing amount of linker was attached to the NPs at increasing reagent concentration during the reaction (Figure 5A). Since some of the target amino groups of Cyt c overlap with the Cyt c-Apaf 1 interaction site, increased levels of SPDP linked to Cyt c NPs reduced the ability of Cyt c to interact with Apaf-1 by about 20% (Figure 5A). However, since a large amount of Cyt c should be delivered to each single target cell in the end (a complete NP with roughly over 100,000 molecules), this was acceptable for our application. Circular dichroism (CD) spectra showed that no major changes in Cyt c

tertiary structure occurred upon modification with SPDP (Figure 5B). We selected the 5 fold-SPDP modification level for PLGA attachment to have sufficient polymer attachment points.

Next, the PLGA polymer was attached to the linker-modified (activated) NPs in order to prevent them to dissolve in aqueous media (e.g., upon reconstitution in buffer). The level of modification accomplished with PLGA-SH was determined by spectrophotometric analysis of the 2-pyridyldithio group released upon synthesis (Figure 4) at 343 nm. The level of modification of the NPs with PLGA was determined to be $2.06 \pm 0.04$ moles per mol of Cyt c under the conditions employed (see Methods Section for details). This construct is referred to from here on as PLGA-S-S-Cyt c NPs.

Finally, we verified whether the constructs obtained would behave in the manner we anticipated – stable in aqueous medium under oxidizing conditions and dissolving after reduction-induced polymer shedding under intracellular conditions. PLGA-S-S-Cyt c NPs were placed in release buffer using glutathione concentrations of 0, 1 μM, and 10 mM and incubated at 37°C simulating extra- and intracellular conditions [28]. At pre-determined times the supernatant was removed and the concentration of dissolved Cyt c determined by measuring the heme absorbance at 530 nm. The supernatant was replaced to maintain sink conditions. The concentration of the released protein was used to construct cumulative release profiles (Figure 6).

The PLGA coated Cyt c NPs released ca. 10% of Cyt c during the experiment, most of it during the first few hours as a "burst release". The amount of burst compares well with most conventional PLGA-based sustained release devices. Release of >80% of Cyt c was accomplished under reducing conditions demonstrating that the system design idea proved correct.

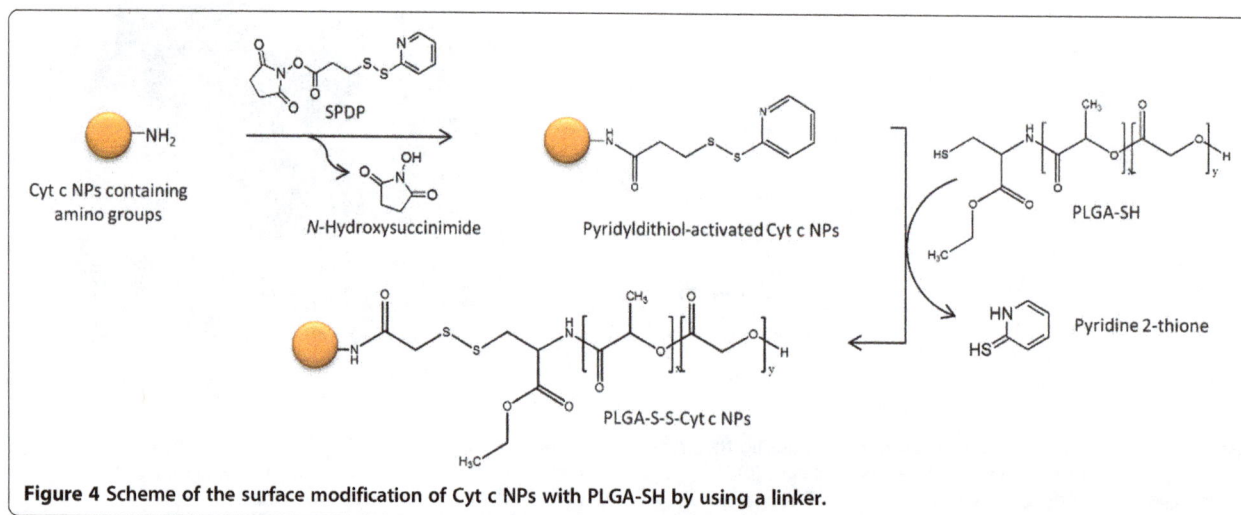

**Figure 4** Scheme of the surface modification of Cyt c NPs with PLGA-SH by using a linker.

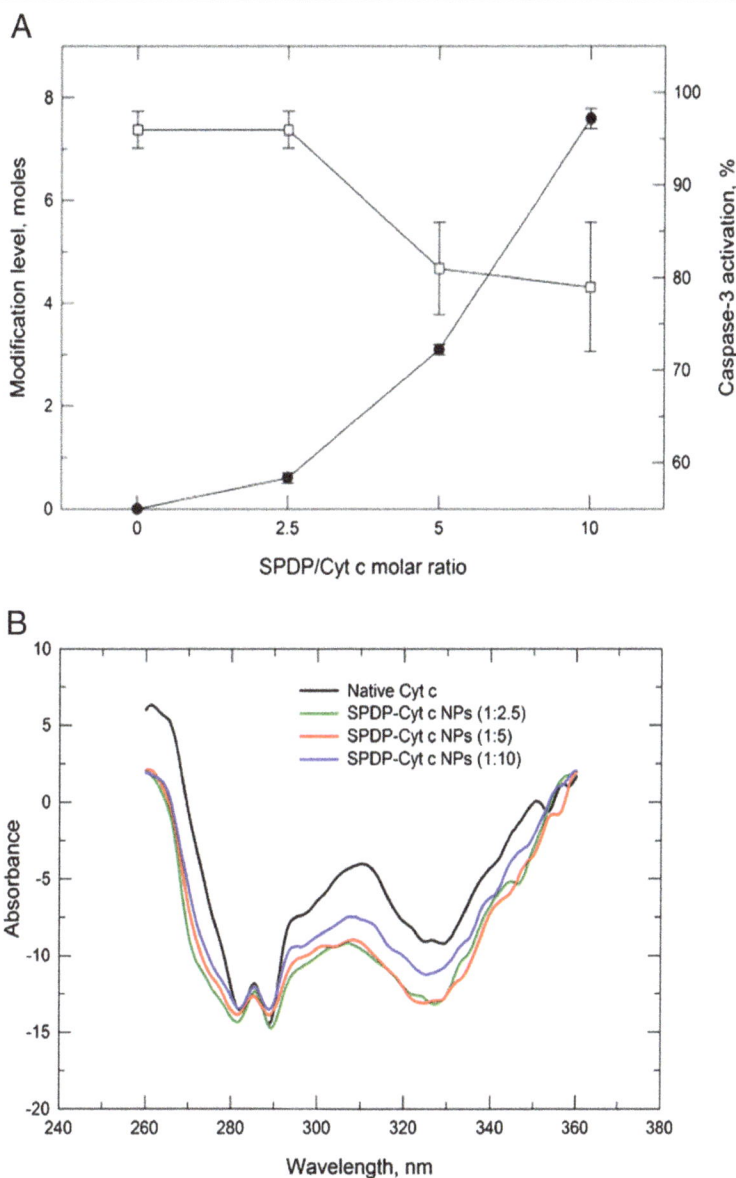

**Figure 5 Characterization of the modification of Cyt c NP surface with the SPDP linker. (A)** SPDP modification level accomplished using increasing reagent concentrations during synthesis (closed symbols) and effect of the modification on *in vitro* caspase-3 activation in a cell-free system (open symbols). **(B)** Near-UV CD spectra of dissolved Cyt c NPs after modification with SPDP.

**PLGA-S-S-Cyt c NPs cytotoxic effects to HeLa cells**

To investigate whether the synthesized NPs would display the desired effect on cancer cells, HeLa cells were incubated with PLGA-S-S-Cyt c NPs at 25, 50, and 100 µg/ml Cyt c concentration for 6 h. Cyt c NPs coated with PLGA induced a significant reduction in cell viability after 6 h of incubation, in particular at the 100 µg/ml Cyt c concentration (Figure 7). Several control experiments were performed. PLGA was conjugated to the SPDP linker and added to HeLa cells at the same concentration as in the corresponding experiments with the highest PLGA-S-S-Cyt c NPs concentration. No significant cytotoxicity was

observed after 6 h (Figure 7) and even 24 h of incubation (data not shown) with PLGA-SPDP. To further confirm that the cell death after treatment with PLGA-S-S-Cyt c NPs was due to Cyt c, we constructed the same PLGA coated NPs delivery using the non-apoptotic protein α-lactalbumin (LA). The LA-based NPs had no effect on cell viability excluding cytotoxic effects of the delivery system *per se* (Figure 7).

Similar to us, Zhao and coworkers designed a smart nanoparticulate DDS for an apoptotic protein (i.e., caspase-3). The protein was released mediated by intracellular reducing conditions. The study showed evidence

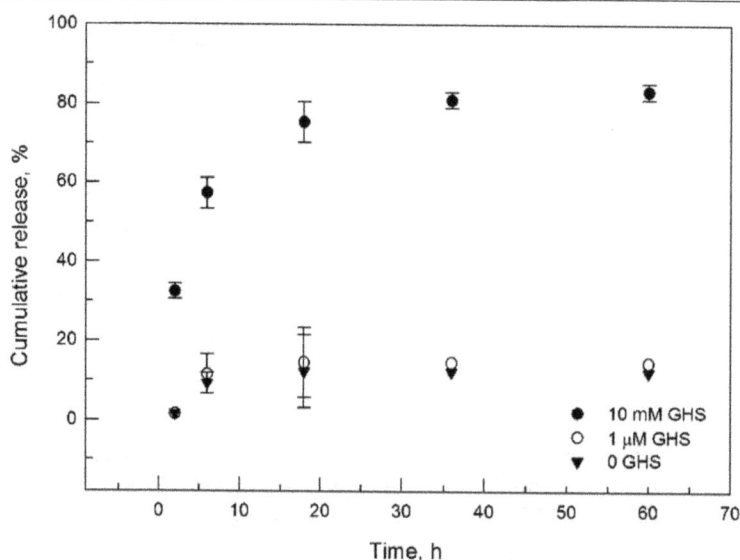

**Figure 6** Sustained release of Cyt c from PLGA-coated NPs after exposure to various conditions.

of cell death after 48 h of incubation [18] in contrast to 6 h for our DDS.

The DDS designed by us herein also proved much more efficient in killing HeLa cells than our previous smart DDS based on silica NPs [15]. Making the drug Cyt c itself the nanoparticulate delivery system eliminated the main problem encountered with the other system, namely, increasing silica toxicity when trying to deliver sufficient Cyt c to the target cells. Similarly, NPs using other materials [29] could present similar issues as our silica-based delivery system.

Another potential advantage of the newly designed DDS is its negative charge. Although positively charged NPs have shown good cell internalization properties [18,30,31], they potentially bind to vascular endothelial cells reducing their tumor accumulation by means of the EPR effect (i.e., the vascular endothelial luminal surface is known to be negatively charged) [2]. In contrast, PLGA has a negative charge at physiological and alkaline pH and it has been demonstrated that it can be internalized by the cell and is able to escape from the endosomes [32]. Therefore, our protein-based NPs could

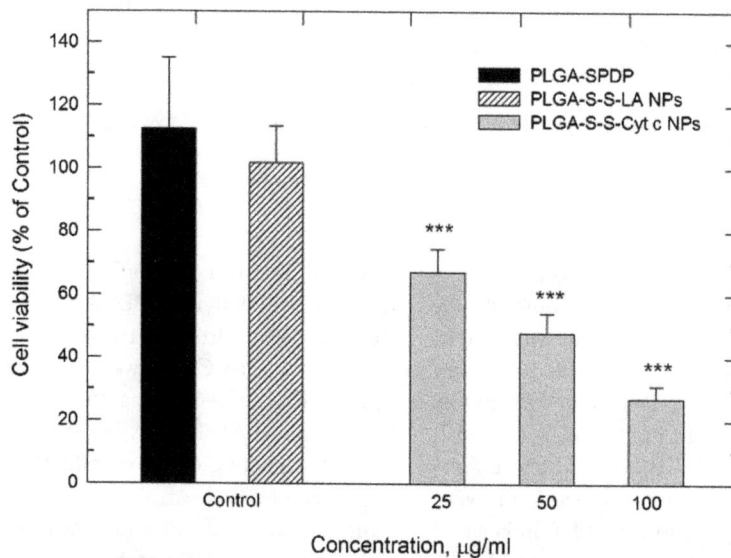

**Figure 7 Viability of HeLa cells treated with PLGA-S-S-Cyt c NPs after 6 h of incubation.** Two controls were included (PLGA-SPDP, PLGA-S-S-LA NPs) and had no effect on HeLa cell viability. Asterisk (***) indicates statistical significance ($p < 0.001$).

be an alternative to the intracellular delivery also of other apoptotic proteins.

## Investigation of apoptosis induction in HeLa cells by the intracellular release of Cyt c

To confirm that the Cyt c-induced cell death observed in the cell viability experiments was indeed due to apoptosis, we assessed the occurrence of nuclear segmentation and chromatin condensation. After HeLa cells were incubated with PLGA-S-S-Cyt c NPs or the PLGA-S-S-LA NPs control for 6 h, the cells were stained with PI and DAPI. Co-localization of DAPI and PI occurred when cells were incubated with Cyt c-based NPs (Figure 8), which points toward nuclear fragmentation and chromatin condensation in the cells indicative of ongoing apoptosis. In contrast, cells without treatment or treated with the PLGA-S-S-LA control showed no indication for dye co-localization and thus apoptosis. These results are in agreement with confocal results by Santra et al [7] who demonstrated that the delivery of Cyt c to the cytoplasm of the cancer cells results in the induction of apoptosis of such cells [7].

## Conclusions

Therapeutic proteins have enormous potential in the treatment and prevention of human diseases but are of limited use because they frequently display low physico-chemical stability. Specifically, when injected proteins usually quickly degrade and have short blood half-lives. Thus, the development of DDS able to protect the protein payload from degradation and in addition enabling targeted and controlled delivery is of considerable interest. Herein, we presented a new method for the delivery of Cyt c, an apoptotic protein, to model cancer cells. The main advantage of the system is that the DDS core consisted of Cyt c NPs and thus the drug itself thus overcoming payload limitations of drug-loaded systems. In order to stabilize the protein NPs we coated them with hydrophobic PLGA using a redox-sensitive linker. Only little Cyt c was released under oxidizing conditions, but 80% was released within hours under intracellular conditions. We furthermore demonstrated that PLGA-S-S-Cyt c NPs were able to induce apoptosis in a human cancer cell line while PLGA-S-S-LA NPs as control did not. Redox-responsive polymer-coated protein-based NPs are a simple and effective method for intracellular delivery of Cyt c for induction of apoptosis. We foresee that this system could potentially be used for the delivery of other proteins. However, we like to point out that our system was not tested in a non-cancerous cell line at this point in time because it does not include targeting ligands. Thus, it would probably show comparable toxicity in cancer and non-cancer cell lines since the EPR effect afforded by the nanoparticle formulation would not be relevant *in vitro* studies. It is also important to acknowledge that particle size is by far not the sole determinant of tumor specificity in *in vivo* studies. For example, immune cells can consume the drug delivery system and some tumors maybe very dense in the core disabling EPR-mediated targeting. To address such questions, experiments ongoing in our

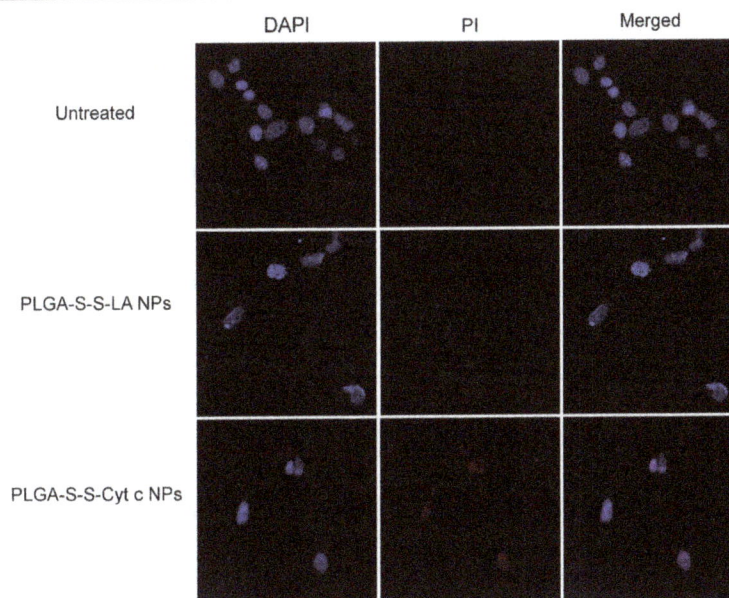

**Figure 8 Study of DAPI and propidium iodine (PI) colocalization, for the detection of apoptotic cells.** Selective induction of apoptosis observed in HeLa cells incubated with the PLGA-S-S-Cyt c NPs. No cellular apoptosis observed in untreated HeLa cells or when incubated with PLGA-S-SLA NPs.

laboratory focus on *in vivo* experiments using animal models. Tumor selectivity and internalization kinetics and mechanism of PLGA-S-S-Cyt c NPs are being scrutinized as well as the decoration of the system with tumor specific ligands. Combining passive targeting with additional ligand-mediating targeting should not only amplify the specificity of the NPs but also facilitate their efficient cellular uptake.

## Experimental procedures

Cytochrome c from equine heart, reduced glutathione ethyl ester, and a protease inhibitor cocktail were from Sigma-Aldrich (St. Louis, MO). Acetonitrile (HPLC grade) was from Fisher (Waltham, MA). Succinimidyl-3-(2-pyridyldithio) propionate (SPDP) was from Proteochem (Denver, CO). Poly (lactide-*co*-glycolide)-SH with a thiol end cap (PLGA-SH, 30,000 Da, copolymer ratio 1:1) was from Akina, Inc (West Lafayette, IN). 4', 6-Diamidino-2-phenylindole (DAPI) and propidium iodide (PI) were purchased from Invitrogen (Grand Island, NY). All the reagents were used without further purification. All other chemicals were from various commercial suppliers and were at least of analytical grade. HeLa cells were purchased from the American Type Culture Collection (Manassas, VA) and grown according to the vendor's instruction.

### Protein nanoprecipitation

Protein nanoparticles were obtained using a similar method as described previously by us [14]. Briefly, Cyt c was solvent-precipitated from nanopure water in presence of the excipient methyl-β-cyclodextrin by adding acetonitrile at a 1:4 volume ratio. Different protein concentrations (5, 10, 20, 40 mg/ml) as well as different mass ratios of protein-to-excipient (1:4, 1:8, 1:20 and 1:40, w/w) were studied. The protein suspension obtained was centrifuged at 6,000 rpm for 15 min, the supernatant discarded, and the pellet vacuum dried for 30 min.

### Determination of the nanoprecipitation yield

Protein concentration and amount of protein aggregates in Cyt c NPs were determined as described by us in detail [33]. In brief, the Cyt c NPs were suspended in 2 ml of potassium phosphate (PBS) buffer for 2 h to dissolve the buffer-soluble fraction. The samples were then subjected to centrifugation at 6,000 rpm for 15 min and the supernatant used to determine the protein concentration. Next, 1 ml of 6 M urea was added to the pellet to dissolve the buffer-insoluble protein fraction. The protein concentration was determined by quantitative ultraviolet (UV) spectrophotometric analysis at 280 nm and also from the heme absorbance at 408 nm. The precipitation yield was calculated from the actual and theoretical quantity of protein recovered after nanoprecipitation and rehydration

(%w/w). Cyt c as obtained from the supplier in PBS or 6 M urea was used to construct a calibration curve. The experiments were performed in triplicate, the results averaged, and the standard deviations calculated.

### Synthesis of SPDP-Cyt c NPs

Following the Cyt c nanoprecipitation, the SPDP linker was added directly to the resulting suspension (i.e., in 80% acetonitrile) to accomplish the NPs surface modification with the linker. Different Cyt c-to-SPDP molar ratios (1:2.5, 1:5, 1:10) were tested. After letting the mixture react for 30 min, the suspension was used to determinate the level of modification with SPDP. To stop the reaction and to remove unreacted SPDP and *N*-hydroxysuccinimide by-products, repeated centrifugation (6,000 rpm, 15 min)/washing cycles were performed. The pyridyldithiol-activated Cyt c (SPDP-Cyt c) NPs pellet was saved for further reaction and analysis. The supernatant was used to determine the concentration of unreacted SPDP by measuring the release of pyridine-2-thione at 343 nm after addition of 10 μL of 15 mg/ml DTT (the absorbance of the supernatant before reaction with DTT was used as blank). The amount of covalent modification of the nanoparticles with the linker was determined from the difference between the initial SPDP concentration and the unreacted SPDP. The experiments were performed in triplicate, the results averaged, and the standard deviations calculated.

### Synthesis of PLGA-Cyt c NPs

HS-PLGA (50 mg) dissolved into 5 ml of Acetonitrile was added to SPDP-Cyt c NPs (5 mg) and reacted at room temperature for 18 h. After reaction, unreacted HS-PLGA and pyridine 2-thione by-products were removed by centrifugation at 6,000 rpm for 10 min. The supernatant was used to determine the level of modification by measuring the concentration of pyridine-2-thione at 343 nm.

### In vitro release of Cyt c

The release of Cyt c from PLGA-S-S-Cyt c NPs was measured similarly as described by us [28]. Briefly, 0.25 mg of PLGA-S-S-Cyt c NPs powder were suspended by sonication in 1 ml of 50 mM PBS with 1 mM EDTA at pH 7.4 and glutathione (GHS) concentrations of 0, 0.001, and 10 mM. Incubation was performed for various times at 37°C, and the NPs were pelleted by centrifugation at 14,000 rpm for 10 min. The supernatant was removed and used to determine the concentration of released Cyt c. The pellet was resuspended in GHS-PBS buffer. The amount of protein released was used to construct cumulative release profiles. The experiments were performed in triplicate, the results averaged, and the standard deviations calculated.

## Dynamic light scattering

Particle sizes of the different formulations of Cyt c NPs were determined by Dynamic light scattering using a DynaPro Titan. The samples were dispersed in DMF and subjected to ultrasonication at 240 W for 30 sec prior to the measurements.

## Scanning electron microscope (SEM)

SEM of the different formulations of Cyt c NPs was performed using a JEOL 5800LV scanning electron microscope at 20 kV. The samples were coated with gold for 10 sec to a thinkness of 10 nm using a Denton Vacuum DV-502A.

## Circular dichroism (CD) spectroscopy

CD spectra were recorded using a JASCO J-1500 High Performance CD spectrometer at room temperature. The protein (Cyt c, Cyt c NPs, or SPDP-Cyt c NPs) was dissolved in nanopure water. CD spectra were acquired from 260 to 350 nm (tertiary structure) at a concentration of 1 mg/ml using a 10 mm quartz cuvette. Each spectrum was obtained by averaging two scans. Spectra of nanopure water blanks were measured prior to the samples and subtracted from the sample spectra.

## Cell-free caspase-3 assay

The cell lysate was obtained as described by us [15]. Briefly, the cell-free reaction was initiated by adding Cyt c or the different Cyt c NP formulations (e.g. 100 µg/mL of Cyt c NPs or SPDP-Cyt c NPs) to freshly purified cytosol (3 mg/ml) in a total reaction volume of 100 µL. The reaction was incubated at 37°C for 150 min. Afterwards the caspase-3 assay was performed following the manufacturer's protocol (CaspACE™ assay; Promega, Madison, WI). The plate was incubated overnight at room temperature and the absorbance at 405 nm was measured in each well using a Thermo Scientific Multiskan FC. All measurements were performed in triplicate.

## Cell culture

HeLa cells were maintained in accordance with the ATCC protocol. Briefly, the cells were cultured in minimum essential medium (MEM) containing 1% L-glutamine, 10% fetal bovine serum (FBS), and 1% penicillin in a humidified incubator with 5% $CO_2$ and 95% air at 37°C. All experiments were conducted before cells reached 25 passages. For the cell viability and confocal microscopy experiments, HeLa cells were seeded in 96-well plates or chambered cover-slides (4 wells), respectively, for 24 h in MEM containing 1% L-glutamine, 10% FBS, and 1% penicillin. Subsequently, cell growth was arrested by decreasing the FBS concentration in the medium to 1% for 18 h. Then, cells were exposed and incubated with the different bioconjugates for 6 h.

## Cell viability assay

Mitochondrial function was measured using the CellTiter 96 aqueous non-radioactive cell proliferation assay from Promega Corporation. HeLa cells (5,000 cells/well) were seeded in 96-well plates as described above. Cells were incubated with serial dilutions of PLGA-S-S-Cyt c NPs (25, 50 and 100 µg/mL of Cyt c) for 6 h. Controls, such as, PLGA-SPDP (80 µg/mL) and NPs of a non-apoptotic protein (i.e. α-lactalbumin at 100 µg/mL) were also tested. After incubation, 20 µL of 3-(4, 5-dimethylthiazol-2-yl)-5-(3-carboxymethoxyphenyl)-2-(4-sulfophenyl)-2H-tetrazolium, inner salt (MTS) and phenazine methosulfate (PMS) was added to each well (333 µg/mL MTS and 25 µM PMS). After 1 h, the absorbance at 492 nm was measured using a microplate reader. HeLa cells treated with 2 µM staurosporin for 6 h were used as positive control and cells without treatment were used as negative control. The relative cell viability (%) was calculated by:

$$Relative\,cell\,viability\,(\%) = \frac{Abs\,test\,sample}{Abs\,control} \times 100$$

$T$-test analysis was used for comparison of two independent groups for cell viability. Difference between control (untreated cells) and experimental groups (i.e., PLGA-SPDP, PLGA-S-S-LA, NPs PLGA-S-S-Cyt c NPs) was considered statistically significant at $p < 0.05$.

## Investigation of apoptosis induction in HeLa cells by the intracellular release of Cyt c

HeLa cells (25,000 cells) were seeded in chambered cover-glass (4-wells) as previously described by us [15]. The cells were incubated with PLGA-S-S-Cyt c NPs at a 25 µg/mL Cyt c concentration at 37°C for 6 h. For detection of apoptosis-dependent nuclear fragmentation, the cells were washed with PBS (1X) and incubated initially with DAPI (300 nM) and thereafter with PI (75 µM) for 5 min each. HeLa cells were then fixed using 3.7% formaldehyde. The coverslips were examined under a Zeiss laser-scanning microscope 510 using a 67× objective. Co-localization of DAPI and PI upon internalization into HeLa cells was determined, which is representative of highly condensed and fragmented chromatin in apoptotic cells [7,34,35]. DAPI was excited at 405 nm and its emission was detected at 420-480 nm. PI was excited at 561 nm and was detected above 600-674 nm.

### Abbreviations

Cyt c: Cytochrome c; Apaf-1: Apoptotic protease activating factor 1; LA: α-lactalbumin; PLGA: Poly (lactic-co-glycolic) acid; EPR: Enhanced permeability and retention; NPs: Nanoparticles; DDS: Drug delivery system; MSN: Mesoporous silica nanoparticles; MEM: Minimum essential medium; FBS: Fetal bovine serum; DAPI: 4',6-diamidino-2-phenylindole; PI: Propidium iodine; MTS: 3-(4, 5-dimethylthiazol-2-yl)-5-(3-carboxymethoxyphenyl)-2-(4-sulfophenyl)-2H-tetrazolium, inner salt; PMS: Phenazine methosulfate; GHS: Glutathione; SPDP: Succinimidyl-3-(2-pyridyldithio) propionate; mβCD: Methyl-β-cyclodextrin; DTT: Dithiothreitol; EDTA: Ethylenediaminetetraacetic acid; SEM: Scanning electron microscopy; CD: Circular dichroism; PBS: Phosphate buffer solution.

## Competing interests

The authors declare they have no competing interests.

## Authors' contributions

MMC carried out all the experimental studies, contributed in the design of the study, analyzed data, and drafted the manuscript. CMF participated in the caspase-3 and cell viability assays, and helped drafting the manuscript. TGR participated in the nanoprecipitation and nanoparticle surface modification studies. YDR participated in the circular dichroism spectroscopy and the dynamic light scattering studies. AM participated in the dynamic light scattering and confocal studies. JM participated in the experimental design of cell culture experiments. MM performed the statistic analysis. KG conceived the study, participated in its design and coordination, and finalized the manuscript. All authors read and approved the final manuscript.

## Acknowledgements

This publication was made possible by grant SC1 GM086240 from the National Institute for General Medical Sciences (NIGMS) at the National Institutes of Health (NIH) through the Support of Competitive Research (SCoRE) Program. Its contents are solely the responsibility of the authors and do not necessarily represent the official views of NIGMS. M.M., C.M.F., T.G., Y.D., and A.M. were supported by fellowships from the NIH Research Initiative for Scientific Enhancement (RISE) Program (2R25GM061151-12), and J.M. received a fellowship from the Institute for Functional Nanomaterials (IFN) at the University of Puerto Rico. The authors are very grateful to Mr. Bismark Madera, M.T., for his outstanding work and dedication in the confocal imaging experiments.

## Author details

[1]Departments of Chemistry, University of Puerto Rico, Río Piedras Campus, San Juan, PR 00931, USA. [2]Department of Biology, University of Puerto Rico, Río Piedras Campus, San Juan, PR 00931, USA. [3]Department of Graduate Studies, University of Puerto Rico, Río Piedras Campus, Río Piedras Campus, San Juan, PR 00931, USA.

## References

1. Danhier F, Feron O, Preat V: **To exploit the tumor microenviroment: passive and active tumor targeting of nanocarriers for anti-cancer drug delivery.** *J Control Release* 2010, **148**:135–146.
2. Fang J, Nakamura H, Maeda H: **The EPR effect: unique features of tumor blood vessels for drug delivery, factors involved, and limitations and augmentation of the effect.** *Adv Drug Deliv Rev* 2011, **63**:136–151.
3. Morachis JM, Mahmoud EA, Almutairi A: **Physical and chemical strategies for therapeutic delivery by using polymeric nanoparticles.** *Pharmacol Rev* 2012, **64**:505–519.
4. Morigi V, Tocchio A, Bellavite-Pellegrini C, Sakamoto JH, Arnone M, Tasciotti E: **Nanotechnology in medicine: from inception to market domination.** *J Drug Deliv* 2012, **2012**:1–7.
5. Lanni JS, Lowe SW, Licitra EJ, Liu JO, Jacks T: **p53-independent apoptosis induced by paclitaxel through an indirect mechanism.** *Proc Natl Acad Sci U S A* 1997, **94**:9679–9683.
6. Ryan KM, Phillips AC, Vousden KH: **Regulation and function of the p53 tumor suppressor protein.** *Curr Opin Cell Biol* 2001, **13**:332–337.
7. Santra S, Kaittanis C, Perez JM: **Cytochrome C encapsulating theranostic nanoparticles: a novel bifunctional system for targeted delivery of therapeutic membrane-impermeable proteins to tumors and imaging of cancer therapy.** *Mol Pharm* 2010, **7**:1209–1222.
8. Yamada Y, Harashima H: **Mitochondrial drug delivery systems for macromolecule and their therapeutic application to mitochondrial diseases.** *Adv Drug Deliv Rev* 2008, **60**:1439–1462.
9. Solá RJ, Griebenow K: **Effects of glycosylation on the stability of protein pharmaceuticals.** *J Pharm Sci* 2009, **98**:1223–1245.
10. Manning MC, Chou DK, Murphy BM, Payne RW, Katayama DS: **Stability of protein pharmaceuticals: an update.** *Pharm Res* 2010, **27**:544–575.
11. Brown LR: **Commercial challenges of protein drug delivery.** *Expert Opin Drug Deliv* 2005, **2**:29–42.
12. Slowing II, Trewyn BG, Lin VSY: **Mesoporous silica nanoparticles for intracellular delivery of membrane-impermeable proteins.** *J Am Chem Soc* 2007, **129**:8845–8849.
13. Solaro R: **Targeted delivery of protein drugs by nanocarriers.** *Materials* 2010, **3**:1928–1980.
14. Morales-Cruz M, Flores-Fernández GM, Morales-Cruz M, Orellano EA, Rodríguez-Martínez JA, Ruiz M, Griebenow K: **Two-step nanoprecipitation for the production of protein-loaded PLGA nanospheres.** *Results Pharma Sci* 2012, **2**:79–85.
15. Méndez J, Morales-Cruz M, Delgado Y, Figueroa CM, Orellano EA, Morales M, Monteagudo A, Griebenow K: **Delivery of chemically glycosylated cytochrome c immobilized in mesoporous silica nanoparticles induces apoptosis in hela cancer cells.** *Mol Pharm* 2014, **11**:102–111.
16. Langer K, Balthasar S, Vogel V, Dinauer N, von Briesen H, Schubert D: **Optimization of the preparation process for human serum albumin (HSA) nanoparticles.** *Int J Pharm* 2003, **257**:169–180.
17. Jahanshahi M, Babaei Z: **Protein nanoparticle: a unique system as drug delivery vehicles.** *Afr J Biotechnol* 2008, **7**:4926–4934.
18. Zhao M, Biswas A, Hu B, Joo K, Wang P, Gu Z, Tang Y: **Redox-responsive nanocapsules for intracellular protein delivery.** *Biomaterials* 2011, **32**:5223–5230.
19. Shive M, Anderson J: **Biodegradation and biocompatibility of PLA and PLGA microspheres.** *Adv Drug Deliv Rev* 1997, **28**:5–24.
20. Lü JM, Wang X, Marin-Muller C, Wang H, Lin PH, Yao Q, Chen C: **Current advances in research and clinical applications of PLGA based nanotechnology.** *Expert Rev Mol Diagn* 2009, **4**:325–341.
21. Nitta SK, Numata K: **Biopolymer-based nanoparticles for drug/gene delivery and tissue engineering.** *Int J Mol Sci* 2013, **14**:1629–1654.
22. Torchilin V: **Tumor delivery of macromolecular drugs based on the EPR effect.** *Adv Drug Deliv Rev* 2011, **63**:131–135.
23. Kim KY: **Nanotechnology platforms and physiological challenges for cancer therapeutics.** *Nanomedicine* 2007, **3**:103–110.
24. Griebenow K, Diaz-Laureano Y, Santos AM, Montañez-Clemente I, Rodriguez L, Vidal M, Barletta G: **Improved enzyme activity and enantioselectivity in organic solvents by methyl-β-cyclodextrin.** *J Am Chem Soc* 1999, **121**:8157–8163.
25. Davies AM, Guillemette JG, Smith M, Greenwood C, Thurgood AG, Mauk AG, Moore GR: **Redesign of the interior hydrophilic region of mitochondrial cytochrome c by site-directed mutagenesis.** *Biochemistry* 1993, **32**:5431–5435.
26. Aachmann FL, Otzen DE, Larsen KL, Wimmer L: **Structural background of cyclodextrin–protein interactions.** *Protein Eng* 2003, **16**:905–912.
27. Cooper A: **Effect of cyclodextrins on the thermal stability of globular proteins.** *J Am Chem Soc* 1992, **114**:9208–9209.
28. Méndez J, Monteagudo A, Griebenow K: **Stimulus-responsive controlled release system by covalent immobilization of an enzyme into mesoporous silica nanoparticles.** *Bioconjug Chem* 2012, **23**:698–704.
29. Besaratinia A, Pfeifer GP: **A review of mechanisms of acrylamide carcinogenicity.** *Carcinogenesis* 2007, **28**:519–528.
30. Yu B, Zhang Y, Zheng W, Fan C, Chen T: **Positive surface charge enhances selective cellular uptake and anticancer efficacy of selenium nanoparticles.** *Inorg Chem* 2012, **51**:8956–8963.
31. Choi SY, Jang SH, Park J, Jeong S, Park JH, Ock KS, Lee K, Yang SI, Joo SW, Ryu PD, Lee SY: **Cellular uptake and cytotoxicity of positively charged chitosan gold nanoparticles in human lung adenocarcinoma cells.** *J Nanopart Res* 2012, **14**:1234.
32. Panyam J, Zhou WZ, Prabha S, Sahoo SK, Labhasetwar V: **Rapid endo-lysosomal escape of poly (DL-lactide-co-glycolide) nanoparticles: implications for drug and gene delivery.** *FASEB J* 2002, **16**:1217–1226.
33. Castellanos IJ, Griebenow K: **Improved α-chymotrypsin stability upon encapsulation in PLGA microspheres by solvent replacement.** *Pharm Res* 2003, **20**:1873–1880.
34. Bratton SB, Salvesen GS: **Regulation of the apaf-1-caspase-9 apoptosome.** *J Cell Sci* 2010, **123**:3209–3214.
35. Shacter E, Williams JA, Hinson RM, Senturker S, Lee YJ: **Oxidative stress interferes with cancer chemotherapy: inhibition of lymphoma cell apoptosis and phagocytosis.** *Blood* 2000, **96**:307–313.

# Transient extracellular application of gold nanostars increases hippocampal neuronal activity

Kirstie Salinas[1†], Zurab Kereselidze[2†], Frank DeLuna[1], Xomalin G Peralta[2] and Fidel Santamaria[1*]

## Abstract

**Background:** With the increased use of nanoparticles in biomedical applications there is a growing need to understand the effects that nanoparticles may have on cell function. Identifying these effects and understanding the mechanism through which nanoparticles interfere with the normal functioning of a cell is necessary for any therapeutic or diagnostic application. The aim of this study is to evaluate if gold nanoparticles can affect the normal function of neurons, namely their activity and coding properties.

**Results:** We synthesized star shaped gold nanoparticles of 180 nm average size. We applied the nanoparticles to acute mouse hippocampal slices while recording the action potentials from single neurons in the CA3 region. Our results show that CA3 hippocampal neurons increase their firing rate by 17% after the application of gold nanostars. The increase in excitability lasted for as much as 50 minutes after a transient 5 min application of the nanoparticles. Further analyses of the action potential shape and computational modeling suggest that nanoparticles block potassium channels responsible for the repolarization of the action potentials, thus allowing the cell to increase its firing rate.

**Conclusions:** Our results show that gold nanoparticles can affect the coding properties of neurons by modifying their excitability.

**Keywords:** Nanoparticle, Uptake, Nanotoxicity, Neurons, Potassium channels, Firing rate

## Background

Several types of nanoparticles, particularly gold, can bind to proteins on the surface of cells [1,2]. Therefore, it is important to determine the effects that such binding has, not only on the metabolism of cells, but also, on their function [3,4]. Although, there is a large body of work on the fatal toxic effects of gold nanoparticles on neurons there is little understanding how these widely used nanoparticles might affect their function, namely their activity and coding properties [5].

We hypothesized that gold nanoparticle-protein interactions alter the electrophysiological properties of neurons, which are mediated by proteins within the neuronal membrane. To test this hypothesis we synthesized

gold nanostars using a silver-seed mediated method we recently developed [6]. We applied the nanoparticles to mouse hippocampal slices while recording the action potential activity of neurons in the CA3 area. Our results show that the firing rate of action potentials of these cells increases by 17% after nanoparticle application. The increase in activity persists after a short nanoparticle application (5 min). The shape of the action potential changes in the area associated with potassium currents, suggesting a preferential effect of these nanoparticles on potassium channels. Overall, our results show that short transient applications of gold nanoparticles have nontoxic functional effects on neurons that should be considered when developing nanotechnology for neurobiology applications.

## Results and discussion

We started our work by synthetizing gold nanoparticles as described in our previous publications [6]. Our method

* Correspondence: fidel.santamaria@utsa.edu
†Equal contributors
[1]UTSA Neurosciences Institute, The University of Texas at San Antonio, San Antonio, Texas 78249, USA
Full list of author information is available at the end of the article

results in star shapes of 180 nm in average width and 70% yield (Figure 1). Energy dispersive X-Ray spectroscopy (EDS) analysis confirms that the nanoparticles are made out of gold (Figure 2). As described in Methods we prepared acute hippocampal slices. A recording electrode was brought in close proximity to the CA3 area of the slice. This section contains the cell bodies of excitatory hippocampal pyramidal cells. A second electrode was also brought close to the first electrode. This electrode could contain regular artificial cerebrospinal fluid (aCSF) or nanoparticle solution (Figure 3).

We recorded the firing rate activity of isolated hippocampal CA3 neurons before and after applying gold nanoparticles (Figure 4A). Our results show a consistent increase in the firing rate of the neurons after nanoparticle application (Figure 4B). Note that nanoparticles were delivered over a period of 5 min with continuous perfusion. Thus, nanoparticles that did not attach to neurons were washed out. Using a fluorescence correlation spectroscopy (FCS) setup we determined that the concentration of the nanoparticles in the pipette was about 3 nM (see Materials and methods and Additional file 1). After being released into the chamber the concentration decays as the nanoparticles diffuse and are carried by the circulating bath and enter the slice tissue, thus, it is expected that a much lower concentration of nanoparticles reaches the cells. The firing rate was averaged for 20 to 50 minutes after nanoparticle application and resulted in an average increase in firing rate of $16.82\% \pm 0.05$ (S.E.M., $p < 0.05$, $n = 8$ experiments, Figure 4C). An identical experiment with a solution free of nanoparticles did not result in a significant increase in firing rate ($2.60\% \pm 0.02$ S.E.M, not significant, $n = 6$ experiments, Figure 4D-F). Similarly, the application of 100 nm diameter latex nano-beads did not cause a significant change in firing rate ($2.29\% \pm 0.03$ S.E.M., not significant, $n = 3$

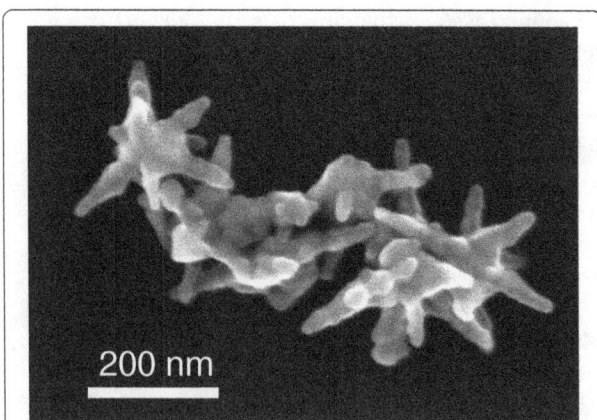

**Figure 1** Scanning electron microscope image of gold nanostars used in this study (imaged using a Hitachi S-5500).

experiments, not shown). Therefore, acute nanoparticle application on neurons results in an increase in firing rate.

In order to elucidate the mechanism through which nanoparticle application affected firing rates we investigated whether there was a change in the shape of the action potentials. For this reason we analyzed the average action potential shape before and after nanoparticle application (Figure 5A). We did not find changes in spike height or spike width. The action potentials rapidly repolarized after about 1.0 ms. However, the current associated with potassium channels appeared smaller than in the action potentials before nanoparticle application (Figure 5B). To quantify differences in the potassium associated current we first determined the time of the minimum voltage deflection in each action potential. The minimum voltage after the action potential peak is the time of maximum potassium current activation [7]. Since the amplitude of the potassium associated current could be affected by noise we decided to integrate the area of this section of the action potentials. Starting from the minimum voltage (maximum potassium activation) we integrated the area of the voltage trace for 0.2 ms, which corresponded to about 30% of the repolarization period (shaded area Figure 5B). We quantified this value in all the experiments before and after the application of nanoparticles. Our results show that there is a significant decrease of $5.0\% \pm 1.8$ S.E.M ($p < 0.05$) in this current (Figure 5C). Thus, our data suggests that our gold nanoparticles preferentially affect potassium channels.

In order to test whether blocking of potassium channels could result in an increase of the firing rate of the recorded neurons, we implemented a computer model using the Hodgkin and Huxley equations (see Materials and methods). We varied the density of non-linear potassium channels (Kdr) while delivering a constant input current (7 nA) until we obtained the change in firing rate observed in the experiments. Our modeling results suggest that it is necessary to change the Kdr by 0.75 (from 36 mS/cm$^2$ to 27 mS/cm$^2$) to obtain an increase of 17.5% in the firing rate (Figure 5D). We then compared the action potential shape from the control and 0.75xKdr simulations (Figure 5E). As in the experimental results the action potential shapes were very similar to each other. However, as in the experiments there was a difference in the strength of the potassium related potential. Since the shape and firing rate of the Hodgkin and Huxley model differ from the recorded action potentials we integrated the same fraction of the voltage trace as in the experiments (30% of the repolarization time after the minimum voltage). Interestingly, our results show that the integral of the potassium voltage is reduced by 4.6% (Figure 5F). Thus, our simulations corroborate our experimental results that blocking of potassium channels by transient application of gold nanoparticles results in increased firing rates for prolonged periods of time.

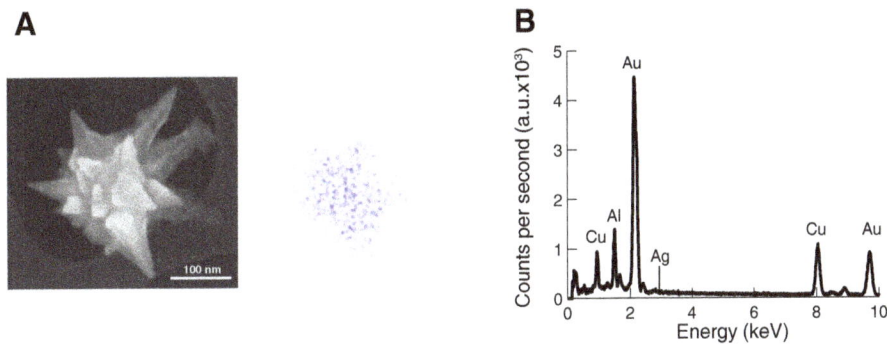

**Figure 2 EDS analysis shows that nanostars are made of gold. (A)** Left: SEM image of gold nanostar. Right: EDS mapping of gold atoms for nanoparticle in left. **(B)** EDS spectrum showing no silver on the surface of the nanoparticles.

## Conclusion

There is increasing evidence that nanoparticles made of different materials and shapes can be fatally toxic to the brain [8-10]. However, little is known about the non-fatal effects on the electrical activity of neurons. We found that a transient extracellular application of gold star-shaped nanoparticles increases the mean firing rate of CA3 hippocampal neurons. A very recent article [11] shows that intracellular injection of gold nanoparticles in the CA1 pyramidal neurons in the hippocampus results in an increase in the excitability of these cells. Our results are consistent with this report and further contribute to suggest that the site of action of the gold nanoparticles is on potassium channels. In our case, given that the reported uptake of gold nanoparticles takes much longer [12] than the effects that we measured, we hypothesize that the nanoparticles are mostly on the surface of the cells. Nanoparticles binding to potassium channels could affect their function by modifying their conformational state through adsorption [13], direct blocking [14] or clustering channels by cross-linking [15].

Our combined experimental and modeling approach suggests that a fraction of potassium channels are rapidly blocked by gold nanoparticles. This blockage causes a faster repolarization of the cell and a subsequent increase in firing rate. It is not known whether this increase in firing rate could cause pathological conditions that involve the hippocampus, such as epilepsy [16]. Increases in neuronal activity in isolated preparations could be compensated by network dynamics [17]. Oscillations in the brain are common [18]; therefore, nanoparticle induced increases in firing rate could occur without a behavioral effect. In any case, it is important to study the neurological and behavioral effects of application of gold nanoparticles in vivo [19].

In our experiments we only used a single concentration (about 3 nM) of the gold nanoparticle solution. This concentration was further decreased when combining the nanoparticle solution with aCSF and releasing it into the space of the chamber. Other factors affect the nanoparticle concentration such as the distance of the nanoparticle application electrode from the slice, the depth of the recorded neuron in the tissue, the movement of the solution in the perfusion chamber, and the diameter of the pipette tip. The uncertainty in controlling the concentration of nanoparticles in the bath did not allow us to perform concentration dependent experiments. A potential future direction to solve this problem could be to

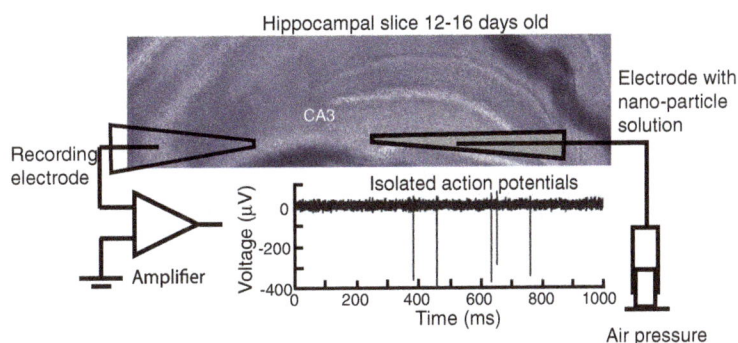

**Figure 3 Experimental setup used to record and deliver star shaped gold nanoparticles to the CA3 region of mouse hippocampal slices.** Transmitted light image obtained using an Olympus BX61WI microscope with a 20× 0.95 N.A. objective.

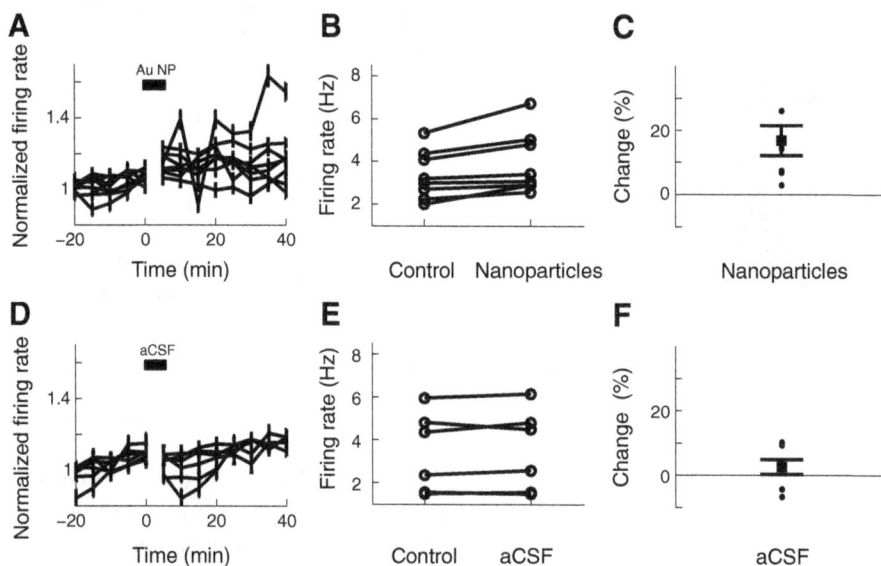

**Figure 4 Transient application of gold nanoparticles increases hippocampal neuronal activity. (A)** Firing rate average from extracellularly recorded CA3 hippocampal neurons before and after gold nanoparticle (Au NP) application. The activity was averaged every 5 min. Bars are S.E.M. Firing rates were normalized to values before application. **(B)** Absolute firing rate of all the experiments in (A) before and after nanoparticle application. Firing rate after application was averaged from t = 20 to t = 50 min. **(C)** Percentage change of firing rate application from B. Error bars are for the S.E.M. **(D-F)** Identical analysis as in A-C when the application pipette only contained artificial cerebrospinal fluid (aCSF) and no nanoparticles.

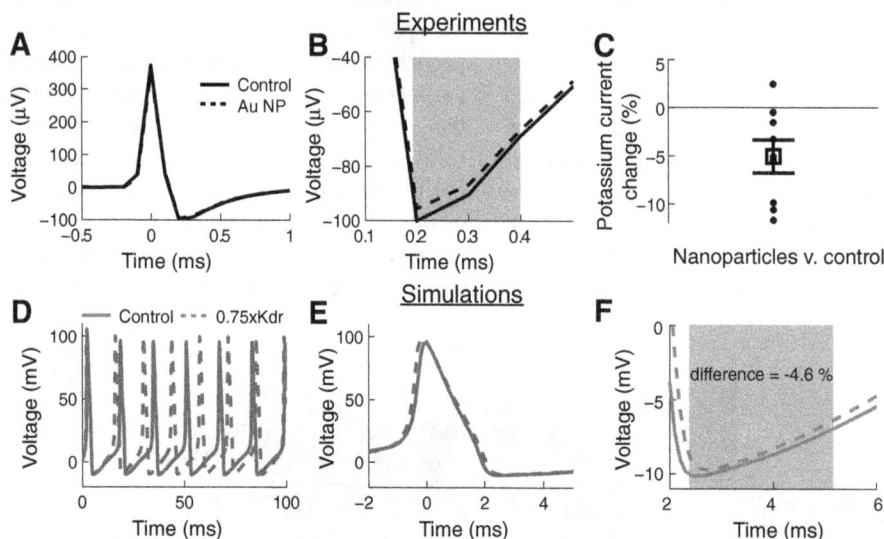

**Figure 5 Transient application of gold nanoparticles reduces the potassium associated current in CA3 hippocampal neurons.** **(A)** Average action potential before (control) and after application of gold nanoparticles (Au NP). The overlay at this scale makes the traces indistinguishable. **(B)** The hyperpolarization region of the action potential (from A) is associated with potassium currents. **(C)** The percent difference in the shaded area integrated from B before and after application of gold nanoparticles for all the experiments (5.0% ± 1.8 S.E.M., n = 8 experiments). **(D-F)** Computer simulation of action potential generation using the Hodgkin and Huxley model. **(D)** The model generated sustained firing rates when stimulated with continuous 7 nA. Varying the density of potassium currents (Kdr) by 0.75 results in an increase in firing rate of 17.5%. **(E)** Action potential overlay comparing the control and 0.75xKdr simulations. **(F)** Integrating the potassium associated voltage in the model shows a decrease in this current of 4.6%. The shaded area in F covers the same fraction of the action potential as in B.

integrate our FCS measuring setup with our nanoparticle experiments to determine their concentration in each experimental condition.

Overall, our work shows that transient exposure of neurons to gold nanoparticles can affect the coding properties of the hippocampus. Thus, these effects have to be taken into account when developing nanomaterials and nanotechnology to study brain function [20].

## Materials and methods

We synthesized gold star-shaped nanoparticles using a method we recently published [6]. Briefly, silver seeds are used as a nucleating agent, upon which growth of the nanoparticle occurred. Our method produced gold nanostars at a 70% yield and a concentration of 2–4 nM. Energy dispersive X-Ray spectroscopy (EDS) was performed with a JEOL JEM-ARM200F (JEOL, JP) to determine the chemical content and relative density of individual nanoparticles. Scanning electron microscopy images were collected with this same microscope or with a Hitachi S-5500.

In order to determine the concentration of nanoparticles we modified a two-photon microscope (Prairie Technologies, Madison, WI) to perform FCS measurements. A photo-multiplier detector was removed and in its place we aligned an optical fiber coupled with a lens. The other end of the optical fiber was connected to an avalanche photo diode (Perkin Elmer, USA). The photo diode was then coupled to an auto-correlator card (Correlator.com, Hong Kong) which was connected to an acquisition computer. A typical experiment collected 10 trials for 20 seconds. Nanoparticle luminescence [21,22] was stimulated with a femtosecond laser Chameleon (Coherent, Santa Clara, CA), at 90 MHz repetition rate with a <150 fs pulse at a wavelength of 760 nm. The luminescence acquisition dichroic had a band pass filter from 584 to 630 nm.

The general form of the auto-correlation function is:

$$G(t) = \frac{1}{V_{eff}[C]} \frac{1}{1 + \frac{t}{t_d}} \frac{1}{\sqrt{1 + \left(\frac{r_o}{z_o}\right)^2 \frac{t}{t_d}}} \quad (1)$$

where $t$ is time, $t_d$ is the auto-correlation time constant, $r_o$ is the waist at the focal point, $z_o$ is the spread along the z-axis, $C$ is the concentration and the effective volume ($V_{eff}$) is the two-photon illumination spot

$$V_{eff} = \pi^{3/2} r_o^2 z_o. \quad (2)$$

We determined the two-photon imaging volume using fluorescent beads of 0.1 μm in diameter (Invitrogen, USA). The measured waist of the focal point is $r_o = 1.25$ μm and the spread along the z-axis is $z_o = 5.40$ μm.

The auto-correlation time constant is related to the diffusion coefficient ($D$) of the particles by:

$$t_d = \frac{r_o^2}{8D}. \quad (3)$$

Finally, the amplitude at zero lag ($G(0)$) is inversely proportional to the average number of particles in the volume ($N$)

$$G(0) = \frac{1}{<N>} = \frac{1}{V_{eff}[C]} \quad (4)$$

from which we can calculate the concentration using equation (2). Fitting equation (1) to our measurements (Additional file 1: Figure S1) shows that the concentration of gold nanoparticles in our solution is $3.44 \pm 0.01$ nM (95% confidence intervals).

We compared the FCS concentration measurements to an estimate based on the content of gold in the formulation and the area and volume of the nanoparticles. We estimated the surface area ($A_S$) and volume ($V_S$) of the nanostars by measuring the radius of the core ($r_c$), the length of the rays ($r_h$), the radius of the base of the ray ($r_b$) and estimating the overlap between a spherical cone and a sphere given by the overlapping length ($\Delta r$) as

$$A_S = 4\pi r_c^2 + n\pi r_b \left( r_b + \left( r_b^2 + r_h^2 \right)^{\frac{1}{2}} \right) - 2n\pi r_b \Delta r \quad (9)$$

$$V_S = \frac{4}{3}\pi r_c^3 + \frac{1}{3} n\pi r_b^2 r_h - \frac{1}{3} n\pi \Delta r \left( 3r_b^2 + \Delta r^2 \right) \quad (10)$$

Where

$$\Delta r = r_c - \left( r_c^2 - r_b^2 \right)^{1/2} \quad (11)$$

and $n$ denotes the number of rays. From the SEM images, we found that the nanostars had an $r_c = 36 \pm 3$ nm, $r_h = 55 \pm 6$ nm, $n = 5.7 \pm 1.3$ peaks with a range of 4–7 peaks, and $r_b = 11 \pm 1$ nm. We calculated the predicted concentration given our synthesis protocol from

$$C = \frac{1}{2} \frac{M_{Au} \cdot V_{Au}}{V_{nps} \cdot N_x} \quad (13)$$

where $M_{Au}$ is the molarity of the gold chloride solution (0.25 mM), $V_{Au}$ is the volume of the gold chloride solution (20 mL), $V_{nps}$ is the volume of the nanoparticle solution (3 mL) and $N_x$ is the number of gold atoms in each shape ($N_V = V_R/d_{Au}^3$ for solid stars and $N_A = A_R/d_{Au}^2$ for hollow stars). The 1/2 comes from diluting the final solution in half. We assumed that the nanoparticles had an FCC crystalline structure and used the lattice constant of gold $d_{Au} = 0.408$ nm to find the number of atoms in each shape. Using this approach we predicted a concentration of stars of 0.12 nM, for solid nanoparticles, and 2.56 nM, for hollow structures (in both cases assuming $n = 5$ peaks). Therefore, the concentration based on a hollow nanoparticle calculation is in very good agreement with the concentration extracted from our experimental FCS measurements.

Artificial cerebrospinal fluid (aCSF) was composed of (in mM): NaCl, 125; KCl, 2.5; $CaCl_2$, 2; $MgCl_2$, 1.3; $NaH_2PO_4$, 1.25; $NaHCO_3$, 26; D-glucose, 20 (Fischer Scientific, USA). C57/BL6NJ mice 14–21 day old were euthanized following a protocol approved by the IACUC of The University of Texas at San Antonio. The brain of the animals was quickly removed and was sectioned in 200 μm thick slices. After incubation for 35 minutes in 37 C the slices were transferred to a chamber and bathed in oxygenated aCSF throughout the experiments [23,24]. The extracellular solution circulated through the chamber at about 2 mL/min.

We fabricated recording electrodes from glass capillary tubes (1–3 MΩ). Electrodes were filled with 0.2 μL of 5 M NaCl. Recordings were obtained using an Alembic VE2 amplifier (Alembic Instruments; Montreal, Canada) together with Axon pclamp software (Molecular Devices; Sunnyvale, CA). The signal was filtered between 1–2 kHz to isolate action potential waveforms and sampled at 10 kHz. Action potentials were recorded and isolated using a combination of voltage thresholds. The data was then imported into Matlab (Natick, MA) to be further analyzed.

Nanoparticles were delivered by pressure injection through a second electrode connected to a micro syringe pump (Harvard Apparatus, Cambridge, MA). This second electrode was placed near the first electrode with a micromanipulator. Approximately 1 microL (μL) of gold nanostars was added during a 5 min window. In other experiments we applied 100 nm diameter latex nanobeads (TetraSpeck nanospheres, Invitrogen).

We also implemented a standard Hodgkin and Huxley model [7]. This model computes the membrane voltage of a neuron. The full model is as follows:

$$C\frac{dV}{dt} = -(\bar{g}_{Na}m^3h(V - E_{Na}) + \bar{g}_k n^4(V - E_k)$$
$$+ \bar{g}_{rest}(V - E_{rest})) + I$$

The passive parameters of the model are the membrane capacitance per unit area ($C = 1$ μF/cm$^2$); the leak resistance ($\bar{g}_{rest} = 0.3$ mS/cm$^2$); and the resting potential ($E_{rest} = 10.6$ mV). The active properties of the model consist of non-linear sodium ($Na = \bar{g}_{Na}m^3h(V - E_{Na})$) and potassium ($K_{dr} = \bar{g}_k n^4(V - E_k)$) channels. The density of the Na current is $\bar{g}_{Na} = 120$ mS/cm$^2$ and $\bar{g}_K = 36$ mS/cm$^2$ for the Kdr current. The reversal potential for potassium current is $E_K = -12$ mV and $E_{Na} = 115$ mV for the Na current. The activation of each current is given by

a set of mass action equations described by the state variables $m$ and $h$ for the Na current, and $n$ for the Kdr current. The value of $m$, $n$, and $h$ is determined by solving the following equation for each one of them ($x = m$, $n$, or $h$):

$$\frac{dx}{dt} = \alpha_x(1 - x) - \beta_x x$$

The reaction rate $\alpha_x$ is called the forward reaction, and $\beta_x$ is the backwards reaction. The forward-backward reaction rate for the activation variables are:

$$\alpha_m = \frac{2.5 - 0.1V}{e^{(2.5 - 0.1v)} - 1}$$
$$\beta_m = 4e^{-V/18}$$
$$\alpha_h = 0.07e^{-V/20}$$
$$\alpha_n = \frac{0.1 - 0.01V}{e^{(1 - 0.1V)} - 1}$$
$$\alpha_m = 0.125e^{-V/80}$$

where $V$ is the voltage from the main equation.

The input to the model was a constant current applied at $t = 50$ ms. The stimulation caused the model to depolarize and generate action potentials continuously. The model was implemented in Matlab and integrated using the Runge–Kutta algorithm.

## Additional file

Additional file 1: "Analysis of fluorescence correlation spectroscopy (FCS) measurements of star shaped nanoparticles". The figure contains two panels showing FCS measurements to determine the concentration of gold nanoparticles in solution and a comparison with theoretical calculations.

## Abbreviations
aCSF: Artificial cerebrospinal fluid; Kdr: Non-linear potassium channels.

## Competing interests
The authors declare that they have no competing interests.

## Authors' contributions
KS participated in the experimental design and carried out the experiments. ZK synthesized the nanoparticles and oversaw their use. FD performed the control latex beads experiments. FS and XGP conceived of the study. XGP participated in the coordination of the project and helped draft the manuscript. FS participated in the design of the project, performed the analysis and drafted the manuscript. All authors read and approved the final manuscript.

## Acknowledgements
This project was supported by the NIH/NIGMS MARC U*STAR GM07717 (KS, FD), the National Institute on Minority Health and Health Disparities RCMI G12MD007591 from the National Institutes of Health (FS and XGP) and the NSF PREM DMR 0934218 (FS, ZK and XGP).

## Author details
[1]UTSA Neurosciences Institute, The University of Texas at San Antonio, San Antonio, Texas 78249, USA. [2]Department of Physics and Astronomy, The University of Texas at San Antonio, San Antonio, Texas 78249, USA.

**References**

1. Chithrani BD, Ghazani AA, Chan WCW: **Determining the size and shape dependence of gold nanoparticle uptake into mammalian cells.** *Nano Lett* 2006, **6:**662–668.

2. Lin C-C, Yeh Y-C, Yang C-Y, Chen C-L, Chen G-F, Chen C-C, Wu Y-C: **Selective binding of mannose-encapsulated gold nanoparticles to type 1 pili in escherichia coli.** *J Am Chem Soc* 2002, **124:**3508–3509.

3. Connor EE, Mwamuka J, Gole A, Murphy CJ, Wyatt MD: **Gold nanoparticles are taken up by human cells but do not cause acute cytotoxicity.** *Small* 2005, **1:**325–327.

4. Alkilany A, Murphy C: **Toxicity and cellular uptake of gold nanoparticles: what we have learned so far?** *J Nanoparticle Res* 2010, **12:**2313–2333.

5. Yang Z, Liu ZW, Allaker RP, Reip P, Oxford J, Ahmad Z, Ren G: **A review of nanoparticle functionality and toxicity on the central nervous system.** *J R Soc Interface* 2010, **7:**S411–S422.

6. Kereselidze Z, Romero VH, Peralta XG, Santamaria F: **Gold nanostar synthesis with a silver seed mediated growth method.** *J Vis Exp* 2012.

7. Koch C: *Biophysics of computation : information processing in single neurons.* New York: Oxford University Press; 1999.

8. de Oliveira GM, Kist LW, Pereira TC, Bortolotto JW, Paquete FL, de Oliveira EM, Leite CE, Bonan CD, de Souza Basso NR, Papaleo RM, Bogo MR: **Transient modulation of acetylcholinesterase activity caused by exposure to dextran-coated iron oxide nanoparticles in brain of adult zebrafish.** *Comp Biochem Physiol C Toxicol Pharmacol* 2014, **162:**77–84.

9. Knudsen KB, Northeved H, Ek PK, Permin A, Andresen TL, Larsen S, Wegener KM, Lam HR, Lykkesfeldt J: **Differential toxicological response to positively and negatively charged nanoparticles in the rat brain.** *Nanotoxicology* 2014, **8:**764–774.

10. Sharma A, Muresanu DF, Patnaik R, Sharma HS: **Size- and age-dependent neurotoxicity of engineered metal nanoparticles in rats.** *Mol Neurobiol* 2013, **48:**386–396.

11. Jung S, Bang M, Kim BS, Lee S, Kotov NA, Kim B, Jeon D: **Intracellular gold nanoparticles increase neuronal excitability and aggravate seizure activity in the mouse brain.** *PLoS One* 2014, **9:**e91360.

12. Chithrani BD, Chan WCW: **Elucidating the mechanism of cellular uptake and removal of protein-coated gold nanoparticles of different sizes and shapes.** *Nano Lett* 2007, **7:**1542–1550.

13. Pan H, Qin M, Meng W, Cao Y, Wang W: **How do proteins unfold upon adsorption on nanoparticle surfaces?** *Langmuir* 2012, **28:**12779–12787.

14. Park KH, Chhowalla M, Iqbal Z, Sesti F: **Single-walled carbon nanotubes are a new class of ion channel blockers.** *J Biol Chem* 2003, **278:**50212–50216.

15. Verma A, Stellacci F: **Effect of surface properties on nanoparticle–cell interactions.** *Small* 2010, **6:**12–21.

16. Ribeiro FM, Paquet M, Cregan SP, Ferguson SSG: **Group I metabotropic glutamate receptor signalling and its implication in neurological disease.** *CNS Neurol Disord Drug Targets* 2010, **9:**574–595.

17. Yang F, Liu ZR, Chen J, Zhang SJ, Quan QY, Huang YG, Jiang W: **Roles of astrocytes and microglia in seizure-induced aberrant neurogenesis in the hippocampus of adult rats.** *J Neurosci Res* 2009.

18. Raghavachari S, Lisman JE, Tully M, Madsen JR, Bromfield EB, Kahana MJ: **Theta oscillations in human cortex during a working-memory task: evidence for local generators.** *J Neurophysiol* 2006, **95:**1630–1638.

19. Sriramoju B, Kanwar RK, Kanwar JR: **Nanomedicine based nanoparticles for neurological disorders.** *Curr Med Chem* 2014. Epub ahead of print.

20. Alivisatos AP, Andrews AM, Boyden ES, Chun M, Church GM, Deisseroth K, Donoghue JP, Fraser SE, Lippincott-Schwartz J, Looger LL, Masmanidis S, McEuen PL, Nurmikko AV, Park H, Peterka DS, Reid C, Roukes ML, Scherer A, Schnitzer M, Sejnowski TJ, Shepard KL, Tsao D, Turrigiano G, Weiss PS, Xu C, Yuste R, Zhuang X: **Nanotools for neuroscience and brain activity mapping.** *ACS Nano* 2013, **7:**1850–1866.

21. Farrer RA, Butterfield FL, Chen VW, Fourkas JT: **Highly efficient multiphoton-absorption-induced luminescence from gold nanoparticles.** *Nano Lett* 2005, **5:**1139–1142.

22. Wang H, Huff TB, Zweifel DA, He W, Low PS, Wei A, Cheng J-X: **In vitro and in vivo two-photon luminescence imaging of single gold nanorods.** *Proc Natl Acad Sci U S A* 2005, **102:**15752–15756.

23. Santamaria F, Wils S, De Schutter E, Augustine GJ: **The diffusional properties of dendrites depend on the density of dendritic spines.** *Eur J Neurosci* 2011, **34:**561–568.

24. Santamaria F, Wils S, De Schutter E, Augustine GJ: **Anomalous diffusion in Purkinje cell dendrites caused by spines.** *Neuron* 2006, **52:**635–648.

# Comparative lung toxicity of engineered nanomaterials utilizing *in vitro*, *ex vivo* and *in vivo* approaches

Yong Ho Kim[1], Elizabeth Boykin[2], Tina Stevens[3], Katelyn Lavrich[1] and M Ian Gilmour[2]*

## Abstract

**Background:** Although engineered nanomaterials (ENM) are currently regulated either in the context of a new chemical, or as a new use of an existing chemical, hazard assessment is still to a large extent reliant on information from historical toxicity studies of the parent compound, and may not take into account special properties related to the small size and high surface area of ENM. While it is important to properly screen and predict the potential toxicity of ENM, there is also concern that current toxicity tests will require even heavier use of experimental animals, and reliable alternatives should be developed and validated. Here we assessed the comparative respiratory toxicity of ENM in three different methods which employed *in vivo*, *in vitro* and *ex vivo* toxicity testing approaches.

**Methods:** Toxicity of five ENM (SiO$_2$ (10), CeO$_2$ (23), CeO$_2$ (88), TiO$_2$ (10), and TiO$_2$ (200); parentheses indicate average ENM diameter in nm) were tested in this study. CD-1 mice were exposed to the ENM by oropharyngeal aspiration at a dose of 100 μg. Mouse lung tissue slices and alveolar macrophages were also exposed to the ENM at concentrations of 22–132 and 3.1-100 μg/mL, respectively. Biomarkers of lung injury and inflammation were assessed at 4 and/or 24 hr post-exposure.

**Results:** Small-sized ENM (SiO$_2$ (10), CeO$_2$ (23), but not TiO$_2$ (10)) significantly elicited pro-inflammatory responses in mice (*in vivo*), suggesting that the observed toxicity in the lungs was dependent on size and chemical composition. Similarly, SiO$_2$ (10) and/or CeO$_2$ (23) were also more toxic in the lung tissue slices (*ex vivo*) and alveolar macrophages (*in vitro*) compared to other ENM. A similar pattern of inflammatory response (e.g., interleukin-6) was observed in both *ex vivo* and *in vitro* when a dose metric based on cell surface area (μg/cm$^2$), but not culture medium volume (μg/mL) was employed.

**Conclusion:** Exposure to ENM induced acute lung inflammatory effects in a size- and chemical composition-dependent manner. The cell culture and lung slice techniques provided similar profiles of effect and help bridge the gap in our understanding of *in vivo*, *ex vivo*, and *in vitro* toxicity outcomes.

**Keywords:** Engineered nanomaterials, Lung toxicity, Alternative toxicity testing

## Background

It is well recognized that nanotechnology has been rapidly growing and advancing over the past 10 years, and will continue to expand in numerous market sectors [1,2]. The advances in nanotechnology, however are accompanied by a need for better understanding of the exposure and toxicity of engineered nanomaterials (ENM) across their life-cycle. Moreover, the enormously diverse and applications of ENM (e.g., shapes, sizes, chemical and surface characteristics) are likely to result in a broad array of exposures and potentially adverse health outcomes. Thus, methods to evaluate and predict the toxicity of ENM are of considerable importance [3]. In particular, more information is needed on the interactions of ENM with lung tissue, since inhalation is a common exposure route and can also lead to potential systemic toxicity [1]. There is already substantial epidemiologic and toxicological evidence that inhaled ENM cause pulmonary effects (e.g., inflammation and/or edema) and/or extrapulmonary or systemic effects (e.g., thrombosis, dysrhythmias, and

* Correspondence: gilmour.ian@epa.gov
[2]Environmental Public Health Division, National Health and Environmental Effects Research Laboratory, United States Environmental Protection Agency, Research Triangle Park, NC, USA
Full list of author information is available at the end of the article

myocardial infarction) [4-7]. In general, nanotoxicology studies of the respiratory tract are performed with *in vivo* (e.g., mice and rats) or *in vitro* (e.g., airway/alveolar epithelial cells, macrophages, and dendritic cells) models. Because of the inherent anatomical complexity of the intact lung which is comprised of about 40 different cell types interpretation of toxicity of ENM in *in vitro* cell culture models is limited as they do not reflect the complex cell-cell contacts and cell-matrix interactions in the tissue. Moreover, despite the need for studying the toxicity of ENM *in vivo*, there is a growing concern that broad toxicity testing will increase the number of animals required. Therefore, developing credible alternative testing methods predictive of *in vivo* ENM toxicity are essential to screen potential hazards and health risks associated with inhalation exposures to these novel materials [2].

Here, we investigated pulmonary toxicity of five ENM: one silicon dioxide ($SiO_2$), two cerium oxide ($CeO_2$), and two titanium dioxide ($TiO_2$) nanomaterials with different primary diameters. $SiO_2$, $CeO_2$, and $TiO_2$ nanomaterials are already widely used in industrial processes and consumer products. $CeO_2$ and $TiO_2$ nanomaterials are the most abundantly produced metal oxide nanomaterials in the U.S. [8] and have been independently tested for adverse health effects *in vitro* and *in vivo*, but not in the same study design [9,10]. $CeO_2$ nanomaterials are of interest because despite having the same crystalline form as the parent compound, the nano-sized material causes more oxidative stress as a result of subtle changes in their surface chemistry [11,12]. $SiO_2$ nanomaterials (particularly the amorphous form), have also recently received attention in biomedical applications, yet their toxicity is not fully understood [13]. In the present study, we conducted acute toxicity tests in mice (*in vivo*), mouse lung tissue slices (*ex vivo*), and mouse alveolar macrophages (*in vitro*) to extrapolate, and compare the results between *ex vivo* or *in vitro* to *in vivo* toxicity testing approaches. Lung tissue slices have shown to preserve almost all cell types and interactions with the microenvironment (i.e., cell-cell or cell-matrix interactions), thus providing the most *in vivo*-like physiologically relevant response. Of all the different types of lung cells, alveolar macrophages are considered to be one of the first lines of a defense against inhaled particles and are primarily responsible for producing pro-inflammatory mediators [14]. The specific aims of this study were to determine the pulmonary toxicity and pro-inflammatory potential of ENM in mice, and compare these effects with the use of *ex vivo* lung slice and *in vitro* cell-based toxicity testing systems.

## Results

### Particle size distributions of ENM

Hydrodynamic diameters of ENM in the various solutions used in this study were determined by dynamic light scattering (Table 1). Diameters of all ENM suspended in water were greater than the specifications provided by the manufacturer, and were even larger when the materials were suspended in culture media. Of all the ENM studied, $TiO_2$ (10) and $SiO_2$ (10) were the most highly agglomerated. Since this clumping behavior controls the density of the ENM agglomerates in suspensions, we estimated the agglomerate density and presented the results in Table 1. $SiO_2$ (10) had the lowest agglomeration density in any solution, indicating that this material was most likely to remain suspended in the solutions and less likely to interact with the cells. Agglomerated $TiO_2$ (200), on the other hand had the highest density which would promote settling and a greater potential to come in contact with the cells on the plate bottom.

### Pulmonary inflammation responses *in vivo*

We monitored concentrations of lactate dehydrogenase (LDH) released into bronchoalveolar lavage fluid (BALF) at 4 hr and 24 hr post-exposure as a biomarker for lung cell injury. None of the ENM, except for $CeO_2$ (88) (at 24 hr post-exposure), significantly increased the concentrations of LDH at any time point compared with saline control groups (Figure 1A). N-acetyl-β-D-glucosaminidase (NAG) and γ-glutamyl transferase (GGT) as biomarkers for lysosomal enzyme and oxidative stress, respectively, were also assessed and were unchanged for any of the ENM (data not shown). Concentrations of albumin and total protein in BALF from the $CeO_2$ (23)-exposed groups were significantly increased at 4 hr and 24 hr post-exposure compared with saline-exposed groups, indicating that this material caused lung edema (Figure 1B and C). As a positive control, LPS increased LDH, albumin, and protein as expected, but did not affect NAG or GGT. The size- and composition-dependent toxicity of ENM was also seen in pulmonary inflammatory cells at 4 hr and 24 hr post-exposure (Figure 2). The $CeO_2$ (23)-exposure groups significantly increased the number of neutrophils (18% and 34% at 4 hrs and 24 hrs, respectively), compared with saline controls. While LPS-exposure groups induced an even stronger neutrophil influx, no other ENM caused significant changes in the neutrophil number. The number of macrophages in BALF was unchanged by any treatment.

Concentrations of pro-inflammatory cytokines (interleukin-6 (IL-6), macrophage inhibitory protein-2 (MIP-2), and tumor necrosis factor-α (TNF-α)) were then monitored in BALF at both time-points (Figure 3). $CeO_2$ (23) significantly increased the concentrations of all three cytokines at 4 hr post-exposure compared with saline control groups. $SiO_2$ (10) significantly increased the concentrations of IL-6 and MIP-2 at 4 hr post-exposure, while $TiO_2$ (10) but not the $TiO_2$ (200) only increased the concentration of MIP-2. These data indicate that the small-sized

**Table 1 Physicochemical properties of engineered nanomaterials (ENM)**

| Chemical | ID | Primary diameter[a] (nm) | Hydrodynamic diameter[b] (nm) | | | | Surface area[e] (m²/g) | Raw density[e] (g/cm³) | Equivalent primary diameter[f] (nm) | Agglomerate density[g] (g/cm³) | | Crystal form |
|---|---|---|---|---|---|---|---|---|---|---|---|---|
| | | | Water | Saline (in vivo) | CM[c] (ex vivo) | CM[c] (in vitro) | | | | CM[c] (ex vivo) | CM[c] (in vitro) | |
| SiO₂ | SiO₂ (10) | 5–15[a] | 401 ± 13 | 574 ± 96 | 458 ± 66 | 342 ± 44 | 590–690[a] | 2.65 | 3.54 | 1.06 ± 0.01 | 1.07 ± 0.01 | Amorphous |
| CeO₂ | CeO₂ (23) | 15–30[a] | 131 ± 55 | 269 ± 91 | 796 ± 46 | 432 ± 133 | 30–50[a] | 7.22 | 20.79 | 1.49 ± 0.03 | 1.88 ± 0.45 | Cerianite |
| CeO₂ | CeO₂ (88) | 70–105[a] | 162 ± 60 | 239 ± 52 | 500 ± 38 | 220 ± 31 | 8–12[a] | 7.22 | 83.16 | 2.78 ± 0.18 | 4.23 ± 0.58 | Cerianite |
| TiO₂ | TiO₂ (10) | 10[a] | 402 ± 16 | 739 ± 10 | 645 ± 3 | 660 ± 62 | 100–130[a] | 3.90 | 12.33 | 1.19 ± 0.00 | 1.19 ± 0.02 | Anatase |
| TiO₂ | TiO₂ (200) | 200[a] | 387 ± 12 | 690 ± 29 | 493 ± 6 | 417 ± 22 | 6.99[d] | 3.90 | 202.92 | 2.65 ± 0.03 | 2.86 ± 0.12 | Anatase |

a provided by the manufacturer.
b determined by dynamic light scattering and expressed as mean ± SEM of multiple measurements.
c CM: culture medium.
d obtained from Sanders et al. [15].
e obtained from the CRC Handbook of Chemistry and Physics [16].
f calculated from equivalent primary diameter $=6/(SSA \times \rho)$ [17,18], where $SSA$ is specific surface area, and $\rho$ is raw nanomaterial density.
g calculated from the Sterling equation [19], agglomerate density $= (1-(1-(d_H/d_{Eq})^{DF-3}))\rho + (1-(d_H/d_{Eq})^{DF-3}))\rho_{media}$ where $d_H$ is hydrodynamic diameter, $d_{Eq}$ is equivalent primary diameter, $DF$ is theoretical fractal dimension (assuming $DF =2.3$ [20]). $\rho$ is raw nanomaterial density, and $\rho_{media}$ is media density (assuming $\rho_{media} = 1$ g/cm³).

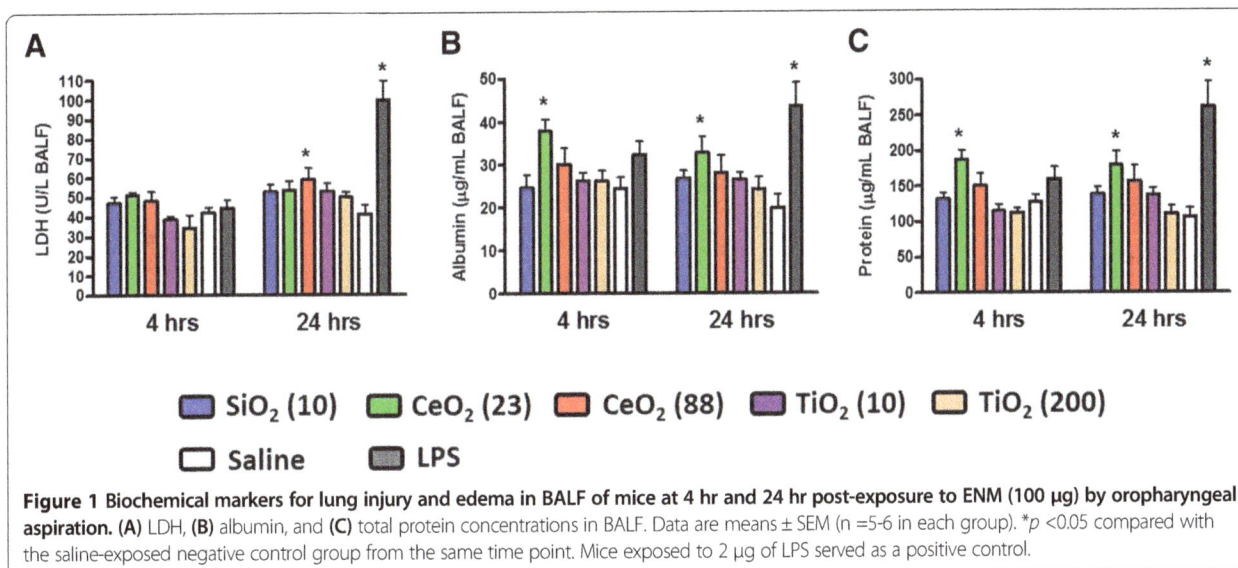

**Figure 1 Biochemical markers for lung injury and edema in BALF of mice at 4 hr and 24 hr post-exposure to ENM (100 μg) by oropharyngeal aspiration. (A)** LDH, **(B)** albumin, and **(C)** total protein concentrations in BALF. Data are means ± SEM (n =5-6 in each group). *$p$ <0.05 compared with the saline-exposed negative control group from the same time point. Mice exposed to 2 μg of LPS served as a positive control.

ENM induced more acute lung inflammation than their larger counterparts, and that chemical composition of ENM was a more important determinant than their size. Based on the cytokine response results, toxicity ranking of ENM approximated $CeO_2$ (23) ≈ $SiO_2$ (10) > $TiO_2$ (10) > $CeO_2$ (88) > $TiO_2$ (200). At 24 hr post-exposure, the cytokine concentrations decreased to saline control values except for $CeO_2$ (23) which maintained elevated levels of IL-6 and TNF-α. Interestingly, the inflammation was not related to uptake of ENM in lung macrophages. The less active $TiO_2$ (10) and $TiO_2$ (200) were avidly taken up by lung macrophages at both time points compared with

other ENM (Additional file 1: Figure S1). Finally, there were no significant changes in circulating white blood cells, red blood cells (RBCs) or RBC indices between the ENM-exposed mice and saline controls (data not shown).

**Pulmonary inflammation responses *ex vivo* and *in vitro***
LDH, GGT, and NAG concentrations in the supernatants from the lung tissue slices at 24 hr post-exposure were unchanged at any of the concentrations tested (data not shown). Only the $SiO_2$ (10) at the highest concentration (132 μg/mL) significantly increased the concentrations of IL-6 and MIP-2 compared with negative controls (Figure 4).

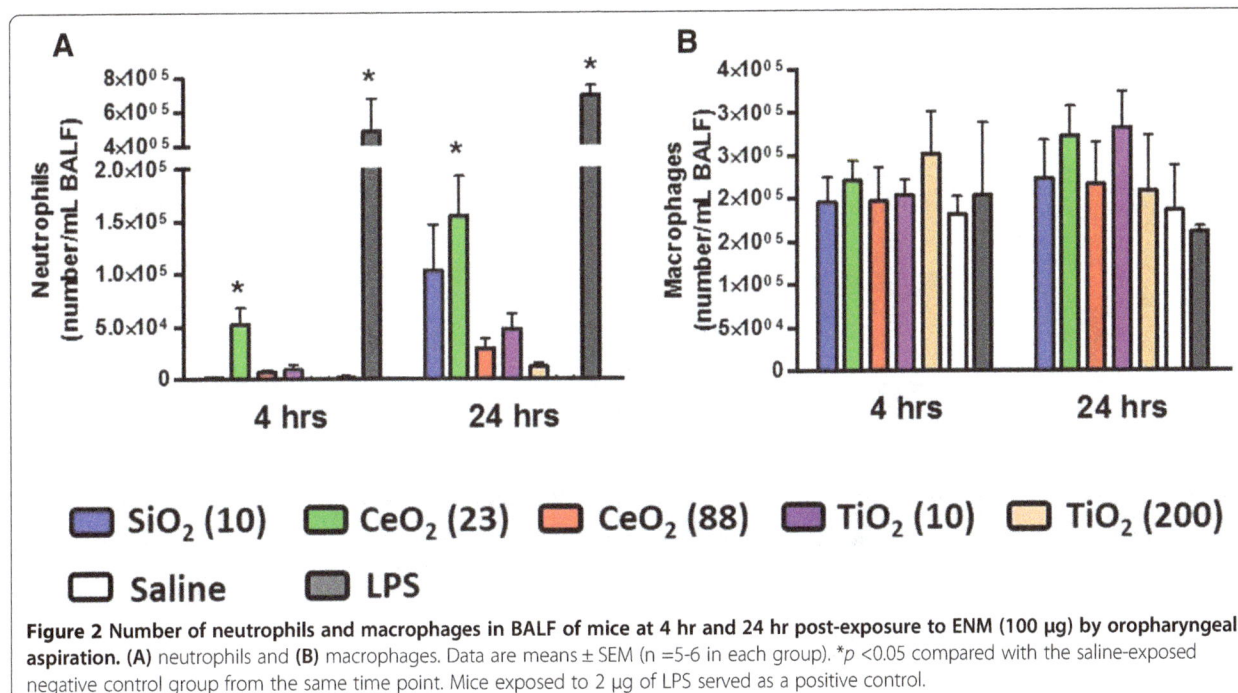

**Figure 2 Number of neutrophils and macrophages in BALF of mice at 4 hr and 24 hr post-exposure to ENM (100 μg) by oropharyngeal aspiration. (A)** neutrophils and **(B)** macrophages. Data are means ± SEM (n =5-6 in each group). *$p$ <0.05 compared with the saline-exposed negative control group from the same time point. Mice exposed to 2 μg of LPS served as a positive control.

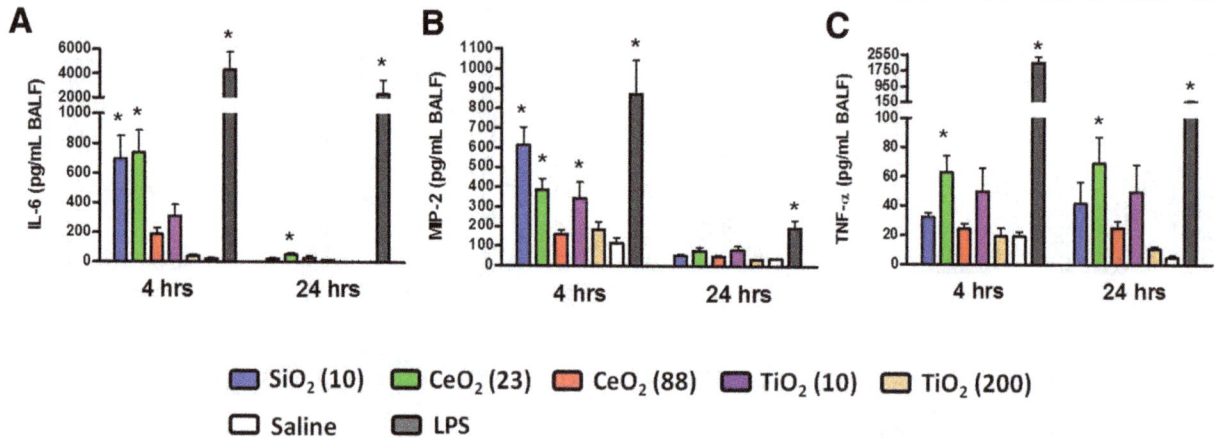

**Figure 3 Cytokine levels in BALF of mice at 4 hr and 24 hr post-exposure to ENM (100 μg) by oropharyngeal aspiration. (A)** IL-6, **(B)** MIP-2, and **(C)** TNF-α concentrations in BALF. Data are means ± SEM (n =5-6 in each group). *$p$ <0.05 compared with the saline-exposed negative control group from the same time point. Mice exposed to 2 μg of LPS served as a positive control.

$CeO_2$ (23) also had increased IL-6 concentration but this was not statistically significant.

Assessment of the cell culture supernatant from ENM-exposed MH-S cells at 24 hr post-exposure revealed that all ENM increased the LDH release in a dose-dependent manner (Figure 5A). $SiO_2$ (10) and $TiO_2$ (10 and 200) appeared to be more and less cytotoxic, respectively, however no apparent size-dependent effects (on cell membrane integrity) were observed. Half-maximal effective concentrations ($EC_{50}$) for the cell membrane integrity of $SiO_2$ (10), $CeO_2$ (23), $CeO_2$ (88), $TiO_2$ (10), and $TiO_2$ (200) were approximately 100, 295, 141, 330, and 384 μg/mL, respectively. Cell viability based on the metabolic

activity of mitochondria was assessed at 24 hr post-exposure (Figure 5B). Similar to the LDH analysis data, we also observed dose-dependent effects of ENM. $EC_{50}$ for the cell viability of $SiO_2$ (10), $CeO_2$ (23), $CeO_2$ (88), $TiO_2$ (10), and $TiO_2$ (200) were approximately 13, 18, 55, 30, and 77 μg/mL, respectively (Additional file 2: Figure S2). Thus, toxicity ranking of ENM based on the $EC_{50}$ for viability was in the order of $SiO_2$ (10) > $CeO_2$ (23) > $TiO_2$ (10) > $CeO_2$ (88) > $TiO_2$ (200). Because the $EC_{50}$ was much higher for LDH, this would indicate that the mitochondrial function was more sensitive to ENM exposure than cell membrane integrity. We also measured cell proliferation based on DNA content at 24 hr post-exposure and found

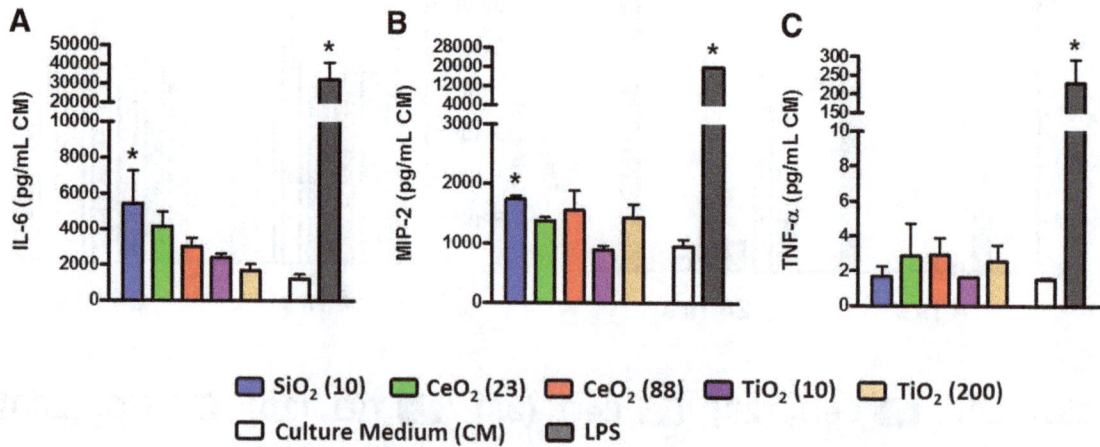

**Figure 4 Cytokine levels in lung tissue slices at 24 hr post-exposure to ENM (132 μg/mL). (A)** IL-6, **(B)** MIP-2, and **(C)** TNF-α concentrations in the culture medium (CM) from the lung tissue slices. Data are means ± SEM (n =3 in each group). *$p$ <0.05 compared with CM-exposed negative control group. Lung tissue slices exposed to 87 ng/mL of LPS served as a positive control.

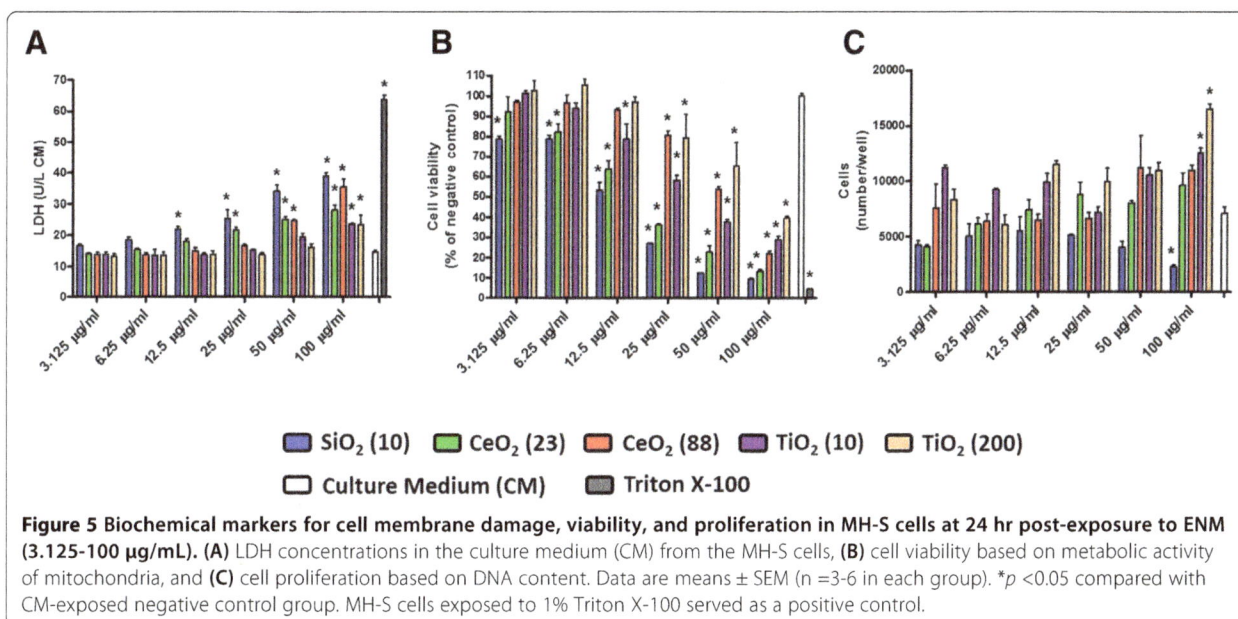

**Figure 5 Biochemical markers for cell membrane damage, viability, and proliferation in MH-S cells at 24 hr post-exposure to ENM (3.125-100 µg/mL). (A)** LDH concentrations in the culture medium (CM) from the MH-S cells, **(B)** cell viability based on metabolic activity of mitochondria, and **(C)** cell proliferation based on DNA content. Data are means ± SEM (n =3-6 in each group). *$p$ <0.05 compared with CM-exposed negative control group. MH-S cells exposed to 1% Triton X-100 served as a positive control.

that cell numbers did not significantly change in any of the ENM-exposed groups except at the high concentration exposure (Figure 5C). At 100 µg/mL concentration, $SiO_2$ (10) significantly decreased MH-S cell numbers, while $TiO_2$ (10) and $TiO_2$ (200) significantly increased the cell numbers. Concentrations of pro-inflammatory cytokine, IL-6, in MH-S cells were measured at 24 hr post-exposure (Figure 6). $SiO_2$ (10) induced more IL-6 production than other ENM which was in line with the IL-6 lung tissue slice response. To provide a more realistic comparison, we converted the nominal mass media concentration (i.e., µg/mL) to mass per unit cell (or tissue) surface area (i.e., µg/cm²) because lung tissue slices have

a larger 3D surface area than the MH-S cells. Taking this into account the exposure dose of 132 µg/mL to the lung slice resulted in a dose of 4.7 µg/cm². Therefore, the IL-6 responses in MH-S cells exposed to 12.5 µg/mL concentration (equivalent to 4.2 µg/cm²) was comparable to those in the lung tissue slices exposed to 132 µg/mL concentration (see the Materials and Methods section for a more detailed calculation).

## Discussion
While much work is being done to better understand the potential toxic effects of ENM on human health, it is still not clear which physico-chemical parameters of ENM are

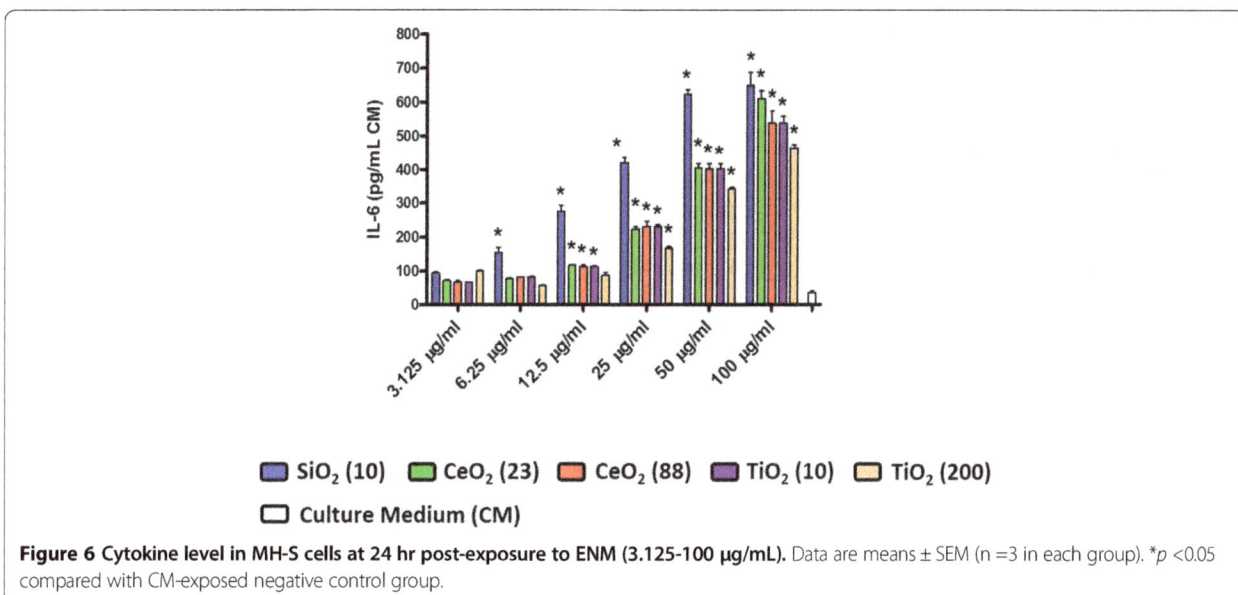

**Figure 6 Cytokine level in MH-S cells at 24 hr post-exposure to ENM (3.125-100 µg/mL).** Data are means ± SEM (n =3 in each group). *$p$ <0.05 compared with CM-exposed negative control group.

most important. Moreover, assessing (or screening) the toxic potential of emerging ENM is likely to increase the numbers of animals required, unless alternative methods are available that consistently reflect the *in vivo* biological effects. Here we utilized three different toxicity testing methods (mice, mouse lung tissue slices, and alveolar macrophages) to investigate the comparative toxicity of five ENM ($SiO_2$ (10), $CeO_2$ (23), $CeO_2$ (88), $TiO_2$ (10), and $TiO_2$ (200)) and determine if the latter two techniques could predict effects seen in animals. We found, in all three different toxicity testing methods, that $SiO_2$ (10) and/or $CeO_2$ (23) had the highest activity on the basis of pro-inflammatory cytokine production. Importantly the mouse lung tissue slices and alveolar macrophages exhibited similar cytokine responses to the distinct ENM when the exposure dose metric was based on cell surface area.

### Size- and chemical composition-dependent lung toxicity of ENM in mice

Numerous studies of nanotoxicology have shown that toxicity of ENM is strongly influenced by two factors: 1) chemical toxicity based on the chemical composition of ENM, and 2) cellular stress caused by the physical properties of ENM [9]. In line with published reports, it was evident that only the smaller-sized ENM caused significant inflammatory effects on mouse lungs, and that the chemical composition was important since stronger effects were noted in $SiO_2$ (10) and $CeO_2$ (23) but not $TiO_2$ (10). Interestingly $TiO_2$ lung macrophage uptake was higher than the other ENM, despite displaying lower toxicity suggesting that the observed inflammatory responses were not dependent on phagocytosis. In support of this, a similar study demonstrated that nanomaterial toxicity was not correlated with particle uptake in the cells [21].

Although further studies are needed to understand the mechanism underlying lung toxicity of ENM, the data also suggest that there was no clear relationship between lung toxicity and degree of ENM agglomeration (i.e., hydrodynamic diameters). Agglomerates of ENM form in biological fluids by loose binding (e.g., van der Waals force) while primary diameters, and not hydrodynamic diameters influence toxicity. In support of this, other researchers have reported that nanoparticle trafficking across lung epithelial cells was correlated with primary diameters and not the hydrodynamic diameters of the agglomerated nanoparticles [22].

Numerous studies have reported that ENM of various crystalline forms and solubility cause varying degrees of lung injury and inflammation. It is generally accepted that insoluble ENM are far less active in producing cellular damage or injury as compared to (partially) soluble ENM of similar size [23-25], although insoluble ENM have the potential to remain in the lungs and other

organs for a long. It also should be noted that while insoluble ENM may not be potent enough to cause cell damage, crystallinity of the ENM (e.g., amorphous or crystalline) might contribute to other toxicological properties [10,26]. In addition, insoluble ENM may cause oxidative stress and lung inflammation depending on their conduction band energy levels [27]. The ENM used in this study ($SiO_2$, $CeO_2$, and $TiO_2$) were insoluble (or poorly soluble) in biological fluids and considered not to release free ions from the nanomaterials to the tissue or cells. Here, the cytokine responses induced by $SiO_2$ (10) and $CeO_2$ (23) were evident in mice at 4 hr post-exposure but receded to control levels at 24 hrs, indicating that the inflammatory response was transient. Others have reported sustained pro-inflammatory cytokine levels at 24 hrs after exposure to $SiO_2$ (amorphous and 14 nm) albeit with 50 mg/kg which is ~15 times higher than the concentration used here [28]. While lung toxicity of $SiO_2$ nanomaterials has been extensively studied [26], there are only a few reports of lung toxicity of $CeO_2$ nanomaterials [9,29-32]. Moreover, these studies have mainly focused on long-term toxicity in mice or rats, demonstrating that intratracheal instillation or inhalation of $CeO_2$ nanomaterials led to severe chronic lung inflammation for up to 28 day post-exposure. Although our findings were limited to the 24 hr time-course, we cannot rule out the possibility of further chronic inflammatory responses, particularly in light of human case studies which report development of lung disease in workers after repeated long-term exposure of $CeO_2$ [33,34]. Similarly, our results showed that $TiO_2$ nanomaterials did not cause significant lung inflammation in mice, consistent with recently published $TiO_2$ toxicity findings performed through multiple interlaboratory comparisons [35].

### Comparing lung toxicity testing in mice to its alternatives

Efforts to reduce the number of animals in toxicity testing have resulted in the development of numerous *ex vivo* and *in vitro* toxicity test methods but the results are still conflicting. This inconsistency could be due to the fact that there are 1) a lack of overall consensus on the relevant dose metric for *in vivo* and *ex vivo/in vitro* studies and 2) inherent limitations to most *in vitro* models such as a lack of complex cell-cell interactions [36]. Here, the mouse lung tissue slices (*ex vivo*) and MH-S cells (*in vitro*) displayed a similar pattern of cytokine response on the basis of the mass per unit surface area of cell or tissue ($\mu g/cm^2$) but not per unit volume of culture medium ($\mu g/mL$), suggesting that cell surface area should be considered in *in vitro* dosimetry when comparing toxicity endpoints from different systems.

It is well documented that nanomaterials form agglomerates in suspension and their fate (or behavior) is governed by different mass transport properties (sedimentation and/

or diffusion), leading to differential exposures of nanomaterials to cells [17,18,20]. The nominal mass media concentration ($\mu$g/mL) in submerged cell-culture conditions assumes that the suspended nanomaterials are completely deposited on the cell surface which may not be always true for all nanomaterials in suspension and may result in misinterpretation of biological response data [37]. In the present study the density of agglomerated ENM in suspension (which influences delivered dose) was associated with the resultant cellular responses *ex vivo* and *in vitro* [17,18,20]. Notably, if the agglomerate density approached the density of the culture medium, the nanomaterials were more likely to remain suspended in the medium (i.e., low delivered dose), leading to a reduced exposure and diminution of biological responses to the nanomaterials [17,18]. In this regard, since the agglomerate densities of $SiO_2$ (10), $CeO_2$ (23) and $TiO_2$ (10) *ex vivo* and *in vitro* were closer to the culture media compared to other ENM, it is likely that the toxic effects were underestimated. In other words, the cytokine responses *ex vivo* and *in vitro* would be expected to increase even more if the cells were exposed to the same delivered dose. Therefore, considering the behavior of ENM agglomerates in submerged cell culture systems (*ex vivo* or *in vitro*) may reduce the disparity between *in vitro* and *in vivo* nanotoxicology outcomes. However, there are limitations to be considered when interpreting *in vitro* cellular responses based on agglomerate density. If ENM are soluble in culture media, their agglomerate density will change over time. Moreover, as described above, in the case of *in vitro* ENM toxicity tests, agglomerations may result in an underestimation of toxicity outcomes (or ranking), while in the case of *in vivo* ENM toxicity tests (via intratracheal instillation or oropharyngeal aspiration technique), agglomeration may cause an overestimation of toxicity outcomes (or ranking) [38]. It is also worth noting that the agglomeration associated with ability of ENM to absorb biological components (e.g., ions, salts, and proteins) in the *in vitro* and *in vivo* system may differently overshadow ENM properties (e.g., chemistry and surface charge), leading to the inconsistent results (*in vitro* versus *in vivo*) [39].

As aforementioned, one of the major challenges faced in cell-based *in vitro* models is that intact lungs are comprised of about 40 different cell types, and *in vitro* models cannot wholly reflect the microenvironment of cell-cell and cell-matrix interactions. Here we utilized the lung tissue slice model which preserves the lung architecture with nearly all cell types. We have previously reported that mouse lung tissue slices incubated with size fractionated particulate matter from a wildfire event displayed similar cytokine responses to those observed in mice [40]. In line with this finding, the lung tissue slice system also showed similar pro-inflammatory responses to ENM as those seen in mice (i.e., pro-inflammatory effects

of $SiO_2$ (10) and $CeO_2$ (23) but not $TiO_2$ (10)). Taken together, the results provide further evidence for particle-mediated biological responses in lung tissue slices and the feasibility of this application to lung toxicity testing. Although several studies have demonstrated toxicity of ENM in lung tissue slices [41,42], this is the first report to our knowledge to compare responses to different size and types of ENM in both mice and mouse lung tissue slices. In addition, the rank order of ENM IL-6 production from the MH-S cells was the same as that observed in both the *ex vivo* and *in vivo* comparisons suggesting that lung macrophages play an important role in this response. In contrast, the response ranking for TNF-$\alpha$ (which is expressed at lower levels in lung macrophages compared to IL-6 [43]) was not the same, suggesting that this biomarker would not be a good readout across the three systems. It should be noted that lung epithelial cells and macrophages differ in pro-inflammatory responses following exposure to ENM [44,45] and that toxicity differs depending on the cell of origin [36], as demonstrated by the observation that cancerous cells are more toxic than their normal precursors.

## Conclusions

We conclude that small-sized ENM, $SiO_2$ (10) and $CeO_2$ (23) but not $TiO_2$ (10), caused acute lung toxicity in mice (*in vivo*). $CeO_2$ (23) had the strongest effect on cytokine (IL-6, TNF-$\alpha$, and MIP-2) release, neutrophil recruitment, and increased protein into the mouse lungs, while the larger $CeO_2$ (88) and $TiO_2$ (200) were less potent, indicating that the effect was dependent on both size and chemical composition of ENM. The rank order of ENM toxicity from both lung tissue slices (*ex vivo*) and alveolar macrophages (*in vitro*) corresponded well to the ranking results from the mice (*in vivo*), suggesting that lung macrophages could replicate this effect. The similar profile of inflammatory response *ex vivo* and *in vitro* was most apparent when the exposure was based on mass per cell surface area. Although we demonstrated a relatively good correlation among the acute lung toxicity endpoints from three different testing methods, further studies are still needed that measure reversibility of effects or progression to long term toxicity. Nevertheless the results provide further evidence for the feasibility of replacing animal lung toxicity testing with cells or lung tissue slices, and provide information about the important parameters (e.g., agglomeration state and exposure dose metric) that will improve interpretation of ENM toxicity in biological systems.

## Materials and methods
### Experimental animals
Adult pathogen-free female CD-1 mice ($\sim$20-25 g and $\sim$30-45 g body weights for pulmonary toxicity and lung tissue

slice studies, respectively) purchased from Charles River Breeding Laboratories (Raleigh, NC). Mice were housed in groups of five in polycarbonate cages with hardwood chip bedding at the U.S. Environmental Protection Agency (EPA) Animal Care Facility accredited by the Association for Assessment and Accreditation of Laboratory Animal Care and were maintained on a 12-hour light to dark cycle at $22.3 \pm 1.1°C$ temperature and $50 \pm 10\%$ humidity. Mice were given access to rodent chow and water *ad libitum* and were acclimated for at least 10 days before the study began. The studies were conducted after approval by the EPA Institutional Animal Care and Welfare Committee.

### Engineered nanomaterials (ENM)

Five ENM were used in this study and designated by their mean primary diameter provided by the manufacturer: $SiO_2$ (10) (silicon dioxide with a primary diameter of 5–15 nm; amorphous; Sigma Aldrich (St. Louis, MO)), $CeO_2$ (23) (cerium oxide with a primary diameter of 15–30 nm; cerianite; NanoAmor (Houston, TX)), $CeO_2$ (88) (cerium oxide with a primary diameter of 70–105 nm; cerianite; Alfa Aesar (Ward Hill, MA)), $TiO_2$ (10) (titanium dioxide with a primary diameter of 10 nm; anatase; Alfa Aesar), and $TiO_2$ (200) (titanium dioxide with a primary diameter of 200 nm; anatase; Acros Organics (Fair Lawn, NJ)). The ENM were suspended in saline for *in vivo* and culture media (see below for further details) for *ex vivo* and *in vitro*, followed by sonication (Sonicator 4000; Misonix Sonicators, Newtown, CT) at 70–80 watts for 10 min and vortex mixing for 1 min to yield a stock solution at a concentration of 2 mg/mL. The ENM suspensions were stored at −80°C until toxicity testing. To explore the effect of solution chemistry on hydrodynamic diameters of ENM, dynamic light scattering (Zetasizer Nano ZS; Malvern Instruments, Malvern, UK) was used at 100 μg/mL ENM concentration in various solutions, such as distilled water, saline, and culture media. Further detailed physicochemical characteristics of ENM are presented in Table 1.

### *In vivo* toxicity of ENM
### Mouse exposure to ENM

Oropharyngeal aspiration was performed on mice anesthetized in a small plexiglass box using vaporized anesthetic isofluorane, following a technique described previously [46]. Briefly, the tongue of the mouse was extended with forceps and 100 μg of ENM in 50 μL saline was pipetted into the oropharynx. Immediately, the nose of the mouse was then covered causing the liquid to be aspirated into the lungs. Similarly, a separate group of mice was instilled with 2 μg of lipopolysaccharide (LPS; *Escherichia coli* endotoxin; 011:B4 containing $10^6$ unit/mg material; Sigma) as a positive control to demonstrate maximal responsiveness to this well characterized inflammatory agent while

additional mice were instilled with 50 μL saline alone as a negative control.

### Bronchoalveolar lavage and hematology

At 4 hr and 24 hr post-exposure, six mice from each treatment group were euthanized with 0.1 mL intraperitoneal injection of Euthasol (diluted 1:10 in saline; 390 mg pentobarbital sodium and 50 mg phenytoin/mL; Virbac AH, Inc., Fort Worth, TX), and blood was collected by cardiac puncture using a 1-mL syringe containing 17 μL sodium citrate to prevent coagulation. The trachea was then exposed, cannulated and secured with suture thread. The thorax was opened and the left mainstem bronchus was isolated and clamped with a microhemostat. The right lung lobes were lavaged three times with a single volume of warmed Hanks balanced salt solution (HBSS; 35 mL/kg mouse). The recovered bronchoalveolar lavage fluid (BALF) was centrifuged at $800xg$ for 10 min at 4°C and the supernatant was stored at both 4°C (for biochemical analysis) and −80°C (for cytokine analysis). The pelleted cells were resuspended in 1 mL HBSS (Sigma). Total BALF cell count of each mouse was obtained by a Coulter counter (Coulter Co., Miami, FL). Additionally, 200 μL resuspended cells were centrifuged in duplicate onto slides using a Cytospin (Shandon, Pittsburgh, PA) and subsequently stained with Diff-Quik solution (American Scientific Products, McGraw Park, PA) for enumeration of macrophages and neutrophils with at least 200 cells counted from each slide. Hematology values including total white blood cells, total red blood cells, hemoglobin, hematocrit, mean corpuscular volume, mean corpuscular hemoglobin concentration, and platelets were measured using a Coulter AcT 10 Hematology Analyzer (Beckman Coulter Inc., Miami, FL).

### Biochemical and cytokine analyses

Concentrations of lactate dehydrogenase (LDH) and γ-glutamyl transferase (GGT) were determined using commercially available kits (Thermo Scientific, Middletown, VA). Albumin and total protein concentrations were measured by the SPQ test system (DiaSorin, Stillwater, MN) and the Coomassie plus protein assay (Pierce Chemical, Rockford, IL) with a standard curve prepared with bovine serum albumin (Sigma), respectively. Activity of N-acetyl-β-D-glucoaminidase (NAG) was determined using a NAG assay kit (Roche Applied Science, Indianapolis, IN). All biochemical assays were modified for use on the KONELAB 30 clinical chemistry spectrophotometer analyzer (Thermo Clinical Lab Systems, Espoo, Finland) as described previously [46]. Concentrations of tumor necrosis factor-α (TNF-α), interleukin-6 (IL-6) and macrophage inhibitory protein-2 (MIP-2) in BALF were determined using commercial multiplexed fluorescent bead-based immunoassays (Milliplex Map Kit, Millpore

Co., Billerica, MA) measured by a Luminex 100 (Luminex Co., Austin, TX) following the manufacturer's protocol. The limits of detection (LOD) of each cytokine were 6.27, 3.28 and 29.14 pg/mL for TNF-α, IL-6 and MIP-2, respectively, and all values below these lowest values were replaced with a fixed value of one-half of the LOD value.

### Ex vivo toxicity of ENM
#### Mouse lung tissue slice preparation and incubation
Lung tissue slices were prepared as previously described [40]. Briefly, mice were euthanized with 0.1 mL intraperitoneal injection of Euthasol (diluted 1:10 in saline; Virbac AH, Inc.). The trachea was exposed and cannulated using a 20G luer stub adapter (Instech Solomon, Plymouth Meeting, PA). The lungs were filled with 1.5% (w/v) low-melting agarose (Sigma) in minimum essential medium (MEM; Simga) at 37°C. The lungs were rinsed with the ice-cold slicing buffer solution (Earle's balanced salt solution (Sigma) supplemented with 15 mM N-(2-hydroxyethyl)piperazine-N'-(2-ethanesulfonic acid) hemisodium salt (HEPES; Sigma)) and removed from the mouse. The lungs were transferred into the ice-cold slicing buffer solution to further solidify the agarose and then the lung lobes were separated using a surgical blade, and the lung tissue cores (8 mm diameter) were prepared using a tissue coring tool (Alabama Research and Development, Munford, AL). Tissue cores were cut into 350 μm thick slices in the ice-cold slicing buffer solution using a specialized vibratome (OTS 5000, FHC Inc., Bowdoinham, ME). The lung tissue slices were then incubated in the wash buffer solution (Dulbecco's modified eagle's medium/nutrient mixture F-12 Ham (Sigma) supplemented with 100 units/mL penicillin (Sigma) and 100 μg/mL streptomycin (Sigma)) under cell culture conditions for 4 hrs. The lung tissue slices were then transferred into a tissue culture treated polystyrene 48-well plate (Corning Inc., Corning, NY) and cultured in the slice incubation medium (the wash buffer solution supplemented with 200 mM L-glutamine (Sigma), 0.1 mM MEM non-essential amino acids (Sigma) and 15 mM HEPES) for up to 6 days at 37°C in a humidified atmosphere of 5% $CO_2$ and 95% air. The lung tissue slices received fresh media every day.

#### Mouse lung tissue slice exposure to ENM
Reconstituted ENM suspensions were sonicated for 2 min, vortexed for 1 min and further diluted with the slice incubation medium to achieve final concentrations of 22, 44, 66, and 132 μg/mL. On day 2 of culture, lung tissue slices were exposed to the ENM for 24 hrs. The initial concentration of 22 μg/mL (total volume of 0.5 mL, therefore of 11 μg of ENM per lung slice) was estimated to be five times higher than the in vivo exposure dose used in this study. If it is assumed that the lung surface area of a 20 g mouse is ~650 cm², 1 cm³ mouse lung tissue has ~800 cm² lung

surface area, and 100% of oropharyngeal instilled ENM is delivered to the lungs, 100 μg of ENM dose in a mouse (~650 cm² lung surface area) is equivalent to 2.2 μg of ENM dose in a mouse lung slice (~14 cm² lung slice surface area) [47]. Moreover, if it is assumed that the lung slice surface area is ~14 cm², the exposure doses of 22, 44, 66, and 132 μg/mL are equivalent to the doses of 0.79, 1.6, 2.3, and 4.7 μg/cm², respectively. Mouse lung tissue slices were exposed to 87 ng/mL LPS which was an equivalent concentration in vivo and served as a positive control. Mouse lung tissue slices exposed to the culture medium alone served as a negative control. At 24 hr post-exposure, lung slice culture fluids were collected, centrifuged at 10,000x$g$ for 5 min, and culture supernatants were stored at both 4°C (for extracellular biochemical analysis) and –80°C (for cytokine analysis). Subsequently, mouse lung tissue slices were homogenized using a tissue homogenizer in a lysis buffer solution containing 0.5% Triton X-100, 150 mM NaCl, 15 mM Tris–HCl (pH 7.4), 1 mM $CaCl_2$ and 1 mM $MgCl_2$ [48]. Homogenates were then centrifuged at 10,000x$g$ for 10 min and supernatants were stored at –80°C (for intracellular biochemical analysis).

#### Biochemical and cytokine analyses
Similar to the in vivo lung inflammation analyses described above, the supernatants of tissue culture fluids and tissue homogenates after exposure to ENM were used to determine the extracellular (LDH and NAG) and intracellular (GGT) biochemical analyses as well as cytokine analysis (IL-6, MIP-2, and TNF-α). Biochemical and pro-inflammatory cytokine analyses were performed using a KONELAB 30 clinical chemistry spectrophotometer analyzer (Thermo Clinical Lab Systems) and multiplexed fluorescent bead-based immunoassays (Milliplex Map Kit) measured by the Luminex 100 (Luminex Co).

### In vitro toxicity of ENM
#### Alveolar macrophage cell culture
The murine alveolar macrophages (MH-S) cells were purchased from ATCC (CRL2019, Manassas, VA) and grown in the following culture medium: RPMI 1640 (Sigma) supplemented with 5% fetal bovine albumin (FBS; Sigma) and 100 units/mL penicillin (Sigma) and 100 μg/mL streptomycin (Sigma) at 37°C in a humidified atmosphere of 5% $CO_2$ and 95% air. MH-S cells at passage 11 yielded 2.4 - 2.9 × $10^6$ cells/mL and were seeded at 3,000 cells per well of a 96-well culture plate.

#### Alveolar macrophage cell exposure to ENM
After 3 days in culture, MH-S cells were exposed to ENM at final concentrations of 3.125, 6.25, 12.5, 25, 50, and 100 μg/mL in the culture medium for 24 hrs. This exposure dose can be converted to the dose based on

cell surface area (assuming MH-S cell surface area is 0.3 cm$^2$). Thus, the exposure doses of 3.125, 6.25, 12.5, 25, 50, and 100 μg/mL are equivalent to the doses of 1.0, 2.1, 4.2, 8.3, 16.7, and 33.3 μg/cm$^2$, respectively. MH-S cells exposed to the culture medium alone served as a negative control and 1% Triton X-100 at 37°C served as a positive control.

### Biochemical and cytokine analyses
After the cells exposed to ENM, the plate was centrifuged at 400xg for 5 min, followed by collection of supernatants to analyze LDH concentrations. The supernatants were also used to determine cytokine production (IL-6). The MH-S cells after centrifugation were then used to evaluate cell proliferation (CyQuant assay; Invitrogen, Eugene, OR). Viability of the MH-S cells exposed to ENM was tested by measuring enzymatic activity based on the cellular cleavage of water-soluble tetrazolium salt (WST-1) to formazan in the cells using a WST-1 assay kit (Roche Applied Science). Biochemical and pro-inflammatory cytokine analyses in this study were also performed using a KONELAB 30 clinical chemistry spectrophotometer analyzer (Thermo Clinical Lab Systems) and multiplexed fluorescent bead-based immunoassays (Milliplex Map Kit) measured by the Luminex 100 (Luminex Co).

### Statistical analysis
Data were expressed as means ± the standard error of the mean (SEM). The results of the ENM-exposed groups were compared to those of the negative control group. Statistical comparison was performed by one-way analysis of variance (ANOVA) with the Newman-Keuls post-hoc test. Statistical analyses were performed using commercial software (GraphPad Prism 6.04, GraphPad Software, Inc., San Diego, CA). If the data did not meet the ANOVA assumptions of either normality or equal variances (Levene's test; $p > 0.05$), the data were transformed. Subsequent to the transformation, the data were checked for requirement compliance and if acceptable, ANOVA proceeded. The statistical significance level was assigned at a probability value of $p < 0.05$.

### Additional files

Additional file 1: Figure S1. Representative BAL cell images at 4 hr and 24 hr post-exposure to ENM. Red arrows indicate ENP uptakes in alveolar macrophages. Original magnification (20x), inset (40x).

Additional file 2: Figure S2. Dose–response curves to determine EC$_{50}$ values for the MH-S cells from WST-1 assay data.

### Abbreviations
ANOVA: Analysis of variance; BALF: Bronchoalveolar lavage fluid; CeO$_2$ (23): Cerium oxide with a primary diameter of 15–30 nm; CeO$_2$ (88): Cerium oxide with a primary diameter of 70–105 nm; CM: Culture medium; EC$_{50}$: Half-maximal effective concentrations; ENM: Engineered nanomaterials; GGT: γ-glutamyl transferase; HBSS: Hanks balanced salt solution; HEPES: N-(2-hydroxyethyl)piperazine-N'-(2-ethanesulfonic acid) hemisodium salt; IL-6: Interleukin-6; LDH: Lactate dehydrogenase; LPS: Lipopolysaccharide; MEM: Minimum essential medium; MIP-2: Macrophage inhibitory protein-2; NAG: N-acetyl-β-D-glucoaminidase; RBC: Red blood cells; SEM: Standard error of the mean; SiO$_2$ (10): Silicon dioxide with a primary diameter of 5–15 nm; TiO$_2$ (10): Titanium dioxide with a primary diameter of 10 nm; TiO$_2$ (200): Titanium dioxide with a primary diameter of 200 nm; TNF-α: Tumor necrosis factor-α; WST-1: Water-soluble tetrazolium salt.

### Competing interests
The authors declare that they have no competing interests.

### Authors' contributions
YHK contributed to the experimental design, carried out the pulmonary assessment and lung tissue slice experiment, performed the data analyses and figure generations, and drafted the manuscript. EB performed the pulmonary assessment and helped with data analyses. TS performed the MH-S cell experiment. KL assisted with the experimental design and pulmonary assessment. MIG conceived and designed the experiment, evaluated the results, and co-wrote the manuscript. All of the authors read and approved the final manuscript.

### Acknowledgements
We thank Debora Andrews, Judy Richards, and Richard Jaskot for technical assistance in toxicologic analyses and Drs. Kevin Dreher and James Samet for their review of this manuscript. The project was supported by grant from the EPA-UNC Toxicology Training Agreement (CR-83515201-0), with the Curriculum in Toxicology, University of North Carolina at Chapel Hill. This paper has been reviewed by the National Health and Environmental Effects Research Laboratory, U.S. Environmental Protection Agency, and approved for publication. Approval does not signify that contents necessarily reflect the views and polices of Agency, nor does the mention of trade names or commercial products constitute endorsement or recommendation for use.

### Author details
$^1$Curriculum in Toxicology, University of North Carolina at Chapel Hill, Chapel Hill, NC, USA. $^2$Environmental Public Health Division, National Health and Environmental Effects Research Laboratory, United States Environmental Protection Agency, Research Triangle Park, NC, USA. $^3$Research Triangle Park Division, National Center for Environmental Assessment, United States Environmental Protection Agency, Research Triangle Park, NC, USA.

### References
1. Jud C, Clift MJ, Petri-Fink A, Rothen-Rutishauser B: Nanomaterials and the human lung: what is known and what must be deciphered to realise their potential advantages? Swiss Med Wkly 2013, 143:w13758.
2. Maynard AD, Aitken RJ, Butz T, Colvin V, Donaldson K, Oberdorster G, Philbert MA, Ryan J, Seaton A, Stone V, Tinkle SS, Tran L, Walker NJ, Warheit DB: Safe handling of nanotechnology. Nature 2006, 444(7117):267–269.
3. Hubbs AF, Mercer RR, Benkovic SA, Harkema J, Sriram K, Schwegler-Berry D, Goravanahally MP, Nurkiewicz TR, Castranova V, Sargent LM: Nanotoxicology–a pathologist's perspective. Toxicol Pathol 2011, 39(2):301–324.
4. Kreyling W, Semmler-Behnke M, Möller W: Health implications of nanoparticles. J Nanoparticle Res 2006, 8(5):543–562.
5. Madl AK, Pinkerton KE: Health effects of inhaled engineered and incidental nanoparticles. Crit Rev Toxicol 2009, 39(8):629–658.
6. Oberdorster G, Oberdorster E, Oberdorster J: Nanotoxicology: an emerging discipline evolving from studies of ultrafine particles. Environ Health Perspect 2005, 113(7):823–839.
7. Phillips JI, Green FY, Davies JC, Murray J: Pulmonary and systemic toxicity following exposure to nickel nanoparticles. Am J Ind Med 2010, 53(8):763–767.
8. Hendren CO, Mesnard X, Dröge J, Wiesner MR: Estimating production data for five engineered nanomaterials as a basis for exposure assessment. Environ Sci Technol 2011, 45(7):2562–2569.
9. Cassee FR, van Balen EC, Singh C, Green D, Muijser H, Weinstein J, Dreher K: Exposure, health and ecological effects review of engineered nanoscale

cerium and cerium oxide associated with its use as a fuel additive. *Crit Rev Toxicol* 2011, **41**(3):213–229.

10. Shi H, Magaye R, Castranova V, Zhao J: Titanium dioxide nanoparticles: a review of current toxicological data. *Part Fibre Toxicol* 2013, **10**:15.

11. Chen J, Patil S, Seal S, McGinnis JF: Rare earth nanoparticles prevent retinal degeneration induced by intracellular peroxides. *Nat Nanotechnol* 2006, **1**(2):142–150.

12. Korsvik C, Patil S, Seal S, Self WT: Superoxide dismutase mimetic properties exhibited by vacancy engineered ceria nanoparticles. *Chem Commun* 2007, **10**:1056–1058.

13. Tang L, Cheng J: Nonporous silica nanoparticles for nanomedicine application. *Nano Today* 2013, **8**(3):290–312.

14. Hiraiwa K, van Eeden SF: Contribution of lung macrophages to the inflammatory responses induced by exposure to air pollutants. *Mediators Inflamm* 2013, **2013**:619523.

15. Sanders K, Degn LL, Mundy WR, Zucker RM, Dreher K, Zhao B, Roberts JE, Boyes WK: In vitro phototoxicity and hazard identification of nano-scale titanium dioxide. *Toxicol Appl Pharmacol* 2012, **258**(2):226–236.

16. Haynes WM: *CRC Handbook of Chemistry and Physics.* 94th edition. Boca Raton, FL: CRC Press/Taylor & Francis Group; 2013.

17. DeLoid G, Cohen JM, Darrah T, Derk R, Rojanasakul L, Pyrgiotakis G, Wohlleben W, Demokritou P: Estimating the effective density of engineered nanomaterials for in vitro dosimetry. *Nat Commun* 2014, **5**:3514.

18. Cohen JM, Teeguarden JG, Demokritou P: An integrated approach for the in vitro dosimetry of engineered nanomaterials. *Part Fibre Toxicol* 2014, **11**:20.

19. Sterling MC Jr, Bonner JS, Ernest AN, Page CA, Autenrieth RL: Application of fractal flocculation and vertical transport model to aquatic sol-sediment systems. *Water Res* 2005, **39**(9):1818–1830.

20. Hinderliter PM, Minard KR, Orr G, Chrisler WB, Thrall BD, Pounds JG, Teeguarden JG: ISDD: a computational model of particle sedimentation, diffusion and target cell dosimetry for in vitro toxicity studies. *Part Fibre Toxicol* 2010, **7**(1):36.

21. Diaz B, Sanchez-Espinel C, Arruebo M, Faro J, de Miguel E, Magadan S, Yague C, Fernandez-Pacheco R, Ibarra MR, Santamaria J, Gonzalez-Fernandez A: Assessing methods for blood cell cytotoxic responses to inorganic nanoparticles and nanoparticle aggregates. *Small* 2008, **4**(11):2025–2034.

22. Fazlollahi F, Kim YH, Sipos A, Hamm-Alvarez SF, Borok Z, Kim KJ, Crandall ED: Nanoparticle translocation across mouse alveolar epithelial cell monolayers: species-specific mechanisms. *Nanomedicine* 2013, **9**(6):786–794.

23. Brunner TJ, Wick P, Manser P, Spohn P, Grass RN, Limbach LK, Bruinink A, Stark WJ: In vitro cytotoxicity of oxide nanoparticles: comparison to asbestos, silica, and the effect of particle solubility. *Environ Sci Technol* 2006, **40**(14):4374–4381.

24. Kim YH, Fazlollahi F, Kennedy IM, Yacobi NR, Hamm-Alvarez SF, Borok Z, Kim KJ, Crandall ED: Alveolar epithelial cell injury due to zinc oxide nanoparticle exposure. *Am J Respir Crit Care Med* 2010, **182**(11):1398–1409.

25. Xia T, Kovochich M, Liong M, Madler L, Gilbert B, Shi H, Yeh JI, Zink JI, Nel AE: Comparison of the mechanism of toxicity of zinc oxide and cerium oxide nanoparticles based on dissolution and oxidative stress properties. *ACS Nano* 2008, **2**(10):2121–2134.

26. Napierska D, Thomassen LC, Lison D, Martens JA, Hoet PH: The nanosilica hazard: another variable entity. *Part Fibre Toxicol* 2010, **7**(1):39.

27. Zhang H, Ji Z, Xia T, Meng H, Low-Kam C, Liu R, Pokhrel S, Lin S, Wang X, Liao YP, Wang M, Li L, Rallo R, Damoiseaux R, Telesca D, Madler L, Cohen Y, Zink JI, Nel AE: Use of metal oxide nanoparticle band gap to develop a predictive paradigm for oxidative stress and acute pulmonary inflammation. *ACS Nano* 2012, **6**(5):4349–4368.

28. Cho WS, Choi M, Han BS, Cho M, Oh J, Park K, Kim SJ, Kim SH, Jeong J: Inflammatory mediators induced by intratracheal instillation of ultrafine amorphous silica particles. *Toxicol Lett* 2007, **175**(1–3):24–33.

29. Aalapati S, Ganapathy S, Manapuram S, Anumolu G, Prakya BM: Toxicity and bio-accumulation of inhaled cerium oxide nanoparticles in CD1 mice. *Nanotoxicology* 2014, **8**(7):786–798.

30. Ma JY, Zhao H, Mercer RR, Barger M, Rao M, Meighan T, Schwegler-Berry D, Castranova V, Ma JK: Cerium oxide nanoparticle-induced pulmonary inflammation and alveolar macrophage functional change in rats. *Nanotoxicology* 2011, **5**(3):312–325.

31. Srinivas A, Rao PJ, Selvam G, Murthy PB, Reddy PN: Acute inhalation toxicity of cerium oxide nanoparticles in rats. *Toxicol Lett* 2011, **205**(2):105–115.

32. Gosens I, Mathijssen LE, Bokkers BG, Muijser H, Cassee FR: Comparative hazard identification of nano- and micro-sized cerium oxide particles based on 28-day inhalation studies in rats. *Nanotoxicology* 2014, **8**(6):643–653.

33. McDonald JW, Ghio AJ, Sheehan CE, Bernhardt PF, Roggli VL: Rare earth (cerium oxide) pneumoconiosis: analytical scanning electron microscopy and literature review. *Mod Pathol* 1995, **8**(8):859–865.

34. Vocaturo G, Colombo F, Zanoni M, Rodi F, Sabbioni E, Pietra R: Human exposure to heavy metals. Rare earth pneumoconiosis in occupational workers. *Chest* 1983, **83**(5):780–783.

35. Bonner JC, Silva RM, Taylor AJ, Brown JM, Hilderbrand SC, Castranova V, Porter D, Elder A, Oberdorster G, Harkema JR, Bramble LA, Kavanagh TJ, Botta D, Nel A, Pinkerton KE: Interlaboratory evaluation of rodent pulmonary responses to engineered nanomaterials: the NIEHS Nano GO Consortium. *Environ Health Perspect* 2013, **121**(6):676–682.

36. Joris F, Manshian BB, Peynshaert K, De Smedt SC, Braeckmans K, Soenen SJ: Assessing nanoparticle toxicity in cell-based assays: influence of cell culture parameters and optimized models for bridging the in vitro-in vivo gap. *Chem Soc Rev* 2013, **42**(21):8339–8359.

37. Teeguarden JG, Hinderliter PM, Orr G, Thrall BD, Pounds JG: Particokinetics in vitro: dosimetry considerations for in vitro nanoparticle toxicity assessments. *Toxicol Sci* 2007, **95**(2):300–312.

38. Baisch BL, Corson NM, Wade-Mercer P, Gelein R, Kennell AJ, Oberdorster G, Elder A: Equivalent titanium dioxide nanoparticle deposition by intratracheal instillation and whole body inhalation: the effect of dose rate on acute respiratory tract inflammation. *Part Fibre Toxicol* 2014, **11**:5.

39. Rivera-Gil P, Jimenez de Aberasturi D, Wulf V, Pelaz B, del Pino P, Zhao Y, de la Fuente JM, Ruiz de Larramendi I, Rojo T, Liang XJ, Parak WJ: The challenge to relate the physicochemical properties of colloidal nanoparticles to their cytotoxicity. *Acc Chem Res* 2013, **46**(3):743–749.

40. Kim YH, Tong H, Daniels M, Boykin E, Krantz QT, McGee J, Hays M, Kovalcik K, Dye JA, Gilmour MI: Cardiopulmonary toxicity of peat wildfire partcualte matter and the predictive utility of precision cut lung slices. *Part Fibre Toxicol* 2014, **11**:29.

41. Neuhaus V, Schwarz K, Klee A, Seehase S, Forster C, Pfennig O, Jonigk D, Fieguth HG, Koch W, Warnecke G, Yusibov V, Sewald K, Braun A: Functional testing of an inhalable nanoparticle based influenza vaccine using a human precision cut lung slice technique. *PLoS One* 2013, **8**(8):e71728.

42. Sauer UG, Vogel S, Aumann A, Hess A, Kolle SN, Ma-Hock L, Wohlleben W, Dammann M, Strauss V, Treumann S, Groters S, Wiench K, van Ravenzwaay B, Landsiedel R: Applicability of rat precision-cut lung slices in evaluating nanomaterial cytotoxicity, apoptosis, oxidative stress, and inflammation. *Toxicol Appl Pharmacol* 2014, **276**(1):1–20.

43. Losa Garcia JE, Rodriguez FM, de Cabo MR M, Garcia Salgado MJ, Losada JP, Villaron LG, Lopez AJ, Arellano JL: Evaluation of inflammatory cytokine secretion by human alveolar macrophages. *Mediators Inflamm* 1999, **8**(1):43–51.

44. Sayes C, Reed K, Subramoney S, Abrams L, Warheit D: Can in vitro assays substitute for in vivo studies in assessing the pulmonary hazards of fine and nanoscale materials? *J Nanoparticle Res* 2009, **11**(2):421–431.

45. Sayes CM, Reed KL, Warheit DB: Assessing toxicity of fine and nanoparticles: comparing in vitro measurements to in vivo pulmonary toxicity profiles. *Toxicol Sci* 2007, **97**(1):163–180.

46. Gilmour MI, McGee J, Duvall RM, Dailey L, Daniels M, Boykin E, Cho SH, Doerfler D, Gordon T, Devlin RB: Comparative toxicity of size-fractionated airborne particulate matter obtained from different cities in the United States. *Inhal Toxicol* 2007, **19**(Suppl 1):7–16.

47. Schmidt-Nielsen K: *Animal Physiology: Adaptation and Environment.* 5th edition. Cambridge, UK: Cambridge University Press; 1997.

48. Whitehead GS, Grasman KA, Kimmel EC: Lung function and airway inflammation in rats following exposure to combustion products of carbon-graphite/epoxy composite material: comparison to a rodent model of acute lung injury. *Toxicology* 2003, **183**(1–3):175–197.

# Enhanced green fluorescent protein-mediated synthesis of biocompatible graphene

Sangiliyandi Gurunathan[1,2*], Jae Woong Han[1], Eunsu Kim[1], Deug-Nam Kwon[1], Jin-Ki Park[3] and Jin-Hoi Kim[1*]

## Abstract

**Background:** Graphene is the 2D form of carbon that exists as a single layer of atoms arranged in a honeycomb lattice and has attracted great interest in the last decade in view of its physical, chemical, electrical, elastic, thermal, and biocompatible properties. The objective of this study was to synthesize an environmentally friendly and simple methodology for the preparation of graphene using a recombinant enhanced green fluorescent protein (EGFP).

**Results:** The successful reduction of GO to graphene was confirmed using UV–vis spectroscopy, and FT-IR. DLS and SEM were employed to demonstrate the particle size and surface morphology of GO and EGFP-rGO. The results from Raman spectroscopy suggest the removal of oxygen-containing functional groups from the surface of GO and formation of graphene with defects. The biocompatibility analysis of GO and EGFP-rGO in human embryonic kidney (HEK) 293 cells suggests that GO induces significant concentration-dependent cell toxicity in HEK cells, whereas graphene exerts no adverse effects on HEK cells even at a higher concentration (100 μg/mL).

**Conclusions:** Altogether, our findings suggest that recombinant EGFP can be used as a reducing and stabilizing agent for the preparation of biocompatible graphene. The novelty and originality of this work is that it describes a safe, simple, and environmentally friendly method for the production of graphene using recombinant enhanced green fluorescent protein. Furthermore, the synthesized graphene shows excellent biocompatibility with HEK cells; therefore, biologically synthesized graphene can be used for biomedical applications. To the best of our knowledge, this is the first and novel report describing the synthesis of graphene using recombinant EGFP.

**Keywords:** Enhanced green fluorescent protein, Graphene oxide, Graphene, Human embryonic kidney 293 cells, Cell viability, Membrane leakage, Oxidative stress

## Background

Graphene has a two-dimensional (2-D) nanostructure with a single layer of carbon atoms and has attracted much interest in recent years because of its unique mechanical, thermal, catalytic, electronic, optical, and biological properties [1-4]. Graphene and graphene-based materials have been widely used in several applications including bio-sensing [5], antibacterial compositions [6-8], drug delivery [9], tissue scaffolds [10], catalysis [11], and energy storage [12]. The production of graphene in large quantities using an environmentally friendly approach is essential but also a significant challenge [13].

Several methods have been established for the synthesis of graphene and its derivatives, including exfoliation of graphite (Gt) [14], flash reduction [15], hydrothermal dehydration [16], mechanical exfoliation [3], epitaxial growth [17], photocatalysis [18], and photodegradation [19]. Although several methods are available for the preparation of graphene, solution-based chemical reduction of graphene oxide (GO) to graphene is considered one of the most efficient methods for low-cost and large-scale production of graphene [13]. Reduction of GO by chemical methods seems to be promising, because of the low cost and potential for large-scale production. Such methods are also appropriate for chemical modification and subsequent processing. However, in chemical methods, the use of hydrazine and hydrazine derivatives as strong reducing agents for the formation of graphene can be toxic or explosive, resulting in challenges for large-scale production. The resulting graphene

* Correspondence: gsangiliyandi@yahoo.com; jhkim541@konkuk.ac.kr
[1]Department of Animal Biotechnology, Konkuk University, 1 Hwayang-Dong, Gwangin-gu, Seoul 143-701, South Korea
[2]GS Institute of Bio and Nanotechnology, Coimbatore, Tamil Nadu 641024, India
Full list of author information is available at the end of the article

also possesses very limited solubility or even irreversible agglomeration during preparation in water and most organic solvents unless capping reagents are used owing to the strong π–π stacking tendency between rGO sheets [20,21]. To overcome the aggregation and solubility problems, several polymers or surfactants have been used, such as poly(N-vinyl-2-pyrrolidone) [22], poly(sodium-4-styrene sulfonate) [23], poly(allylamine) [24], and potassium hydroxide [25]. Recently, Akhavan et al. [26] demonstrated a possible route for inexpensive mass production of high-quality graphene sheets from natural and industrial carbonaceous wastes.

The toxicity of GO and graphene has been studied in various cell types such as neuronal cells [27], lung epithelial cells [28], fibroblasts [29], primary mouse embryonic fibroblast cells [30], and cancer cells [31], and the results vary across cell and material types.

Surface modification of graphene has been reported to alter its toxicity [31], with reduced GO and carboxylated graphene reported to be less toxic than GO or native graphene [32]. Single-layer GO sheets were found to be internalized and sequestered in cytoplasmic, membrane-bound vacuoles in human lung epithelial cells and fibroblasts, with toxicity induced at concentrations above 20 µg/mL after 24 h [27,29]. Sanchez et al. [4] reported that graphene-family nanomaterials (GFNs) can be either benign or toxic to cells, and that the biological responses depend on layer number, lateral size, stiffness, hydrophobicity, surface functionalization, and concentration. In addition, the biocompatibility and cytotoxicity depend on the type of reducing agent used for the functionalization of GO.

Graphene has been used as a possible biocompatible nanocarrier for delivering drugs [33] and also as a functional biomaterial. Sun et al. [9] reported that non-toxic PEGylated nano-graphene oxide could deliver water-insoluble cancer drugs. Fan et al. [34] showed that graphene/chitosan composites were biocompatible in L929 cells and that the absence of metallic impurities in graphene sheets makes them potential candidates as scaffolds for tissue engineering. Furthermore, Chen et al. [35] reported that graphene oxide (GO)/ultra-high-molecular-weight polyethylene (GO/UHMWPE) composites showed remarkably enhanced hardness and slightly improved yield strength compared with pure UHMWPE. The addition of small amounts of GO did not affect the attachment and proliferation of MC3T3-E1 osteoblasts cultured on GO/UHMWPE composite surfaces, indicating its excellent biocompatibility. Akhavan et al. [36] reported size-dependent cyto- and genotoxic effects of reduced graphene oxide nanoplatelets (rGONPs) rGONPs on cells. A cell viability test showed significant cell death on treatment with 1.0 µg/mL rGONPs with an average lateral dimension (ALD) of $11 \pm 4$ nm, whereas rGO sheets an ALD of $3.8 \pm 0.4$ µm exhibited a significant

cytotoxic effect only at the high concentration of 100 µg/mL after 1 h of exposure time. Akhavan et al. [37] demonstrated the size-dependent cytotoxic and genotoxic effects of reduced graphene oxide nanoplatelets on human mesenchymal stem cells (hMSCs). Furthermore, Akhavan et al. [38] used ginseng extract-reduced GO to differentiate stem cells. Park et al. [39] used graphene-as a substrate to promote human neural stem cell adhesion and differentiation into neurons. Lee et al. [40] reported that the strong non-covalent binding ability of graphene allows it to act as a pre-concentration platform for osteogenic inducers, which accelerate the differentiation of mesenchymal stem cells (MSCs) growing on it toward the osteogenic lineage. Akhavan et al. [37] used graphene nanogrids as two-dimensional selective templates for accelerated differentiation of human MSCs (hMSCs) isolated from umbilical cord blood into osteogenic lineages. The biocompatible and hydrophilic graphene nanogrids showed high actin cytoskeleton expression coinciding with the patterns of the nanogrids. Akhavan and Ghaderi [41] introduced a reduced graphene oxide (rGO)/TiO$_2$ heterojunction film as a biocompatible flash photo stimulator for the effective differentiation of hNSCs into neurons. Graphene nanogrids on a SiO$_2$ matrix containing TiO$_2$ nanoparticles (NPs) were also applied as a photocatalytic stimulator to accelerate the differentiation of human neural stem cells (hNSCs) into two-dimensional neural networks [42].

Several environmentally friendly methods have been developed using various biomolecules such as ascorbic acid [43], amino acids [44], glucose [45], and bovine serum albumin [46] as reducing agents or stabilizers. In addition, microorganisms have also used to reduce GO, including *Shewanella* [47], *Escherichia coli* [48,49], *Pseudomonas aeruginosa* [8], *Bacillus marisflavi* [50], and *Ganoderma* spp [21]. Some purified proteins have also been used for synthesis of graphene, such as melatonin [51], l-glutathione [52], and humanin [53]. Recently, the synthesis of graphene has been increased significantly because of the wide range of resources and availability of simple, cost-effective, and environmentally friendly approaches. The major problem encountered during the synthesis of nanoparticles using biomass is the isolation and purification of the nanoparticles from the biomass, which requires many downstream processing steps including sonication and ultracentrifugation to attain maximum yield [54]. Moreover, endotoxin may be present in the nanoparticles, which may limit the use of the nanoparticles in medical applications [55]. Therefore, this study attempted to use a recombinant protein.

Recombinant enhanced green fluorescent protein (EGFP) (Gene Bank Accession no. U57607) is a protein composed of 293 amino acid residues (32.7 kDa) that has an isoelectric point of 6.2 and exhibits bright green fluorescence when exposed to light in the blue to ultraviolet range.

EGFP has been widely used as a biological reporter to identify tissue and cells with target gene expression [56,57]. Previous studies showed no obvious detrimental effects of EGFP and no toxicity, i.e., it is biologically inert [58,59]. In addition, EGFP was selected here as a reducing and stabilizing agent for synthesis of graphene because it is a natural protein from the jellyfish *Aequorea victoria* and has been proven to be an excellent biological reporter [60]. Thus, without any other toxic reagents added, the raw material and reaction products are all environmentally friendly, which should increase the efficiency and large-scale synthesis of graphene. Additionally, EGFP contains five cysteine amino acid residues, each containing a thiol group that can be oxidized to form the disulfide derivative cysteine, which functions as a nucleophile [61]. Protons have high binding affinity to oxygen-containing groups, such as hydroxyl and epoxide groups on GO, resulting in the formation of $H_2O$ molecules [27,62]. The unique chemical structure of EGFP makes it not only an ideal reducing agent but also an effective capping agent. Therefore, we addressed the following objectives: first, the development of a simple, dependable, and environmentally friendly approach for synthesis of graphene using recombinant EGFP; second, the characterization of GO and EGFP-reduced GO; and finally, the evaluation of cellular responses of GO and EGFP-rGO in human embryonic kidney 293 cells.

## Results and discussion
### Synthesis and characterization of EGFP-rGO
As shown in Figure 1, EGFP-rGO was synthesized by a two-step process, including an oxidation step and an

EGFP-based reduction step. In the first step, graphene oxide was formed by the oxidation of graphite crystals according to a modification of the Hummers method [63]; the crystals were dispersible in water. In the second step, a stable black aqueous suspension was obtained through a chemical deoxidization process by using EGFP as both a reducer and a stabilizer. Similarly, Wang et al. [13] reported a simple method of reduction of GO to rGO using the natural polymer heparin as both a reducing agent and a stabilizer to produce a stable aqueous suspension of heparin-rGO sheets. Fan et al. [34] fabricated biocompatible graphene-reinforced chitosan composites in which chitosan was significantly reinforced by the addition of a small amount of graphene sheets. The graphene/chitosan composites were biocompatible in the L929 fibrosarcoma cell line.

The reduction of GO was confirmed using UV–vis absorption spectroscopy. As shown in Figure 1, the absorption peak of the GO dispersion was located at 230 nm with a shoulder peak at about 300 nm, which was consistent with previous reports [13,27,62]. After the reduction process, the peak was red-shifted to 258 nm and the absorbance was increased dramatically in the entire spectral region. This result suggests that GO was reduced by EGFP and that the aromatic structure of graphene may be restored. Further evidence showed that the UV–vis absorption spectrum of GO was characterized by the $\pi-\pi^*$ of the C = C plasmon peak at approximately 230 nm and a shoulder at approximately 300 nm that is often attributed to $n-\pi^*$ transitions of the carbonyl groups [62,64]. With reduction by EGFP, the plasmon peak gradually red-shifted to approximately 258 nm,

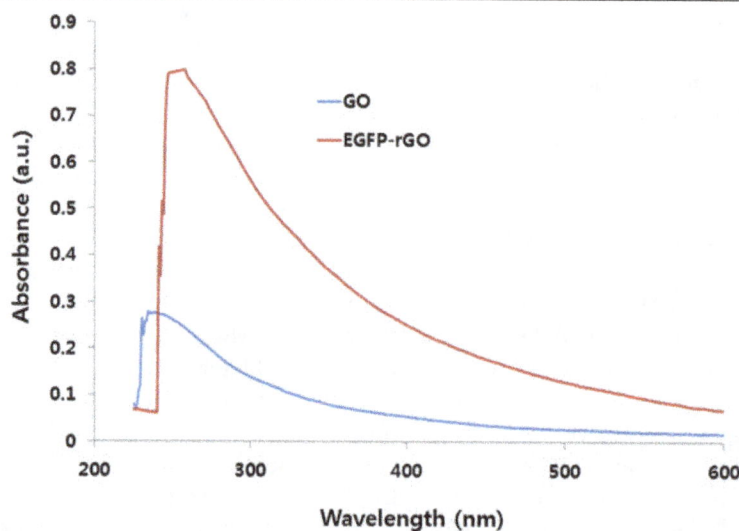

**Figure 1 Synthesis and characterization of GO and EGFP-rGO by ultraviolet–visible spectroscopy.** Spectra of GO exhibited a maximum absorption peak at approximately 230 nm, which corresponds to a π-π transition of aromatic C–C bonds. The absorption peak for reduced GO was red-shifted to 258 nm. At least three independent experiments were performed for each sample and reproducible results were obtained. Data from a representative experiment are shown.

indicating the restoration of sp$^2$ carbon and possible rearrangement of atoms [65]. Similar trends were also observed for the reduction of GO by L-ascorbic acid [43,66], L-cysteine [62], melatonin [51], heparin [13], dopamine [67], and humanin [53].

## FTIR spectra of GO and EGFP-rGO

The reduction of oxygen-containing functional groups of GO by EGFP was confirmed by FT-IR spectroscopy. Figure 2 shows the FT-IR spectra of GO and EGFP-rGO. The presence of different types of oxygen-containing groups in graphene oxide was confirmed at 3440 cm$^{-1}$ (O-H stretching vibrations), 1725 cm$^{-1}$ (stretching vibrations from C = O), 1225 cm$^{-1}$ (C-OH stretching vibrations), and 1070 cm$^{-1}$ (C-O stretching vibrations), as reported earlier [68,69]. In addition, the substitution of hydroxyl groups on the GO surface by carboxyl groups was confirmed by the CH$_2$-stretching vibration at 2,920 cm$^{-1}$ (lower spectrum) [70]. In contrast, the FT-IR spectrum of graphene completely differs from that of GO. The FTIR peak of EGFP-rGO showed O-H stretching vibrations, stretching vibrations from C = O, C-OH stretching vibrations, and C-O stretching vibrations at 3440, 1725, 1225, and 1070 cm$^{-1}$, respectively, indicating that GO was significantly reduced by the deoxygenation procedure. The intensities of absorption peaks corresponding to oxygen functional groups decreased and these functional groups almost disappeared. Altogether, these results clearly confirm that the oxygen-containing groups were removed during reduction using EGFP. These changes in EGFP-rGO compared with GO in FT-IR spectra were identical with those of earlier reports that used various reducing agents such as hydrazine [14], vitamin C [66], L-cysteine [62], heparin [13], and humanin [53].

## XRD analysis of GO and EGFP-rGO

To further characterize the crystal structures, the XRD patterns of the exfoliated GO and EGFP-rGO were studied. The characteristic peak of GO appears at 11.7°, corresponding to a d-spacing of 0.76 nm resulting from the formation of hydroxyl, epoxy, and carboxyl groups (Figure 3). In contrast to GO, EGFP-rGO showed no peaks at 11.7°, which indicates that most of the oxygen functional groups of GO were removed. Compared with pristine graphite (2θ = 26.4°), the diffraction peak of exfoliated GO moved to 11.7° (002) with a layer-to-layer distance (d-spacing) of 0.76 nm. This value was larger than the d-spacing of pristine graphite (0.34 nm) because of the introduction of numerous oxygenated functional groups on the carbon sheets [13]. After the exfoliated GO was reduced by EGFP, the peak at 11.7° disappeared, but a new diffraction peak appeared at 2θ = 25.8° with a d-spacing of 0.36 nm, which was closer to the typical (002) diffraction peak of graphite (2θ = 26.4°, d-spacing of 0.34 nm). The higher interlayer spacing value of exfoliated GO resulted from the introduction of numerous oxygenated functional groups on the carbon sheets [7,21,48]. The data obtained from this experiment suggest that EGFP played an important role in the deoxygenation of GO and also that the reduction of GO by EGFP was consistent with earlier reports using various reducing agents including vitamin C [66], L-cysteine [62], heparin [13], and humanin [53].

## Size distribution analysis of GO and EGFP-rGO

Size distribution analysis was performed to elucidate the state of GO and EGFP-rGO in an aqueous solution using DLS measurement [71] with a concentration of 250 μg/mL. The average hydrodynamic diameter (AHD)

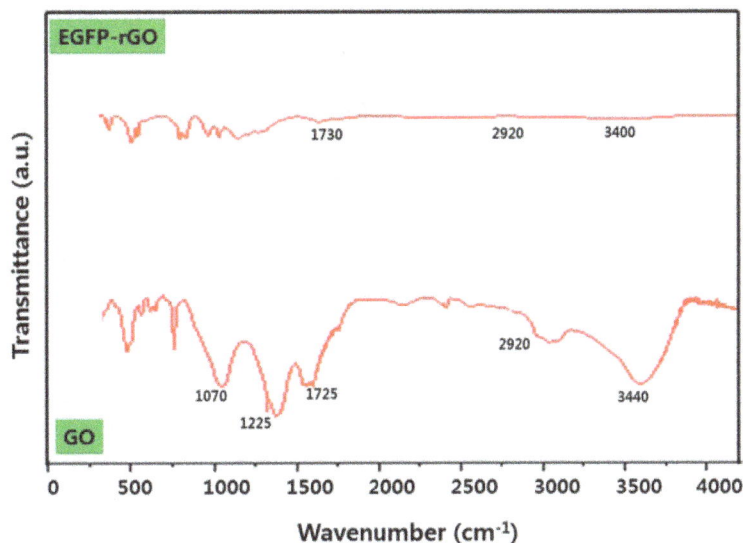

**Figure 2** Fourier transform infrared spectroscopy spectra of GO and EGFP-rGO.

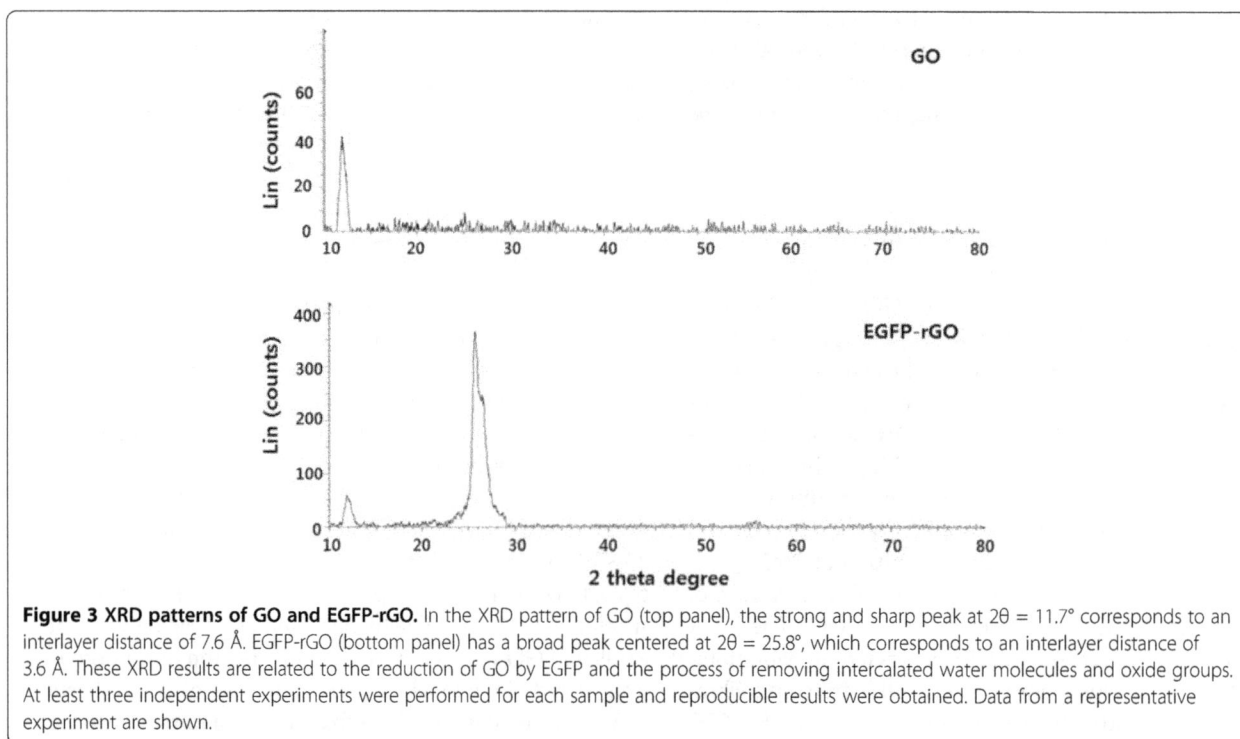

**Figure 3 XRD patterns of GO and EGFP-rGO.** In the XRD pattern of GO (top panel), the strong and sharp peak at 2θ = 11.7° corresponds to an interlayer distance of 7.6 Å. EGFP-rGO (bottom panel) has a broad peak centered at 2θ = 25.8°, which corresponds to an interlayer distance of 3.6 Å. These XRD results are related to the reduction of GO by EGFP and the process of removing intercalated water molecules and oxide groups. At least three independent experiments were performed for each sample and reproducible results were obtained. Data from a representative experiment are shown.

of GO and EGFP-rGO was 2288 ± 20 nm and 2607 ± 32 nm, respectively (Figure 4). However, after the reduction of GO with EGFP, the AHD increased and was relatively larger than that of GO. This obvious change of size distribution suggests that EGFP not only acted as a reducing agent to prepare rGO but also functionalized on the surface of the resulting rGO. Similar results were observed for heparin and biopolymer-functionalized reduced graphene oxide [13,72]. Graphene nanoplates functionalized with isocyanate showed the effective hydrodynamic diameter size of 560 ± 60 nm. Lammel et al. [73] reported that the hydrodynamic diameter of GO functionalized with carboxyl graphene nanoplatelets increased from 385 to 1,110 nm. Liu et al. [74] reported that aqueously dispersed graphite (Gt), graphite oxide (GtO), graphene oxide (GO), and reduced graphene oxide (rGO) had sizes of 5,250, 4,420, 560, and 2,930 nm, respectively. A similar trend was observed for GO reduced by *Pseudomonas aeruginosa* [8], *Bacillus marisflavi* [50], *Ginkgo biloba* [70], and *Ganoderma* spp [21]. The size of EGFP-rGO was slightly larger than that of GO, indicating that EGFP not only acted as a reducing agent but also was functionalized on the surfaces of the resulting rGO, leading to an increased size [75]. Similarly, Wang et al. [13] found that the average size of heparin-reduced graphene oxide was larger than that of GO under the same experimental conditions. Altogether, our data and data from other groups suggest that EGFP used as a reducing agent plays an important role in increasing the size of rGO.

## Surface properties of GO and EGFP-rGO

Zeta potential is an important factor for characterizing the dispersion stability of colloids because the magnitude and sign of the effective surface charge associated with the double layer around the colloid, and it directly influences the electrostatic interaction between different graphene sheets [76,77]. Zeta potential measurements were carried out in aqueous solutions of the GO and EGFP-rGO in function of pH is important to determine the surface charge of the sheets (Figure 5). The results show that GO sheets are highly negative charged with an average −29.7 mV at pH range between 2 and 10. This value is attributed to the presence of oxygen species at the surface of GO. On the contrary, EGFP-rGO, shows positive zeta potential values for the same pH range, which is suggest that the lower charge density of this type of graphene. Interestingly, recombinant proteins treated GO sheets resulted in the reduction and almost complete elimination of the oxygen functionalities at the surface of graphene materials.

## Surface morphology analysis of GO and EGFP-rGO by SEM

The surface morphology of the GO and EGFP-rGO samples was analyzed using SEM. As shown in Figure 6A, the GO samples contain several layers of sheets, and further the sheets are aggregated and crumpled sheets are closely associated with each other to form a continuous conducting network. The edges of the GO sheets appeared crumpled, folded, and closely restacked with one

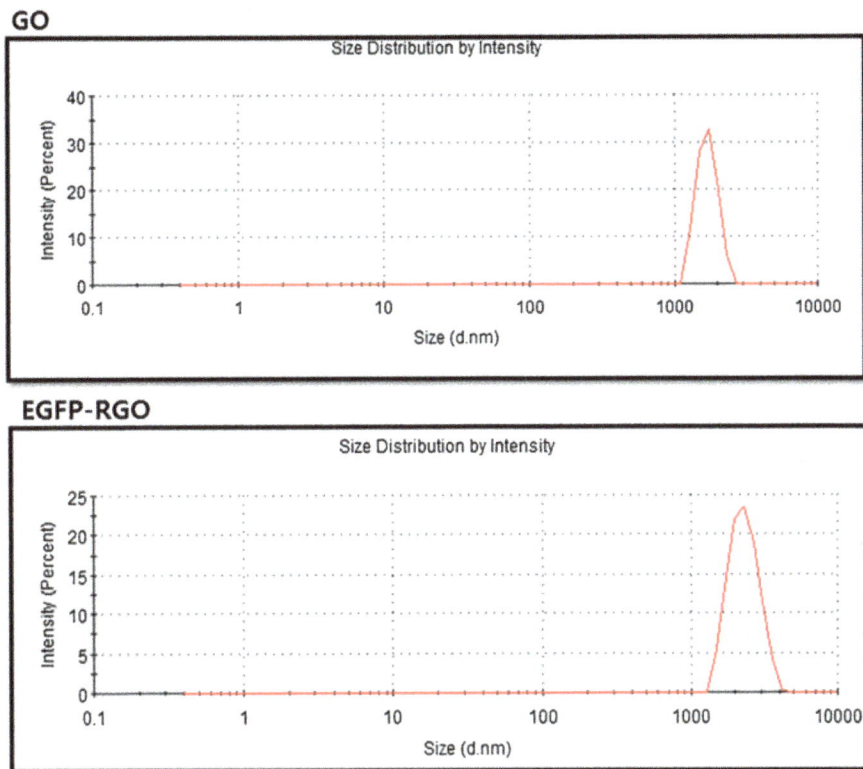

**Figure 4 Size distribution analysis of GO and EGFP-rGO.** Aqueous dispersions of GO and EGFP-rGO were characterized by DLS analysis using a particle size analyzer at the scattering angle θ = 90°. The data show the average values from triplicate measurements. The sample concentrations were all 250 µg/mL.

another because of the oxidation process [78]. Jeong et al. [79] reported that at higher concentrations, the surfaces of GO sheets have a soft-carpet-like morphology, possibly because of the presence of residual $H_2O$ molecules and hydroxyl or carboxyl groups attached to GO. In contrast to GO, on SEM the EGFP-rGO samples resemble transparent and rippled silk waves (Figure 6B). He and Gao [80] reported that Gt appears to pile up in thick cakes, whereas GO is exfoliated into thin large flakes with wavy wrinkles. Previously, we observed on

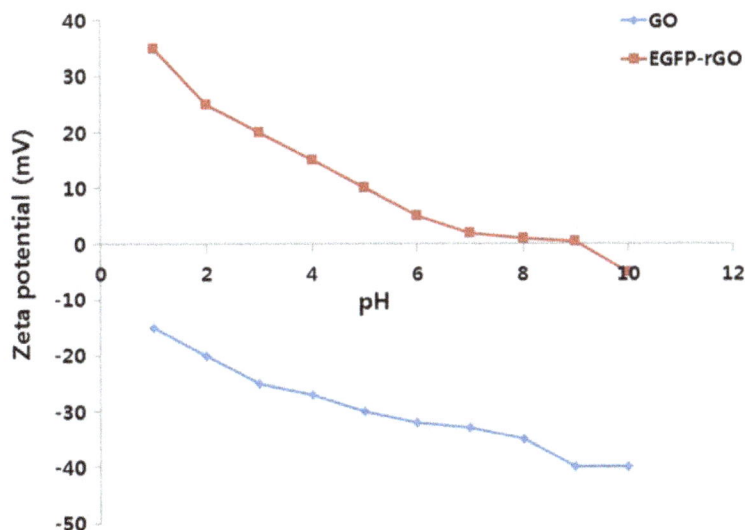

**Figure 5** Zeta potential of as-prepared GO and EGFP-rGO as a function of pH, in aqueous dispersions at a concentration of ~0.05 mg ml$^{-1}$.

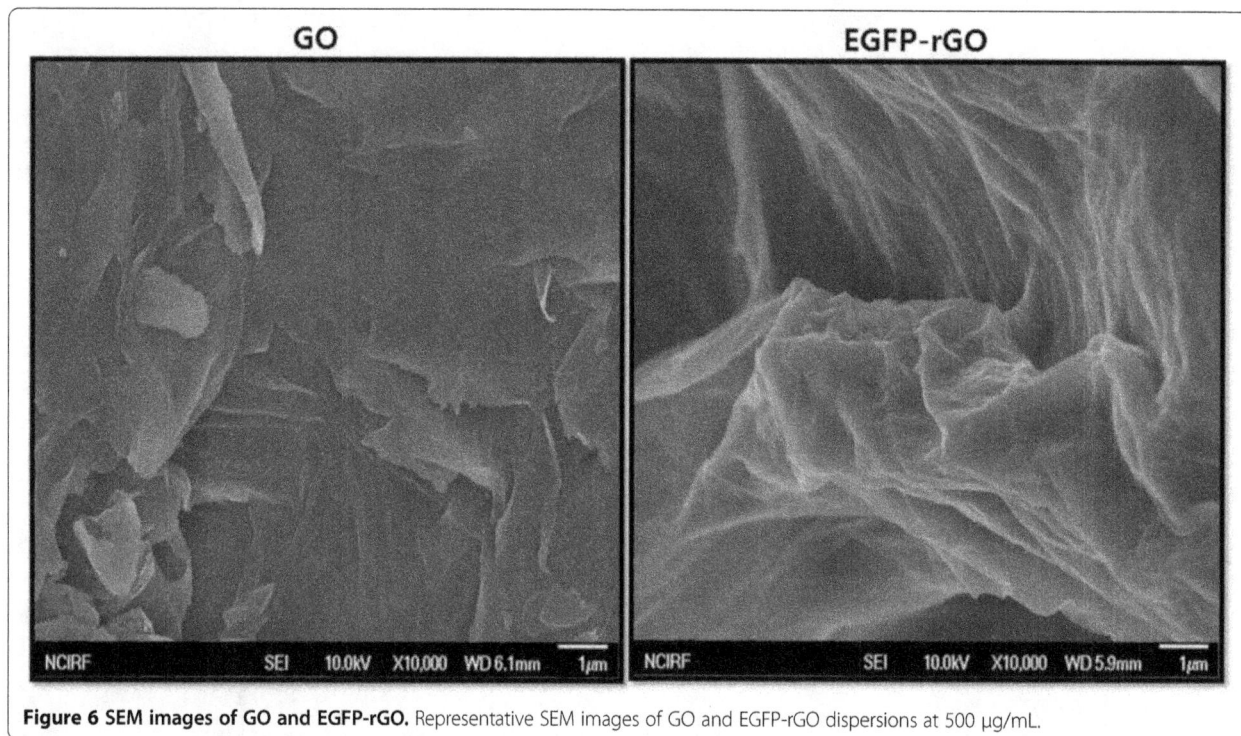

**Figure 6 SEM images of GO and EGFP-rGO.** Representative SEM images of GO and EGFP-rGO dispersions at 500 µg/mL.

SEM that GO consisted of individual sheets closely associated with each other, with a silky and leaf-like structure, whereas *Ginkgo biloba* extract-reduced GO (Gb-rGO) sheets showed thin layers of nanosheets and were mainly comprised of larger, wavy forms [70]. The graphene sheets were found to possess a curled morphology consisting of a thin, wrinkled, paper-like structure, with fewer layers (approximately four layers) and a large specific surface area [81]. Graphene nanosheets were functionalized with long chains and polymers, resulting in coarse and hairy surfaces with blurry edges of the flakes [80]. Previously, we reported using SEM that GO was present as multilayered, wavy, folded flakes, whereas fungal extract-reduced graphene oxide showed several layers stacked on top of one another similarly to sheets of paper, with a silky, wrinkled, and flower-like curling morphology [70]. This difference in morphology between the folded, stacked structure of GO and transparent and rippled silk wave structure of graphene suggests that EGFP played an important role in the reduction of GO to graphene. The data obtained from this study suggest that synthesis of graphene using biological molecules was similar to that of graphene sheets prepared from Gt powder through oxidation followed by rapid thermal expansion in a nitrogen atmosphere [81].

### Raman spectroscopy analysis of GO and EGFP-rGO

Raman spectroscopy is used to characterize the structural electronic properties of graphite and graphene-based materials [21,82,83]. Raman spectra are also used to measure induced enormous structural changes during chemical oxidation of pristine graphite and the reduction of GO to rGO [83]. In the Raman spectra, the G band resulting from first-order scattering of the $E_{2g}$ phonons of $sp^2$ carbon atoms and the D band originating from the breathing mode of k-point photons of $A_{1g}$ symmetry are the two main characteristic features of graphene-based materials [84-86]. In the Raman spectrum of GO, the G band is broadened and shifted to 1615 cm$^{-1}$. In addition, the D band at 1359 cm$^{-1}$ becomes prominent, indicating a reduction in the size of the in-plane $sp^2$ domains, possibly because of extensive oxidation-induced defects in the sheets (Figure 7). The Raman spectrum of the rGO reduced by EGFP also contains both G and D bands located at 1607 and 1351 cm$^{-1}$, respectively; however, the D/G intensity ratio (2.149) is increased compared to that in GO upon reduction. This change suggests a decrease in the average size of the $sp^2$ domains upon reduction of the exfoliated GO [14,84].

The major effects of deoxygenation are the restoration of the $sp^2$ network and the introduction of small and isolated aromatic domains, and these effects are responsible for the observed increase in the ID/IG ratio in rGO [66,83,86,87]. Wang et al. [82] suggested that the G band is broadened and shifted upward to 1,595 cm$^{-1}$, and the increased intensity of the D band at 1,350 cm$^{-1}$ could be attributed to the significant decrease in the size of the in-plane $sp^2$ domains resulting from oxidation and ultrasonic exfoliation, in addition to the partially ordered graphite crystal structure of graphene nanosheets. The

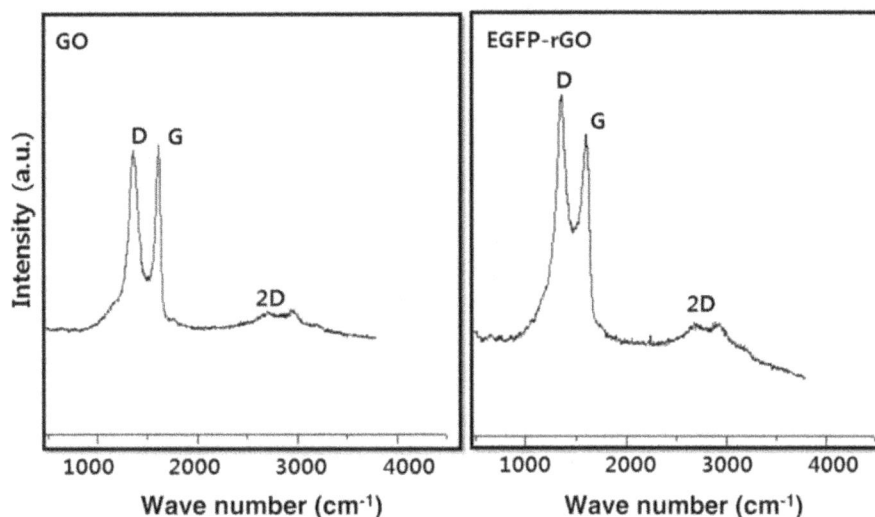

**Figure 7** **Raman spectroscopy analyses of GO and EGFP-rGO samples.** Raman spectra were obtained using a laser excitation of 532 nm at a power of, 1 mW. The figure shows representative Raman spectra of GO and EGFP-rGO samples after removal of the fluorescent background. The intensity ratios of the D-peak to the G-peak were 1.8 and 2.149 for GO and EGFP-rGO, respectively. At least three independent experiments were performed for each sample and reproducible results were obtained.

Raman spectra of graphene-based materials also show a two-dimensional (2D) band that is sensitive to the stacking of graphene sheets. It is well known that the two-phonon (2D) Raman scattering of graphene-based materials is useful to differentiate monolayer graphene from multilayer graphene as it is highly sensitive to the stacking of graphene layers [14,88,89]. Another characteristic of single-layer graphene is the relatively strong Raman intensity of the 2D band with respect to the G-band [90]. Usually, a Lorentzian peak for the 2D band of monolayer graphene sheets is observed at 2,679 $cm^{-1}$, whereas this peak is broadened and shifted to a higher wave number in the case of multilayer graphene [14,88,89]. We observed the 2D band at 2699 $cm^{-1}$, which is the same as the previously reported peak position for single-layer graphene [90,91]. Thus, our sample could consist of single-layer graphene flakes.

It should be noted that this ratio is higher than those reported for rGO produced using various reducing agents such as L-cysteine [62], dextran [92], baker's yeast [93], DTT [83,94], and NaBH$_4$ [95]. The Raman spectroscopy analyses described here agree with those of previous studies that used various biomolecules and organisms to reduce GO to graphene, such as L-cysteine [62], Baker's yeast [93], heparin [13], *Escherichia coli* [48], *P. aeruginosa* [8], Humanin [53], *Ganoderma* spp [21], and *Ginkgo biloba* [70].

### Biocompatibility of GO and EGFP-rGO

The HEK cell line has been extensively used as an expression tool for recombinant proteins [96]. Therefore, we used the HEK cell line as a model system to study the effect of GO and EGFP-rGO. Figure 8 shows the biocompatibility of EGFP-rGO in HEK cells assessed using the WST assay. GO exhibited concentration-dependent toxicity compared to untreated control cells, whereas EGFP-rGO-treated cells showed no significant toxicity when compared to untreated cells. Several studies have shown interactions between dispersed graphene and GO sheets in various cell types such as monolayer cultures of neuronal cells [27], lung epithelial cells [28], fibroblasts [30,47], and human breast cancer cells [21]. Single-layer GO sheets were found to be internalized and sequestered in cytoplasmic, membrane-bound vacuoles by human lung epithelial cells or fibroblasts, and they induced toxicity at concentrations above 20 μg/mL after 24 h [28,29,94,97]. Limited literature is available on the biocompatibility of graphene [4]. GFNs have been suggested to be useful as biosensors [98], tissue scaffolds [10], carriers for drug delivery and gene therapy [99], antibacterial agents [7,8], and bio-imaging probes [27] because of their unique features over other types of nanomaterials, including their high specific surface area, which allows high-density bio-functionalization and drug loading. The results from our study indicate that EGFP-rGO can be used as a biocompatible material. Altogether, the results from our study and those from other groups suggest that EGFP-rGO can be used in various biomedical applications.

### Effect of EGFP-rGO on LDH leakage

LDH (lactate dehydrogenase) is present in all types of cells and LDH leakage is a useful index for cytotoxicity on the basis of loss of membrane integrity, a hallmark of necrosis [100]. Based on the percentage of the maximum LDH release, in the present study EGFP-rGO was

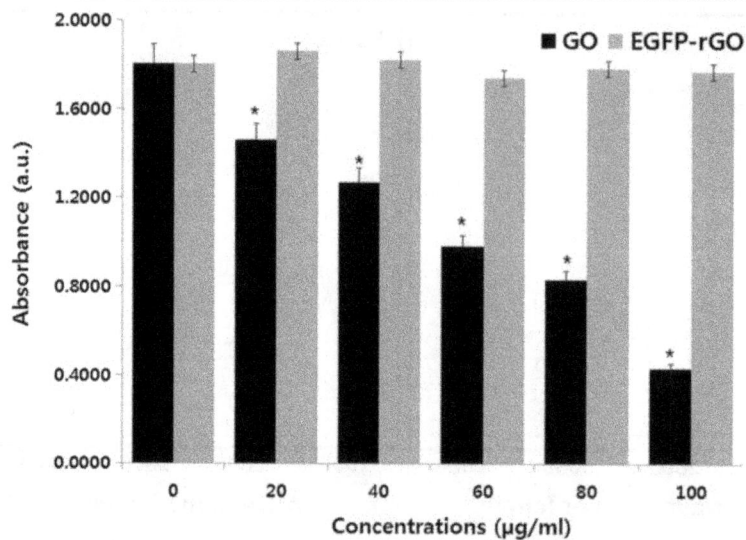

**Figure 8 Effects of GO and EGFP-rGO on cell viability of human embryonic kidney 293 cells.** Cell viability of human kidney cells was determined using WST-8 assay after 24 hours exposure to different concentrations of GO or EGFP-rGO. The results represent the means of three separate experiments, and error bars represent the standard error of the mean. GO- and EGFP-rGO-treated groups showed statistically significant differences from the control group by the Student's *t*-test (P < 0.05).

considered non-toxic to cells, whereas GO showed toxicity to the cells in a concentration-dependent manner when compared to untreated cells (Figure 9). Significant LDH release was observed after 24 h of exposure to GO at higher concentrations, whereas graphene had no effect on the release of LDH. Thus, the LDH assay results were consistent with the cell-viability assay results. The toxicity of graphene materials depends on their size, shape, composition, surface charge, and surface chemistry,

in addition to the target cell type [101]. Zhang et al. [27] observed that graphene aggregates/agglomerates that had sedimented onto the surface of rat PC12 cells caused an increase in LDH leakage only at the highest exposure concentration (100 μg/mL). Our earlier findings also suggest that at higher concentrations, TEA-rGO has no significant toxicity in mouse embryonic fibroblast cells [102]. Therefore, EGFP-derived graphene is also biocompatible.

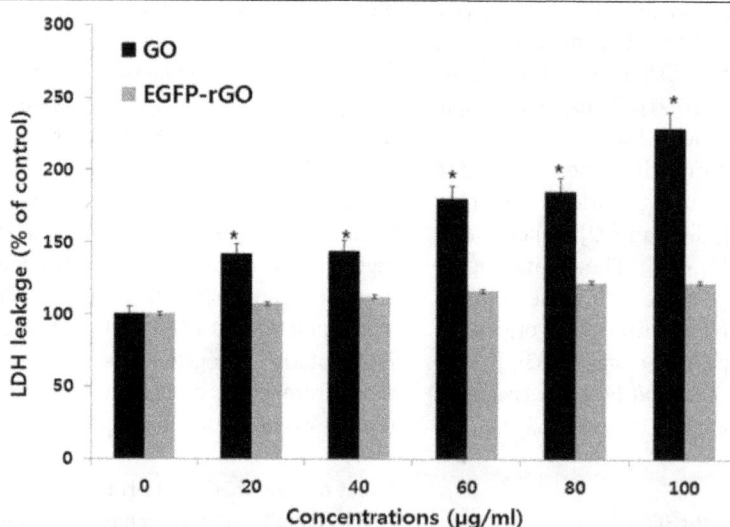

**Figure 9 Effects of GO and EGFP-rGO on lactate dehydrogenase activity in human embryonic kidney 293 cells.** Lactate dehydrogenase activity was measured at 490 nm, using the cytotoxicity detection lactate dehydrogenase kit. The results represent the means of three separate experiments, and error bars represent the standard error of the mean. GO- and EGFP-rGO-treated groups showed statistically significant differences from the control group by the Student's *t*-test (P < 0.05).

## Effects of EGFP-rGO on oxidative stress

The DCF assay was performed to investigate the toxicity of nanomaterials attributable to ROS generation. Following exposure of HEK cells for 24 h to GO and EGFP-rGO, the state of oxidative stress in the cells was observed. As shown in Figure 10, the ROS generation increased in a concentration-dependent manner as the concentration of GO was increased, whereas EGFP-rGO had no significant impact, even at high concentrations, when treated cells were compared to untreated cells. These results were consistent with the results from the WST-8 assay and LDH assay, suggesting that toxicity in cells exposed to GO may result from oxidative stress mediated by ROS generation. It was previously shown that exposure to multiwalled carbon nanotubes (MWCNTs) resulted in a concentration-dependent cytotoxicity in cultured human embryonic kidney cells, which was associated with increased oxidative stress [103]. Zhang et al. [104] reported that surface functionalization (e.g., PEGylation) of single-walled carbon nanotubes (SWCNTs) reduced the ROS-mediated toxicological response in PC-12 cells. Induction of oxidative stress is considered to be one of the principal mechanisms underlying nanomaterial toxicity [105]. Lammel et al. [73] demonstrated that GO and carboxyl graphene nanoplatelets (CXYG) induce the generation of intracellular ROS in a concentration- and time-dependent manner in the human hepatocellular carcinoma cell line HepG2. GO-mediated cell death is caused by increased intracellular ROS levels originating from mitochondrial damage [73]. Stern et al. [106] suggest that several nanomaterials cause cell death through autophagy and lysosomal dysfunction. Qu et al. [107] reported that

ROS production was independent of surface modification on QDs and that ROS did not account for the cytotoxicity of QD-PEG-NH$_2$ particles in J774A.1 cells. Recently, Wu et al. [108] investigated the toxicity of graphene oxide in *Caenorhabditis elegans* at adult day 10 and found that prolonged exposure to 0.1 mg/L GO did not induce the noticeable intestinal autofluorescence or intestinal ROS production compared with the control; however, prolonged exposure to 10–100 mg/L GO resulted in intestinal autofluorescence and intestinal ROS production. Chong et al. [109] assessed the effect of graphene quantum dots (GQD) using various measures such as cell viability, cell apoptosis and necrosis, and LDH and ROS levels, and found that over 95% and 85% of HeLa cells and A549 cells, respectively, remained alive after 24 h of incubation with GQD-PEG, even when the GQD concentration increased to 160 μg/mL. Furthermore, they suggested that the low cytotoxicity resulted from PEGylation or the inherent properties of the GQD sample. Graphene nanoparticles, depending on the synthesis method, can exhibit different morphologies, chemical properties, and physical properties. Earlier studies also suggest that graphene nanoparticles show diverse responses in cells and tissues depending on their morphology and synthesis method [110].

## Effect of EGFP-rGO on cell morphology

Biocompatibility is important for the development of new nanomaterials for biological and biomedical applications [50]. In addition to the biochemical assays described above, we evaluated the morphology of the cells treated with GO and EGFP-rGO. The effect of EGFP-rGO on cell morphology was determined using higher

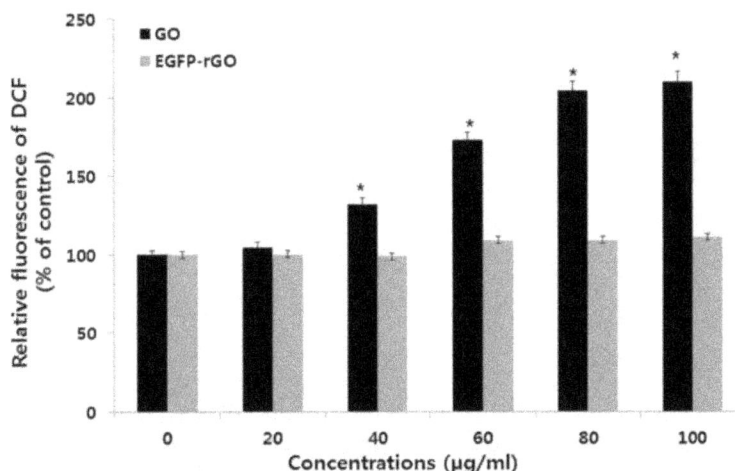

**Figure 10 Effects of GO and EGFP-rGO on generation of ROS in human embryonic kidney 293 cells.** The relative fluorescence of 2',7'-dichlorofluorescein was measured using a spectrofluorometer with excitation at 485 nm and emission at 530 nm. The results represent the means of three separate experiments and the error bars represent the standard error of the mean. Treated groups GO, showed statistically significant differences from the control group, as determined by Student's t-test ($P < 0.05$).

concentrations of GO and EGFP-rGO (100 µg/mL), and the cells were seeded at the same density of $1 \times 10^4$ cells per plate. After 24 and 48 h of incubation, we observed the morphology of cells, and surprisingly, EGFP-rGO had no apparent effect; the cells were healthy (Figure 11); conversely, GO-treated cells were unhealthy, and the structure of the cells was contracted (Figure 11). Cheng et al. [67] reported that biopolymer-functionalized rGO exhibits an ultralow hemolysis ratio and significant cytocompatibility in human umbilical vein endothelial cells (HUVECs), even at a high concentration of 100 µg/mL. Talukdar et al. [71] evaluated the effect of various types of graphene materials such as graphene nano-onions (GNOs), graphene oxide nanoribbons (GONRs), and graphene oxide nanoplatelets (GONPs) on the viability and differentiation of human mesenchymal stem cells (MSCs). They found that the cytotoxic effect was concentration-dependent but not time-dependent. In our study, concentrations lower than 50 µg/mL showed no significant differences compared to untreated controls. Our data suggest that EGFP-rGO at up to 100 µg/mL has no effect on cell viability, LDH, ROS generation, or on cell morphology. Our earlier studies demonstrated both cytotoxicity and biocompatibility of graphene materials in various cell types. Altogether, our findings and those of other research groups suggest that the cytotoxicity or biocompatibility of graphene materials is dependent on physicochemical properties such as the density of functional groups, size, and conductivity, in addition to the type of reducing agents used for the deoxygenation

of GO, degree of functionalization, and cell type [50,75]. Finally, graphene materials prepared using recombinant EGFP could be useful for potential biomedical applications.

## Conclusion

Commonly, the reduction of GO using chemical reducing agents is harmful to human health and the environment, and aggregation is another problem that occurs during the reduction process. Here, we show the synthesis of biocompatible graphene using recombinant EGFP. EGFP is one of the most widely used tools in biology because of its stability and lack of toxicity. In the present study, we explored the potential application of EGFP for a different purpose other than the tagging usually reported in the literature. We have developed a simple, dependable, and environmentally friendly method for the fabrication of reduced GO. Our findings suggest that GO induced significant concentration-dependent decreases in the viability of HEK cells, whereas graphene exerted no toxic effects on HEK cells at a concentration of 100 µg/mL. Therefore, it is concluded that the use of a biological substrate in a simple and environmentally friendly approach for synthesis of graphene resulted in significant deoxygenation of suspended GO suspensions, thus providing a suitable substitute for chemical reducing agents and potentially enabling biomedical applications of graphene-based materials. This work may provide additional insight into graphene synthesis.

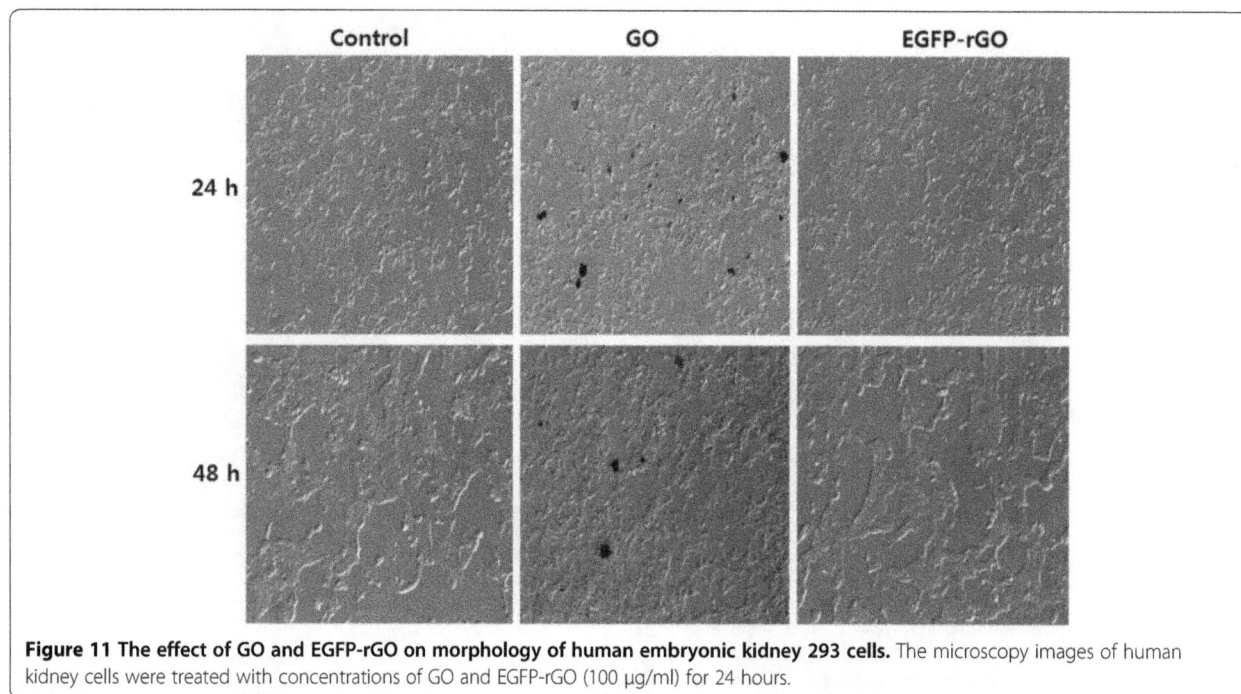

**Figure 11 The effect of GO and EGFP-rGO on morphology of human embryonic kidney 293 cells.** The microscopy images of human kidney cells were treated with concentrations of GO and EGFP-rGO (100 µg/ml) for 24 hours.

## Materials and methods

### Materials

Gt powder, NaOH, KMnO4, NaNO$_3$ anhydrous ethanol, 98% H$_2$SO$_4$, 36% HCl, and 30% H$_2$O$_2$ aqueous solution were purchased from Sigma-Aldrich (St Louis, MO, USA). Penicillin-streptomycin solution, trypsin-ethylene-diaminetetraacetic acid solution, Dulbecco's Modified Eagle Medium (DMEM), and 1% antibiotic-antimycotic solution were obtained from Gibco (Life Technologies, Carlsbad, CA, USA). Fetal bovine serum and the in vitro toxicology assay kit were purchased from Sigma-Aldrich. Enhanced green fluorescent protein was purchased from Bio-vision (Cat.No. 4999–100; Milpitas, California, USA).

### Synthesis of GO

GO was synthesized as described previously [21,57]. In a typical synthesis process, natural Gt powder (2 g) was added to cooled (0°C) H$_2$SO$_4$ (350 mL), and then KMnO$_4$ (8 g) and NaNO$_3$ (1 g) were added gradually while stirring. The mixture was transferred to a 40°C water bath and stirred for 60 min. Deionized water (250 mL) was slowly added and the temperature was increased to 98°C. The mixture was maintained at 98°C for a further 30 minutes and the reaction was terminated by the addition of deionized water (500 mL) and 30% H$_2$O$_2$ solution (40 mL). The color of the mixture changed to brilliant yellow, indicating the oxidation of pristine Gt to Gt oxide. The mixture was then filtered and washed with diluted HCl to remove metal ions. Finally, the product was washed repeatedly with distilled water until pH 7.0 was achieved, and the synthesized Gt oxide was further sonicated by ultrasonication for 30 min.

### Preparation of EGFP-rGO

Reduction of GO was performed as described previously [21,41] with suitable modifications. Using GO as a precursor, EGFP-rGO was prepared using EGFP as both a reducing agent and a stabilizer. In a typical procedure, reduced GO (rGO) was obtained from the reaction of EGFP with GO. A mixed aqueous solution containing EGFP (100 µg/mL) and GO (1 mg/mL) was ultrasonicated for 15 min, and the mixture was maintained at 40°C for 1 h. The mixture was then cooled to room temperature and ultrasonicated for a further 15 min. After being vigorously stirred for 5 min, the mixture was stirred in a water bath (90°C) for 1 h. The resulting stable black dispersion was then centrifuged and washed with water three times. A homogenous EGFP-rGO suspension was obtained without aggregation. Finally, the obtained EGFP-rGO sheets were redispersed in water before further use.

### Characterization of GO and EGFP-rGO

GO and EGFP-rGO were characterized according to methods described previously [41]. UV-visible spectra were recorded using a WPA Biowave II spectrophotometer (Biochrom, Cambridge, UK). The particle sizes of the GO and EGFP-rGO dispersions were measured using a Zetasizer Nano ZS90 instrument (Malvern Instruments, Worcestershire, UK). X-ray diffraction (XRD) analyses were performed in a Bruker D8 DISCOVER X-ray diffractometer (Bruker AXS GmBH, Karlsruhe, Germany). The X-ray source was 3 kW with a Cu target, and high-resolution XRD patterns were measured using a scintillation counter ($\lambda = 1.5406°A$). The XRD was run at 40 kV and 40 mA, and samples were recorded at 2$\theta$ values between 5° and 80°. The dried powder of GO and EGFP-rGO was diluted with potassium bromide and the Fourier transform infrared spectroscopy (FTIR) (Perkin Elmer Inc., USA) and spectrum GX spectrometry were recorded within the range of 500–4000 cm$^{-1}$. A JSM-6700 F semi-in-lens field emission scanning electron microscope was used to acquire SEM images. The solid samples were transferred to a carbon tape held in an SEM sample holder, and then the analyses were performed at an average working distance of 6 mm. Raman spectra of GO and EGFP-rGO were measured using a WITEC Alpha300 laser with a wavelength of 532 nm. Calibration was initially performed using an internal silicon reference at 500 cm$^{-1}$ and gave a peak position resolution of less than 1 cm$^{-1}$. The spectra were measured from 500 to 4500 cm$^{-1}$. All samples were deposited onto glass slides in powdered form without using any solvent.

### Cell culture and exposure of cells to GO and EGFP-rGO

Human embryonic kidney 293 cells were cultured in DMEM supplemented with 10% FBS and 100 U/mL penicillin-streptomycin in a humidified incubator maintained at 37°C and 5% CO$_2$. At approximately 75% confluence, cells were harvested using 0.25% trypsin and subcultured in 75 cm$^2$ flasks, 6-well plates, or 96-well plates depending on the intended use. Cells were allowed to attach to the substratum for 24 h prior to treatment. The medium was replaced three times per week, and cells were passaged at subconfluency. Cells were prepared in 100 µL aliquots at a density of $1 \times 10^5$/mL and plated in 96-well plates. After the cells were cultured for 24 h, the medium was replaced with medium containing GO or EGFP-rGO at different concentrations (0–100 µg/mL). After incubation for an additional 24 h, cells were analyzed for viability, lactate dehydrogenase (LDH) release, and reactive oxygen species (ROS) generation. Cells not exposed to GO or EGFP-rGO served as the control. Further, morphology of cells treated with GO or EGFP-rGO or untreated was examined using an OLYMPUS IX71 microscope (Japan) using appropriate filter sets.

## Cell-viability assay

The WST-8 assay was performed as described previously
[29]. Typically, $1 \times 10^4$ cells were seeded in a 96-well
plate and cultured in DMEM supplemented with 10%
FBS at 37°C under 5% $CO_2$. After 24 h, the cells were
washed with 100 µL of serum-free DMEM two times
and incubated with 100 µL of different concentrations of
GO or EGFP-rGO suspensions in serum-free DMEM.
After 24 h of exposure, the cells were washed twice with
serum-free DMEM, and 15 µL of WST-8 solution was
added to each well containing 100 µL of serum-free
DMEM. After 1 h of incubation at 37°C under 5% $CO_2$,
80 µL of the mixture was transferred to another 96-well
plate because residual GO or EGFP-rGO can affect the
absorbance values at 450 nm. The absorbance of the
mixture solutions was measured at 450 nm using a micro plate reader. Cell-free control experiments were performed to determine whether GO and EGFP-rGO react
directly with the WST-8 reagents. Typically, 100 µL of
GO or EGFP-rGO suspensions with different concentrations (0–100 µg/mL) were added to a 96-well plate and
10 µL of WST-8 reagent solution was added to each
well; the mixture was incubated at 37°C under 5% $CO_2$
for 1 h. After incubation, the GO or EGFP-rGO was
centrifuged and 100 µL of the supernatant was transferred to another 96-well plate. The optical density was
measured at 450 nm.

## Membrane integrity

The cell membrane integrity of human embryonic kidney 293 cells was evaluated by determining the activity
of lactate dehydrogenase (LDH) leaking out of the cells
according to the manufacturer's instructions (in vitro
toxicology assay kit, TOX7, Sigma, USA) and also as described previously [21]. Briefly, the cells were exposed to
various concentrations of GO and EGFP-rGO (0–
100 µg/mL) for 24 h, and then 100 µL per well of each
cell-free supernatant was transferred in triplicate into
wells in a 96-well plate, and 100 µL of the LDH assay reaction mixture was added to each well. After 3 h of incubation under standard conditions, the optical density of
the color generated was determined at a wavelength of
490 nm using a micro plate reader.

## Determination of ROS

ROS were estimated according to a method described
previously [43]. Intracellular ROS were measured based
on the intracellular peroxide-dependent oxidation of
2',7'-dichlorodihydrofluorescein diacetate (DCFH-DA,
Molecular Probes, USA) to form the fluorescent compound 2',7'-dichlorofluorescein (DCF), as previously described. Cells were seeded onto 24-well plates at a
density of $5 \times 10^4$ cells per well and cultured for 24 h.
After washing twice with PBS, fresh medium containing

different concentrations of GO or EGFP-rGO (0–
100 µg/mL) was added and the cells were incubated for
24 h. The cells were then supplemented with 20 µM
DCFH-DA, and incubation continued for 30 min at 37°C.
The cells were rinsed with PBS, 2 mL of PBS was added to
each well, and the fluorescence intensity was determined
using a spectrofluorometer (Gemini EM) with excitation
at 485 nm and emission at 530 nm.

## Statistical analyses

All assays were carried out in triplicate and the experiments were repeated at least three times. The results are
presented as means ± SD. All experimental data were
compared using the Student's t test. A p value less than
0.05 was considered statistically significant.

### Competing interests
The authors declare that they have no competing interests.

### Authors' contributions
SG conceived the idea and participated in the design, preparation of
graphene, and writing of the manuscript. JWH performed the
characterization of graphene. EK, JKP, DNK participated in culturing,
biocompatibility, and other biochemical assays. SG and JHK participated in
the coordination of the study. All authors read and approved the final
manuscript.

### Acknowledgments
This work was supported by the KU-Research Professor Program of Konkuk
University. Dr Sangiliyandi Gurunathan was supported by a Konkuk
University KU-Full-time Professorship. This work was also supported by the
Woo Jang-Choon project (PJ007849).

### Author details
[1]Department of Animal Biotechnology, Konkuk University, 1 Hwayang-Dong,
Gwangin-gu, Seoul 143-701, South Korea. [2]GS Institute of Bio and
Nanotechnology, Coimbatore, Tamil Nadu 641024, India. [3]Animal Biotechnology
Division, National Institute of Animal Science, Suwon 441-350, Korea.

### References
1. Novoselov KS, Jiang Z, Zhang Y, Morozov SV, Stormer HL, Zeitler U, Maan
   JC, Boebinger GS, Kim P, Geim AK: Room-temperature quantum Hall effect
   in graphene. Science 2007, 315:1379.
2. Rao CN, Sood AK, Subrahmanyam KS, Govindaraj A: Graphene: the new
   two-dimensional nanomaterial. Angew Chem Int Ed Engl 2009,
   48:7752–7777.
3. Dreyer DR, Park S, Bielawski CW, Ruoff RS: The chemistry of graphene
   oxide. Chem Soc Rev 2010, 39:228–240.
4. Sanchez VC, Jachak A, Hurt RH, Kane AB: Biological interactions of
   graphene-family nanomaterials: an interdisciplinary review. Chem Res
   Toxicol 2012, 25:15–34.
5. Lu CH, Yang HH, Zhu CL, Chen X, Chen GN: A graphene platform for
   sensing biomolecules. Angew Chem Int Ed Engl 2009, 48:4785–4787.
6. Hu W, Peng C, Luo W, Lv M, Li X, Li D, Huang Q, Fan C: Graphene-based
   antibacterial paper. ACS Nano 2010, 4:4317–4323.
7. Akhavan O, Ghaderi E: Toxicity of graphene and graphene oxide
   nanowalls against bacteria. ACS Nano 2010, 4:5731–5736.
8. Gurunathan S, Han JW, Dayem AA, Eppakayala V, Kim JH: Oxidative stress-
   mediated antibacterial activity of graphene oxide and reduced graphene
   oxide in Pseudomonas aeruginosa. Int J Nanomedicine 2012, 7:5901–5914.
9. Sun XM, Liu Z, Welsher K, Robinson JT, Goodwin A, Zaric S, Dai HJ: Nano-
   graphene oxide for cellular imaging and drug delivery. Nano Res 2008,
   1:203–212.

10. Nayak TR, Andersen H, Makam VS, Khaw C, Bae S, Xu X, Ee PL, Ahn JH, Hong BH, Pastorin G, Ozyilmaz B: Graphene for controlled and accelerated osteogenic differentiation of human mesenchymal stem cells. ACS Nano 2011, 5:4670–4678.

11. Song Y, Qu K, Zhao C, Ren J, Qu X: Graphene oxide: intrinsic peroxidase catalytic activity and its application to glucose detection. Adv Mater 2010, 22:2206–2210.

12. Wang L, Lee K, Sun YY, Lucking M, Chen ZF, Zhao JJ, Zhang SBB: Graphene oxide as an ideal substrate for hydrogen storage. ACS Nano 2009, 3:2995–3000.

13. Wang Y, Zhang P, Liu CF, Zhan L, Li YF, Huang CZ: Green and easy synthesis of biocompatible graphene for use as an anticoagulant. Rsc Adv 2012, 2:2322–2328.

14. Stankovich S, Dikin DA, Piner RD, Kohlhaas KA, Kleinhammes A, Jia Y, Wu Y, Nguyen ST, Ruoff RS: Synthesis of graphene-based nanosheets via chemical reduction of exfoliated graphite oxide. Carbon 2007, 45:1558–1565.

15. Cote LJ, Cruz-Silva R, Huang JX: Flash reduction and patterning of graphite oxide and its polymer composite. J Am Chem Soc 2009, 131:11027–11032.

16. Zhou Y, Bao QL, Tang LAL, Zhong YL, Loh KP: Hydrothermal dehydration for the "green" reduction of exfoliated graphene oxide to graphene and demonstration of tunable optical limiting properties. Chem Mater 2009, 21:2950–2956.

17. Hass J, de Heer WA, Conrad EH: The growth and morphology of epitaxial multilayer graphene. J Phys Condens Matter 2008, 20:323202 (27pp).

18. Akhavan O, Ghaderi E: Photocatalytic reduction of graphene oxide nanosheets on TiO2 thin film for photoinactivation of bacteria in solar light irradiation. J Phys Chem C 2009, 113:20214–20220.

19. Akhavan O, Abdolahad M, Esfandiar A, Mohatashamifar M: Photodegradation of graphene oxide sheets by TiO2 nanoparticles after a photocatalytic reduction. J Phys Chem C 2010, 114:12955–12959.

20. Akhavan O, Choobtashani M, Ghaderi E: Protein degradation and RNA efflux of viruses photocatalyzed by graphene-tungsten oxide composite under visible light irradiation. J Phys Chem C 2012, 116:9653–9659.

21. Gurunathan S, Han J, Park JH, Kim JH: An in vitro evaluation of graphene oxide reduced by Ganoderma spp. in human breast cancer cells (MDA-MB-231). Int J Nanomedicine 2014, 9:1783–1797.

22. Tang LAL, Lee WC, Shi H, Wong EYL, Sadovoy A, Gorelik S, Hobley J, Lim CT, Loh KP: Highly wrinkled cross-linked graphene oxide membranes for biological and charge-storage applications. Small 2012, 8:423–431.

23. Stankovich S, Piner RD, Chen XQ, Wu NQ, Nguyen ST, Ruoff RS: Stable aqueous dispersions of graphitic nanoplatelets via the reduction of exfoliated graphite oxide in the presence of poly(sodium 4-styrenesulfonate). J Mater Chem 2006, 16:155–158.

24. Min K, Han TH, Kim J, Jung J, Jung C, Hong SM, Koo CM: A facile route to fabricate stable reduced graphene oxide dispersions in various media and their transparent conductive thin films. J Colloid Interface Sci 2012, 383:36–42.

25. Fan XB, Peng WC, Li Y, Li XY, Wang SL, Zhang GL, Zhang FB: Deoxygenation of exfoliated graphite oxide under alkaline conditions: a green route to graphene preparation. Adv Mater 2008, 20:4490–4493.

26. Akhavan O, Bijanzad K, Mirsepah A: Synthesis of graphene from natural and industrial carbonaceous wastes. Rsc Adv 2014, 4:20441–20448.

27. Zhang Y, Ali SF, Dervishi E, Xu Y, Li Z, Casciano D, Biris AS: Cytotoxicity effects of graphene and single-wall carbon nanotubes in neural phaeochromocytoma-derived PC12 cells. ACS Nano 2010, 4:3181–3186.

28. Chang Y, Yang ST, Liu JH, Dong E, Wang Y, Cao A, Liu Y, Wang H: In vitro toxicity evaluation of graphene oxide on A549 cells. Toxicol Lett 2011, 200:201–210.

29. Wang K, Ruan J, Song H, Zhang JL, Wo Y, Guo SW, Cui DX: Biocompatibility of graphene oxide. Nanoscale Res Lett 2011, 6:8.

30. Gurunathan S, Han JW, Eppakayala V, Dayem AA, Kwon DN, Kim JH: Biocompatibility effects of biologically synthesized graphene in primary mouse embryonic fibroblast cells. Nanoscale Res Lett 2013, 8:393.

31. Lu CH, Zhu CL, Li J, Liu JJ, Chen X, Yang HH: Using graphene to protect DNA from cleavage during cellular delivery. Chem Commun 2010, 46:3116–3118.

32. Sasidharan A, Panchakarla LS, Chandran P, Menon D, Nair S, Rao CNR, Koyakutty M: Differential nano-bio interactions and toxicity effects of pristine versus functionalized graphene. Nanoscale 2011, 3:2461–2464.

33. Pan YZ, Bao HQ, Sahoo NG, Wu TF, Li L: Water-Soluble Poly(N-isopropylacrylamide)-Graphene sheets synthesized via click chemistry for drug delivery. Adv Funct Mater 2011, 21:2754–2763.

34. Fan HL, Wang LL, Zhao KK, Li N, Shi ZJ, Ge ZG, Jin ZX: Fabrication, mechanical properties, and biocompatibility of graphene-reinforced chitosan composites. Biomacromolecules 2010, 11:2345–2351.

35. Chen YF, Qi YY, Tai ZX, Yan XB, Zhu FL, Xue QJ: Preparation, mechanical properties and biocompatibility of graphene oxide/ultrahigh molecular weight polyethylene composites. Eur Polym J 2012, 48:1026–1033.

36. Akhavan O, Ghaderi E, Akhavan A: Size-dependent genotoxicity of graphene nanoplatelets in human stem cells. Biomaterials 2012, 33:8017–8025.

37. Alzhavan O, Ghaderi E, Shahsavar M: Graphene nanogrids for selective and fast osteogenic differentiation of human mesenchymal stem cells. Carbon 2013, 59:200–211.

38. Akhavan O, Ghaderi E, Abouei E, Hatamie S, Ghasemi E: Accelerated differentiation of neural stem cells into neurons on ginseng-reduced graphene oxide sheets. Carbon 2014, 66:395–406.

39. Park SY, Park J, Sim SH, Sung MG, Kim KS, Hong BH, Hong S: Enhanced differentiation of human neural stem cells into neurons on graphene. Adv Mater 2011, 23:H263–+.

40. Lee WC, Lim CHYX, Shi H, Tang LAL, Wang Y, Lim CT, Loh KP: Origin of enhanced stem cell growth and differentiation on graphene and graphene oxide. ACS Nano 2011, 5:7334–7341.

41. Akhavan O, Ghaderi E: Flash photo stimulation of human neural stem cells on graphene/TiO2 heterojunction for differentiation into neurons. Nanoscale 2013, 5:10316–10326.

42. Akhavan O, Ghaderi E: Differentiation of human neural stem cells into neural networks on graphene nanogrids. J Mater Chem B 2013, 1:6291–6301.

43. Fernandez-Merino MJ, Guardia L, Paredes JI, Villar-Rodil S, Solis-Fernandez P, Martinez-Alonso A, Tascon JMD: Vitamin C is an ideal substitute for hydrazine in the reduction of graphene oxide suspensions. J Phys Chem C 2010, 114:6426–6432.

44. Gao J, Liu F, Liu YL, Ma N, Wang ZQ, Zhang X: Environment-friendly method to produce graphene that employs vitamin C and amino acid. Chem Mater 2010, 22:2213–2218.

45. Zhu CZ, Guo SJ, Fang YX, Dong SJ: Reducing sugar: New functional molecules for the green synthesis of graphene nanosheets. ACS Nano 2010, 4:2429–2437.

46. Liu JB, Fu SH, Yuan B, Li YL, Deng ZX: Toward a universal "adhesive nanosheet" for the assembly of multiple nanoparticles based on a protein-induced reduction/decoration of graphene oxide. J Am Chem Soc 2010, 132:7279–+.

47. Wang GM, Qian F, Saltikov C, Jiao YQ, Li Y: Microbial reduction of graphene oxide by Shewanella. Nano Res 2011, 4:563–570.

48. Gurunathan S, Han JW, Eppakayala V, Kim JH: Microbial reduction of graphene oxide by Escherichia coli: a green chemistry approach. Colloid Surface B 2013, 102:772–777.

49. Akhavan O, Ghaderi E: Escherichia coli bacteria reduce graphene oxide to bactericidal graphene in a self-limiting manner. Carbon 2012, 50:1853–1860.

50. Gurunathan S, Han JW, Eppakayala V, Kim JH: Green synthesis of graphene and its cytotoxic effects in human breast cancer cells. Int J Nanomedicine 2013, 8:1015–1027.

51. Esfandiar A, Akhavan O, Irajizad A: Melatonin as a powerful bio-antioxidant for reduction of graphene oxide. J Mater Chem 2011, 21:10907–10914.

52. Pham TA, Kim JS, Kim JS, Jeong YT: One-step reduction of graphene oxide with L-glutathione. Colloid Surface A 2011, 384:543–548.

53. Gurunathan S, Han J, Kim JH: Humanin: a novel functional molecule for the green synthesis of graphene. Colloid Surface B 2013, 111:376–383.

54. Deepak V, Umamaheshwaran PS, Guhan K, Nanthini RA, Krithiga B, Jaithoon NMH, Gurunathan S: Synthesis of gold and silver nanoparticles using purified URAK. Colloid Surface B 2011, 86:353–358.

55. Vallhov H, Qin J, Johansson SM, Ahlborg N, Muhammed MA, Scheynius A, Gabrielsson S: The importance of an endotoxin-free environment during the production of nanoparticles used in medical applications. Nano Lett 2006, 6:1682–1686.

56. Tsien RY: The green fluorescent protein. Annu Rev Biochem 1998, 67:509–544.

57. Godwin AR, Stadler HS, Nakamura K, Capecchi MR: Detection of targeted GFP-Hox gene fusions during mouse embryogenesis. Proc Natl Acad Sci U S A 1998, 95:13042–13047.

58. Heim R, Prasher DC, Tsien RY: Wavelength mutations and posttranslational autoxidation of green fluorescent protein. Proc Natl Acad Sci U S A 1994, 91:12501–12504.

59. Rafat M, Cleroux CA, Fong WG, Baker AN, Leonard BC, O'Connor MD, Tsilfidis C: PEG-PLA microparticles for encapsulation and delivery of Tat-EGFP to retinal cells. *Biomaterials* 2010, 31:3414–3421.

60. Li X, Zhang G, Ngo N, Zhao X, Kain SR, Huang CC: Deletions of the Aequorea victoria green fluorescent protein define the minimal domain required for fluorescence. *J Biol Chem* 1997, 272:28545–28549.

61. Brocklehurst K, Little G: Reactivities of the various protonic states in the reactions of papain and of L-cysteine with 2,2'- and with 4,4'- dipyridyl disulphide: evidence for nucleophilic reactivity in the un-ionized thiol group of the cysteine-25 residue of papain occasioned by its interaction with the histidine-159-asparagine-175 hydrogen-bonded system. *Biochem J* 1972, 128:471–474.

62. Chen D, Li L, Guo L: An environment-friendly preparation of reduced graphene oxide nanosheets via amino acid. *Nanotechnology* 2011, 22:325601.

63. Hummers WS, Offeman RE: Preparation of graphitic oxide. *J Am Chem Soc* 1958, 80:1339–1339.

64. Luo ZT, Lu Y, Somers LA, Johnson ATC: High yield preparation of macroscopic graphene oxide membranes. *J Am Chem Soc* 2009, 131:898–+.

65. Eda G, Chhowalla M: Chemically derived graphene oxide: towards large-area thin-film electronics and optoelectronics. *Adv Mater* 2010, 22:2392–2415.

66. Zhang JL, Yang HJ, Shen GX, Cheng P, Zhang JY, Guo SW: Reduction of graphene oxide via L-ascorbic acid. *Chem Commun* 2010, 46:1112–1114.

67. Cheng C, Nie SQ, Li S, Peng H, Yang H, Ma L, Sun SD, Zhao CS: Biopolymer functionalized reduced graphene oxide with enhanced biocompatibility via mussel inspired coatings/anchors. *J Mater Chem B* 2013, 1:265–275.

68. Xu YX, Bai H, Lu GW, Li C, Shi GQ: Flexible graphene films via the filtration of water-soluble noncovalent functionalized graphene sheets. *J Am Chem Soc* 2008, 130:5856–+.

69. Choi EY, Han TH, Hong JH, Kim JE, Lee SH, Kim HW, Kim SO: Noncovalent functionalization of graphene with end-functional polymers. *J Mater Chem* 2010, 20:1907–1912.

70. Gurunathan S, Han JW, Park JH, Eppakayala V, Kim JH: Ginkgo biloba: a natural reducing agent for the synthesis of cytocompatible graphene. *Int J Nanomedicine* 2014, 9:363–377.

71. Talukdar Y, Rashkow JT, Lalwani G, Kanakia S, Sitharaman B: The effects of graphene nanostructures on mesenchymal stem cells. *Biomaterials* 2014, 35:4863–4877.

72. Cheng C, Li S, Nie SQ, Zhao WF, Yang H, Sun SD, Zhao CS: General and biomimetic approach to biopolymer-functionalized graphene oxide nanosheet through adhesive dopamine. *Biomacromolecules* 2012, 13:4236–4246.

73. Lammel T, Boisseaux P, Fernandez-Cruz ML, Navas JM: Internalization and cytotoxicity of graphene oxide and carboxyl graphene nanoplatelets in the human hepatocellular carcinoma cell line Hep G2. *Part Fibre Toxicol* 2013, 10:27.

74. Liu SB, Zeng TH, Hofmann M, Burcombe E, Wei J, Jiang RR, Kong J, Chen Y: Antibacterial activity of graphite, graphite oxide, graphene oxide, and reduced graphene oxide: membrane and oxidative stress. *ACS Nano* 2011, 5:6971–6980.

75. Gurunathan S, Han JW, Kim JH: Green chemistry approach for the synthesis of biocompatible graphene. *Int J Nanomedicine* 2013, 8:2719–2732.

76. Yang F, Liu YQ, Gao LA, Sun J: pH-Sensitive highly dispersed reduced graphene oxide solution using lysozyme via an in situ reduction method. *J Phys Chem C* 2010, 114:22085–22091.

77. Zhou NL, Gu H, Tang FF, Li WX, Chen YY, Yuan J: Biocompatibility of novel carboxylated graphene oxide-glutamic acid complexes. *J Mater Sci* 2013, 48:7097–7103.

78. Prasanna K, Natarajan R, Kaveripatnam S, Dhathathreyan KS: Functionalized exfoliated graphene oxide as supercapacitor electrodes. *Sci Res Pub* 2012, 2:59–66.

79. Jeong HK, Lee YP, Lahaye RJWE, Park MH, An KH, Kim IJ, Yang CW, Park CY, Ruoff RS, Lee YH: Evidence of graphitic AB stacking order of graphite oxides. *J Am Chem Soc* 2008, 130:1362–1366.

80. He HK, Gao C: General approach to individually dispersed, highly soluble, and conductive graphene nanosheets functionalized by nitrene chemistry. *Chem Mater* 2010, 22:5054–5064.

81. Lian PC, Zhu XF, Liang SZ, Li Z, Yang WS, Wang HH: Large reversible capacity of high quality graphene sheets as an anode material for lithium-ion batteries. *Electrochim Acta* 2010, 55:3909–3914.

82. Wang YY, Ni ZH, Shen ZX, Wang HM, Wu YH: Interference enhancement of Raman signal of graphene. *Appl Phys Lett* 2008, 92:043121.

83. Vernekar AA, Mugesh G: Hemin-functionalized reduced graphene oxide nanosheets reveal neuroxynitrite reduction and isomerization activity. *Chem-Eur J* 2012, 18:15122–15132.

84. Tuinstra F, Koenig JL: Raman spectrum of graphite. *J Chem Phys* 1970, 53:1126–1130.

85. Ferrari AC, Robertson J: Resonant Raman spectroscopy of disordered, amorphous, and diamondlike carbon. *Phys Rev B* 2001, 64:075414.

86. Fan ZJ, Kai W, Yan J, Wei T, Zhi LJ, Feng J, Ren YM, Song LP, Wei F: Facile synthesis of graphene nanosheets via Fe reduction of exfoliated graphite oxide. *ACS Nano* 2011, 5:191–198.

87. Lin ZY, Yao YG, Li Z, Liu Y, Li Z, Wong CP: Solvent-assisted thermal reduction of graphite oxide. *J Phys Chem C* 2010, 114:14819–14825.

88. Stankovich S, Dikin DA, Dommett GHB, Kohlhaas KM, Zimney EJ, Stach EA, Piner RD, Nguyen ST, Ruoff RS: Graphene-based composite materials. *Nature* 2006, 442:282–286.

89. Akhavan O, Ghaderi E, Aghayee S, Fereydooni Y, Talebi A: The use of a glucose-reduced graphene oxide suspension for photothermal cancer therapy. *J Mater Chem* 2012, 22:13773–13781.

90. Sim Y, Park J, Kim YJ, Seong MJ, Hong S: Synthesis of graphene layers using graphite dispersion in aqueous surfactant solutions. *J Korean Phys Soc* 2011, 58:938–942.

91. Green AA, Hersam MC: Solution phase production of graphene with controlled thickness via density differentiation. *Nano Lett* 2009, 9:4031–4036.

92. Kim YK, Kim MH, Min DH: Biocompatible reduced graphene oxide prepared by using dextran as a multifunctional reducing agent. *Chem Commun* 2011, 47:3195–3197.

93. Khanra P, Kuila T, Kim NH, Bae SH, Yu DS, Lee JH: Simultaneous bio-functionalization and reduction of graphene oxide by baker's yeast. *Chem Eng J* 2012, 183:526–533.

94. Gurunathan S, Han JW, Dayem AA, Eppakayala V, Park MR, Kwon DN, Kim JH: Antibacterial activity of dithiothreitol reduced graphene oxide. *J Ind Eng Chem* 2013, 19:1280–1288.

95. Shin HJ, Kim KK, Benayad A, Yoon SM, Park HK, Jung IS, Jin MH, Jeong HK, Kim JM, Choi JY, Lee YH: Efficient reduction of graphite oxide by sodium borohydride and its effect on electrical conductance. *Adv Funct Mater* 2009, 19:1987–1992.

96. Thomas P, Smart TG: HEK293 cell line: a vehicle for the expression of recombinant proteins. *J Pharmacol Toxicol Methods* 2005, 51:187–200.

97. Hu WB, Peng C, Lv M, Li XM, Zhang YJ, Chen N, Fan CH, Huang Q: Protein corona-mediated mitigation of cytotoxicity of graphene oxide. *ACS Nano* 2011, 5:3693–3700.

98. Kuila T, Bose S, Khanra P, Mishra AK, Kim NH, Lee JH: Recent advances in graphene-based biosensors. *Biosens Bioelectron* 2011, 26:4637–4648.

99. Feng LZ, Liu ZA: Graphene in biomedicine: opportunities and challenges. *Nanomedicine* 2011, 6:317–324.

100. Hong SW, Lee JH, Kang SH, Hwang EY, Hwang YS, Lee MH, Han DW, Park JC: Enhanced neural cell adhesion and neurite outgrowth on graphene-based biomimetic substrates. *Biomed Res Int* 2014, Article ID 212149, 8 pages.

101. Yang K, Li YJ, Tan XF, Peng R, Liu Z: Behavior and toxicity of graphene and its functionalized derivatives in biological systems. *Small* 2013, 9:1492–1503.

102. Gurunathan S, Han JW, Eppakayala V, Kim JH: Biocompatibility of microbially reduced graphene oxide in primary mouse embryonic fibroblast cells. *Colloids Surf B: Biointerfaces* 2013, 105:58–66.

103. Reddy ARN, Reddy YN, Krishna DR, Himabindu V: Multi wall carbon nanotubes induce oxidative stress and cytotoxicity in human embryonic kidney (HEK293) cells. *Toxicology* 2010, 272:11–16.

104. Zhang YB, Xu Y, Li ZG, Chen T, Lantz SM, Howard PC, Paule MG, Slikker W, Watanabe F, Mustafa T, Biris AS, Ali SF: Mechanistic toxicity evaluation of uncoated and PEGylated single-walled carbon nanotubes in neuronal PC12 cells. *ACS Nano* 2011, 5:7020–7033.

105. Shvedova AA, Pietroiusti A, Fadeel B, Kagan VE: Mechanisms of carbon nanotube-induced toxicity: Focus on oxidative stress. *Toxicol Appl Pharmacol* 2012, 261:121–133.

106. Stern ST, Adiseshaiah PP, Crist RM: Autophagy and lysosomal dysfunction as emerging mechanisms of nanomaterial toxicity. *Part Fibre Toxicol* 2012, 9:20.

107. Qu G, Wang X, Wang Z, Liu S, Jiang G: **Cytotoxicity of quantum dots and graphene oxide to erythroid cells and macrophages.** *Nanoscale Res Lett* 2013, **8**:198.
108. Wu Q, Zhao Y, Zhao G, Wang D: **microRNAs control of in vivo toxicity from graphene oxide in Caenorhabditis elegans.** *Nanomedicine* 2014, doi:10.1016/j.nano.2014.04.005.
109. Chong Y, Ma Y, Shen H, Tu X, Zhou X, Xu J, Dai J, Fan S, Zhang Z: **The in vitro and in vivo toxicity of graphene quantum dots.** *Biomaterials* 2014, **35**:5041–5048.
110. Shen JF, Hu YH, Li C, Qin C, Ye MX: **Synthesis of Amphiphilic graphene nanoplatelets.** *Small* 2009, **5**:82–85.

# 11

# Plants and microbes assisted selenium nanoparticles: characterization and application

Azamal Husen[1]* and Khwaja Salahuddin Siddiqi[2]

**Abstract**

Selenium is an essential trace element and is an essential component of many enzymes without which they become inactive. The Se nanoparticles of varying shape and size may be synthesized from Se salts especially selenite and selenates in presence of reducing agents such as proteins, phenols, alcohols and amines. These biomolecules can be used to reduce Se salts in vitro but the byproducts released in the environment may be hazardous to flora and fauna. In this review, therefore, we analysed in depth, the biogenic synthesis of Se nanoparticles, their characterization and transformation into t- Se, m-Se, Se-nanoballs, Se-nanowires and Se-hollow spheres in an innocuous way preventing the environment from pollution. Their shape, size, FTIR, UV–vis, Raman spectra, SEM, TEM images and XRD pattern have been analysed. The weak forces involved in aggregation and transformation of one nano structure into the other have been carefully resolved.

**Keywords:** Nanotechnology, Selenium, Plant extracts, Microbes, Biofabrication

## Introduction

Selenium was known as a notorious element until it was recognized by Schwarz and Foltz in 1957 as an essential trace element for both plants and mammals. Normally Se is available as selenate and selenite oxoanions. The reduction of soluble $Se^{4+}$ and $Se^{6+}$ by microbes to insoluble non toxic elemental Se is an effective way to remove it from contaminated soil, water and drainage [1]. Se is one of the chalcogens occurring as selenate $SeO_4^{2-}$, selenite $SeO_3^{2-}$ and selenide $Se^{2-}$ which may be reduced to atomic state by a precursor containing an appropriate reducing agent. Biogenic synthesis of Se nanoparticles is frequently achieved by reduction of selenate/selenite in presence of bacterial proteins or plant extracts containing phenols, flavonoids amines, alcohols, proteins and aldehydes. The deficiency of Se is known to be associated with over 40 diseases in man [2,3]. At low dosage it can stimulate the growth of the plant whereas at high dosages it can cause damage to it [4-6]. Se has also been shown to be effective against cancer [7,8]. Their compounds in the form of selenocysteine and selenomethionine are metabolized in biological system [7,8].

A variety of microorganisms, enzymes and fungi, besides plant extracts have been used to synthesise Se nanoparticles of different size and morphology. Se itself is used in rectifier, solar cells, photocopier and semiconductor. In addition, they exhibit biological activity owing to their interaction with the proteins and other biomolecules present in the bacterial cells and plant extracts, containing functional groups such as ›NH, C = O, COO and C-N [9]. Se-nanobelts have been synthesised on large scale with an approximate diameter of 80 nm and length up to 5 µm [10]. Se exists in many crystalline and amorphous forms but the shape, size and structure of the nanoparticles depend on the concentration, temperature, nature of biomolecules and pH of the reaction mixture. The properties of Se nanoparticles varies with size and shape for instance, Se nanospheres have high biological activity and low toxicity while Se nanowires of t-Se have high photoconductivity [11]. Various methods have been employed to produce large scale Se-nanowires and trigonal selenium (t-Se) [12,13]. Pulse laser ablation, electro-kinetic technique, hydrothermal treatment, vapour deposition methods [10,14-16] generally used for production of Se nanoparticles on large scale require either sophisticated instruments or specific chemicals which are time consuming and uneconomical. Such methods often employ toxic

* Correspondence: adroot92@yahoo.co.in
[1]Department of Biology, College of Natural and Computational Sciences, University of Gondar, P.O. Box 196, Gondar, Ethiopia
Full list of author information is available at the end of the article

chemicals or high temperature and high pressure which further pollute the environment.

In order to circumvent the effect of toxic chemicals in the fabrication of nanoparticles, biogenic protocol is generally followed [17,18]. Scientists have developed benign and harmless methods for the fabrication of nanoparticles using plant extracts, bacteria and fungi. For instance, *Capsicum annum*, *Escherichia coli* and *Bacillus subtilis* [19-21] have recently been used to produce nanoparticles. Over 16 different species of bacteria and Arechaea have been found to reduce colourless selenate and selenite to red elemental Se of different shape and size [22,23]. Plants and microbes act as producers and protectors of the environment when they are properly used. Pure element in its, atomic state may be produced by many bacteria [24,25] mainly due to the chemicals present in them or protein exuded by them. We have limited knowledge of the mechanism of the formation of Se nanoparticles by microbes and plant extract, nevertheless for a better understanding attempts are being made to explore the chemical reactions occurring in these media. Many bacterial strains have been found to reduce selenate/selenite to Se nanoparticles in different environment [26] even in sewage and sludge under both aerobic and anaerobic conditions [27-29]. It has been suggested that substantial quantity of soluble toxic selenate/selenite is reduced by bacterial strain to produce non toxic insoluble Se nanoparticles, although in doing so many such microbes would die. The production of $Se^0/Te^0$ by two anaerobic bacteria *Sulfurospirillum barnesii* and *Bacillus selenireducens* has been demonstrated by Oremland et al. and Baesman et al. [25,24].

The main objective of this review is to identify the plant extracts and bacterial strains involved in the biosynthesis of Se nanoparticles. Also the characterization and identification of Se-nanoballs, nanorods, nanowires and hollow spheres have been undertaken with a view to update the nanobiotechnology of Se nanoparticles and their application in diverse areas.

## Se nanoparticles from plants, characterization and application

There is a fine line between optimum limit/or deficiency and excess of Se in living system which may cause toxicity. It is known that the Se nanoparticles prepared from biological material are less toxic than the bulk Se nanoparticles prepared from chemicals. The biomolecules present in the extract act both as reducing agent and stabilizers of Se nanoparticles. Bacteria, algae, dry fruits and plant extracts are used to produce nanoparticles. Green synthesis of selenium nanoparticles from selenious acid was achieved by dried extract of raisin (*Vitis vinifera*) [30]. Like other biological materials, raisin also contains sugar, flavonoids and phenols in addition to

minerals such as iron, potassium and calcium [19,31]. A change from colourless to deeply brick-red colour indicated the formation of nanoparticles. The formation of Se-nanoballs was examined at different interval of time. It took nearly 6 min to start conversion of Se ion to Se nanoparticles which was indicated by a decrease in Se ion concentration in the solution. The nature of nanoparticles was analysed by TEM images. It showed that the diameter of nanoballs ranges between 3 and 18 nm. They were found to be encapsulated with a thin polymorphic layer. The formation of Se nanoparticles was confirmed from the energy dispersive x-ray spectroscopy. The Se nanoballs were identified from their characteristic absorption peaks at 1.37KeV, 11.22KeV and 12.49KeV [32]. The morphology of Se nanoparticles can be analysed by x-ray diffraction (XRD) analysis. The broad diffraction peak suggests the presence of amorphous nature of Se nanoparticles [33]. Their particle size [34] has been found to be of the order of 12 nm.

Sharma et al. [30] have characterized Se nanoballs fabricated from *V. vinifera* by FTIR spectral studies. The spectrum exhibited two sharp absorption peaks at $3420$ cm$^{-1}$ attributed to OH and, the second peak at $1620$ cm$^{-1}$ to C-H vibration of the organic molecules. A distinct peak at $1375$ cm$^{-1}$ has been assigned to phenolic OH. The other peaks of medium intensity are due to – CH$_3$ and OCH$_3$ groups associated with the biopolymers, present in the *V. vinifera* extract acting as reducing agent and stabilizer for the Se nanoballs. Since lignin is a component of all vegetables, fruits and cell wall, it can be extracted from them and the compounds present in them may be identified. In the present work, phenolic group has been identified which generally acts as reducing agent and, it is oxidised to ketone during the redox process. However, the extract also contains fairly substantial amount of reducing sugars and therefore, they also help in the reduction and formation of Se nanoballs. These authors have given a flow diagram for Se nanoparticles synthesis but it does not reveal the chemical changes which occur as a consequence of redox reactions. We now propose the following scheme Figure 1 based on the general synthetic route.

Although, biochemicals may often be used for the synthesis of nanomaterials, the biogenic synthetic route is frequently applied due to its ease and simplicity and, also because no hazardous and toxic residues are released in the environment [35,36]. In general, a variety of Se nanoparticles are produced when $H_2SeO_3$ is treated with plant extracts for instance, α-Se nanoparticles have been fabricated from *Capsicum annum* extract in aqueous medium at low pH and at ambient temperature [19]. The light green extract of *C. annum* turns pale after 5 h of the addition of $H_2SeO_3$, and then gradually turned red after 12 h (Figure 2a). This red colour is the characteristic signature

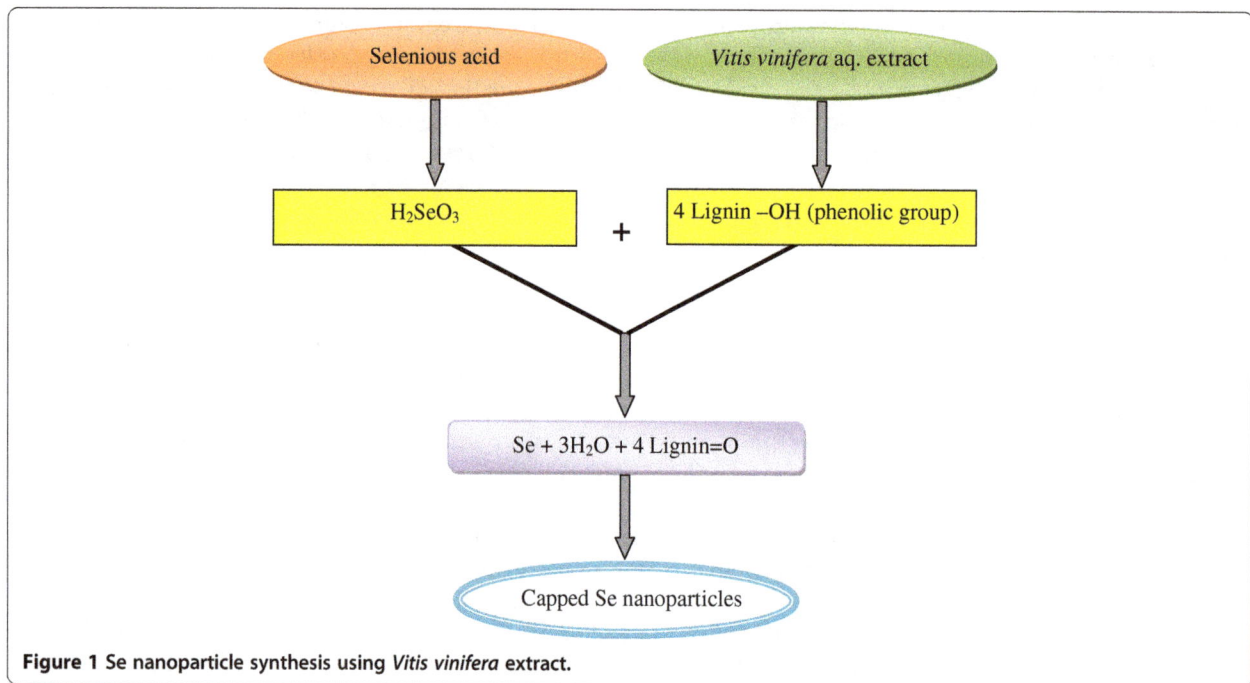

**Figure 1** Se nanoparticle synthesis using *Vitis vinifera* extract.

of α-Se in the x-ray photoemission spectroscopy (XPS) which is due to excitation of their surface plasmon vibration [37]. Its XPS spectrum (Figure 2b) shows a sharp peak at 54.4 eV which corresponds to the elemental selenium [38]. The XRD pattern of the Se nanoparticles shows a broad peak at $2\theta$ angles of 15-35$^0$ (Figure 2c) which suggests that the nanoparticles are not crystalline. Their Raman spectrum displayed a resonance peak at 263.7 cm$^{-1}$ which (Figure 2d) further confirms the formation of α-Se nanoparticles [39]. An additional peak at 474 cm$^{-1}$ has been attributed to the protein vibration which is mixed with amorphous Se.

SEM and TEM images of the α-Se nanoparticles showed that they consist of nanorods and nanoballs laced with C. annum protein which makes them slightly irregular in shape. The length and diameter of rods and nanoballs range between 1–3 μm and 0.4 μm, respectively. A closer look at the highly magnified field emission scanning electron microscopy (FESEM) image suggests that rod like nanoparticles are actually aggregates of spherical particles with protein coating, making the surface rough and uneven. It is quite likely that proteins are held together by hydrogen bonding and Se nanoparticles are held by van der Waals forces.

When the pH of the reaction mixture is lowered to 2 the time to produce α-Se nanoparticles increases. It has been observed from their FESEM images that a variety of polygonal Se nanoparticles are produced with size varying from 200–500 nm. It is of interest to note that some hollow spherical particles were also produced with a pore diameter of 160 nm. Li et al. [19] have hypothesized

that hollow spheres are formed as a consequence of rise in temperature when the reaction product is placed in an electric field. Although, the melting point of α-Se nanoparticles is not very low (~490 K) even then this temperature is seldom achieved in such system, so that it may melt and produce hollow spheres. It is to be noted that even if microwave energy is supplied without rise in temperature only the outer surface of α-Se nanoballs, made of protein layer would melt, because organic materials have inherently lower melting point than metalloid Se. However, if these α-Se particles also melt with electronic impact even then the hollow sphere would not be produced because the lattice would rupture resulting in the formation of irregular sheets and dot like structures. It is more likely, that hollow spheres of α-Se are also formed along with solid nanoballs and polygonal structures during the synthesis of nanoparticles.

A comparison of FTIR spectrum of pure *C. annum* extract with the reaction mixture (*C. annum* extract + $SeO_3^{2-}$) showed many peaks at 1652, 1542 and 1241 cm$^{-1}$ corresponding to amide I, II and III bands owing to $\upsilon(C=O)$, $\upsilon(N\text{-}H)$ and $\upsilon(C\text{-}N)$ respectively [36]. These bands slightly shift after the formation of nanoparticles. The UV–vis spectrum of the *C. annum* protein (washed with SDS-PAGE gel) with molecular weight of 30 kDa, showed peak (210 nm) corresponding to peptide bonds and amino residues (280 nm). As these are reducing agents they help in the formation of nanoparticles. It has also been confirmed from cyclic voltammogram that the redox reaction occurs between - 0.7 and 0.9 V [19].

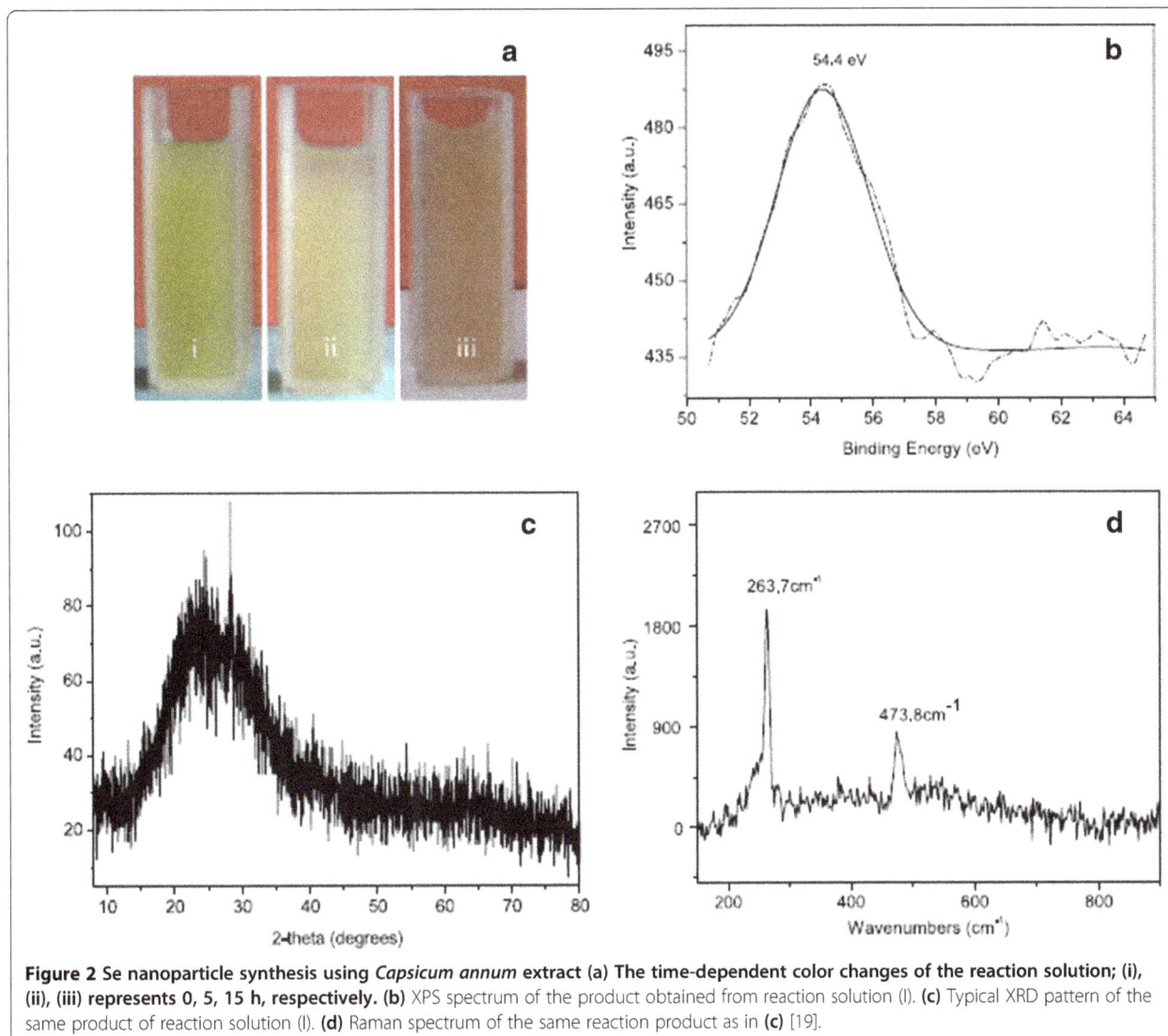

**Figure 2** Se nanoparticle synthesis using *Capsicum annum* extract **(a) The time-dependent color changes of the reaction solution; (i), (ii), (iii) represents 0, 5, 15 h, respectively. (b)** XPS spectrum of the product obtained from reaction solution (I). **(c)** Typical XRD pattern of the same product of reaction solution (I). **(d)** Raman spectrum of the same reaction product as in **(c)** [19].

Inorganic Se (selenite or selenate) also occur as seleno-methionine, selenocysteine, selenocystathione, methyl selenol, dimethyl selenide and selenium methyl seleno-cysteine. Absorption of Se depends on its morphology and solubility in aqueous medium. Sodium salts of Se are generally soluble in water. One form may change into another to suit the basic requirements of binding to certain functional groups such as proteins. Different plants absorb Se in different quantities for instance; wheat accumulates Se proportional to its availability in the soil while *Astragalus* grown in the same soil had manifold excess of the element in it. Broccoli is known as Se accumulator. Finley [40], showed from an experiment, on broccoli grown in sodium selenate laden soil, that it accumulated $\sim 10^3$ μg Se/gm dry weight of the plant tissue. Broccoli is known to contain fairly [41] large quantity of selenium as methyl selenocysteine,

perhaps due to its greater solubility in aqueous medium. However, it is strange that Se from broccoli does not accumulate efficiently in man or rats [42,43] because its major part as selenium methyl selenocysteine is perhaps metabolised to methyl selenol [44]. It has been demonstrated experimentally that, methyl substituted forms of Se is an effective anticancer agent than the other derivatives of organo-Se compounds [45]. Garlic with as much Se as 1000 μg/gm of dry weight has been grown [46] and found to contain Se as selenium methyl selenocysteine but when the Se concentration falls below 200 μg/gm, it is found as q-glutamyl selenium methyl selenocysteine. Although Se from high Se garlic is a chemoprotective agent [47] it is particularly selective against breast cancer in rats [48,49] induced by 7, 12-dimethyl benzene (a) anthracene. Even the aqueous solution of garlic is chemoprotective [50]. Glucosinolate as secondary plant compound known to induce phase II enzymes [42] is chemoprotective against bladder cancer. Se fed experimental rats in the forms of selenite, selenate, selenomethionine did not acumulate in most rats which means it is either not absorbed or excreted through urine.

Trigonal Se nanowires and nanotubes have been synthesized in absolutely ecofriendly way. The Se nanowires of 70–100 nm width and length in several μm were prepared in absolute ethanol at room temperature while trigonal Se nanotubes of diameter 180–350 nm were obtained in aqueous medium at 85°C (358 K). It was observed that amorphous Se nanoparticles were formed in the beginning and subsequently transformed into nanowires and tubes [51].

Stable Se nanoparticles in colloidal form were prepared from *Terminalia arjuna* extract in aqueous medium. They were characterized by UV–vis, energy dispersive X-ray analysis (EDAX), transmission electron microscopy (TEM), FTIR and XRD analysis. The colloidal solution had an absorbance maximum at 390 nm. Its IR spectrum showed peaks corresponding to O-H, NH, C = O and C-O stretches suggesting the presence of hydroxyl, amino, ketonic and carbonyl functional groups in the extract which act both as stabilizer and capping agent for the Se nanoparticles [52]. The Se nanoparticles synthesized from fenugreek seed extract in aquous medium at room temperature are between 50–150 nm. They have been found to be active against human breast cancer cells [53].

Se nanoparticles of approximately 35 nm have been synthesized from gum arabic which remain stable in solution for about 30 days. The gum arabic was found to be a better stabilizer for Se nanoparticles than the hydrolysed gum arabic [54]. The Se nanoparticles synthesized from lemon leaf extract exhibited an absorption maximum at 395 nm in the UV–vis region. Initially, the mixture of leaf extract and $SeO_3^{2-}$ remains colourless but after stirring and incubating it for 24 hr at 30°C, it turns red [55]. The photoluminescence spectra exhibited excitation peak at 395 and emission peak at 525 nm (Figure 3). It has been found from TEM image that the size of particles ranges between 60–80 nm. They are polydispersed in colloidal solution but crystalline in nature [55]. The FTIR spectra of the samples with and without Se nanoparticles were compared to examine the changes in the functional groups of the biomolecules. The broad peak at 3415 due to υ(NH) shifts to 3418 $cm^{-1}$ but new peaks appear at 2930 and 3456 $cm^{-1}$ in the colloidal solution containing Se nanoparticles. The region 1500–1800 $cm^{-1}$ is due to various amide bands which split into some new bands in colloidal solution. However, after reduction of the $Na_2SeO_3$ to Se nanoparticles by the biomolecules in the extract containing functional groups such as alcohol, aldehyde, phenol etc., they are oxidized to the following species:

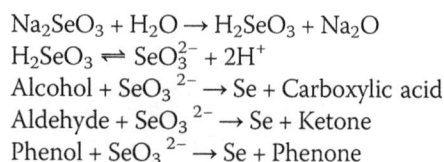

$$Na_2SeO_3 + H_2O \rightarrow H_2SeO_3 + Na_2O$$
$$H_2SeO_3 \rightleftharpoons SeO_3^{2-} + 2H^+$$
$$Alcohol + SeO_3^{2-} \rightarrow Se + Carboxylic\ acid$$
$$Aldehyde + SeO_3^{2-} \rightarrow Se + Ketone$$
$$Phenol + SeO_3^{2-} \rightarrow Se + Phenone$$

It has also been detected from gel electrophoresis that Se nanoparticles prevented DNA damage when cells were exposed to UVB [55].

Polyphenol gallic acid nanoparticles from plant have been used to fabricate Se nanoparticles since the gallic acid nanoparticles may behave differently than the bulk gallic aicd. Since gallic acid can quantitatively coordinate with the Se ions, another reducing agent dithioerthreitol was added to gallic acid-coated with Se ions. A change in colour was taken as an indication of the formation of nanoparticles which was confirmed by UV–vis and emission spectroscopy [56]. A slightly different method has also been employed by Ingole et al. [34] to prepare Se nanoparticles from glucose. $Na_2SeSO_3$ prepared from Se powder was treated with glucose solution according to the following chemical reactions:

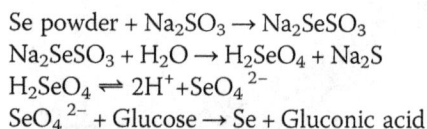

$$Se\ powder + Na_2SO_3 \rightarrow Na_2SeSO_3$$
$$Na_2SeSO_3 + H_2O \rightarrow H_2SeO_4 + Na_2S$$
$$H_2SeO_4 \rightleftharpoons 2H^+ + SeO_4^{2-}$$
$$SeO_4^{2-} + Glucose \rightarrow Se + Gluconic\ acid$$

The colourless solution in the beginning becomes yellow then orange and finally turns red which does not change even after heating for over 1 h. These changes have been ascribed to the changes in size of Se nanoparticles.

### Se nanoparticles from microbes, characterization and application

Microorganisms reduce the toxic, selenate and selenite oxoanions into non toxic elemental selenium which is insoluble in water. Continuous use of water or edible plants from Se rich soil can cause skin lesion and early hair fall. Effort is therefore, made to reduce Se compounds to elemental Se with the help of bacteria. It is a simple process of detoxification of selenites/selenates to Se nanoparticles as the reverse reaction is too slow to produce Se compounds [57].

Fast (forward reaction)
$$Se(IV)/Se\ (VI) \rightleftharpoons Se(o)$$

Slow (backward reaction)

Due to their unique property Se nanoparticles are photovoltaic and semiconductor, antioxidant and chemoprotective agents [58]. Since Se nanoparticles inhibit the growth of *Staphylococcus aureus* it can be used as a

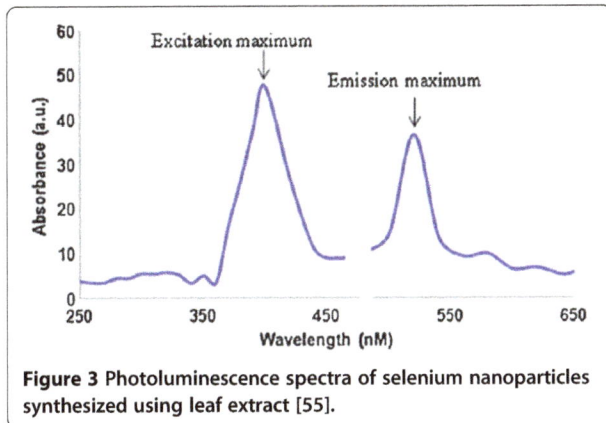

**Figure 3** Photoluminescence spectra of selenium nanoparticles synthesized using leaf extract [55].

medicine against *S. aureus* infection. Different concentration of Se starting from 65 to 230 mg/L of Se(IV) were allowed to interact with different types of microbes. Appearance of red colour was taken as sign of reduction of Se(IV) to Se(0) as shown in forward reaction above. However, there was no decolouration later, indicating the absence of any species causing oxidation of Se(IV) → Se(VI) (Figure 4). The redox process is time and concentration dependent. When bacterial culture was grown in presence of 40–100 mg/L selanate, no change in colour was observed even after long time. It appears as if the bacteria are resistant to Se(VI) reduction. However, such bacterial culture may be used to reduce soluble and toxic Se(IV) to

non toxic and insoluble Se nanoparticles. It is also indicative of bioremediation of Se from selenites.

Amporphous Se nanoparticles have been synthesized from sodium selenite in presence of *Shewanella* sp. HN-41 in aqueous medium under anaerobic conditions taking care of reaction time, selenite concentration and biomass of *Shewanella* [59]. Different types of the Se nanoparticles are synthesized using protein, peptides and several other reducing agents [1,60,61]. Nanowires and nanorods have been fabricated from *Rhizobium selenireducens* sp., *Dechlorosoma* sp., *Pseudomonas* sp., *Paracoccus* sp., *Enterobactor* sp., *Thaurea* sp., *Sulfurospirillium* sp., *Desulfovibrio* sp., and *Shewanella* sp., [61-63]. It has been reported that the particle size is decreased in presence of $O_2$. It is obvious that oxygen will promote oxidation of Se (forward reaction) as a consequence of which the redox step becomes slower producing smaller Se nanoparticles. Selenite reductase is also helpful in the synthesis of Se nanoparticles. A wide range of selenite concentration starting from 0.01, 0.05, 0.15, 0.25, 0.050, 0.75 and 1 mM were used to study the effect of concentration, size and morphology. Average particle size for all the above concentrations was nearly $103 \pm 5.1$ nm. For large quantity of nanoparticles the selenite concentration not exceeding 0.1 mM is needed. It has been observed from TEM and SAED image that Se nanoparticles are amorphous [59]. Extracellular synthesis [64] of fairly smaller particles of the order of 47 nm from the fungus, *Aspergillus terreus* was done in 60 min.

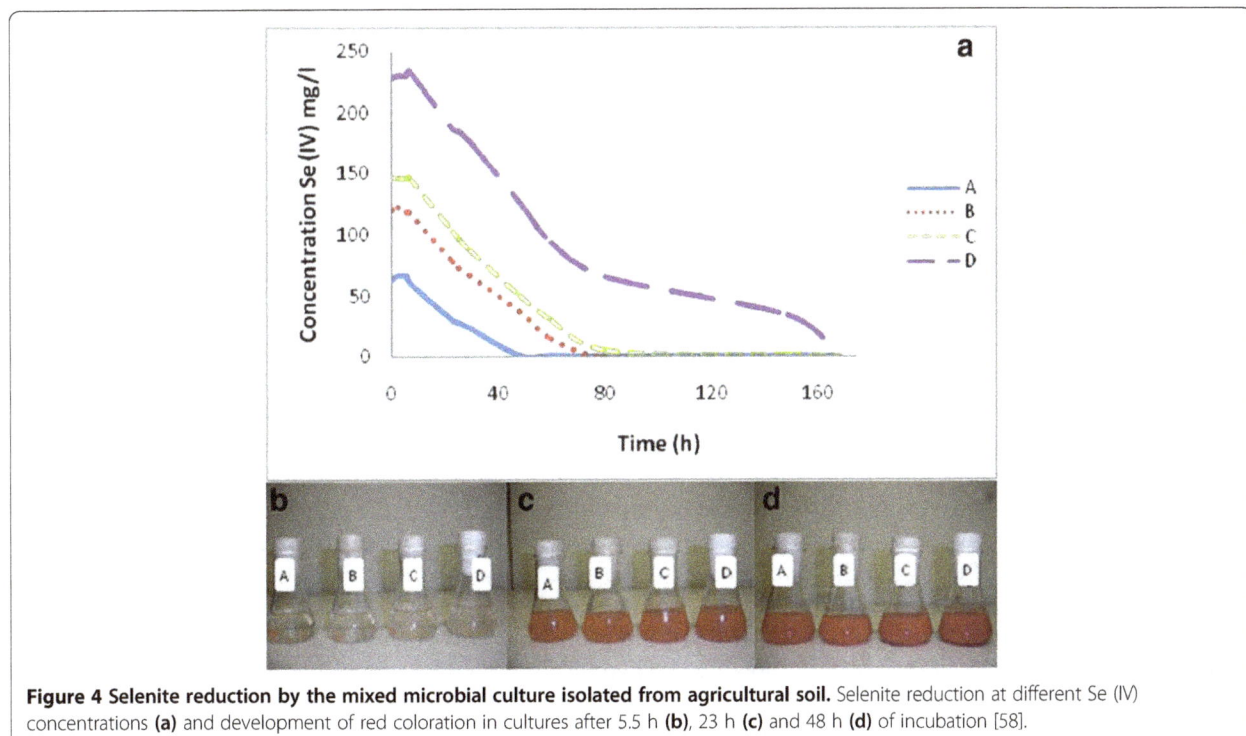

**Figure 4 Selenite reduction by the mixed microbial culture isolated from agricultural soil.** Selenite reduction at different Se (IV) concentrations **(a)** and development of red coloration in cultures after 5.5 h **(b)**, 23 h **(c)** and 48 h **(d)** of incubation [58].

Microbes like *Klebsiella pneumoniae* [31] and *Pseudomonas alcaliphila* [65] have also been used to synthesize Se nanoparticles in good yield. When $Na_2SeO_3$ was added to the activated culture of *P. alcaliphila* the reaction started immediately but completed after 48 h [65]. A gradual colour change with time was observed in the following order:

Grey → Red → Intense Red
0 h 6 h 48 h

The characteristic red colour of Se nanoparticles [37] was detected spectrophotometrically and has been ascribed to the excitation of the surface plasmon vibration of the monoclinic Se. It has been noticed that particle size is directly proportional to reaction time and it ranges between 50–500 nm [21]. Field emission scanning electron microscopic (FESEM) image shows nanoparticles of varying size and shape.

The FTIR spectra of the samples with and without Se nanoparticles showed that the intensity of the spectral peaks containing Se nanoparticles is drastically diminished [65] which suggests strong interaction between Se atoms and the protein molecules present in the *P. alcaliphila*. This is to be noted that the interaction between Se nanoparticles and protein is simply electrostatic because the intensity of sample containing Se atoms was decreased followed by an increase in $\upsilon(NH)$ from 3421 to 3435 $cm^{-1}$. The Raman spectra also support the formation of trigonal selenium (t-Se) and monoclinic selenium (m-Se) by the appearance of peaks at 234 and 254 $cm^{-1}$, respectively (Figure 5). A peak at 235 $cm^{-1}$ is mainly due to chain like structure of t-Se. As the peak at 234 $cm^{-1}$ appears after 48 h of inoculation, it is considered as the transformation of one form of Se into other. The FESEM images which show the accumulation of nanorods on the nanoballs. The size can be controlled by PVP at different time of incubation of nanospheres ranging from 20–200 nm.

On the basis of above studies a possible mechanism for the formation and transformation of Se nanoballs to Se nanorods has been proposed. Zhang et al. [65] presume that $SeO_3^{2-}$ is reduced to elemental selenium by the protein excreted by the *P. alcaliphila* and, their aggregation gives Se nanoparticles of varying size [13]. It is true that protein acts as reducing agent for $SeO_3^{2-}$ but it is available as excretion from *P. alcaliphila* is not convincing. The excretion contains toxins, pyrogens and traces of protein but they may not be sufficient for reduction of selenite. Authors also suggest that large m-Se nanoballs are not stable in solution and they dissolve to form Se atoms. A fraction of dissolved Se atoms crystallize as t-Se forming nanorods [66]. It is not rue because Se in atomic state is not soluble in a solvent but stays in colloidal form. It is a misconception. However,

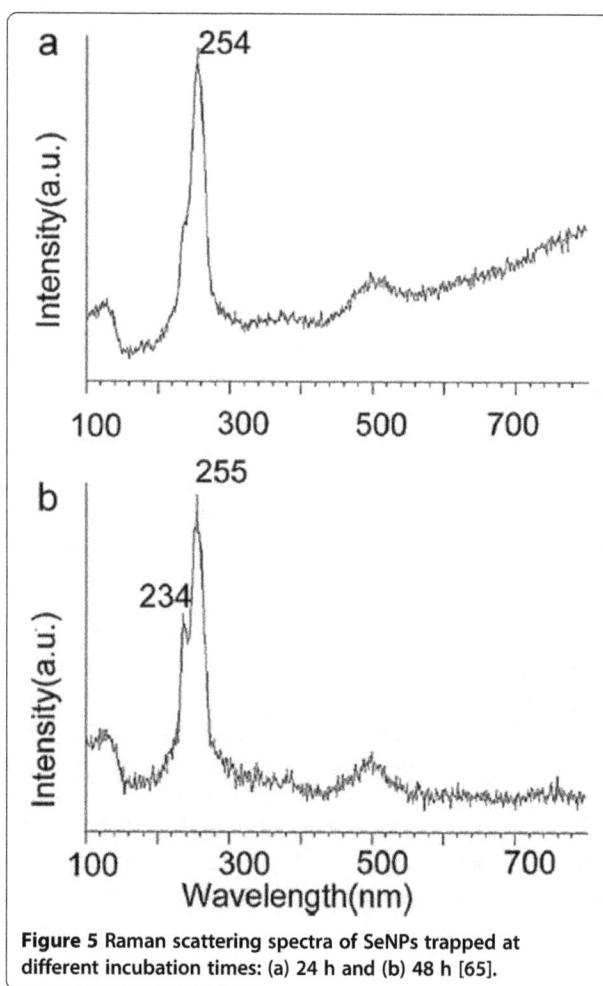

**Figure 5** Raman scattering spectra of SeNPs trapped at different incubation times: (a) 24 h and (b) 48 h [65].

PVP controls the size of Se nanoparticles. If the Se nanoparticles without PVP are left for 2–3 weeks they form aggregates of different shapes and size. Since $SeO_3^{2-}$ in ionic form is toxic to bacterial culture, the bacteria in selenite solution may therefore, die and the disintegrated protein may then act as reducing agent for selenite. The large m-Se nanoparticle can not dissolve in solution to give Se atom forming Se-nanorods. It is quite likely that they may be segregated and rearranged into nanorods.

Se nanoparticles synthesized from sodium selenite and glutathione in aqueous medium had been tested for its growth inhibition efficacy against *Staphylococcus aureus* [67]. *It was found that growth of S. aureus was inhibited in presence of Se nanoparticles within 3–4 h with as low concentration as 7–15 μg/ml which suggests that Se nanoparticles may be used against bacterial infections.*

Biogenic Se nanoparticles fabricated from protein produced by *E. coli* have been compared with those synthesized from chemical reaction via redox mechanism [68]. There are specific types of protein produced by *E. coli* (AdhP, Idh, OmpC, AceA) which are associated with Se nanoparticles. They are also responsible for their uniform size and distribution. One of the proteins (AdhP) was found to bind strongly to Se nanoparticles. *E. coli* was found to produce Se nanoparticles from 2 mM of $SeO_3{}^{2-}$ in about 48 h which was distinguished by the change in colour from colourless to dark red. The amorphous Se nanoparticles thus produced were between 10–90 nm. Since the bacterial growth continued even in presence of selenite, it is confirmed that selenite is not toxic to *E. coli* at this concentration. The enzymes alcohol dehydrogenase, propanol-preferring (ADHP), ACEP (Isocitrate lyase), ENO, KPYKI, IDH and GLPK require metal ions as cofactor for their activity while the enzymes DCEP, ASTC and TNAA require pyridoxal phosphate as coenzyme without which their activity is lost. The authors have not distinguished between cofactor and co-enzyme, they have termed both as metallic cofactor and non-metallic cofactor which is misleading. The cofactors in the enzyme are metal ions bonded through a coordinate covalent bond and can accept lone pair of electrons from the donor atoms in the enzyme into their vacant orbital to form the bond. The Se nanoparticle is in the elemental state and no metal in atomic state can bind to protein or any electron donating species. It is therefore; proper to use the word association of Se nanoparticles to protein rather than bonding. The size of Se nanoparticles produced in presence of the protein, alcohol dehydrogenase propanol-preferring (AdhP) were much smaller than those produced in their absence. However, since *E. coli* contains many other proteins than only pAdhP (purified protein), the decrease in Se nanoparticle size may be the cumulative effect of all proteins available in *E. coli*.

## Conclusion

Bioreduction of selenate or selenite from microorganism such as bacteria, fungi and plant extract have become the favourite pursuit of biologist, chemist and engineers. It is expected that in future the metal would be extracted by biomineralization because they produce the purest form of the element. Many raw materials like waste vegetables, fruit peels and leather cuttings may be utilized to produce elemental metal/metalloid from their oxide, halide, nitrate, sulphide and carbonates. Generally, protein, phenol, alcohol, flavonoid or sugar are required for the reduction of $SeO_3^{-2}$, $SeO_4^{-2}$ and at least one of the above organic molecules is present in microbes and plant extracts. They may therefore, be exploited for the biotransformation of selenate and selenide to elemental Se of various shape and size. Since the reduced metals or metalloids are insoluble in aqueous medium they can be easily sequestered. Growth inhibition of some of the bacterial stains occurs during the redox process which suggests that selenite/selenate may be used against infection caused by such microbes.

## Competing interests
The authors declare that they have no competing interests.

## Authors' contributions
AH gathered the research data. AH and KSS analysed these data findings and wrote this review paper. Both authors read and approved the final manuscript.

## Acknowledgements
The authors are thankful to the publishers for the permission to adopt their figures for this review.

## Author details
[1]Department of Biology, College of Natural and Computational Sciences, University of Gondar, P.O. Box 196, Gondar, Ethiopia. [2]Department of Chemistry, College of Natural and Computational Sciences, University of Gondar, P.O. Box 196, Gondar, Ethiopia.

## References
1.  Dungan RS, Frankenberger T Jr: Microbial transformations of selenium and the bioremediation of seleniferous environments. *Biorem J* 1999, **3**:171–188.
2.  Tapiero H, Townsend DM, Tew KD: The antioxidant role of selenium and seleno-compounds. *Biom Pharmaco* 2003, **57**:134–144.
3.  Cox DN, Bastiaans K: Understanding Australian consumers' perceptions of selenium and motivations to consume selenium enriched foods. *Food Qua Pref* 2007, **18**:66–76.
4.  Turakainen M, Hartikainen H, Seppanen MM: Effects of selenium treatments on potato (*Solanum tuberosum* L.) growth and concentrations of soluble sugars and starch. *J Agric Food Chem* 2004, **52**:5378–5382.
5.  Hartikainen H, Xue T, Piironen V: Selenium as an antioxidant and pro-oxidant in ryegrass. *Plant Soil* 2000, **225**:193–200.
6.  Lyons GH, Genc Y, Soole K, Stangoulis JCR, Liu F, Graham RD: Selenium increases seed production in *Brassica*. *Plant Soil* 2009, **318**:73–80.
7.  Ip C, Thompson HJ, Zhu Z, Ganther HE: In vitro and in vivo studies of methylseleninic acid: evidence that a monomethylated selenium metabolite is critical for cancer chemoprevention. *Cancer Res* 2000, **60**:2882–2886.
8.  Miller S, Walker SW, Arthur JR, Nicol F, Pickard K, Lewin MH, Howie AF, Beckett GJ: Selenite protects human endothelial cells from oxidative damage and induces thioredoxin reductase. *Clin Sci* 2001, **100**:543–550.
9.  Zhang Y, Zhang J, Wang HY, Chen HY: Synthesis of selenium nanoparticles in the presence of polysaccharides. *Mater Lett* 2004, **58**:2590–2594.
10. Xie Q, Dai Z, Huang WW, Zhang W, Ma DK, Hu XK, Qian YT: Large-scale synthesis and growth mechanism of single-crystal Se nanobelts. *Crystal Growth Des* 2006, **6**:1514–1517.
11. Liu MZ, Zhang SY, Shen YH, Zhang ML: Seleniumnanoparticles prepared from reverse microemulsion process. *Chin Chem Lett* 2004, **15**:1249.
12. Quintana M, Haro-Poniatowski E, Morales J, Batina N: Synthesis of selenium nanoparticles by pulsed laser ablation. *App Surf Sci* 2002, **195**:175–186.
13. Gates B, Mayers B, Cattle B, Xia Y: Synthesis and characterization of uniform nanowires of trigonal selenium. *Adv Fun Mat* 2002, **12**:219–227.
14. Wang MCP, Zhang X, Majidi E, Nedelec K, Gates BD: Electrokinetic assembly of selenium and silver nanowires into macroscopic fibers. *ACS Nano* 2010, **4**:2607–2614.
15. Yang L, Shen Y, Xie A, Liang J: Oriented attachment growth of three-dimensionally packed trigonal selenium microspheres into large area wire networks. *Eur J Inor Chem* 2007, **2007**:4438–4444.
16. Filippo E, Manno D, Serra A: Characterization and growth mechanism of selenium microtubes synthesized by a vapor phase deposition route. *Crystal Growth Des* 2010, **10**:4890–4897.
17. Husen A, Siddiqi KS: Carbon and fullerene nanomaterials in plant system. *J Nanobiotechno* 2014, **12**:16.
18. Husen A, Siddiqi KS: Phytosynthesis of nanoparticles: concept, controversy and application. *Nano Res Lett* 2014, **9**:229.

19. Li SK, Shen YH, Xie AJ, Yu XR, Zhang XZ, Yang LB, Li CH: **Rapid, room-temperature synthesis of amorphous selenium/protein composites using** *Capsicum annuum* **L. extract.** *Nanotechno* 2007, 18:405101–405109.

20. Gurunathan S, Kalishwaralal K, Vaidyanathan R, Venkataraman D, Pandian SRK, Muniyandi J, Hariharan N, Eom SH: **Biosynthesis, purification and characterization of silver nanoparticles using** *Escherichia coli.* *Coll Surf B* 2009, 74:328–335.

21. Wang T, Yang L, Zhang B, Liu J: **Extracellular biosynthesis and transformation of selenium nanoparticles and application in H₂O₂ biosensor.** *Coll Surf B* 2010, 80:94–102.

22. Oremland RS, Stolz J: **Dissimilatory reduction of selenate and arsenate in nature.** In *Environmental Metal–Microbe Interaction.* Edited by Lovley DR. Washington, DC: ASM Press; 2000:25.

23. Stolz JF, Oremland RS: **Bacterial respiration of arsenic and selenium.** *FEMS Micro Rev* 1999, 23:615–627.

24. Baesman SM, Bullen TD, Dewald J, Zhang D, Curran S, Islam FS, Beveridge TJ, Oremland RS: **Formation of tellurium nanocrystals during anaerobic growth of bacteria that use Te oxyanions as respiratory electron acceptors.** *Appl Environ Microbiol* 2007, 73:2135–2143.

25. Oremland RS, Herbel MJ, Blum JS, Langley S, Beveridge TJ, Ajayan PM, Sutto T, Ellis AV, Curran S: **Structural and spectral features of selenium nanospheres produced by Se-respiring bacteria.** *Appl Environ Microbiol* 2004, 70:52–60.

26. Narasingarao P, Haggblom MM: **Identification of anaerobic selenate-respiring bacteria from aquatic sediments.** *Appl Environ Microbiol* 2007, 73:3519–3527.

27. Lortie L, Gould WD, Rajan S, McCready RG, Cheng KJ: **Reduction of Selenate and Selenite to Elemental Selenium by a** *Pseudomonas stutzeri* **Isolate.** *Appl Environ Microbiol* 1992, 58:4042–4044.

28. Oremland RS, Blum JS, Culbertson CW, Visscher PT, Miller LG, Dowdle P, Strohmaier FE: **Isolation, growth, and metabolism of an obligately anaerobic, selenate-respiring bacterium, strain SES-3.** *Appl Environ Microbiol* 1994, 60:3011–3019.

29. Sabaty M, Avazeri C, Pignol D, Vermeglio A: **Characterization of the reduction of selenate and tellurite by nitrate reductases.** *Appl Environ Antiontiol* 2001, 67:5122–5126.

30. Sharma G, Sharma AR, Bhavesh R, Park J, Ganbold B, Nam JS, Lee SS: **Biomolecule-mediated synthesis of selenium nanoparticles using dried** *Vitis vinifera* **(raisin) extract.** *Molecules* 2014, 19:2761–2770.

31. Fesharaki PJ, Nazari P, Shakibaie M, Rezaie S, Banoee M, Abdollahi M, Shahverdi AR: **Biosynthesis of selenium nanoparticles using** *Klebsiella pneumoniae* **and their recovery by a simple sterilization process.** *Braz J Microbiol* 2010, 41:461–466.

32. Dhanjal S, Cameotra SS: **Aerobic biogenesis of selenium nanospheres by** *Bacillus cereus* **isolated from coalmine soil.** *Microb Cell Fact* 2010, 9:52.

33. Ingole AR, Thakare SR, Khati NT, Wankhade AV, Burghate DK: **Green synthesis of selenium nanoparticles under ambient condition.** *Chalcogenide Lett* 2010, 7:485–489.

34. Klug HP, Alexander LE: *X-ray Diffraction Procedures for Polycrystalline and Amorphous Materials.* New York, NY, USA: Wiley; 1967.

35. Mukherjee P, Senapati S, Mandal D, Ahmad A, Khan MI, Kumar R, Sastry M: **Extracellular synthesis of gold nanoparticles by the fungus** *fusarium oxysporum.* *Chem Bio Chem* 2002, 5:461–463.

36. Mukherjee P, Ahmad A, Sastry M, Kumar R: **Bioreduction of AuCl₄⁻ ions by the fungus,** *Verticillium* **sp. and surface trapping of the gold nanoparticles formed.** *Angew Chem Int Ed* 2001, 40:3585–3588.

37. Lin ZH, Wang CRC: **Evidence on the size-dependent absorption spectral evolution of selenium nanoparticles.** *Mater Chem Phys* 2005, 92:591–594.

38. Xi GC, Xiong K, Zhao QB, Qian YT: **Nucleation–dissolution–recrystallization: a new growth mechanism for t-Selenium nanotubes.** *Cryst Growth Des* 2006, 6:577–582.

39. Cao XB, Xie Y, Zhang SY, Li FQ: **Ultra-thin trigonal selenium nanoribbons developed from series-wound beads.** *Adv Mater* 2004, 16:649–653.

40. Finley J: **Selenium from broccoli is metabolized differently than Se from selenite, selenate or selenomethionine.** *J Agric Food Chem* 1998, 46:3702–3707.

41. Roberge MT, Borgerding AJ, Finley JW: **Speciation of selenium compounds from high selenium broccoli is affected by the extracting solution.** *J Agric Food Chem* 2003, 51:4191–4197.

42. Finley J: **The retention and distribution by healthy young men of stable isotopes of selenium consumed as selenite, selenate or hydroponically-grown broccoli are dependent on the chemical form.** *J Nutr* 1999, 129:865–871.

43. Finley JW, Davis C, Feng Y: **Selenium from high-selenium broccoli protects rats from colon cancer.** *J Nutr* 2000, 130:2384–2389.

44. Ganther HE, Lawrence JR: **Chemical transformations of selenium in living organisms: improved forms of selenium for cancer prevention.** *Tetrahedron* 1997, 53:12299–12310.

45. Ip C, Ganther HE: **Activity of methylated forms of selenium in cancer prevention.** *Cancer Res* 1990, 50:1206–1211.

46. Ip C, Lisk D: **Characterization of tissue selenium profiles and anticarcinogenic responses in rats fed natural sources of selenium-rich products.** *Carcinogen* 1994, 15:573–576.

47. Ip C, Lisk D: **Enrichment of selenium in allium vegetables for cancer prevention.** *Carcinogen* 1994, 9:1881–1885.

48. Ip C, Lisk J, Stoewsand G: **Mammary cancer prevention by regular garlic and selenium enriched garlic.** *Nutr Cancer* 1992, 17:279–286.

49. Ip C, Lisk DJ: **Efficacy of cancer prevention by highselenium garlic is primarily dependent on the action of selenium.** *Carcinogen* 1995, 16:2649–2652.

50. Lu J, Pei H, Ip C, Lisk DJ, Ganther H, Thompson HJ: **Effect of an aqueous extract of selenium-enriched garlic on in vitro markers and in vivo efficacy of cancer prevention.** *Carcinogen* 1996, 17:1903–1907.

51. Chen H, Shin DW, Nam JG, Kwon KW, Yoo JB: **Selenium nanowires and nanotubes synthesized via a facile template-free solution method.** *Mat Res Bull* 2010, 45:699–704.

52. Prasad KS, Selvaraj K: **Biogenic synthesis of selenium nanoparticles and their effect on As(III)-induced toxicity on human lymphocytes.** *Biol Trace Elem Res* 2014, 157:275–283.

53. Ramamurthy C, Sampath KS, Arunkumar P, Kumar MS, Sujatha V, Premkumar K, Thirunavukkarasu C: **Green synthesis and characterization of selenium nanoparticles and its augmented cytotoxicity with doxorubicin on cancer cells.** *Bioprocess Biosyst Eng* 2013, 36:1131–1139.

54. Kong H, Yang J, Zhang Y, Fang Y, Nishinari K, Phillips GO: **Synthesis and antioxidant properties of gum arabic-stabilized selenium nanoparticles.** *Int J Biol Macromol* 2014, 65:155–162.

55. Prasad KS, Patel H, Patel T, Patel K, Selvaraj K: **Biosynthesis of Se nanoparticles and its effect on UV-induced DNA damage.** *Coll Surf B* 2013, 103:261–266.

56. Stacey B, Sarker N, Dowdell A, Banerjee I: **The spontaneous formation of selenium nanoparticles on gallic acid assemblies and their antioxidant properties.** *Ford Under Res J* 2011, 1:41–46.

57. Bajaj M, Schmidt S, Winter J: **Formation of Se (0) nanoparticles by** *Duganella* **sp. and** *Agrobacterium* **sp. isolated from Se-laden soil of North-East Punjab, India.** *Microbial Cell Fact* 2012, 11:64.

58. Chen T, Wong YS, Zheng W, Bai Y, Huang L: **Selenium nanoparticles fabricated in** *Undaria pinnatifida* **polysaccharide solutions induce mitochondria-mediated apoptosis in A375 human melanoma cells.** *Coll Surf B* 2008, 67:26–31.

59. Tam K, Ho CT, Lee JH, Lai M, Chang CH, Rheem Y, Chen W, Hur HG: **Growth mechanism of amorphous selenium nanoparticles synthesized by** *Shewanella* **sp. HN-41.** *Biosci Biotech Bioch* 2010, 74:696–700.

60. Hunter WJ, Kuykendall LD, Manter DK: *Rhizobium selenireducens* **sp. nov.: a selenite-reducing a-Proteobacteria isolated from a bioreactor.** *Curr Microbiol* 2007, 55:455–460.

61. Rathgeber C, Yurkova N, Stackebrandt E, Beatty JT, Yurkov V: **Isolation of tellurite- and selenite-resistant bacteria from hydrothermal vents of the juan de fuca ridge in the pacific ocean.** *Appl Environ Microbiol* 2002, 68:4613–4622.

62. Morita M, Uemoto H, Watanabe A: **Reduction of selenium oxyanions in wastewater using two bacterial strains.** *Eng Life Sci* 2007, 7:235–240.

63. Klonowska A, Heulin T, Vermeglio A: **Selenite and tellurite reduction by** *Shewanella oneidensis.* *Appl Environ Microbiol* 2005, 71:5607–5609.

64. Zare B, Babaie S, Setayesh N, Shahverdi AR: **Isolation and characterization of a fungus for extracellular synthesis of small selenium nanoparticles.** *Nanomed J* 2013, 1:13–19.

65. Zhang W, Chen Z, Liu H, Zhang L, Gaoa P, Li D: **Biosynthesis and structural characteristics of selenium nanoparticles by** *Pseudomonas alcaliphila.* *Coll Surf B* 2011, 88:196–201.

66. Gates B, Yin Y, Xia Y: **A solution-phase approach to the synthesis of uniform nanowires of crystalline selenium with lateral dimensions in the range of 10–30 nm.** *J Am Chem Soc* 2000, **122**:12582–12583.
67. Tran PA, Webster TJ: **Selenium nanoparticles inhibit** *Staphylococcus aureus* **growth.** *Int J Nanome* 2011, **6**:1553–1558.
68. Dobias J, Suvorova EI, Bernier-Latmani R: **Role of proteins in controlling selenium nanoparticle size.** *Nanotechno* 2011, **22**:195605.

# 12

# Facile synthesis of fluorescent Au/Ce nanoclusters for high-sensitive bioimaging

Wei Ge[1], Yuanyuan Zhang[1], Jing Ye[1], Donghua Chen[2], Fawad Ur Rehman[1], Qiwei Li[1], Yun Chen[1], Hui Jiang[1] and Xuemei Wang[1*]

## Abstract

**Background:** Tumor-target fluorescence bioimaging is an important means of early diagnosis, metal nanoclusters have been used as an excellent fluorescent probe for marking tumor cells due to their targeted absorption. We have developed a new strategy for facile synthesis of Au/Ce nanoclusters (NCs) by doping trivalent cerium ion into seed crystal growth process of gold. Au/Ce NCs have bright fluorescence which could be used as fluorescent probe for bioimaging.

**Results:** In this study, we synthesized fluorescent Au/Ce NCs through two-step hydrothermal reaction. The concentration range of 25–350 μM, Au/Ce NCs have no obvious cell cytotoxicity effect on HeLa, HepG2 and L02 cells. Furthermore, normal cells (L02) have no obvious absorption of Au/Ce NCs. Characterization of synthesized Au/Ce NCs was done by using TEM, EDS and XPS. Then these prepared Au/Ce NCs were applied for *in vitro/in vivo* tumor-target bioimaging due to its prolonged fluorescence lifetime and bright luminescence properties.

**Conclusions:** The glutathione stabilized Au/Ce NCs synthesized through hydrothermal reaction possess stable and bright fluorescence that can be readily utilized for high sensitive fluorescence probe. Our results suggest that Au/Ce NCs are useful candidate for *in vitro/in vivo* tumor bioimaging in potential clinical application.

**Keywords:** Fluorescence bioimaging, Au/Ce nanoclusters, Probe, Tumor

## Background

Cancer is still a serious threat to human health and its effective treatment is still a big challenge. Its early diagnosis provides opportunity for the effective treatment and hence can improve the survival rate. The early diagnosis have been researched extensively through finding biomarkers [1,2] and *in situ* fluorescent bio-imaging [3,4]. Fluorescence imaging has been introduced as an important bioimaging tool and *in vivo* fluorescence imaging can improve the visibility of infected site.

Recently, many fluorescent composites have been developed as sensitive optical imaging probes, including fluorescent dyes [5,6], quantum dots [7-10], metal nanoparticles [11-13] and up-converted nanomaterials [14,15]. Precious metals like gold and silver nanomaterials have been paid much attention for application in a wide range of

biomedical field due to their excellent biocompatibility and physicochemical properties [16-18]. Folic acid conjugated AuNCs@SiO$_2$ nanoprobes (AuNCs@SiO$_2$-FA) with good biocompatibility have been designed and applied into fluorescent imaging of gastric cancer cells [19]. Meanwhile, various kinds of metal nanoclusters have been synthesized by coating biological ligands such as BSA, PEG, DHLA and GSH, etc., and accordingly, different kinds of biomedical function have been developed to meet the practical and clinical needs [17]. In addition, lanthanide-doped luminescent nanomaterials have been also explored for some disease diagnosis [20,21]. Ce$^{3+}$ and Eu$^{3+}$ co-doped multifunctional nanoparticles like NaGdF$_4$:Ce$^{3+}$, Eu$^{3+}$ NPs have been also explored for bioimaging [21]. Moreover, some studies reported the combining of various detection mediums like quantum dots and magnetic nanoparticles for multi-mode imaging [22]. In view these observations, a new strategy for facile synthesis of fluorescent Au/Ce nanoclusters (NCs) has been explored in this contribution by doping trivalent cerium ion into seed crystal growth process of gold. Through

\* Correspondence: xuewang@seu.edu.cn
[1]State Key Lab of Bioelectronics (Chien-Shiung Wu Lab), Department of Biological Science and Medical Engineering, Southeast University, Nanjing 210096, China
Full list of author information is available at the end of the article

hydrothermal synthesis of these glutathione stabilized nanoclusters, it is possible to utilize the biocompatible and fluorescent Au/Ce NCs to realize high-sensitive *in vitro/in vivo* tumor-target bioimaging.

## Results and discussion
### Synthetic strategy
Considering the multiple roles played by glutathione (GSH) in cell survival and metabolic functions, GSH functionalized nanocomposites were prepared and tested. This study represents promising tools for a wide variety of investigations in biomedical field. As shown in Figure 1, the facile synthesis of GSH protected Au/Ce NCs could be readily realized to obtain well dispersible Au/Ce nanoclusters. It is noted that after mixing a certain concentration of GSH and $HAuCl_4$, one step hydrothermal reaction assisted the synthesis of Au nanoclusters, where GSH acted as a reductant and stabilizing agent. Afterwards, a certain amount of $CeNO_3$ was added to the as-synthesized solution that facilitated the realization of stable ultra-small Au/Ce nanoclusters. The end-products of fluorescent Au/Ce nanoclusters were well separated and purified by ethanol centrifugal.

### Characterization of Au/Ce NCs and Au NCs
Transmission Electron Microscopy (TEM) images of GSH capped Au/Ce NCs illustrated the perfect dispersion of the nanocomposites without any aggregation. Figure 2 showed typical images of purified Au NCs and Au/Ce NCs that can disperse well in ultrapure water.

This typical TEM image of the resulting Au NCs and Au/Ce NCs evidenced their high mono-dispersion and relatively uniform sizes. As shown in Figure 2, TEM characterization demonstrates that 90% of the Au NCs (Figure 2A) and Au/Ce NCs (Figure 2C) ranged between 1.2–2.2 nm in diameter (i.e., with narrow size distribution), while the high resolution image (HRTEM) of Au NCs showed clear crystal of metallic structure. HRTEM (Figure 2A, inset) illustrated that the gold nanoclusters kept their interplanar Au–Au spacing at ca. 0.2 nm.

Moreover, X-ray photoelectron spectroscopy (XPS) and energy dispersive X-ray spectroscopy (EDS) were used to investigate the valence of gold and cerium in the Au/Ce NCs after the formation of relevant nanoclusters. The EDS analysis indicated that Au and Ce elements co-exist in the composites without other elemental impurity present in the prepared Au/Ce NCs (Figure 3A). As shown in Figure 3B, two peaks located at the binding energy of 83.9 and 87.7 eV were observed, which were consistent with the emission of 3d photoelectrons from Au (0), while cerium's 3d orbital spectrum with prominent $Ce^{3+}$ peaks at 884.8 and 904.2 eV, respectively, as earlier reported [23].

Fluorescence and UV–Vis absorption spectroscopy were further utilized to characterize the optical properties of the GSH capped Au/Ce NCs. As shown in Figure 4A, fluorescence emission peak of $Ce^{3+}$ solution appeared at 350 nm, while two apparent emission bands of Au/Ce NCs characteristic peak of gold and cerium located at ca. 570 nm and 360 nm, respectively. The peak of trivalent cerium red shifted for about 10 nm, which

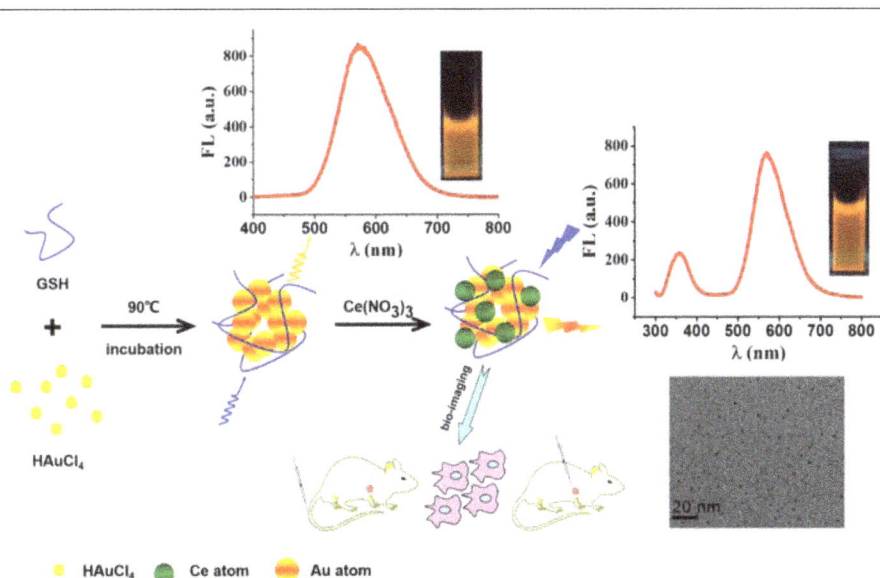

**Figure 1 Illustration of the synthesis of fluorescent GSH–Au/Ce NCs.** Mixed with a certain concentration of glutathione and $HAuCl_4$ until the solution became colorless, then the mixture was placed in a water bath with 90°C for two hours, and a certain concentration of $Ce(NO_3)_3$ aqueous solution was added immediately, the mixture was blending and heating in water bath with 90°C for another one hour. The separated and purified Au/Ce NCs were applied in bioimaging and measured by fluorescence spectrometer and transmission electron microscope.

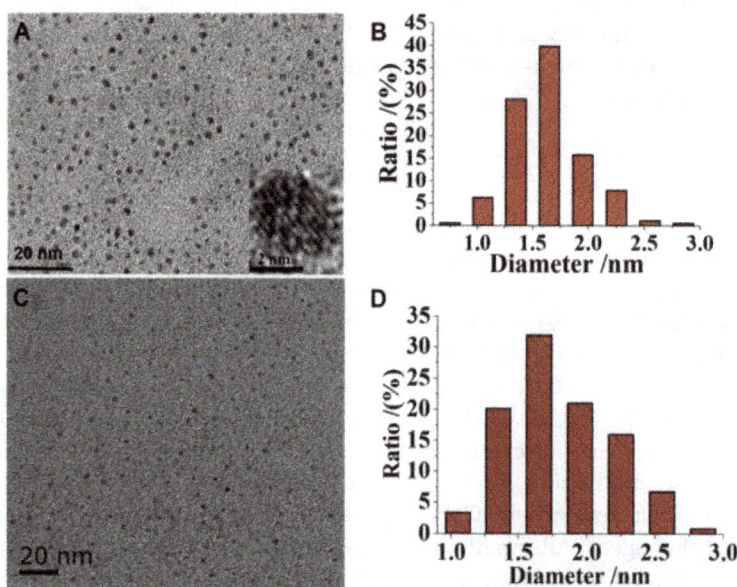

**Figure 2 Transmission Electron Microscope (TEM) images. (A)** Typical image of Au NCs. **(C)** Typical image of Au/Ce NCs. Inset in image **(A)**: high resolution image with the crystallinity of the metallic structure. **(B)** The size distribution histogram of Au NCs. **(D)** The size distribution histogram of Au/Ce NCs.

may attributed to the forming of Au/Ce NCs. In addition, curve a showed UV–Vis absorption peak of $Ce^{3+}$ aqueous solution; it is evident that UV–Vis absorption peak appeared at ca. 290 nm, while the relevant absorption peak of Au/Ce NCs was almost smeared out due to the

formation of the hybrid Au/Ce NCs, thereby suggesting the successful formation of Au/Ce nanoclusters. Based on these observations, we believe that $Ce^{3+}$ ions doped in the lattice of gold during seed crystal growth process affected the optical properties of Au/Ce NCs.

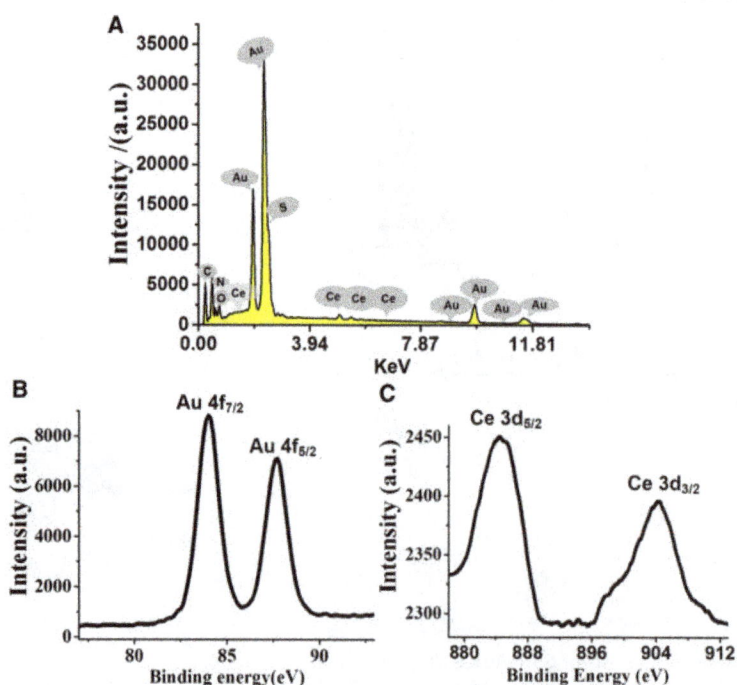

**Figure 3 Elemental analysis of Au/Ce NCs. (A)** EDS of the Au/Ce NCs formed by hydrothermal synthesis. **(B)** and **(C)** were X-ray photoelectron spectra (XPS) evidencing the Au 4f and Ce 3d photoelectron emission from the Au/Ce NCs, respectively.

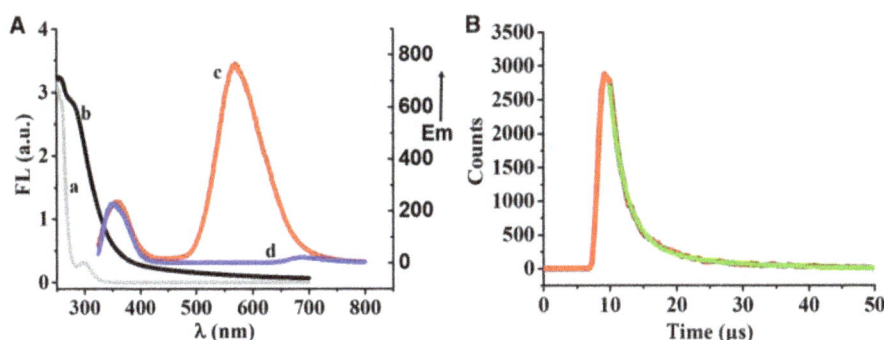

**Figure 4 Optical characterization of Au/Ce NCs. (A)** UV–Vis absorption spectroscopy and fluorescence emission spectrum: Curve a and curve b were UV–Vis absorption spectroscopy of $Ce^{3+}$ aqueous solution and Au/Ce NCs, respectively. Curve c and curve d were fluorescence emission spectrum of Au/Ce NCs and $Ce^{3+}$ aqueous solution, respectively, excitation wavelength was 290 nm. **(B)** Fluorescence lifetime analysis of Au/Ce NCs, excitation was 430 nm.

Several noble metal clusters have shown remarkably long fluorescence lifetimes of 1 or 2 orders of magnitude higher than organic dyes and quantum dots [24]. As shown in Figure 4B, the fluorescence lifetime determined by time-correlated single photon counting indicates that Au/Ce NCs exhibit a luminescence lifetime with a biexponential decay; a short component $\tau_1$ ($2.34 \pm 0.05$ µs, 87.2%) and a long component $\tau_2$ ($11.25 \pm 0.30$ µs, 12.8%), respectively. The microsecond time scale long lifetime component maybe due to the electron transfer from the clusters to the ligand where a redox process might occur and charge-separated trap [25], hence, making it possible to avoid the interference of short-lived background fluorescence. The apparent long fluorescence lifetime of Au/Ce NCs makes it possible for their future bio-application in monitoring some important biological process through biosensing or bioimaging.

### Application of Au/Ce NCs in bioimaging

Based on the above observations, we thus examined the cytotoxic activity of Au/Ce NCs against cancer cell lines (HeLa and HepG2 cells) and normal cells (L02) using MTT assays. MTT assay revealed that Au/Ce NCs have good compatibility even at relatively high concentrations (Figure 5).

The bright green cellular fluorescence of Au/Ce NCs inside HeLa cancer cells appeared to be adequate for use in the *in vivo* bio-imaging of relevant live tumor cells. It is evidenced in Figure 6 that the Au/Ce NCs were well distributed in the cells so that the relevant edges and morphologies of the cells were neatly delineated. Importantly, the fluorescence intensity increased with increase of incubation time and concentration of Au/Ce NCs (Figure 6A–C). The relative fluorescence intensity variations are further confirmed by a comparison of the quantitative variations in the fluorescence intensities across both cell types, as shown in Figure 6D. Similarly, we also

explored that HepG2 and L02 cells support the bio-labeling of Au/Ce NCs. Figure 7 showed that HepG2 treated with Au/Ce NCs displayed clear fluorescence. In contrast, little or almost no intracellular fluorescence was observed in control group involving L02 cells, which showed that there was no obvious fluorescence for normal cells subjected to the same incubation conditions (i.e., in the presence of Au/Ce NCs) as provided for the HeLa cancer cells. Therefore, Au/Ce NCs can be readily applied into *in vitro* / *in vivo* bio-imaging of relevant tumor cells.

On the basis of aforementioned *in vitro* results, we now wish to establish the feasibility of *in vivo* bio-imaging of tumors by fluorescence based on fluorescent Au/Ce NCs. For this purpose, we relied on a xenograft tumor model of Cervical carcinoma. As shown in Figure 8, subcutaneous injection of Au/Ce NCs around xenograft tumors allowed the clear observation of bright fluorescence around the tumor after 24 hours while the fluorescence in the mouse injected with intravenous injection Au/Ce NCs solution through the tail was also observed by *in vivo* fluorescence. No obvious toxic effects were observed during the experimental trail, suggesting that Au/Ce NCs solutions can be administered for high-sensitive *in vivo* tumor-targeted bioimaging.

### Conclusions

In summary, a novel strategy of facile Au/Ce nanoclusters synthesis by doping trivalent cerium ion into seed crystal growth process of gold has been developed. The EDS, UV–Vis absorption spectroscopy, fluorescence and XPS characterization validates that the glutathione stabilized Au/Ce NCs via hydrothermal synthesis possess stable and bright fluorescence. The as-prepared GSH–Au/Ce NCs have strong fluorescence and long fluorescence lifetime, which could be readily utilized for high sensitive tumor-target bioimaging.

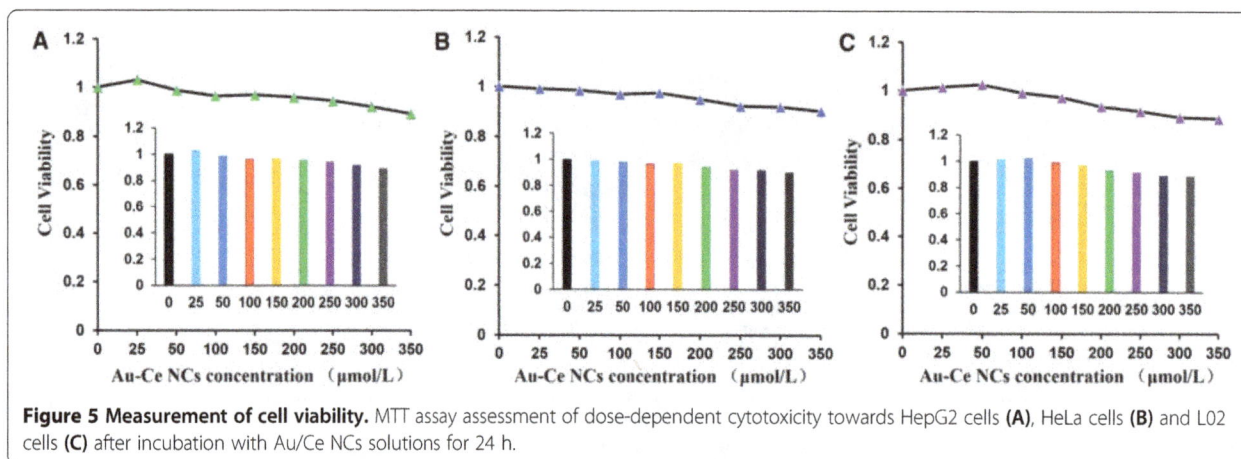

**Figure 5 Measurement of cell viability.** MTT assay assessment of dose-dependent cytotoxicity towards HepG2 cells **(A)**, HeLa cells **(B)** and L02 cells **(C)** after incubation with Au/Ce NCs solutions for 24 h.

## Methods

### Chemicals, materials and cells

Auric chloride acid ($HAuCl_4 \cdot 6H_2O$), L-glutathione (GSH) and Cerium nitrate ($Ce(NO_3)_3 \cdot 6H_2O$) along with all chemicals (analytical reagent, AR) were purchased from Sinopharm Chemical Reagent Co., Ltd. DMEM medium, fetal bovine serumwere (FBS) and penicillin were purchased from SunShineBio Technology Co., Ltd (Nanjing, China). Ultrapure water (18.2MU cm; Milli-Q, Millipore) was used as lyticagent of all aqueous solutions reagent. Tumor cells like HeLa cells, HepG2 cells and normal cells like L02 cells (purchased from Cell Bank of Chinese Academy of Sciences, Shanghai) were applied in our study. All of them were cultured in DMEM medium supplemented with 10%

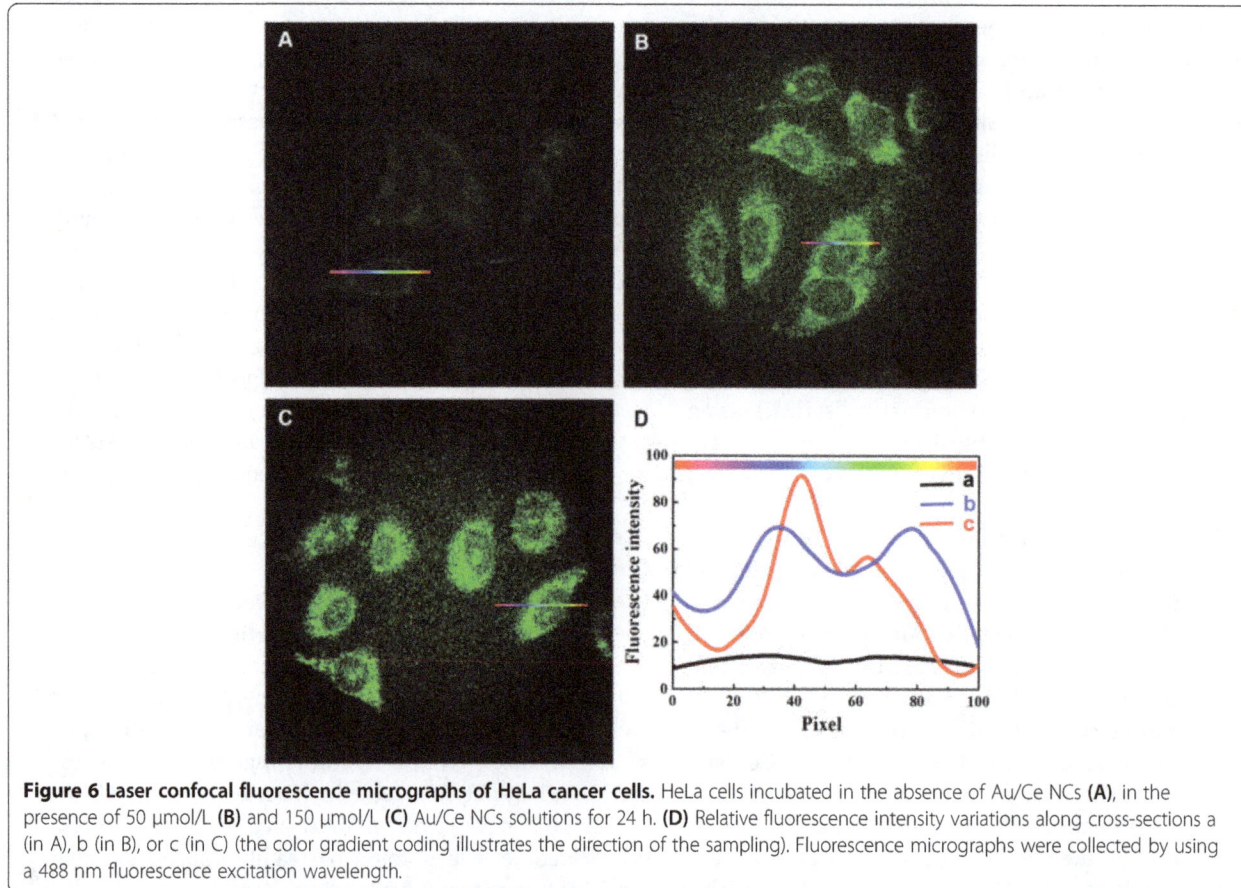

**Figure 6 Laser confocal fluorescence micrographs of HeLa cancer cells.** HeLa cells incubated in the absence of Au/Ce NCs **(A)**, in the presence of 50 μmol/L **(B)** and 150 μmol/L **(C)** Au/Ce NCs solutions for 24 h. **(D)** Relative fluorescence intensity variations along cross-sections a (in A), b (in B), or c (in C) (the color gradient coding illustrates the direction of the sampling). Fluorescence micrographs were collected by using a 488 nm fluorescence excitation wavelength.

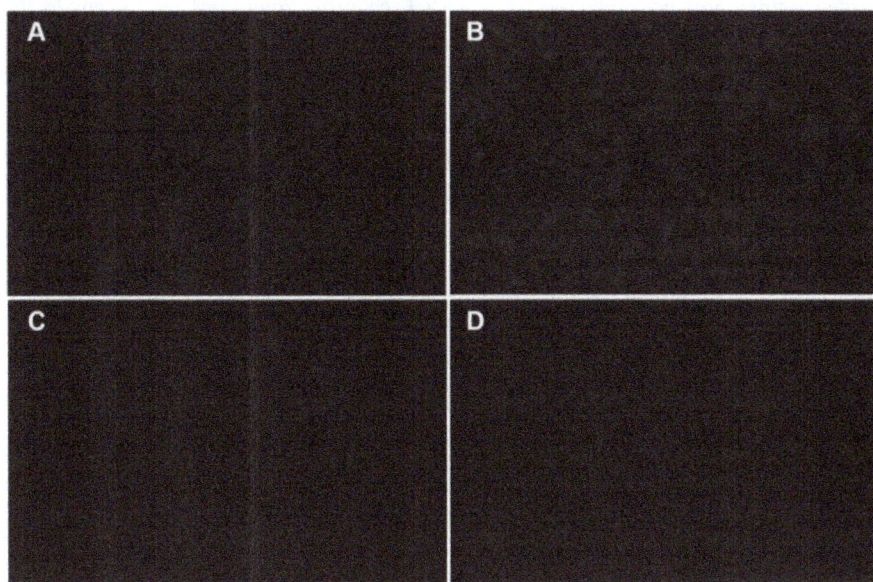

**Figure 7** **Laser confocal fluorescence micrographs of HepG2 cells and L02 cells.** **(A)** HepG2 cells without Au/Ce NCs treatment. **(B)** HepG2 cells treated with 150 μmol/L Au/Ce NCs solutions for 24 h. **(C)** L02 cells without Au/Ce NCs treatment. **(D)** L02 cells treated with 150 μmol/L Au/Ce NCs solutions for 24 h. The micrographs of HepG2 cells were acquired by 20× IR coated objective. The micrographs of L02 cells were collected by 63× IR coated objective. The excitation wavelength was 488 nm.

fetal bovine serum and 1% penicillin at 37°C in a carbon dioxide cell incubator with 5% $CO_2$ and 95% relative humidity.

## Characterizations

Transmission Electron Microscopy (TEM) images were recorded using a JEM-2100 microscope (JEOS, Japan) to characterize the size and size distribution, a certain concentration of sample solution was spotted on carbon coated copper grid (300 meshes) and was dried in desiccator at room temperature. Energy dispersive X-ray spectroscopy (EDS) analyses was done in a Zeiss Ultra Plus scanning electron microscopic (SEM). The valence state of gold and cerium atoms in the Au/Ce nanoclusters was investigated by a PHI 5000 VersaProbe X-ray photoelectron spectrometer (XPS), briefly, samples were droped on a silicon wafer and dried in laboratory ambience to form evenly spread film. Thermo BioMate 3S UV–visible spectrophotometer was used for the UV–Vis absorption measurements, spectra were typically measured in the range of 200–700 nm. Photoluminescence spectra were carried out using SHIMADZU RF-5301 PC

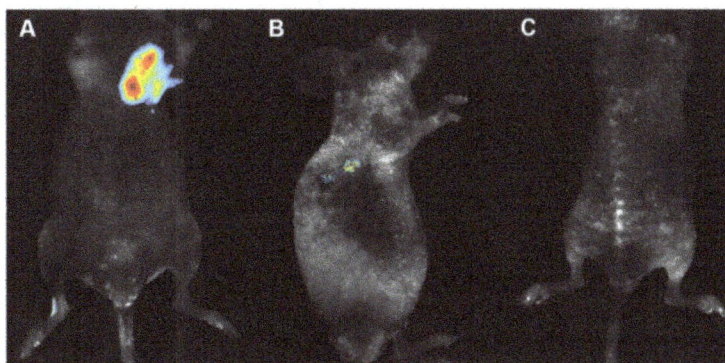

**Figure 8** **Representative xenograft tumor nude mice models of Cervical carcinoma *in vivo* imaging.** **(A)** *In vivo* fluorescence imaging 24 h after a subcutaneous injection of 5 mmol/L Au/Ce NCs solution near the tumor. **(B)** *In vivo* fluorescence imaging 24 h after a intravenous injection 5 mmol/L Au/Ce NCs solution through the tail. **(C)** Control nude mice without tumor after a intravenous injection equivalent PBS through the tail. Fluorescent Au/Ce NCs were observed inside the tumors using a 455 nm excitation wavelength.

instrument. Cells fluorescence imaging were collected by laser scanning confocal microscope Carl Zeiss LSM710 (Zeiss, Germany). Nude mice *in vivo* imaging were carried on vivo multispectral imaging system (Maestro EX).

### Preparation of glutathione stabilized Au nanoclusters and Au/Ce nanoclusters

Precisely prepared 2.4 mM glutathione aqueous solution, measured out 8.5 µL pre-prepared 1 M HAucl$_4$ aqueous solution and added into 5 mL 2.4 mM GSH solution, and shaked vigorously until the solution becomes colorless. The final concentration of HAuCl$_4$ was 1.7 mM. Then the mixture was placed in a water bath with 90°C for two hours to obtain low fluorescence Au seed crystal and added 5 µL 1 M Ce(NO$_3$)$_3$ aqueous solution immediately. The mixture was blending and heating in water bath with 90°C for another one hour. A illustration need to be added that the synthesis of pure Au nanoclusters were performed in same condition without adding cerium ion. The products were separed and purified by centrifugation at 10000 rpm for 10 min with 1:4 ethanol. The precipitation were redispersed in deionised water or phosphate buffer solution (PBS, pH 7.2) for further application.

### Cell growth inhibition study by MTT assay

Briefly, 100 µL medium with $2 \times 10^3$ Cells/well were plated in 96-well plates, after about eight hours incubation, cells were treated with 100 µL various concentrations of Au/Ce NCs medium solution. Each concentration set up five repetition. After treatment for 24 hours, 20 µL MTT solution (5 mg/ml) was added to each well. And cells were incubated for another four hours, then the supernatant was removed and 150 µL DMSO was added per well. Samples were then shaked well for 10 min and the optical density (OD) was read at a wavelength of 490 nm by microplate reader (MK3, ThermoFisher). All experiments were performed in triplicate.

### Construction of the xenografted tumor mouse model

BALB/c female athymic nude mice, age-matched (four weeks of age) and weight-matched (18–22 g), were purchased from Peking University Health Science Center. All experiments involving mice were approved by the National Institute of Biological Science and Animal Care Research Advisory Committee of Southeast University, and experiments were conducted following the guidelines of the Animal Research Ethics Board of Southeast University. The mice were randomly assigned to groups for experimental purposes. They were maintained in clean facilities with a 12-hour light/dark cycle and received water and food through a semi-barrier system. Subcutaneous tumor models were generated by the subcutaneous inoculation (0.10 mL volume containing $5 \times 10^7$ cells/mL media) of HeLa cells in the right side of their armpit or nearby sites using a 1-mL syringe with a 25 G needle. Tumor growth was monitored until a palpable size for next applications.

### *In vitro* and *in vivo* bioimaging study

For cellular imaging, HeLa cells and L02 were treated with a certain concentrations of Au/Ce NCs solutions and incubated at 37°C for 24 h. The cells were washed three times with PBS before fluorescence imaging. A 488-nm excitation laser beam (Andor Revolution XD) was focused using a 63× IR coated objective (Nikon). Similarly, HepG2 cells were treated with 150 µmol/L Au/Ce NCs solutions, the cells were washed three times with PBS before fluorescence imaging. A 488-nm excitation laser beam (Andor Revolution XD) was focused using a 20× IR coated objective (Nikon).

For *in vivo* bio-imaging of Au/Ce NCs in the tumor location complex solution was administered into the solid tumor mouse model through local injection or intravenous injection through the tail. The mice were fully anesthetized by gaseous 5% isoflurane anesthesia. The *in vivo* bio-images were acquired on Cri Maestro *in vivo* imaging system. After incubation for 24 h, fluorescent Au/Ce nanoclusters were observed inside the tumors by *in vivo* fluorescence imaging using a 455 nm excitation wavelength. In comparison, the negative control groups, which received an equivalent volume of Phosphate Buffered Saline (PBS), did not exhibit any apparent fluorescence. The ROI (regions of interest) analysis was measured under the assistance of CRi Maestro Image software. The studies were approved by the National Institute of Biological Science and Animal Care Research Advisory Committee of Southeast University, while experiments conducted the guidelines of the Animal Research Ethics Board of Southeast University.

**Competing interests**
The authors declare that they have no competing interests.

**Authors' contributions**
WG and XMW designed the study and wrote the manuscript. WG synthesized Au/Ce nanoclusters and performed the relevant biological experiments; YYZ and HJ, FUrR, QWL helped to complete a part of data processing and discussion; JY, DHC, YC helped for the animal studies. All authors read and approved the final manuscript.

**Acknowledgements**
This work is supported by National Natural Science Foundation of China (81325011), National High Technology Research and Development Program of China (2012AA022703) and the National Basic Research Program (2010CB732404) as well as Suzhou Science & Technology Major Project (ZXY2012028).

**Author details**
[1]State Key Lab of Bioelectronics (Chien-Shiung Wu Lab), Department of Biological Science and Medical Engineering, Southeast University, Nanjing 210096, China. [2]School of Chemistry and Chemical Engineering, Southeast University, Nanjing 211189, China.

## References

1. Jie GF, Liu P, Zhang SS. Highly enhanced electrochemiluminescence of novel gold/silica/CdSe-CdS nanostructures for ultrasensitive immunoassay of protein tumor marker. Chem Commun. 2010;46:1323–5.
2. Wong GL, Chan HL, Tse YK, Chan HY, Tse CH, Lo AO, et al. On-treatment alpha-fetoprotein is a specific tumor marker for hepatocellular carcinoma in patients with chronic hepatitis B receiving entecavir. Hepatology. 2014;59:986–95.
3. Gao SP, Chen DH, Li QW, Ye J, Jiang H, Amatore C, et al. Near-infrared fluorescence imaging of cancer cells and tumors through specific biosynthesis of silver nanoclusters. Sci Rep. 2014;4:4384.
4. Wang JL, Zhang G, Li QW, Jiang H, Liu CY, Amatore C, et al. *In vivo* self-bio-imaging of tumors through in situ biosynthesized fluorescent gold nanoclusters. Sci Rep. 2013;3:1157.
5. Luo SL, Zhang EL, Su YP, Cheng TM, Shi CM. A review of NIR dyes in cancer targeting and imaging. Biomaterials. 2011;32:7127–38.
6. Jin YH, Ye FM, Zeigler M, Wu CF, Chiu DT. Near-infrared fluorescent dye-doped semiconducting polymer dots. ACS Nano. 2011;5:1468–75.
7. Hong GS, Robinson JT, Zhang YJ, Diao S, Antaris AL, Wang QB, et al. In vivo fluorescence imaging with $Ag_2S$ quantum dots in the second near-infrared region. Angew Chem Int Ed. 2012;124:9956–9.
8. Liu YL, Ai KL, Yuan QH, Lu LH. Fluorescence-enhanced gadolinium-doped zinc oxide quantum dots for magnetic resonance and fluorescence imaging. Biomaterials. 2011;32:1185–92.
9. Feugang JM, Youngblood RC, Greene JM, Fahad AS, Monroe WA, Willard ST, et al. Application of quantum dot nanoparticles for potential non-invasive bio-imaging of mammalian spermatozoa. J Nanobiotechnol. 2012;10:45.
10. Gérard VA, Maguire CM, Bazou D, Gun'ko YK. Folic acid modified gelatine coated quantum dots as potential reagents for in vitro cancer diagnostics. J Nanobiotechnol. 2011;9:50.
11. Wu X, He XX, Wang KM, Xie C, Zhou B, Qing ZH. Ultrasmall near-infrared gold nanoclusters for tumor fluorescence imaging *in vivo*. Nanoscale. 2010;2:2244–9.
12. Shang L, Dong SJ, Nienhaus GU. Ultra-small fluorescent metal nanoclusters: synthesis and biological applications. Nano Today. 2011;6:401–18.
13. Shang L, Dorlich RM, Brandholt S, Schneider R, Trouillet V, Bruns M, et al. Facile preparation of water-soluble fluorescent gold nanoclusters for cellular imaging applications. Nanoscale. 2011;3:2009–14.
14. Shen J, Zhao L, Han G. Lanthanide-doped upconverting luminescent nanoparticle platforms for optical imaging-guided drug delivery and therapy. Adv Drug Deliv Rev. 2013;65:744–55.
15. Yin WY, Zhou LJ, Gu ZJ, Tian G, Jin S, Yan L, et al. Lanthanide-doped $GdVO_4$ upconversion nanophosphors with tunable emissions and their applications for biomedical imaging. J Mater Chem. 2012;22:6974–81.
16. Liu CL, Wu HT, Hsiao YH, Lai CW, Shih CW, Peng YK, et al. Insulin-directed synthesis of fluorescent gold nanoclusters: preservation of insulin bioactivity and versatility in cell imaging. Angew Chem Int Ed. 2011;50:7056–60.
17. Dreaden EC, Alkilany AM, Huang XH, Murphy CJ, El-Sayed MA. The golden age: gold nanoparticles for biomedicine. Chem Soc Rev. 2012;41:2740–79.
18. Li JJ, Zhong XQ, Cheng FF, Zhang JR, Jiang LP, Zhu JJ. One-pot synthesis of aptamer-functionalized silver nanoclusters for cell-type-specific imaging. Anal Chem. 2012;84:4140–6.
19. Zhou ZJ, Zhang CL, Qian QR, Ma JB, Huang P, Zhang X, et al. Folic acid-conjugated silica capped gold nanoclusters for targeted fluorescence/X-ray computed tomography imaging. J Nanobiotechnol. 2013;11:17.
20. Chen GY, Qiu HL, Prasad PN, Chen XY. Upconversion Nanoparticles: Design, Nanochemistry, and Applications in Theranostics. Chem Rev. 2014;114:5161–214.
21. Hao SW, Chen GY, Yang CH. Sensing Using Rare-Earth-Doped Upconversion Nano-particles. Theranostics. 2013;3:331–45.
22. Zhou L, Li ZH, Ju EG, Liu Z, Ren JS, Qu XG. Aptamer-directed synthesis of multifunctional lanthanide-doped porous nanoprobes for targeted imaging and drug delivery. Small. 2013;9:4262–8.
23. Wang K, Ruan J, Qian QR, Song H, Bao CC, Zhang XQ, et al. BRCAA1 monoclonal antibody conjugated fluorescent magnetic nanoparticles for *in vivo* targeted magnetofluorescent imaging of gastric cancer. J Nanobiotechnol. 2011;9:23.
24. McCormack RN, Mendez P, Barkam S, Neal CJ, Das S, Seal S. Inhibition of nanoceria's catalytic activity due to $Ce^{3+}$ site-specific interaction with phosphate ions. J Phys Chem C. 2014;118:18992–9006.
25. Guével XL, Trouillet V, Spies C, Li K, Laaksonen T, Auerbach D, et al. High photostability and enhanced fluorescence of gold nanoclusters by silver doping. Nanoscale. 2012;4:7624–31.

# Novel metal allergy patch test using metal nanoballs

Tomoko Sugiyama[1], Motohiro Uo[2][*], Takahiro Wada[2], Toshio Hongo[2], Daisuke Omagari[3], Kazuo Komiyama[3], Hitoshi Sasaki[4], Heishichiro Takahashi[5], Mikio Kusama[1] and Yoshiyuki Mori[1]

## Abstract

**Background:** Patch tests are often used in the clinical diagnosis of metal allergies. In currently available patch tests, high concentrations of metal salt solutions are used. However, diagnosis accuracy can be influenced not only by acute skin reactions to high concentrations of metal salt, but also by skin reactions to other components present in the patch or to pH changes. In this study, we developed Ni nanoparticles (termed "nanoballs") for use in patch-test solutions.

**Findings:** Highly soluble, spherical Ni nanoballs were prepared using plasma electrolysis. The Ni released from the nanoballs permeated through a dialysis membrane, and the nanoball-containing solution's pH was maintained constant. Ni ions were released slowly at low concentrations in a time-dependent manner, which contrasted the rapid release observed in the case of a commercial patch test. Consequently, in the new test system, reactions caused by high concentrations of metal salts were avoided.

**Conclusions:** By exploiting the high specific surface area of Ni nanoballs, we obtained an effective dissolution of Ni ions that triggered Ni allergy in the absence of direct contact between the nanoballs and mouse skin. This novel patch system can be applied to other metals and alloys for diagnosing various types of metal-induced contact dermatitis.

**Keywords:** Nanoparticle, Nickel, Metal allergy, Patch test, Elemental distribution

## Background

Metal-allergy patch tests are routinely used in the clinical diagnosis of metal-induced contact dermatitis. Currently available patch tests use high concentrations of various metal salts in aqueous solution. They contain a reservoir sheet that allows the test solution to permeate into the skin to induce a local allergic reaction (Figure 1a). A patch test solution typically contains a metal salt under an acidic condition [1]. Current patch tests can cause pustular or follicular reactions because of the high concentration of metal salts. In addition, false positive or negative reactions [2] and skin irritation often occur [3-5]. Metal allergies, however, are often triggered by metal ions that are continuously eroded from metallic materials under neutral pH conditions, and in this case, the metal ion concentration is typically low. Factors such as pH, metal-ion

concentration, and dosage rates differ considerably between a genuine metal-allergic reaction and that which occurs in a patch test.

An ideal patch test must reproduce the metal erosion that occurs in metallic equipment (Figure 1b). Therefore, historically, large metal particles have been used as the metal source in the patch test [6]. However, their dissolution rates are inadequate for triggering an allergic reaction because of their low specific surface area [7,8].

To address the aforementioned limitations of patch tests, in this study, we prepared 40–50-nm-diameter Ni nanoballs (Figure 1c) through plasma electrolysis at the surface of electrodes [9]. Because of their size, we expected the Ni nanoballs to exhibit a high rate of Ni dissolution, which is suitable for a patch test. In our novel patch system, we used a Ni-nanoball suspension enclosed in a dialysis tube, which we call a "Ni-nanoball pack." Ni ions released from the nanoball pack permeated through the dialysis membrane into the skin of mice, and thus direct contact between the Ni nanoballs and the skin was avoided. We evaluated this Ni-permeation behavior *in vitro*

* Correspondence: uo.abm@tmd.ac.jp
[2]Advanced Biomaterials Department, Graduate School of Medical and Dental Sciences, Tokyo Medical and Dental University, 1-5-45 Yushima, Bunkyo-ku, Tokyo 113-8549, Japan
Full list of author information is available at the end of the article

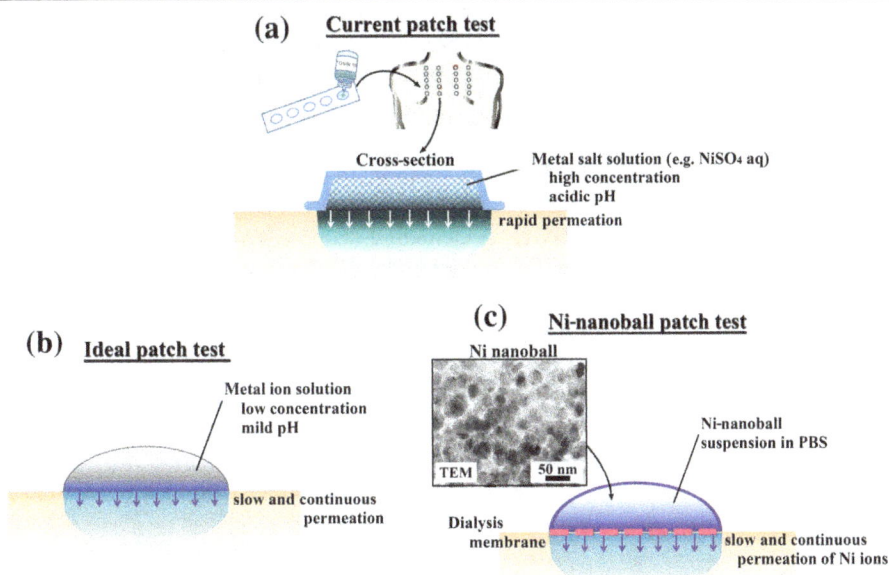

**Figure 1 Various patch-test schemes.** A currently available commercial patch test **(a)**, an ideal patch test **(b)**, and the novel patch test designed using Ni nanoballs **(c)**.

and also *in vivo* by using mouse skin. The Ni distribution in the skin was measured using synchrotron radiation-excited X-ray fluorescence (SR-XRF) analysis and particle-induced X-ray emission (PIXE), which are highly sensitive analysis methods. Furthermore, we examined the chemical state of the Ni nanoballs and the permeated Ni in the skin by performing X-ray absorption fine structure (XAFS) analysis.

## Methods

### Preparation of Ni-nanoball suspensions and an *in vitro* test of Ni-ion release

Ni nanoballs were synthesized according to the method described by Toriyabe *et al.* [9] Ni nanoballs were dispersed in distilled water (DW) and 0.1 M phosphate buffer solution (PBS) at pH =5.8 to reach a concentration of 500 ppm

of Ni nanoballs. In the *in vitro* test of Ni-ion release, the experimental setup used for examining Ni permeation through the dialysis membrane was prepared as shown in Figure 2a. The inner cylinder was filled with 500 μL of Ni-nanoball suspension in DW and PBS (pH =5.8) and immersed in 2500 μL of DW for 1 h to 7 days. Evaluation of the permeated Ni in the outer DW solution was based on a colorimetric reaction using 2-(5-Nitro-2-pyridylazo)-5-[N-n-propyl-N-(3-sulfopropyl)amino]phenol disodium salt dehydrate (Nitro-PAPS), which results in the Ni complex and shows a characteristic absorption around 568 nm (Additional file 1: Figure S1). Then, 500 μL of the outer solution was mixed with 2.0 mL of Nitro-PAPS aqueous solution (20 ppm), and the Ni concentration was estimated by the absorbance at 568 nm using the optical absorption (UV-Vis) spectrometer based on Yamashita *et al.*

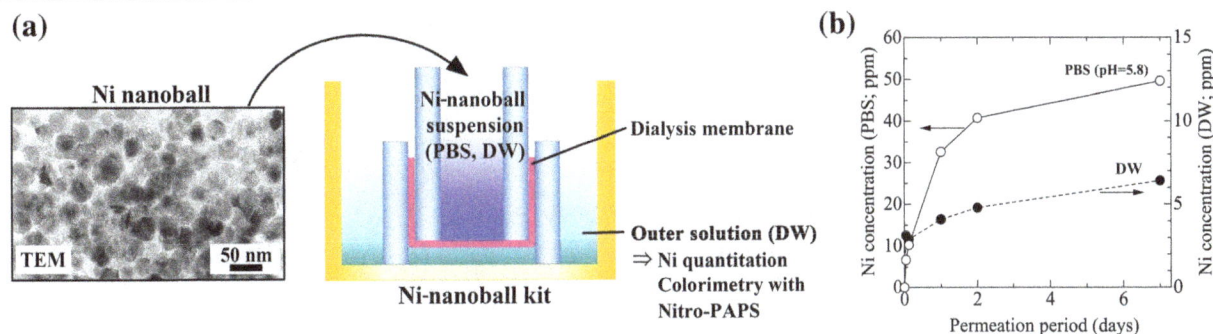

**Figure 2 *In vitro* Ni release from Ni-nanoball suspensions through a dialysis membrane.** Experimental setup **(a)** and time-dependent release of Ni **(b)**.

[10]. The standard Ni solutions (0.2-10 ppm) were prepared by diluting a 1 mg/mL $Ni(NO_3)_2$ solution with DW.

### *In vivo* Ni-permeation test designed for mouse skin

C57BL/6 mice (45–75 weeks old) were used in this study. All animal protocols were approved by the Animal Ethics Review Board of the Dental Hospital of Nihon University School of Dentistry, Tokyo, Japan, and conformed to the guidelines of the National Institutes of Health. The skin on the back of mice was depilated under general anesthesia. Approximately 20 mL of the Ni-nanoball suspension was placed into the dialysis tube and both ends were clamped. The Ni-nanoball pack was fixed onto the depilated back skin of the mice by using film dressing. The application period ranged from 30 min to 24 h, after which the mice were sacrificed. The Ni-nanoball pack was then removed and the skin was gently wiped, and the part of the skin that contacted the pack was excised and used for preparing frozen sections (20-μm thick) according to the method reported by Kubo *et al.* [11]. The sections were placed on Kapton film, dried and subjected to elemental-distribution analysis. As a control, a commercial patch test for Ni allergy (5% w/v $NiSO_4$ aq.) was applied to mice using the

same method. Adjacent specimen slices were stained with hematoxylin and eosin (H-E) and used for histopathological diagnosis. Detailed information on the chemicals, materials, and equipments used are presented in Additional file 1: Table S1.

### Elemental-distribution and chemical-state analyses

Elemental distribution analysis of the entire specimen was performed using SR-XRF. The specimen was irradiated with micro-focused X-ray and the specimen stage was scanned two-dimensionally. The fluorescence X-ray was detected at each point and the elemental distribution images were processed. The high-resolution elemental distribution was analyzed using micro-PIXE analysis, which involved exposure to the micro-focused proton beam with raster scanning over the target area of the specimen. The characteristic X-rays were detected, and the elemental distribution image was processed. The chemical states of the Ni contained in Ni nanoballs and the Ni in mouse skin were examined using XAFS analysis. Detailed conditions of the elemental analyses are shown in Additional file 1: Table S2.

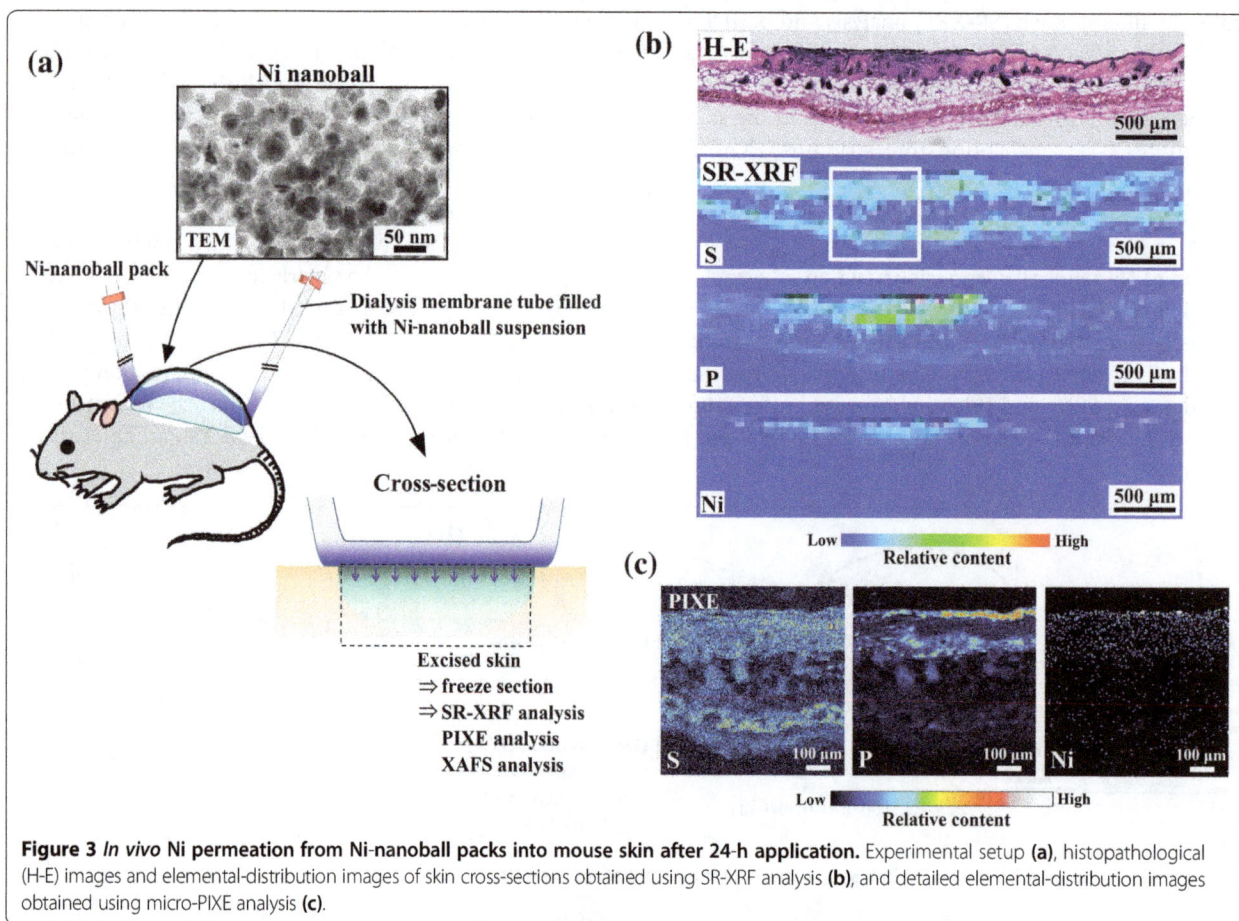

**Figure 3** *In vivo* **Ni permeation from Ni-nanoball packs into mouse skin after 24-h application.** Experimental setup **(a)**, histopathological (H-E) images and elemental-distribution images of skin cross-sections obtained using SR-XRF analysis **(b)**, and detailed elemental-distribution images obtained using micro-PIXE analysis **(c)**.

## Results and discussion

The time-dependent release of Ni *in vitro* is shown in Figure 2b. We observed a continuous release of Ni. The Ni-dissolution rate of Ni nanoballs dispersed in PBS (pH 5.8) was higher than that of nanoballs dispersed in DW. This result agrees with the previous findings of previous studies showing that Ni solubility is increased in acidic solutions [12,13]. The pH of the commercial Ni patch-test solution was 3.8, which is lower than that of natural human skin, where the pH is typically below 5.0 [14]. Thus, we do not expect the Ni-nanoball suspension at pH 5.8 to irritate human skin. Furthermore, the Ni-release rate in our test system could be controlled by adjusting the pH.

Figure 3b shows H-E-stained cross-sections of mouse skin (histopathological analysis) and their SR-XRF images of elemental distribution obtained after treatment with the Ni-nanoball pack for 24 h. The Ni that permeated from the patch was clearly observed on the surface side of the skin. Figure 3c shows the detailed elemental distribution of S, P, and Ni in the areas that exhibited high Ni accumulation (white squares in Figure 3b), as assessed using micro-PIXE analysis. The Ni concentration was high in the epidermis and spread into the dermis layer beyond the basal layer. These results indicate a clear internal permeation of Ni. Furthermore, the H-E-stained images of the same area showed a slight inflammatory response at the

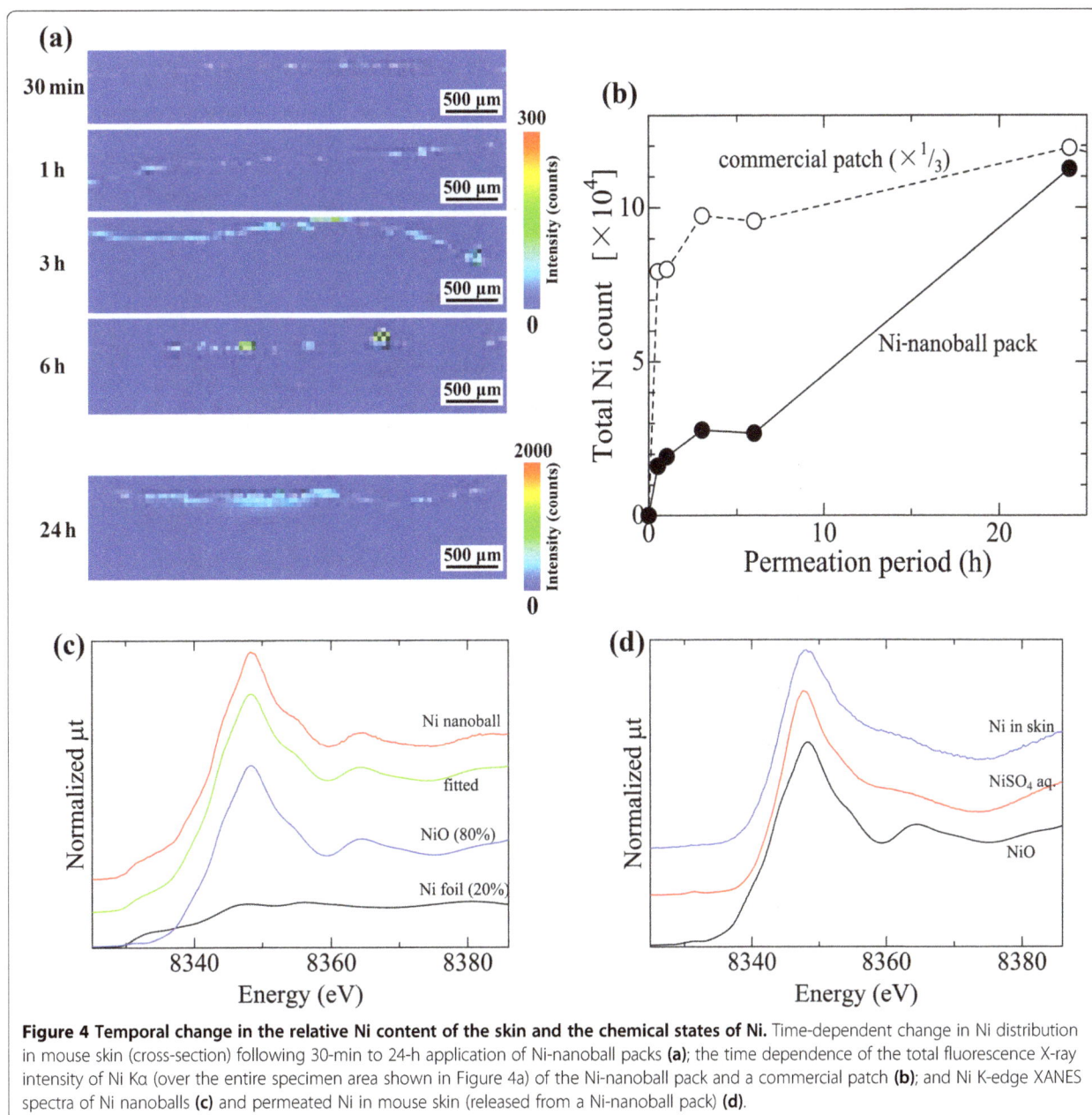

**Figure 4 Temporal change in the relative Ni content of the skin and the chemical states of Ni.** Time-dependent change in Ni distribution in mouse skin (cross-section) following 30-min to 24-h application of Ni-nanoball packs (a); the time dependence of the total fluorescence X-ray intensity of Ni Kα (over the entire specimen area shown in Figure 4a) of the Ni-nanoball pack and a commercial patch (b); and Ni K-edge XANES spectra of Ni nanoballs (c) and permeated Ni in mouse skin (released from a Ni-nanoball pack) (d).

epidermis. The regions of inflammation overlapped with localized areas of Ni and P accumulation. Because the phosphate originates from inflammatory cells, this colocalization of Ni and P suggests that the Ni that permeates from the patch and penetrates the skin induces local inflammation.

Figure 4 shows the temporal change in the relative Ni content of the skin. Figure 4a shows the images of Ni distribution in the skin cross-section, and Figure 4b shows the total fluorescence X-ray intensity of Ni Kα over the entire specimen area (shown in Figure 4a). When the Ni-nanoball pack was used, the Ni content in the skin increased linearly in a time-dependent manner until 24 h. Figure 4b shows the result obtained when we applied the commercial Ni-allergy patch to the skin by using the same method. The concentration of the Ni that permeated from the commercial patch increased drastically after application for only 30 min, and the permeated-Ni content was approximately three times higher than that after application for 24 h. This extremely high dose of Ni is likely to lower the accuracy of the commercial patch test as a result of the potential side reactions that it might cause. By contrast, the comparatively slow and time-dependent permeation of Ni from the Ni-nanoball pack is likely to be optimal for eliciting a Ni allergic reaction.

Figure 4 also shows the Ni K-edge XAFS spectra of Ni nanoballs (Figure 4c) and the permeated Ni in skin (Ni released from the nanoballs) (Figure 4d). The spectrum of the Ni nanoballs closely fits that of a mixture of metallic Ni (20%) and NiO (80%). This suggests that the Ni nanoballs were almost oxidized during storage. Conversely, the spectrum of the permeated Ni in skin (derived from the Ni-nanoball pack) was similar to that of $NiSO_4$ aq., but distinct from that of NiO, a major component of Ni nanoballs. Therefore, the dissolved Ni ions were successfully released from the Ni nanoballs and they permeated the skin through the dialysis membrane.

During a genuine allergic reaction to metals, metal erosion and skin permeation occur in a low and continuous dose. We have shown that our Ni-nanoball pack releases Ni ions slowly from the suspension under mildly acidic conditions. Recently, the health risks posed by nanomaterials have become a growing concern [15-18]. Thus, in this study, we exploited the high specific surface area of Ni nanoballs in order to obtain an effective dissolution of Ni ions and used this to trigger Ni allergy without direct contact between the nanoballs and the skin.

## Conclusions

A high dose of metals and counter-ions and an acidic pH are factors that lower the accuracy of currently available metal-allergy patch tests. These factors were addressed effectively in this study, and the test system presented here could potentially serve as a state-of-the-art patch

test. Another limitation of current patch tests is that they are available only for a few metal species, and the accuracy of these tests is still in question. The facile process used in this study for preparing Ni nanoballs can be applied to most pure metals and alloys, and then the method can be readily used for diagnosing other metal allergies. By conducting further studies on the ability to induce allergy in animal and human skin, the accuracy of the diagnosis obtained using our novel patch test designed for metal allergy could be potentially enhanced.

## Additional file

Additional file 1: Table S1. Detailed information on chemicals, materials, and equipment. Table S2. Detailed experimental conditions used for elemental analysis. Figure S1. Optical absorption spectra of Nitro-PAPS solution mixed with Ni ion solutions of various concentrations (a) and the calibration curve of Ni with the absorbance at 568 nm (b). Linear correlation was observed between 0 to 2.0 ppm of Ni. Quantitation of Ni was carried within this range by diluting the sample solution adequately.

### Abbreviations
SR-XRF: Synchrotron radiation-excited X-ray fluorescence; PIXE: Particle-induced X-ray emission; XAFS: X-ray absorption fine structure; PBS: Phosphate buffer solution; Nitro-PAPS: 2-(5-Nitro-2-pyridylazo)-5-[N-n-propyl-N-(3-sulfopropyl)-amino]phenol disodium salt dehydrate.

### Competing interests
The authors declare that they have no competing interests.

### Authors' contributions
TS and MU conceived and designed the experiments. HS and HT produced and measured the Ni nanoballs. TS, MU, and DO performed the experiments. DO and KK performed the histopathological analysis. TS, MU, and TW performed the SR-XRF and PIXE analyses. TS and MU co-wrote the manuscript. KK, MK, and YM edited the manuscript. All authors discussed the results and commented on the manuscript. All authors read and approved the final manuscript.

### Acknowledgements
The SR-XRF measurements were performed with the approval of the Photon Factory Program Advisory Committee (Proposal Nos. 2011G011 and 2014G017). This work was financially supported by the Japan Society for the Promotion of Science KAKENHI (Grant No. 23390438 to M. Uo). T. Sugiyama is supported by a research fellowship for young scientists from the Japan Society for the Promotion of Science.

### Author details
[1]Department of Dentistry, Oral and Maxillofacial Surgery, Jichi Medical University, 3311-1 Yakushiji, Shimotsuke, Tochigi 329-0498, Japan. [2]Advanced Biomaterials Department, Graduate School of Medical and Dental Sciences, Tokyo Medical and Dental University, 1-5-45 Yushima, Bunkyo-ku, Tokyo 113-8549, Japan. [3]Department of Pathology, Nihon University School of Dentistry, 1-8-13 Kanda-Surugadai, Chiyoda-ku, Tokyo 131-8310, Japan. [4]Nakayamagumi Co. Ltd., North 19, East 1, Higashi-ku, Sapporo, Hokkaido 065-8610, Japan. [5]Graduate School of Engineering, Hokkaido University, North 13, West 8, Sapporo, Hokkaido 060-8628, Japan.

### References
1.  Endo K, Ohno H, Kawashima I, Yamane Y, Yanagi T: Reliability of patch testing for metal hypersensitivity: Stability of metal ions and solution pH. *Shika Zairyo Kikai* 2005, **24**:82.
2.  Mitchell J, Maibach HI: Managing the excited skin syndrome: patch testing hyperirritable skin. *Contact Dermatitis* 1997, **37**:193–199.

3.  Mowad CM: **Patch testing: pitfalls and performance.** *Curr Opin Allergy Clin Immunol* 2006, **6:**340–344.

4.  Diepgen TL, Coenraads PJ: **Sensitivity, specificity and positive predictive value of patch testing: the more you test, the more you get? ESCD Working Party on Epidemiology.** *Contact Dermatitis* 2000, **42:**315–317.

5.  Fischer T, Rystedt I: **False-positive, follicular and irritant patch test reactions to metal salts.** *Contact Dermatitis* 1985, **12:**93–98.

6.  Thyssen JP, Menné T, Schalock PC, Taylor JS, Maibach HI: **Pragmatic approach to the clinical work-up of patients with putative allergic disease to metallic orthopaedic implants before and after surgery.** *Br J Dermatol* 2011, **164:**473–478.

7.  Midander K, Pan J, Wallinder IO, Heim K, Leygraf C: **Nickel release from nickel particles in artificial sweat.** *Contact Dermatitis* 2007, **56:**325–330.

8.  Larese F, Gianpietro A, Venier M, Maina G, Renzi N: **In vitro percutaneous absorption of metal compounds.** *Toxicol Lett* 2007, **170:**49–56.

9.  Toriyabe Y, Watanabe S, Yatsu S, Shibayama T, Mizuno T: **Controlled formation of metallic nanoballs during plasma electrolysis.** *Appl Phys Lett* 2007, **91:**041501.

10. Yamashita S, Abe A, Noma A: **Sensitive, direct procedures for simultaneous determinations of iron and copper in serum, with use of 2-(5-nitro-2-pyridylazo)-5-(N-propyl-N-sulfopropylamino)phenol (nitro-PAPS) as ligand.** *Clin Chem* 1992, **38:**1373–1375.

11. Kubo A, Ishizaki I, Kubo A, Kawasaki H, Nagao K, Ohashi Y, Amagai M: **The stratum corneum comprises three layers with distinct metal-ion barrier properties.** *Sci Rep* 2013, **3:**1731. doi:10.1038/srep01731.

12. Palmer DA, Bénézeth P, Xiao C, Wesolowski DJ, Anovitz LM: **Solubility measurements of crystalline NiO in aqueous solution as a function of temperature and pH.** *J Solution Chem* 2011, **40:**680–702.

13. Plyasunova NV, Zhang Y, Muhammed M: **Critical evaluation of thermodynamics of complex formation of metal ions in aqueous solutions. IV. Hydrolysis and hydroxo-complexes of $Ni^{2+}$ at 298.15 K.** *Hydrometallurgy* 1998, **48:**43–63.

14. Lambers H, Piessens S, Bloem A, Pronk H, Finkel P: **Natural skin surface pH is on average below 5, which is beneficial for its resident flora.** *Int J Cosmetic Sci* 2006, **28:**359–370.

15. Auffan M, Rose J, Bottero JY, Lowry GV, Jolivet JP, Wiesner MR: **Towards a definition of inorganic nanoparticles from an environmental, health and safety perspective.** *Nat Nanotechnol* 2009, **4:**634–641.

16. Kertész Z, Szikszai Z, Gontier E, Moretto P, Surlève-Bazeille JE, Kiss B, Juhász I, Hunyadi J, Kiss ÁZ: **Nuclear microprobe study of $TiO_2$-penetration in the epidermis of human skin xenografts.** *Nucl Instrum Methods Phys Res, Sect B Beam Interact with Mater Atoms* 2005, **231:**280–285.

17. Larese Filon F, Crosera M, Timeus E, Adami G, Bovenzi M, Ponti J, Maina G: **Human skin penetration of cobalt nanoparticles through intact and damaged skin.** *Toxicol In Vitro* 2013, **27:**121–127.

18. Larese FF, D'Agostin F, Crosera M, Adami G, Renzi N, Bovenzi M, Maina G: **Human skin penetration of silver nanoparticles through intact and damaged skin.** *Toxicology* 2009, **255:**33–37.

# Development of antimicrobial biomaterials produced from chitin-nanofiber sheet/silver nanoparticle composites

Vinh Quang Nguyen[1,2], Masayuki Ishihara[2*], Jun Kinoda[3], Hidemi Hattori[2], Shingo Nakamura[2], Takeshi Ono[4], Yasushi Miyahira[4] and Takemi Matsui[1]

## Abstract

**Background:** Chitin nanofibers sheets (CNFSs) with nanoscale fiber-like surface structures are nontoxic and biodegradable biomaterials with large surface-to-mass ratio. CNFSs are widely applied as biomedical materials such as a functional wound dressing. This study aimed to develop antimicrobial biomaterials made up of CNFS-immobilized silver nanoparticles (CNFS/Ag NPs).

**Materials and methods:** CNFSs were immersed in suspensions of Ag NPs ($5.17 \pm 1.9$ nm in diameter; mean $\pm$ SD) for 30 min at room temperature to produce CNFS/Ag NPs. CNFS/Ag NPs were characterized by transmission electron microscopy (TEM) and then tested for antimicrobial activities against *Escherichia* (*E.*) *coli*, *Pseudomonas* (*P.*) *aeruginosa*, and H1N1 influenza A virus, three pathogens that represent the most widespread infectious bacteria and viruses. Ultrathin sectioning of bacterial cells also was carried out to observe the bactericidal mechanism of Ag NPs.

**Results:** The TEM images indicated that the Ag NPs are dispersed and tightly adsorbed onto CNFSs. Although CNFSs alone have only weak antimicrobial activity, CNFS/Ag NPs showed much stronger antimicrobial properties against *E. coli*, *P. aeruginosa*, and influenza A virus, with the amount of immobilized Ag NPs onto CNFSs.

**Conclusions:** Our results suggest that CNFS/Ag NPs interacting with those microbes exhibit stronger antimicrobial activities, and that it is possible to apply CNFS/Ag NPs as anti-virus sheets as well as anti-infectious wound dressings.

**Keywords:** Antimicrobial biomaterials, Chitin nanofiber sheets, Silver nanoparticles, Wound dressings, Anti-virus sheets

## Background

Chitin/chitosan is second most abundant natural nontoxic biomaterial, and is produced from the exoskeleton of sea food, shellfish, crabs, shrimps, insects, edible mushrooms, and sea weed algae [1]. The advantages in biochemical activities of chitin/chitosan-based materials include: anti-infectious activity [2]; stimulation of angiogenesis/wound repair; and stabilization/activation of growth factors [3-7]. Since chitin nanofiber sheets (CNFSs) are biodegradable and exhibit large surface-to-mass ratios, CNFSs are widely

applied in pharmaceuticals as composite materials. The favorable properties of CNFS-based materials are enhanced as sizes of their fibers are decreased across the range of 1-100 nm [8]. In case of cosmetic dermatology, chitin nano-fibrils do not only protect corneocytes and intracorneal lamellae, but also helping to maintain cutaneous homeostasis. In addition, CNFSs neutralize the activity of free radicals and trap them in their structure, thereby regulating correct cell turnover [9]. The positive charges on the surface of the fibers, along with the chelating capacity of the acetamido groups of the chitin/chitosan molecule, play important roles in adsorption of heavy metals [10,11]. Our previous study demonstrated that chitin powder with nano-scale fiber-like surface structures can adsorb Ag NPs more

* Correspondence: ishihara@ndmc.ac.jp
[2]Research Institute, National Defense Medical College, 3-2 Namiki, Tokorozawa, Saitama 359-1324, Japan
Full list of author information is available at the end of the article

efficiently than chitin powder with flat/smooth film-like surface structures [12]. CNFSs have attracted much attention for application as components of pharmaceutics such as drug carriers, textile materials, sutures, and scaffold materials for tissue engineering [13,14].

Ag NPs have strong antimicrobial activity against most microorganisms, including bacteria, fungi, and viruses. In recent publications, we demonstrated that chitin/chitosan/Ag NP composites have enhanced antimicrobial activities against microbial pathogens, including bacteria (*E. coli*), fungi (*Aspergillus niger*), and virus (H1N1 influenza A virus) [15-17]. The bactericidal activity of Ag NPs is believed to result from Ag NP interactions with the cell wall, permitting Ag NPs to penetrate the membrane, thereby leading to the cell death [18]. In addition, silver ions released from Ag NP surface are thought to bind to sulfhydryl groups, leading to protein denaturation [19]. Furthermore, silver ions have been shown to penetrate through ion channels without causing damage to the cell membranes, where the ions denature the ribosome and suppress the expression of enzymes and proteins essential to ATP production [20]. Silver ions can interact with the bases in DNA causing loss of replication [21,22]. Ag NPs also induce the formation of free radicals, which in turn damage the membrane and cause cell death, formation of bactericidal reactive oxygen species (ROS), and lactate dehydrogenase activity involved in the respiratory chain [23]. Bacterial DNA can be affected by ROS, resulting in the production of superoxide anion ($O_2^-$), hydroxyl radical (OH$^-$), and singlet oxygen ($^1O_2$) with subsequent oxidative damage [23,24].

Questions have been raised about the safety of prolonged use of Ag NPs in humans and animals. Generally, silver does not adversely affect mammalian cell viability. Hence,

silver has been incorporated into various materials and used in antimicrobial materials to protect from infectious contamination [25]. Recently, *in vitro* studies in human cells have reported that Ag NP exposure induces metabolic arrest rather than cell death, and that human cells have a greater resistance to the toxic effects of Ag NPs in comparison with those from other organisms [26,27]. Furthermore, previous studies have revealed that while Ag NP-containing chitosan-based wound dressings are cytotoxic *in vitro*, such wound dressings perform satisfactorily *in vivo* [28]. The aim of the present work is to evaluate the bactericidal (against *E. coli* and *P. aeruginosa*) and antiviral (against influenza virus H1N1) activities of CNFS/Ag NPs, for potential biomedical applications such as wound dressings and antivirus sheets.

## Results

### Characterization of Ag NPs and CNFS/Ag NPs

Ag NPs were synthesized by autoclaving (at 121°C and 20 kPa) a mixture of only three components: silver-containing glass powder, glucose, and water [17,29]. TEM images showed that the Ag NPs were spherical with the average particle size of 5.17 ± 1.92 nm (mean ± SD). The results of UV-Vis analysis of the Ag NPs suspension revealed that the peak at 390.5 nm is representative of the Ag NPs in this study (Figure 1).

The surface morphology of the CNFS has been characterized using SEM imaging. The CNFS has a nanoscale fiber-like surface structure (Figure 2A). TEM observation of CNFS/Ag NPs revealed that the Ag NPs were stably adsorbed to the surface of CNFS (Figure 2B). Based on comparison of absorbance values of Ag NP suspension before and after reaction with CNFS, along with the equation for the standard curve of absorbance at 390.5 nm as a

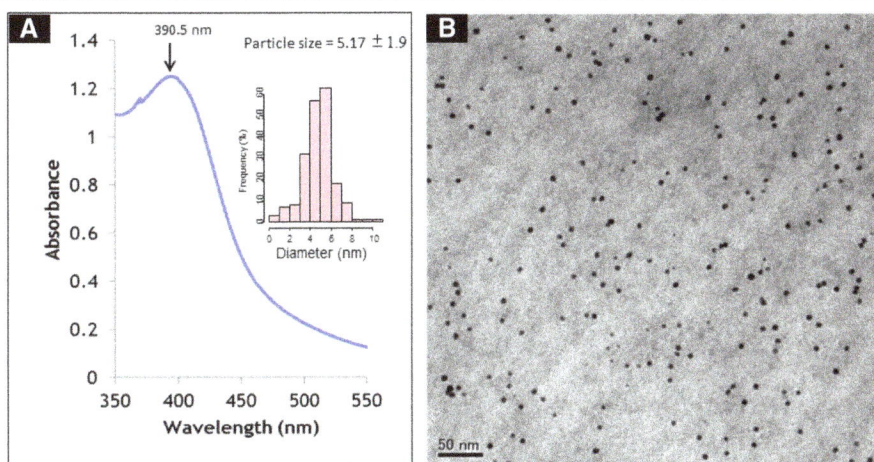

**Figure 1 Absorption of Ag NPs to ANSF. (A)** The UV-visible absorption spectrum of Ag NPs. The peak of absorbance at the wavelength of 390.5 nm is indicated. The inset figure is the particle size distribution histogram of the Ag NPs. **(B)** TEM micrograph of Ag NPs. Scale bar represents 50 nm.

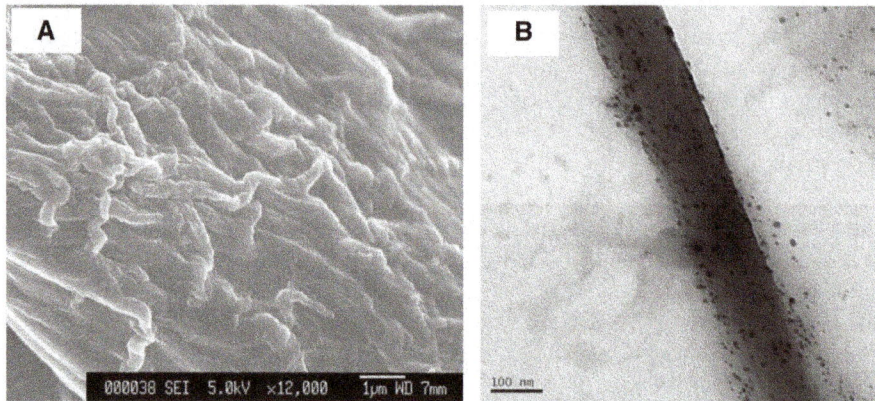

**Figure 2 SEM and TEM images of CNFS.** SEM image of CNFS; scale bar represents 1 μm **(A)**. TEM image of CNFS/Ag NPs composite sheet; scale bar represents 100 nm **(B)**.

function of the concentration of Ag NPs in suspension, we estimated that Ag NPs were immobilized on CNFS at 8. 45 μg per cm$^2$ (Figure 3A and Figure 3B).

### Bactericidal activity of CNFS/Ag NPs

To establish the bactericidal properties of CNFS/Ag NPs, the sheet were tested with two infectious bacteria: *E. coli* and *P. aeruginosa*. The inhibition zone of bacterial growth around CNFS/Ag NPs and CNFS alone against *E. coli* and *P. aeruginosa* are shown in Figure 4. There was no zone of growth inhibition around CNFS alone for either *E. coli* or *P. aeruginosa*. With CNFS/Ag NPs (8.5 μg/ml), there were clear zones of inhibition of ≈ 30 mm diameter (for *E. coli*) and ≈ 25 mm diameter (for *P. aeruginosa*) around CNFS/Ag NPs after 24 h incubation (Figure 4).

Bactericidal tests of CNFS/Ag NPs were performed against *E. coli* and *P. aeruginosa* by counting the viable bacterial colonies after treatment with different concentrations of Ag NPs immobilized on CNFS (2.3, 3.8, 8.5 μg/1 cm$^2$ CNFS). Samples of *E. coli* were completely eradicated when exposed to CNFS contained 8.5 μg/ml of Ag NPs. The high concentration Ag NPs immobilized on CNFS gave significant decreases of cell number in log 10 CFU/ml, while CNFS alone gave only a little decrease against *E. coli* and *P. aeruginosa*. Thus, the bactericidal activity of CNFS/Ag NPs increased with increased Ag NP loading (Figure 5).

### Antiviral activity of CNFS/Ag NPs

In order to confirm the antiviral activity of CNFS/Ag NPs, the CNFSs carrying various amounts of immobilized Ag NPs were evaluated for antiviral activity for human influenza A virus (A/PR/8/34 (H1N1)). The high concentration Ag NPs immobilized on CNFS gave significant decreases

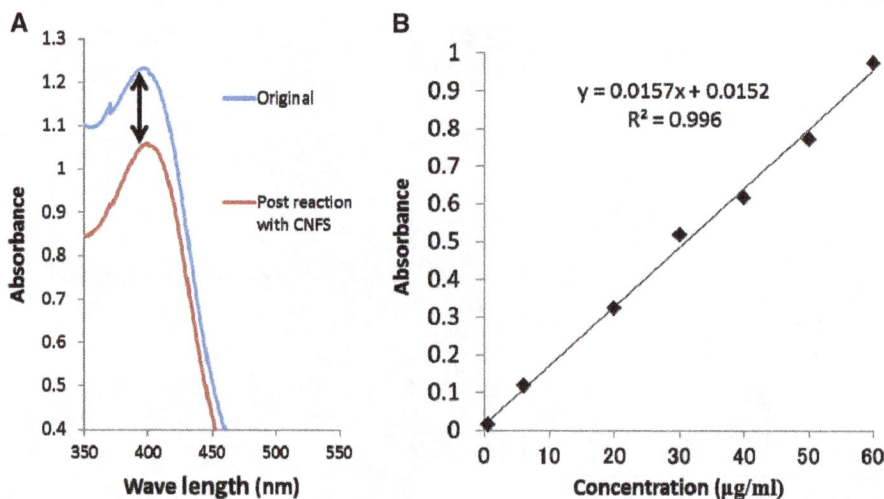

**Figure 3 The absorbance spectra of Ag NPs.** The absorbance spectra of original Ag NP suspension (blue line) and suspension after reaction with 1 cm$^2$ CNFS (red line) **(A)**. The relationship between absorbance at 390.5 nm and the concentration of Ag NPs in the suspension **(B)**.

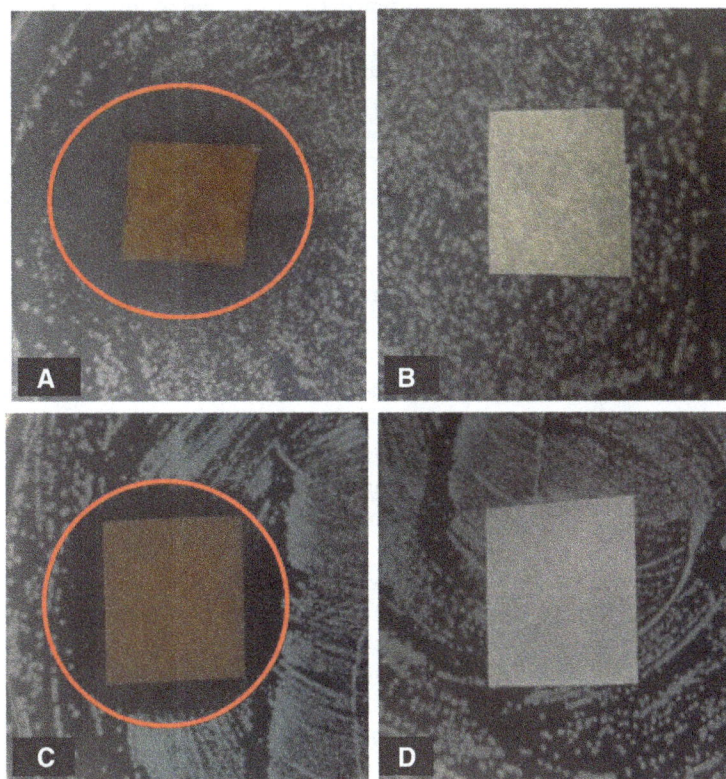

**Figure 4** Antimicrobial activity of CNFS/Ag NPs against *E. coli* (A); CNFS alone against *E. coli* (B); CNFS/Ag NPs against *P. aeruginosa* (C); and CNFS alone against *P. aeruginosa* (D). The CNFS/Ag NPs showed inhibition zone of bacterial growth as shown with red circles (A and C), although the CNFS alone exhibited no detectable inhibitory activity against either bacterial species (B and D).

of virus number in log 10 CFU/ml, while CNFS alone gave only a little decrease against influenza A virus. At concentration of Ag NPs of 8.5 µg/1 cm$^2$ chitin sheet, there was a reduction of greater than 2 log10 (100-fold) corresponding to reduction of viral titers by approximate 99% (Figure 6). This mentions that the antiviral activities of the CNFS/Ag NPs sheet were due to the interaction between virions and Ag NPs. The viruses may be adsorbed and immobilized on the CNFS/Ag NPs sheet. Therefore, inceasing amount of nAg on the chitin sheet makes more viruses adsored and immobilized on the CNFS/Ag NPs sheet.

**Ultrathin sectioning of bacterial cells**

The mechanism(s) of bactericidal activity of Ag NPs remain poorly understood. Ultrathin sectioning was carried out to obtain further understanding of the bactericidality and the interaction of the Ag NPs with bacterial cells. After 1 h treatment with a suspension of Ag NPs, the cytoplasmic components of *E. coli* and *P. aeruginosa* were coagulated, leading to vacant spaces within the cells. Gross inspection of the TEM images revealed non-homogeneity of the cytoplasm in the Ag NP-treated cells compared with the controls (Figure 7). The density of cytoplasmic components in treated cells was obviously decreased compared with the control. Plasma membranes of treated

cells were detached from the cell wall, leaving open spaces between the membrane and cell wall. Furthermore, DNA was condensed (Figure 7).

**Discussion**

Several kinds of materials are used for wound dressings, including cotton, chitin, chitosan, alloskin, pigskin, and other biologic-based materials. The various materials are commonly used in clinical settings, but these dressings often have some disadvantages such as low antimicrobial activity, allergenicity, toxic effects, and poor adhesiveness [8-10]. In the present study, we developed a potential wound dressing composed of Ag NPs (5.17 ± 1.9 nm in diameter) immobilized on CNFS to remedy some of the disadvantages of current wound dressings. CNFS was combined with Ag NPs, which act as a barrier to microorganisms, thereby limiting cross contamination. The Ag NPs were homogeneously dispersed and tightly immobilized on CNFS. The CNFS/Ag NPs showed strong antimicrobial activity against *E. coli*, *P. aeruginosa*, and influenza A virus.

The Ag NPs used in this research were produced using environmental-friendly materials and processes to control the size of Ag NPs, yielding Ag NPs of ≈ 5 nm in diameter. Components of the NPs included silver-containing glass

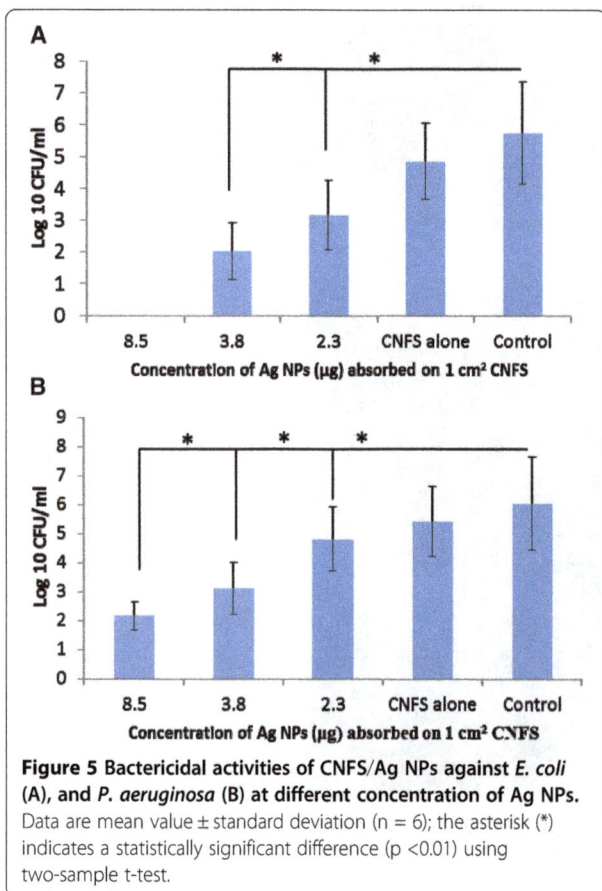

**Figure 5** Bactericidal activities of CNFS/Ag NPs against *E. coli* (A), and *P. aeruginosa* (B) at different concentration of Ag NPs. Data are mean value ± standard deviation (n = 6); the asterisk (*) indicates a statistically significant difference (p <0.01) using two-sample t-test.

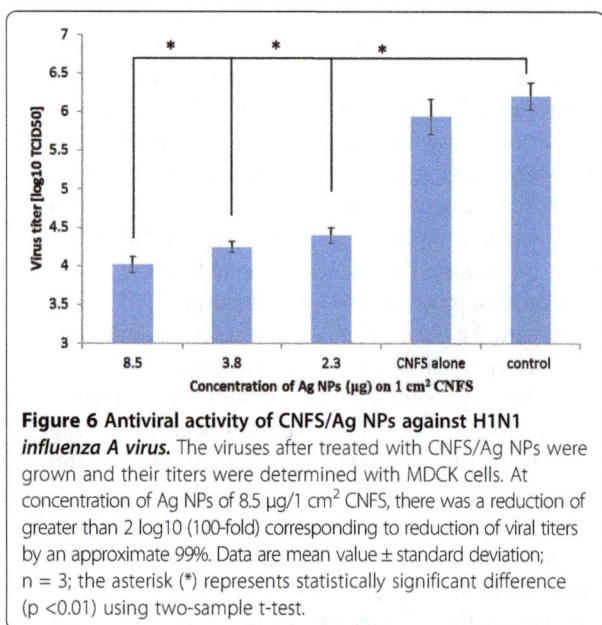

**Figure 6** Antiviral activity of CNFS/Ag NPs against H1N1 *influenza A virus.* The viruses after treated with CNFS/Ag NPs were grown and their titers were determined with MDCK cells. At concentration of Ag NPs of 8.5 μg/1 cm² CNFS, there was a reduction of greater than 2 log10 (100-fold) corresponding to reduction of viral titers by an approximate 99%. Data are mean value ± standard deviation; n = 3; the asterisk (*) represents statistically significant difference (p <0.01) using two-sample t-test.

powder, glucose, and water. Silver-containing glass powder often is used in osteal or dental applications as an anti-microbial agent; glucose has the advantage of being an environmentally friendly agent [29]. In previous work, we have demonstrated that the antimicrobial activity of Ag NPs depended on particle size [12,15,17]. Ag NPs of small size (3-10 nm diameter) have strong antimicrobial activity against *E. coli*, *P. aeruginosa*, and influenza A virus, and it is hypothesized that Ag NPs with smaller particle sizes have larger available surface areas for interaction with microorganisms [17,29].

The CNFS used in this study has a nanoscale fiber-like surface structure, with corresponding increases in the available surface area for adsorption of Ag NPs. In addition, the advantages in biochemical activities of chitin/chitosan-based materials include anti-infectious activity [2], stimulation of angiogenesis/wound repair, and stabilization/activation of growth factors [3-7]. Recent studies show that the application of CNFS to skin improved the epithelial granular layer and increased granular density, suggesting the potential use of CNFS as a component of skin-protective formulations [30]. CNFS also has been shown to inhibit mucosal inflammation by suppressing the MPO-positive cells such as leukocytes [31].

This study used CNFS, a commercially available wound dressing, in combination with Ag NPs to provide stronger antimicrobial ability. The composite thus might serve as a new biocompatible wound dressing with reduced danger of cross contamination. CNFS/Ag NPs showed much stronger bactericidal activity against *E. coli* and *P. aeruginosa*, with clear zones of inhibition around the sheet. We also observed antiviral activity against influenza A virus, presumably due to the interaction between virions and Ag NPs immobilized on CNFS. Therefore, increasing amounts of Ag NPs on CNFS may further increase the number of virions immobilized on CNFS/Ag NPs, yielding increased virucidal activity. Our results suggest that CNFS/Ag NPs can be applied not only as functional wound dressings, but also as antimicrobial agents, including antivirus sheets.

CNFSs containing 8.5 μg Ag NPs/1 cm² sheet (7.3 ± 0.1 mg) completely eradicated *E. coli*. Several potential mechanisms have been reported for the bactericidal activities of Ag NPs [21,22]. Those studies showed that Ag NP exposure resulted in decreased density of cytoplasmic components, condensation of bacterial DNA, and disorganization of the cytoplasmic membrane with detachment of the plasma membrane from the cell wall. These phenomena suggested that Ag NPs induced a loss of integrity of the cytoplasm and membranes, causing malfunction of organelles and membranes, and leading to cell death. Alternatively, bacterial DNA could be affected by ROS, resulting in the production of superoxide anions ($O_2^-$) with subsequent oxidative damage [23]. This study is required to carry out biochemical analyses to confirm those mechanism.

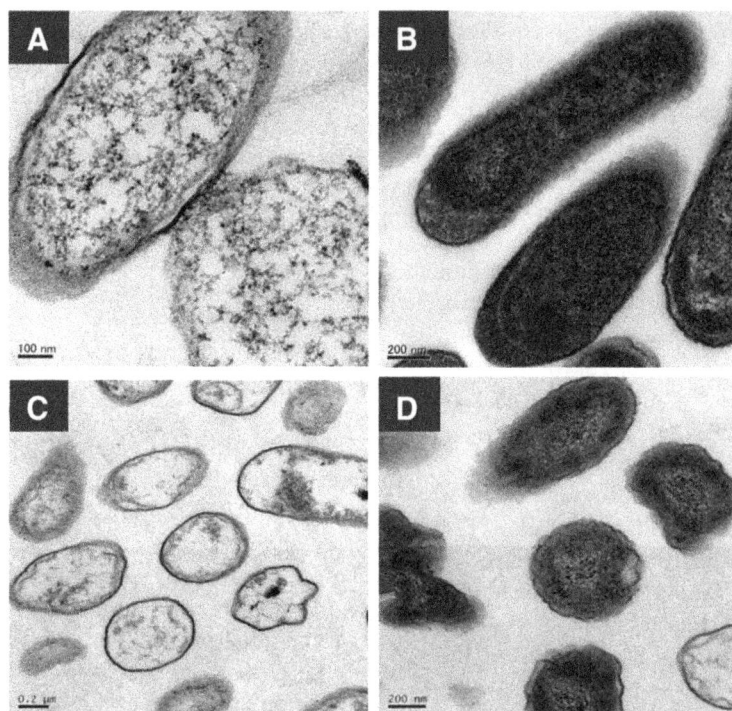

**Figure 7 Representative TEM images of morphology and structure of *E. coli* and *P. aeruginosa*. (A)** *E. coli* treated with Ag NPs; **(B)** normal *E. coli*; **(C)** *P. aeruginosa* treated with Ag NPs; **(D)** normal *P. aeruginosa*. Exposure to Ag NPs resulted in damage to the structure of bacterial cell membranes, condensed DNA, and coagulated cytoplasmic components. Normal bacterial cells were smooth, exhibiting intact surfaces and undamaged structures of inner membranes. Scale bars are as indicated.

The effect of the size of Ag NPs on antiviral activity suggests that various viruses interact selectively with smaller (≤10 nm diameter) Ag NPs, as previously reported for HIV-1 [32] and hepatitis B viruses [33]. We also previously reported a size-dependence for the effect of free Ag NPs on antiviral activity against influenza A virus [29]. In the context of anti-influenza A virus activity, further spatial restriction due to the CNFS would be expected to prevent or weaken the interaction between virions and Ag NPs. Although the virus was not completely eradicated when exposed to CNFS/Ag NPs, the adsorption of Ag NPs onto CNFS provided stronger antivirus activity. Thus, the interaction between the virions and the Ag NPs is expected to be further increased with increasing amounts of Ag NPs on the CNFSs.

## Conclusions

Our TEM image analysis indicated that Ag NPs are dispersed and tightly adsorbed on CNFS. Our antimicrobial assays further demonstrated that Ag NPs immobilized on CNFS provide much higher antimicrobial activities against *E. coli*, *P. aeruginosa*, and influenza A virus. Thus, we propose that CNFS/Ag NPs might find use as anti-influenza sheets as well as anti-bacterial wound dressing sheets.

## Materials and methods

### Materials

Silver-containing glass powder (BSP21, Ag content: 1 wt%; average grain size: 10 μm) was purchased from Kankyo Science (Kyoto, Japan). CNFSs (degree of deacetylation: ≈ 30%) used in this study were obtained as the commercial product (BeschitinW, Unichika Ltd., Tokyo, Japan). D-Glucose was purchased from Wako Pure Chemical Industries, Ltd. (Osaka, Japan). All chemicals were used as received.

### Preparation of Ag NPs

A suspension of size-controlled Ag NPs was prepared as previously described [29]. Briefly, 0.5 g of Ag-containing glass powder was dispersed in 50 ml of an aqueous solution of 0.8 wt% glucose in a 100 ml glass vial. The mixture was autoclaved at 121°C and 200 kPa for 20 min and then gradually cooled to room temperature; the mixture then was centrifuged at 1500 g for 10 min. The resulting brown supernatant containing the Ag NP suspension was stored in the dark at 4°C. Transmission electron microscopy (TEM) specimens were prepared by casting a small drop of a suspension of Ag NPs onto a carbon-coated copper grid; excess solution was then removed using filter paper and the specimens were dried at room temperature. TEM images were obtained using a JEOL JEM-1010 microscope

(Nihon Electronics Inc., Tokyo, Japan) operated at 80 kV. The diameter size of Ag NPs from TEM image was determined using ImageJ 1.45 software (http://rsb.info.nih.gov/ij).

### Preparation of CNFS/Ag NPs

CNFS (1 cm × 1 cm) was submerged in a 1.5 ml ClickFit polypropylene microcentrifuge tube (TreffLab AG, Degersheim, Switzerland) containing 1 ml of Ag NP suspension (at about 30 µg/ml) and shaken well for 30 min using a shaker (Mild Mixer PR-36; TAITEC, Tokyo, Japan). The post-reaction supernatant was analyzed using a UV-visible spectrometer (Jasco V-630, Tokyo, Japan) to measure the amount of unreacted Ag NPs as a peak of absorbance at wavelength of 390.5 nm. The CNFS/Ag NP composites were washed twice with distilled water. The washed composites were air dried up on a clean bench for 1 h and used in bactericidal assays on the same day. TEM inspection confirmed that the Ag NPs were homogeneously dispersed and immobilized on the CNFS, which had become brown in color. The concentrations of Ag NPs immobilized on the CNFS were calculated based on the UV-Vis spectra of Ag NPs before and after mixing with CNFS, using a standard curve of Ag NPs generated for a previous publication [15].

Scanning electron microscopy (SEM) specimens of the CNFS/Ag NP composites were mounted on metal mounts with double-sided adhesive tape and coated with gold plasma to enhance conductivity using a plasma multi-coater PMC-5000 (Meiwafosis Co., Ltd., Tokyo, Japan). The surface morphology of coated samples was examined by JSM-6340 F microscope (JEOL, Tokyo, Japan) operated at 5 kV. The TEM image of CNFS/Ag NPs were carried out by cutting the composite sheet into very small pieces and then resuspending in 200 µl distilled water; 5 µl of the resulting suspension was observed by TEM with the JEOL JEM-1010 microscope.

### Bactericidal activity of CNFS/Ag NPs

A culture of E. coli strain DH5α (Takara Co., Kyoto, Japan) was stored at -80°C in Luria-Bertani (LB) broth containing 50% sterile glycerol. Overnight cultures were prepared by growing a single E. coli colony overnight at 37°C in 5 ml of LB medium. On the next day, 200 µl of the overnight culture was inoculated into 2 ml of LB medium and incubated at 37°C for 6 h or until the optical density at 600 nm ($OD_{600}$) reached 0.260. The E. coli culture then was diluted 4-fold with LB broth, and 30 µl of the diluted suspension were spread on LB agar (ForMedium Ltd., Hunstanton, England). CNFS/Ag NPs and CNFS alone then were placed onto the surface of the inoculated agar plate, which was incubated at 37°C overnight. Growth inhibition zones around the sheets were measured on the subsequent day using a centimeter scale.

The experiment on P. aeruginosa was carried out using essentially the same technique as for E. coli. The P. aeruginosa strain ATCC 27853 (American Type Culture Collection (ATCC), Manassas, USA) was stored at -80°C in Luria-Bertani (LB) broth containing 50% sterile glycerol. The cell suspension of P. aeruginosa was prepared as follows: 20 µl of stock suspension was plated onto Pseudomonas Isolation agar (Neogen Ltd. Michigan, USA) and incubated at 37°C for 18 h. Colonies then were resuspended by placing 2 ml of LB broth on the plate surface and gently shaking the plate by hand for few minutes. The resulting suspension was pipetted to a new tube for the next experiments (or stocked at -80°C with glycerol). Following adjustment to $OD_{600}$ of 0.26, this suspension was diluted a further 4-fold, and 30 µl was spread onto nutrient agar (Nissui Pharmaceutical CO., LTD, Tokyo, Japan). CNFS/Ag NPs and CNFS alone then were placed onto the surface of the inoculated agar plate, which was incubated at 37°C overnight. Growth inhibition zones around the sheets were measured on the subsequent day using a centimeter scale.

Forty µl of the diluted cultures of E. coli and P. aeruginosa were dropped onto each CNFS composite harboring immobilized Ag NPs at various concentrations (8.5, 3.8, 2.3, and 0 µg/1 cm$^2$ sheet (7.3 ± 0.1 mg)). All of the sheets were incubated at 37°C for 1 h, and each sheet then was immersed/washed in 1 ml LB medium. The resulting wash suspensions were subjected to 10-fold serial dilutions, and 50 µl samples of diluted suspensions were plated (to 90 × 15 mm petri plates of LB agar (for E. coli) or nutrient agar (for P. aeruginosa)). Plates were incubated at 37°C for 24 h, and viable cells were enumerated.

### Evaluation of the antiviral activity of CNFS /Ag NPs

Antiviral activity of CNFS/Ag NPs was evaluated against H1N1 influenza A virus as described previously [17,30]. Fifty µl of viral suspension (about 10$^5$ TCID$_{50}$/ml) was added onto CNFS/Ag NPs consisting of various amounts of Ag NPs (8.5, 3.8, 2.3, and 0 µg) immobilized on 1 cm$^2$ CNFS (7.3 ± 0.1 mg). The virus-inoculated composites were placed in an empty petri dish and incubated at room temperature for 1 h to facilitate the interaction between the viruses and the CNFS/Ag NPs. The sheets then were individually transferred to 1.5 ml tubes, each of which received 450 µl phosphate-buffered saline (PBS) and 1 min of vortexing. Following centrifugation at 6400 g for 5 min, the supernatants were transferred to new tubes, then subjected to eleven 2-fold serial dilutions in PBS. Fifty µl of each diluted supernatant was added to the individual wells of a 96-well plate containing MDCK cells. The samples were incubated at 37°C and 5% CO$_2$ for 1 h to allow virus adsorption to the cells. Aliquots of growth medium (50 µl DMEM medium containing 0.4% BSA and 5 µg/ml trypsin) were added to each well. After

5 days of incubation, another 50 µl of DMEM medium containing 0.4% BSA was added to each well. Seven days post-infection, surviving cells were fixed with methanol (200 µl/well, two times), and stained with 50 µl of 5% Giemsa stain solution. Cells counts (stained (uninfected) and unstained (infected)) were determined, and viral titers (in $TCID_{50}$/ml) were calculated according to method of Reed and Muench [17,30].

## Ultrathin sectioning of bacterial cells

In order to understand the bactericidal activities of silver nanoparticles, ultrathin sectioning was carried out to observe ultrastructural changes in bacterial cells. Two ml of Ag NPs suspension (6 µg/ml) was placed on the surfaces of agar plates containing colonies of *E .coli* or *P. aeruginosa*. After 1 h the colonies were recovered and fixed overnight (minimum of 2 h) at 4°C with 2% glutaraldehyde and 2% paraformaldehyde in 0.1 M phosphate buffer, pH 7.4. The fixed samples were washed overnight (minimum of 2 h) at 4°C in 0.1 M phosphate buffer, then post-fixed for 2 h at 4°C in 1% $OsO_4$ in 0.1 M phosphate buffer. The samples then were dehydrated by using a series of alcohol solutions at increasing concentration (50, 75, 95% at 20 min each, followed by 2 passages in 100% ethanol for 30 min each). Samples were infiltrated at room temperature by immersion in propylene oxide (2 × 30 min), 1 : 1 mixture of propylene oxide and epoxy resin (1 h), 1 : 2 mixture of propylene oxide and epoxy resin (overnight), and epoxy resin only (minimum 4 h). The samples then were embedded with epoxy resin in a Beem capsule and polymerized in an oven at 37°C/12 h, 45°C/24 h, and 60°C/48 h. The polymerized samples were first semi-thin sectioned at 1.5 µm with glass knives using UltraCut S and stained with Toluidine Blue. Ultrathin sections were obtained with an ultramicrotome (UltraCut S, Reichert) with ultrathin slices 60 to 90 nm in thickness. Ultrathin slices were recovered on a 3.0 mm-diameter 200-mesh copper grid and stained with uranyl acetate for 20 min and lead acetate for 1 min. The grids were examined by TEM (JEM-1010).

## Statistical analyses

Statistical analyses were carried out using StatMate III, Macintosh Version (ATMS Co., Tokyo, Japan). Statistical significance was assumed when p <0.01. Where relevant, values are provided as mean ± SD.

## Abbreviations

CNFSs: Chitin nanofibers sheets; CNFS/Ag NPs: CNFS-immobilized silver nanoparticles; TEM: Transmission electron microscopy; SEM: Scanning electron microscopy; UV-Vis: Visible-Ultraviolet spectrums.

## Competing interests

The authors declare that they have no competing interests.

## Authors' contributions

VQN designed the research, performed the experiments, and drafted the manuscript and the figures. MI and TM supervised and coordinated the study and approved the manuscript. TO and YMi supervised, guided and performed the viral study. JK assisted VQN and performed some parts of experiments. HH and SN guided the study and assisted to draft the manuscript and the figures. All authors read and approved the final manuscript.

## Acknowledgment

The authors also thank Ms. Y Ichiki from Laboratory Center of National Defense Medical College for her help with the electron microscopy experiments.

## Author details

[1]Faculty of System Design, Tokyo Metropolitan University, 6-6 Asahigaoka, Hino, Tokyo 191-0065, Japan. [2]Research Institute, National Defense Medical College, 3-2 Namiki, Tokorozawa, Saitama 359-1324, Japan. [3]Department of Oral and Maxillofacial Surgery, National Defense Medical College, 3-2 Namiki, Tokorozawa, Saitama 359-8513, Japan. [4]Department of Global Infectious Diseases and Tropical Medicine, National Defense Medical College, 3-2 Namiki, Tokorozawa, Saitama 359-8513, Japan.

## References

1. Ishihara M, Fujita M, Kishimoto S, Hattori H, Kanatani Y: **Biological, Chemical, and Physical Compatibility of Chitosan and Biopharmaceuticals.** In *Chitosan-Based Systems for Biopharmaceuticals.* Edited by Samento B, Neves JD. West Sussex, UK: John Wiley&Sons, Ltd; 2012:93–107.
2. Muzzarelli RAA, Morganti P, Morganti G, Palombo P, Palombo M, Biagini G, Mattioli BM, Belmonte M, Giantomassi F, Orlandi F, Muzzarelli C: **Chitin nanofibrils/chitosan glycolate composites as wound medicaments.** *Carbohydr Polym* 2007, **70**:274–284.
3. Ishihara M, Nakanishi K, Ono K, Sato M, Saito Y, Yura H, Matsui T, Hattori H, Uenoyama M, Kurita A: **Photocrosslinkable chitosan as a dressing for wound occlusion and accelerator in healing process.** *Biomaterials* 2002, **23**(3):833–840.
4. Hattori H, Amano Y, Nogami Y, Takase B, Ishihara M: **Hemostasis for severe hemorrhage with photocrosslinkable chitosan hydrogel and calcium alginate.** *Ann Biomed Eng* 2010, **38**(12):3724–3732.
5. Kiyozumi T, Kanatani Y, Ishihara M, Saitoh D, Shimizu J, Yura H, Suzuki S, Okada Y, Kikuchi M: **Medium (DMEM/F12)-containing chitosan hydrogel as adhesive and dressing in autologous skin grafts and accelerator in the healing process.** *J Biomed Mater Res* 2006, **79B**(1):129–136.
6. Ishihara M, Fujita M, Obara K, Hattori H, Nakamura S, Nambu M, Kiyosawa T, Maehara T: **Controlled releases of FGF-2 and paclitaxel from chitosan hydrogels and their subsequent effects on wound repair, angiogenesis, and tumor growth.** *Curr Drug Deliv* 2006, **3**(4):351–358.
7. Masuoka K, Ishihara K, Asazuma T, Hattori H, Matsui T, Takase B, Kanatani Y, Fujita M, Saito Y, Yura H, Fujikawa K, Nemoto N: **Interaction of chitosan with fibroblast growth factor-2 and its protection from inactivation.** *Biomaterials* 2005, **26**(16):3277–3284.
8. Jayakumar R, Prabaharan M, Nair SV, Tamura H: **Novel chitin and chitosan nanofibers in biomedical applications.** *Biotechnol Advances* 2010, **28**:142–150.
9. Morganti P, Morganti G: **Chitin nanofibrils for advanced cosmeceuticals.** *Clinic Dermatol* 2008, **26**(4):334–340.
10. Dutta AK, Kawamoto N, Sugino G, Izawa H, Morimoto M, Saimoto H, Ifuku S: **Simple preparation of chitosan nanofibers from dry chitosan powder by the star burst system.** *Carbohydro Polym* 2013, **98**:1198–1202.
11. Lee M-Y, Park JM, Yang J-W: **Micro precipitation of lead on the surface of crab shell particles.** *Process Biochem* 1997, **32**(8):671–677.
12. Nguyen VQ, Ishihara M, Nakamura S, Hattori H, Ono T, Miyahira Y, Matsui T: **Interaction of Silver Nanoparticles and Chitin Powder with Different Sizes and Surface Structures: The Correlation with Antimicrobial Activities.** *J Nanomater* 2013, **2013**:9. art no.467534.
13. Min SK, Lee SC, Hong SD, Chung CP, Park WH, Min BM: **The effect of a laminin-5-derived peptide coated onto chitin microfibers on re-epithelialization in early-stage wound healing.** *Biomaterials* 2010, **31**(17):4725–4730.

14. Pillai CKS, Sharma CP: **Electrospinning of Chitin and Chitosan Nanofibres.** *Trends Biomater Artif Organ* 2009, **22**(3):179–201.

15. Nguyen VQ, Ishihara M, Mori Y, Nakamura S, Kishimoto S, Hattori H, Fujita M, Kanatani Y, Ono T, Miyahira Y, Matsui T: **Preparation of Size-Controlled Silver Nanoparticles and Chitin-Based Composites and Their Antimicrobial Activities.** *J Nanomater* 2013, **2013**:7. art no. 693486.

16. Nguyen VQ, Ishihara M, Mori Y, Nakamura S, Kishimoto S, Fujita M, Hattori H, Kanatani Y, Ono T, Miyahira Y, Matsui T: **Preparation of size-controlled silver nanoparticles and chitosan-based composites and their anti-microbial activities.** *Biomed Mater Eng* 2013, **23**(6):473–483.

17. Mori Y, Ono T, Miyahira Y, Nguyen VQ, Matsui T, Ishihara M: **Antiviral activity of silver nanoparticle/chitosan composites against H1N1 influenza A virus.** *Nanoscale Res Lett* 2013, **8**(1):88–93.

18. Tran HV, Tran LD, Ba CT, Vu HD, Nguyen TN, Pham DG, Nguyen PX: **Synthesis, characterization, antibacterial and antiproliferative activities of monodisperse chitosan-based silver nanoparticles.** *Colloids and Surfaces A: Physicochem Engineer Aspects* 2010, **360**(1–3):32–40.

19. Sotiriou GA, Pratsinis SE: **Antibacterial Activity of Nanosilver Ions and Particles.** *Environ Sci Technol* 2010, **44**(14):5649–5654.

20. Yamanaka M, Hara K, Kudo J: **Bactericidal actions of a silver ion solution on Escherichia coli, studied by energy-filtering transmission electron microscopy and proteomic analysis.** *Appl Environ Microbiol* 2005, **71**(11):7589–7593.

21. Sondi I, Salopek-Sondi B: **Silver nanoparticles as antimicrobial agent: a case study on E. coli as a model for Gram-negative bacteria.** *J Colloid Interface Sci* 2004, **275**(1):177–182.

22. Kim S-H, Lee H-S, Ryu D-S, Choi S-J, Lee D-S: **Antibacterial activity of silver-nanoparticles against Staphylococus and Escherichia coli.** *Korean J Microbiol Biotechnol* 2011, **39**:770–785.

23. Pellieux C, Dewilde A, Pierlot C, Aubry J-M: **Bactericidal and virucidal activities of singlet oxygen generated by thermolysis of naphthalene endoperoxides.** *Methods Enzymol* 2000, **319**:197–207.

24. Abdelgawad AM, Hudson SM, Rojas OJ: **Antimicrobial wound dressing nanofiber mats from multicomponent (chitosan/silver-NPs/polyvinyl alcohol) systems.** *Carbohydr Polym* 2014, **100**:166–178.

25. AshaRani PV, Low Kah Mun G, Hande MP, Valiyaveettil S: **Cytotoxicity and genotoxicity of silver nanoparticles in human cells.** *ACS Nano* 2009, **3**(2):279–290.

26. De Lima R, Seabra AB, Duran N: **Silver nanoparticles: a brief review of cytotoxicity and genotoxicity of chemically and biogenically synthesized nanoparticles.** *J Appl Toxicol* 2012, **32**(11):867–879.

27. Hackenberg S, Scherzed A, Kessler M, Hummel S, Technau A, Froelich K, Ginzkey C, Koehler C, Hagen R, Kleinsasser N: **Silver nanoparticles: evaluation of DNA damage, toxicity and functional impairment in human mesenchymal stem cells.** *Toxicol Let* 2011, **201**(1):27–33.

28. Ong S-Y, Wu J, Moochhala SM, Tan M-H, Lu J: **Development of a chitosan-based wound dressing with improved hemostatic and antimicrobial properties.** *Biomaterials* 2008, **29**(32):4323–4332.

29. Mori Y, Tagawa T, Fujita M, Kuno T, Suzuki S, Matsui T, Ishihara M: **Simple and environmentally friendly preparation and size control of silver nanoparticles using an inhomogeneous system with silver-containing glass powder.** *J Nanopart Res* 2011, **13**(7):2799–2806.

30. Ito I, Osaki T, Ifuku S, Saimoto H, Takamori Y, Kurozumi S, Imagawa T, Azuma K, Tsuka T, Okamoto Y, Minami S: **Evaluation of the effects of chitin nanofibrils on skin function using skin models.** *Carbohydr Polym* 2014, **101**:464–470.

31. Azuma K, Osaki T, Wakuda T, Ifuku S, Saimoto H, Tsuka T, Imagawa T, Okamoto Y, Minami S: **Beneficial and preventive effect of chitin nanofibrils in a dextran sulfate sodium-induced acute ulcerative colitis model.** *Carbohydr Polym* 2012, **87**:1399–1403.

32. Elechiguerra J, Burt JL, Morones JR, Camacho-Bragado A, Gao X, Lara HH, Yacaman M: **Interaction of silver nanoparticles with HIV-1.** *J Nanobiotechnol* 2005, **3**(6):1–10.

33. Lu L, Sun RW, Chen R, Hui CK, Ho CM, Luk JM, Lau GK, Che CM: **Silver nanoparticles inhibit hepatitis B virus replication.** *Antivir Ther* 2008, **13**(2):253–262.

# One pot light assisted green synthesis, storage and antimicrobial activity of dextran stabilized silver nanoparticles

Muhammad Ajaz Hussain[1*], Abdullah Shah[1], Ibrahim Jantan[2], Muhammad Nawaz Tahir[3], Muhammad Raza Shah[4], Riaz Ahmed[5] and Syed Nasir Abbas Bukhari[2*]

## Abstract

**Background:** Green synthesis of nanomaterials finds the edge over chemical methods due to its environmental compatibility. Herein, we report green synthesis of silver nanoparticles (Ag NPs) mediated with dextran. Dextran was used as a stabilizer and capping agent to synthesize Ag NPs using silver nitrate ($AgNO_3$) under diffused sunlight conditions.

**Results:** UV–vis spectra of as synthesized Ag nanoparticles showed characteristic surface plasmon band in the range from ~405-452 nm. Scanning electron microscopy (SEM) and atomic force microscopy (AFM) studies showed spherical Ag NPs in the size regime of ~50-70 nm. Face centered cubic lattice of Ag NPs was confirmed by powder X-ray diffraction (PXRD). FT-IR spectroscopy confirmed that dextran not only acts as reducing agent but also functionalizes the surfaces of Ag NPs to make very stable dispersions. Moreover, on drying, the solution of dextran stabilized Ag NPs resulted in the formation of thin films which were found stable over months with no change in the plasmon band of pristine Ag NPs. The antimicrobial assay of the as synthesized Ag NPs showed remarkable activity.

**Conclusion:** Being significantly active against microbes, the Ag NPs can be explored for antimicrobial medical devices.

**Keywords:** Ag nanoparticles, Storage of nanoparticles, Diffused sun light, Antimicrobial activity

## Introduction

Ag NPs have wide variety of applications, e.g., opto-electrical [1,2], microbiocidal [3], nanorobotics [4] and medicinal [5]. However, clustering of Ag NPs on storage and in physiological media [6-9] is a major limitation in their biomedical applications. Chemical methods for the synthesis of Ag NPs have harmful effects on environment as well as on human health [9]. Due to said reasons, nowadays, polysaccharides and polypeptides [10] have attracted the vigil eye of researchers for the biosynthesis of Ag NPs as they can act as reducing, capping and stabilizing agents [11-13]. Recently, polysaccharide based Ag NPs have been prepared by adding NaOH [14] but use of such corrosive reagent has harmful effects on

environment as well as on human health. Therefore, it is promising to fabricate Ag NPs that could sustain themselves for longer period of time using environmentally benign molecules like biopolymers.

In this report, we have explored the green synthesis of Ag NPs using dextran as co-reducing as well as capping ligand without using any environmentally hostile ingredient like NaOH and $NaBH_4$. Dextran was choice because it is cheaper, non-toxic, biocompatible, efficient reducing and self-capping agent, *in situ* stabilizer of nanoparticles and environment friendly. The as synthesized NPs can be stored within matrix of dextran in the form of thin films without changing the optical properties over months. Moreover, the as-prepared Ag NPs were tested as antimicrobial probes against *S. aureus* (ATCC 25923), *E. coli* (ATCC 25922), *B. subtilis* (ATCC 6633), *S. epidermidis* (ATCC 12228), *P. aeruginosa* (ATCC 27853) and fungal strains *Actinomycetes and A. niger*.

* Correspondence: majaz172@yahoo.com; snab@ukm.edu.my
[1]Department of Chemistry, University of Sargodha, Sargodha 40100, Pakistan
[2]Drug and Herbal Research Centre, Faculty of Pharmacy, Universiti Kebangsaan Malaysia, Jalan Raja Muda Abdul Aziz, Kuala Lumpur 50300, Malaysia
Full list of author information is available at the end of the article

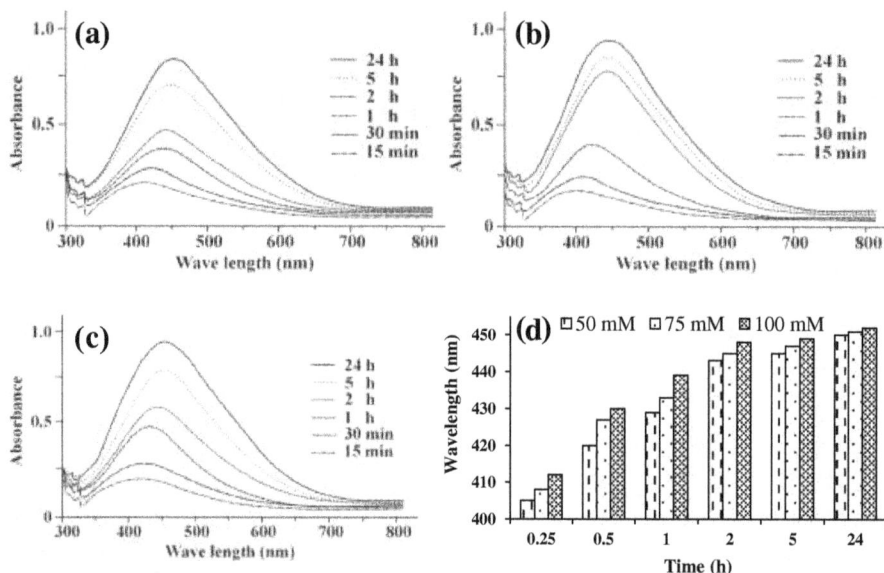

**Figure 1** UV–vis spectra of Ag NPs prepared in dextran: 50 (a), 75 (b) and 100 mmol (c) and graph showing effect of reaction time and concentration on absorbance (d).

## Experimental

### Materials and measurements

Dextran (molar mass 40000) was obtained from Sigma Aldrich, Germany. AgNO$_3$ (99.98%) from Merck, Germany was used as silver precursor. Deionized water was used for preparation of all solutions. UV–vis analyses were performed on UV-1700 PharmaSpec (Shimadzu, Japan). FT-IR spectra were recorded on IR Prestige-21 (Shimadzu, Japan). The samples (microtomes) were analyzed by SEM Plano (Wetzlar, Germany) using carbon stubs (carbon adhesive Leit-Tabs No. G 3347). The sizes and shapes of NPs were analyzed using AFM, Multimode, Nanoscope IIIa, Veeco, (California, USA) in tapping mode. Powder X-ray diffraction measurements were carried out (over a range of 5-100°, 2$\Theta$) on an Xpert Pro MPD, (PANalytical, The Netherlands) diffractometer equipped with monochromatic X-rays.

### Sample preparation of AgNO$_3$ and dextran

AgNO$_3$ solutions (50, 75 and 100 mmol) were prepared by dissolving AgNO$_3$ (0.85, 1.27 and 1.7 g, respectively) in deionized water. Concentrated solution of dextran was freshly prepared by dissolving dextran in deionized water (10 mL).

### Synthesis of Ag NPs mediated by dextran

Freshly prepared AgNO$_3$ (50 mmol, 2 mL) solution was added to the dextran solution (2 mL). The reaction mixture was exposed to diffused sunlight and color change was monitored over a period of 24 h by using UV–vis spectrophotometer. The same procedure was adopted for AgNO$_3$ (75 and 100 mmol) solutions, respectively.

### Thin film formation of dextran loaded with Ag NPs

Concentrated aq. solution of dextran loaded with Ag NPs (100 mmol) was kept in a petri dish for drying under air and stored.

### Atomic force microscopy (AFM)

The samples were prepared by dissolving thin films in deionized water and dispersing them on freshly cleaved sheet of mica substrate. AFM images were recorded at ambient temperature and repeated with different concentrations of the samples.

### Scanning electron microscope (SEM)

Surface of dextran thin films was analyzed by SEM to study geometry of embedded Ag NPs.

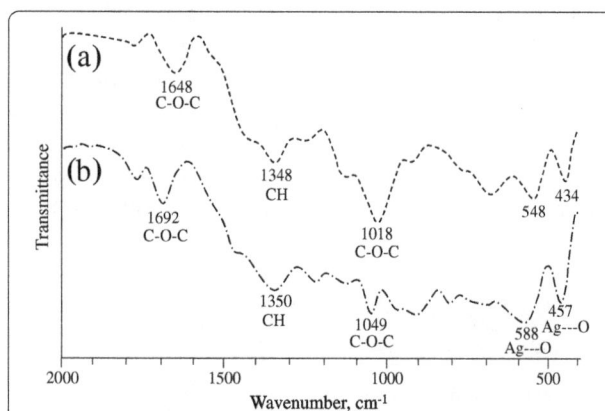

**Figure 2** FT-IR Spectra of dextran (a) and Ag NPs (50 mmol) loaded in dextran thin film (b).

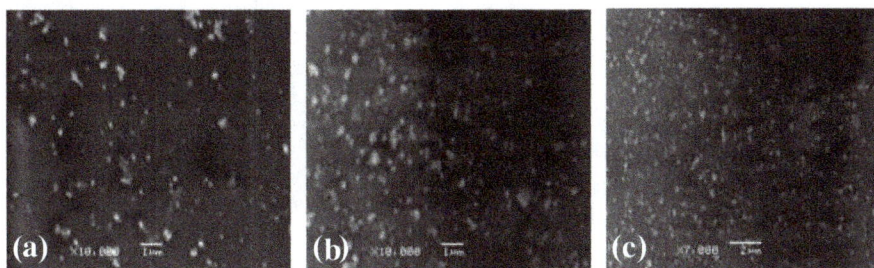

**Figure 3** SEM images of Ag NPs (50–70 nm) embedded in dextran thin films of 50 (a), 75 (b) and 100 mmol (c) AgNO$_3$ solution.

### Antimicrobial activity of Ag NPs

The test organism *S. aureus* (ATCC 25923), *E. coli* (ATCC 25922), *B. subtilis* (ATCC 6633), *S. epidermidis* (ATCC 12228), *P. aeruginosa* (ATCC 27853) and fungal strains *Actinomycetes and A. niger* were used for testing the antimicrobial activity of Ag NPs. The bacterial and fungal strains were procured from Microbiology Labs of Agriculture University, Faisalabad, Pakistan. Mueller Hinton Agar Media (Oxoid Ltd., England) was used for bacterial growth and Sabouraud Dextrose Agar (Hardy Diagnostics, USA) was used for fungal growth. Inoculums were prepared by transferring the microorganism culture in both tubes having 10 mL of respective broth media (Mueller Hinton broth for bacterial culture and Sabouraud Dextrose broth for fungal culture) and were inoculated for 24 h at 37°C for bacteria and 27-30°C for fungi. Seven days old culture of fungal strain was washed and suspended in normal saline solution. Then filtered through glass wool aseptically and incubated at 28°C. The tubes were shaken periodically to accelerate the growth of microorganisms. The turbidity of inoculums was adjusted by 0.5 Mc Farland Standard.

Antimicrobial assay of Ag NPs against different bacterial and fungal strains was conducted by disc diffusion method. *In vitro* antimicrobial activity was screened by using Mueller Hinton Agar plates for bacterial strains. Inoculum (0.1 mL) was spread uniformly on plates. Ag NPs solution was loaded on 6 mm discs of Whatman No. 1 filter paper. Loaded discs were placed on the surface of medium and plates were incubated for 24 h at 37°C. Pure DMSO (15–20 mL) loaded disc was used as negative control. At the end of incubation period, inhibition zones were measured in millimeters. These studies were performed in triplicate.

Similarly, antifungal activity of Ag NPs was screened on Sabouraud Dextrose Agar plates by using disc diffusion method and plates were incubated at 27-30°C for 36–48 h. After incubation period, zones of inhibition were measured.

### Results and Discussion

AgNO$_3$ (50, 75 and 100 mmol) solutions mixed with concentrated dextran solution were colorless in the beginning but turned light brown after 10 min indicating the nucleation of Ag NPs. The color changed to ruby red after 60 min while chocolate red color was observed after 24 h indicating the completion of growth process.

UV–vis absorption bands appeared ranging from ~405-450, 408–451 and 412–452 nm for nanoparticles synthesized using 50, 75 and 100 mmol dextran-AgNO$_3$ solutions, respectively and the corresponding UV–vis spectra of dextran-Ag NPs are shown in Figure 1a,b and c. The all reactions were monitored for 24 h at different time intervals. The red shift was observed in UV–vis absorptions for dextran-Ag NPs by increasing reaction time. Increase in absorption coefficient was also observed by increasing the concentration of AgNO$_3$ solution from 50–100 mmol. The increase in wavelength of absorption may be attributed to increase in size of Ag NPs. It is noteworthy that no absorption band was observed in the spectrum when sample was stored in dark. The reaction progressed on exposing the sample to diffused sunlight. Graphical representation of increase in absorption of Ag NPs solutions with increase in reaction time and AgNO$_3$ concentration (50, 75 and 100 mmol) is depicted in Figure 1d.

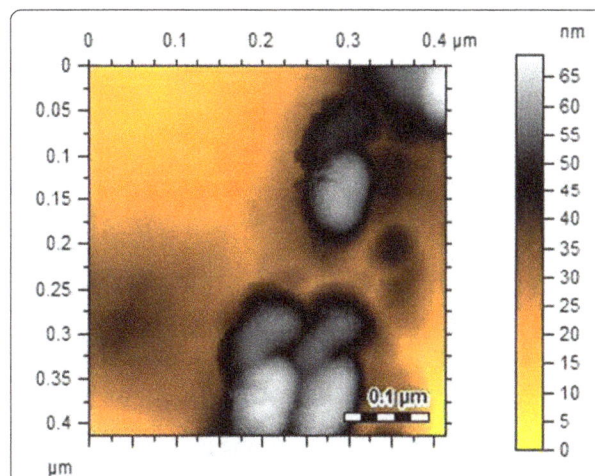

**Figure 4** AFM images of Ag NPs (50–70 nm) embedded in dextran thin films prepared from 100 mmol AgNO$_3$ solution.

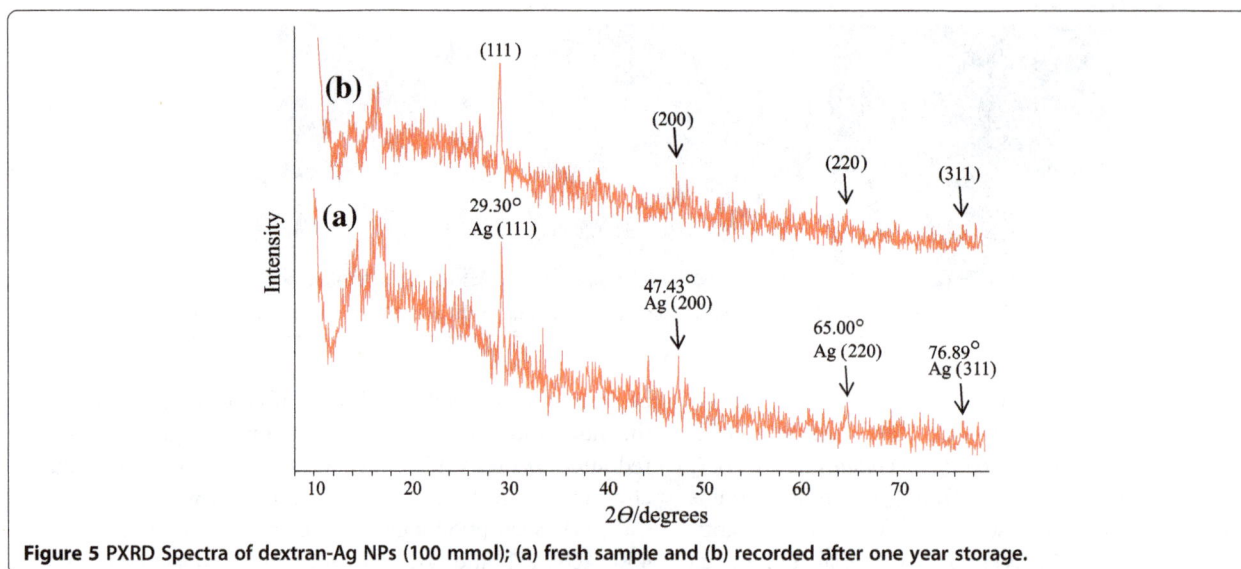

**Figure 5** PXRD Spectra of dextran-Ag NPs (100 mmol); (a) fresh sample and (b) recorded after one year storage.

FT-IR (using pellet mixed with KBr) spectra of dextran and dextran-Ag NPs (50 mmol solution) were recorded to confirm interaction between dextran and Ag$^+$ ions. Peaks at 434 and 548 cm$^{-1}$ in pure dextran were shifted to 457 and 588 cm$^{-1}$ in dextran-Ag NPs due to Ag--O excitation [15]. It is obvious from FT-IR spectra that there exist significant Van der Waal interactions between the chain of dextran and Ag NPs as all of the signals of dextran were shifted to somewhat higher positions (Figure 2).

Microtomes of dextran thin films loaded with Ag NPs observed by SEM showed spherical Ag NPs with uniform distribution (Figure 3). Dextran-Ag NPs film was dissolved in Milli-Q water and studied by AFM as well. The AFM images also witnessed the results of SEM that the NPs were found spherical (50–70 nm, Figure 4).

The crystalline nature of the as synthesized Ag NPs using dextran was confirmed via XRD analysis. As shown in Figure 5, there are four distinct reflections in the diffractogram at 29.3° (111), 47.43° (200), 65.05° (220), and 76.89° (311). These characteristics reflections show crystallographic planes of face centered cubic

structure of the Ag NPs. Same sample was stored in the form of thin film for one year and PXRD was re-recorded to confirm the structural stability of the Ag NPs. Similarity of PXRD pattern in sample before and after one year (Figure 4) indicated that Ag NPs are quite stable on storage in thin film of dextran. In this way we could successfully avoid agglomeration of Ag NPs on storage in solid state. So, this novel method for long term storage of Ag NPs can be further exploited for potential biomedical applications and optoelectronic devices.

There was no difference in SEM images of the sample (100 mmol) before and after one year storage (Figure 6) in thin films. The synthesized thin films were foldable, ruby red in color and almost optically transparent as demonstrated by digital photograph (see Figure 6). Likewise, after one year storage of thin films under dark were re-dissolved in water. The UV–vis spectroscopic analysis of aqueous solution (100 mmol) of thin films (see Figure 6) after one year showed an absorption band centered at 446 nm. The concordant absorbance therefore indicated no change in size and morphology of the stored Ag NPs

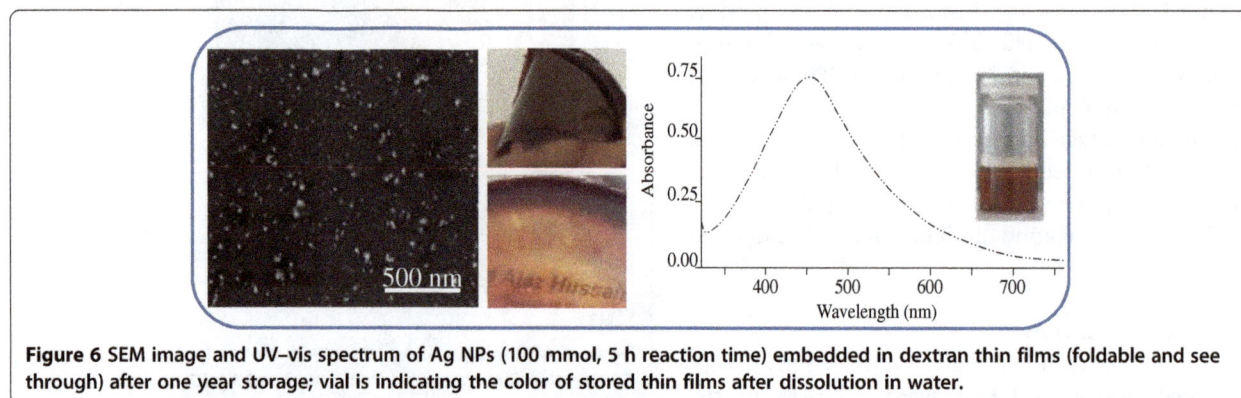

**Figure 6** SEM image and UV–vis spectrum of Ag NPs (100 mmol, 5 h reaction time) embedded in dextran thin films (foldable and see through) after one year storage; vial is indicating the color of stored thin films after dissolution in water.

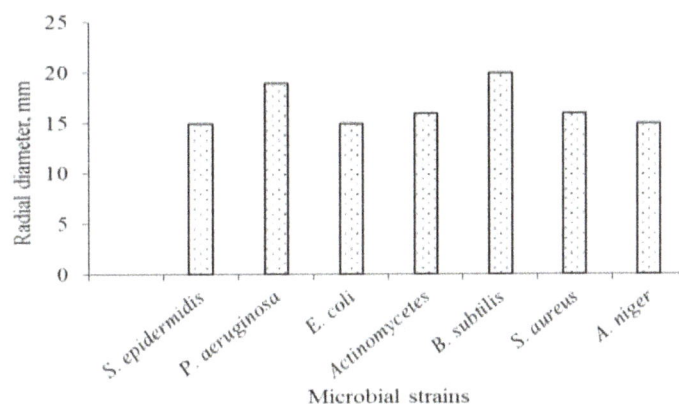

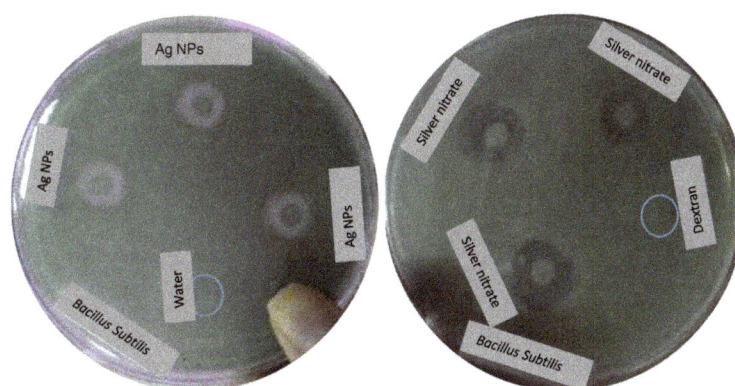

**Figure 7** Graph indicating inhibitory zone (radial diameter) of Ag NPs (50 *mmol*) *vs.* different microbial strains whereas plates indicating that deionized water and dextran did not show any activity however, AgNO₃ solution (0.01 M) was found active against *Bacillus subtilis* strains.

in thin films. So, the present method appeared highly efficient for the long term storage of Ag NPs in dextran thin films without agglomeration.

Solution of Ag NPs showed significant antimicrobial activity against different bacterial (*S. aureus, E. coli, B. subtilis, S. epidermidis, P. aeruginosa)* and fungal strains *Actinomycetes and A. niger* as depicted in Figure 7. The inhibiting zones vs. microbial strains for Ag NPs solution of 50 mmol concentration and antibacterial activity of silver nanoparticles (Ag NPs) against *Bacillus subtilis* are also shown as a typical example (see Figure 7). It was observed that deionized water and dextran do not show any activity however, AgNO₃ solution (0.01 M) was found active against mentioned strains. All of the experiments were carried out in triplicate and mean values have been reported. The prepared pristine Ag NPs can be used as effective therapeutic tools.

## Conclusions

We report on the diffused sun light assisted green synthesis of dextran stabilized Ag NPs without use of any hazardous and costly reducing agent or any extra functionalizing ligand. The as synthesized nanoparticles can be stored in solid state over months without imparting any change in the physical or optical properties. Being significantly active against microbes, the Ag NPs can be exploited for antimicrobial medical devices.

**Competing interests**
The authors declare that they have no competing interests.

**Authors' contributions**
MAH contributed in conception, design and acquisition of data. MAH and IJ have given final approval of the version to be published. AS synthesized the Ag nanoparticles and performed UV spectroscopic analysis. IJ and MNT contributed in spectral analysis and interpretation of experimental data. MRS performed and interpreted the AFM analysis. RA performed SEM and critical revision of scientific contents. SNAB performed antimicrobial analysis and involved in revising the manuscript critically for important intellectual content. All authors read and approved the final manuscript.

**Author details**
[1]Department of Chemistry, University of Sargodha, Sargodha 40100, Pakistan. [2]Drug and Herbal Research Centre, Faculty of Pharmacy, Universiti Kebangsaan Malaysia, Jalan Raja Muda Abdul Aziz, Kuala Lumpur 50300, Malaysia. [3]Institute of Inorganic and Analytical Chemistry, Johannes Guttenberg University of Mainz, Duesbergweg 10-14, Mainz 55128, Germany. [4]International Center for Chemical and Biological Sciences, University of Karachi, Karachi 75270, Pakistan. [5]Centre for Advanced Studies in Physics (CASP), GC University, Lahore 54000, Pakistan.

**References**
1.  Raveendran P, Fu J, Wallen SL: Completely "green" synthesis and stabilization of metal nanoparticles. *J Am Chem Soc* 2003, **125**:13940–13941.
2.  Carsin H, Wassermann D, Pannier M, Dumas R, Bohbot S: A silver sulphadiazine-impregnated lipidocolloid wound dressing to treat second-degree burns. *J Wound Care* 2004, **13**:145–148.
3.  Hayward RC, Saville DA, Aksay IA: Electrophoretic assembly of colloidal crystals with optically tunable micropatterns. *Nature* 2000, **404**:56–59.
4.  Haberzettl CA: Nanomedicine: destination or journey. *Nanotechnol* 2002, **13**:9–13.
5.  Ong C, Lim JZZ, Ng C-T, Li JJ, Yung L-YL, Bay B-H: Silver nanoparticles in cancer: therapeutic efficacy and toxicity. *Curr Med Chem* 2013, **20**:772–781.
6.  El-Nour KMMA, Eftaiha A, Al-Warthan AA, Ammar RAA: Synthesis and applications of silver nanoparticles. *Arab J Chem* 2010, **3**:135–140.
7.  Tahir MN, Eberhardt M, Zink N, Therese HA, Kolb U, Theato P, Tremel W: From single molecules to nanoscopically structured functional materials: au nanocrystal growth on TiO2 nanowires controlled by surface-bound silicatein. *Angew Chem Int Ed* 2006, **45**:4803–4809.
8.  Tahir MN, Zink N, Eberhardt M, Therese HA, Kolb U, Faiss S, Janshoff A, Kolb U, Theato P, Tremel W: Hierarchical assembly of TiO2 nanoparticles on WS2 nanotubes achieved through multifunctional polymeric ligands. *Small* 2007, **3**:829–834.
9.  Tahir MN, Andre R, Sahoo JK, Jochum FD, Theato P, Natalio F, Berger R, Branscheid R, Kolb U, Tremel W: Hydrogen peroxide sensors for cellular imaging based on horse radish peroxidase reconstituted on polymer-functionalized TiO(2) nanorods. *Nanoscale* 2011, **3**:3907–3914.
10. Bar H, Bhui DK, Sahoo GP, Sarkar P, De SP, Misra A: Green synthesis of silver nanoparticles using latex of *Jatropha curcus*. *Colloid Surface A* 2009, **339**(1–3):134–139.
11. Lou C-W, Chen A-P, Lic T-T, Lin J-H: Antimicrobial activity of UV-induced chitosan capped silver nanoparticles. *Mater Lett* 2014, **128**:248–252.
12. Oluwafemi OS, Vuyelwa N, Scriba M, Songca SP: Green controlled synthesis of monodispersed, stable and smaller sized starch-capped silver nanoparticles. *Mater Lett* 2013, **106**:332–336.
13. Long Y, Ran X, Zhang L, Guo Q, Yang T, Gao J, Cheng H, Cheng T, Shi C, Su Y: A method for the preparation of silver nanoparticles using commercially available carboxymethyl chitosan and sunlight. *Mater Lett* 2013, **112**:101–104.
14. Bankura KP, Maity D, Mollick MMR, Mondal D, Bhowmick B, Bain MK, Chakraborty A, Sarkar J, Acharya K, Chattopadhyay D: Synthesis, characterization and antimicrobial activity of dextran stabilized silver nanoparticles in aqueous medium. *Carohydr Polym* 2012, **89**:1159–1165.
15. Shameli K, Ahmad MB, Jazayeri SD, Sedaghat S, Shabanzadeh P, Jahangirian H, Mahdavi M, Abdollahi Y: Synthesis and characterization of polyethylene glycol mediated silver nanoparticles by the green method. *Int J Mol Sci* 2012, **13**:6639–6650.

# Interfacial film stabilized W/O/W nano multiple emulsions loaded with green tea and lotus extracts: systematic characterization of physicochemical properties and shelf-storage stability

Tariq Mahmood[1,2]*, Naveed Akhtar[1] and Sivakumar Manickam[3]

## Abstract

**Background and aims:** Multiple emulsions have excellent encapsulating potential and this investigation has been aimed to encapsulate two different plant extracts as functional cosmetic agents in the W/O/W multiple emulsions and the resultant system's long term stability has been determined in the presence of a thickener, hydroxypropyl methylcellulose (HPMC).

**Methods:** Multiple W/O/W emulsions have been generated using cetyl dimethicone copolyol as lipophilic emulsifier and a blend of polyoxyethylene (20) cetyl ether and cetomacrogol 1000® as hydrophilic emulsifiers. The generated multiple emulsions have been characterized with conductivity, pH, microscopic analysis, phase separation and rheology for a period of 30 days. Moreover, long term shelf-storage stability has been tested to understand the shelf-life by keeping the generated multiple emulsion formulations at $25 \pm 10°C$ and at $40 \pm 10\%$ relative humidity for a period of 12 months.

**Results:** It has been observed that the hydrophilic emulsifiers and HPMC have considerably improved the stability of multiple emulsions for the followed period of 12 months at different storage conditions. These multiple emulsions have shown improved entrapment efficiencies concluded on the release rate of conductometric tracer entrapped in the inner aqueous phase of the multiple emulsions.

**Conclusion:** Multiple emulsions have been found to be stable for a longer period of time with promising characteristics. Hence, stable multiple emulsions loaded with green tea and lotus extracts could be explored for their cosmetic benefits.

**Keywords:** Green tea, Lotus, Nano, Multiple, Emulsions, Stability

## Introduction

In these days the consumers worldwide are looking for personal care products, in specific the cosmetic products that assure multiple benefits. Besides, they also expect the latest advancements of technology to be incorporated into these innovative product formulations. Facing these trends, formulators strive to develop highly differentiated multifunctional product formulations that focus not only on the treatment but also on their aesthetics. A significant number of novel products are available in the markets that have been incorporated with a variety of new generation active ingredients. During the incorporation of these emerging actives, a range of formulation challenges are normally encountered that include control in the stability and the complications of combining several actives into a single cosmetic formulation [1]. Although, multiple emulsions were described in 1925;

* Correspondence: tariqmahmood750@gmail.com
[1]Department of Pharmacy, Faculty of Pharmacy and Alternative Medicine, The Islamia University of Bahawalpur, Bahawalpur 63100, Pakistan
[2]School of Pharmacy, The University of Faisalabad, Faisalabad 37610, Pakistan
Full list of author information is available at the end of the article

but much attention was not given until the report which was published in the late 1960s [2]. Recently, emulsions have established growing interest as vehicles to deliver the drugs efficiently to the body [3] and more importantly cavitation technique has been exploited to generate the nanoemulsions incorporated with the active constituents [4-7].

Complex water-in-oil-in-water (W/O/W) multiple emulsions consist of water-in-oil (W/O) dispersed into the continuous aqueous phase and stabilized with hydrophilic emulsifiers. Due to the unique structure and properties, these multiple emulsions are interesting carrier systems for various drug delivery approaches [8] especially for the controlled release of actives [9]. These multiple emulsions have been generated using cavitation technique [10-13]. In cosmetics, multiple emulsions play an important role to prepare skin care products with prolonged action [1].

However, compared with simple emulsions that consist of only two phases, much more destabilization processes need to be taken into consideration for the multiple emulsions. Four possible processes lead to the instability of W/O/W emulsions (a) coalescence of the internal aqueous droplets (b) coalescence of the oil droplets (c) rupture of the oil film resulting in the loss of the internal aqueous droplets, and (d) passage of water and water-soluble substances between the inner and outer aqueous phases. Thus, for the stabilization of multiple emulsions, following strategies have been proposed; (a) the use of high viscous oils to prevent the diffusion of water and water-soluble substances between the inner and outer aqueous phases, (b) the polymerization of interfacially adsorbed surfactant molecules, and (c) the gelation of oily or aqueous phases of the emulsions [14].

Green tea is among the most important plant extracts unfolded as cosmeceuticals [15] and it is now the subject of focus owing to its proven antioxidant properties and for its ability to repair photo-damage and phototoxicity caused by UV. It is also useful against a variety of skin disorders [16]. *Nelumbo nucifera* (lotus) extract has been observed to show potent antioxidant and anti-tyrosinase activities and has been found to have higher potential to be further developed as functional cosmetic agent [17,18].

The current study investigates the development of a stable cosmetic multiple emulsion loaded with green tea and lotus extracts which serve as functional cosmetic agents. Moreover, the influence of additives on the resultant aesthetic properties of these multiple emulsions have been investigated by subjecting the multiple emulsions for long term stability under varying conditions of temperature and humidity. Furthermore, it is the first time that a blend of hydrophilic emulsifiers, polyoxyethylene (20) cetyl ether and cetomacrogol 1000® has been employed in this study.

## Materials and methods
### Materials
To develop the multiple emulsions, paraffin oil was used as an oil phase (Merck, Germany). The lipophilic emulsifier Abil® EM 90 was supplied by Franken (Franken, Germany). Polyoxyethylene (20) cetyl ether (Brij 58® - Merck, Germany) and Cetomacrogol 1000® were used as hydrophilic emulsifiers. The thickener used was hydroxypropyl methylcellulose (HPMC, German Grade) and AnalaR grade $MgSO_4.7H_2O$ (BDH, Poole, England) was used as a conductometric tracer. Standardized green tea and lotus extracts were used as functional cosmetic agents to be encapsulated in the inner phase of primary emulsion.

### Preparation of extracts
#### Extraction of green tea leaves
50 g of grounded and dried green tea leaves were first subjected to a continuous hot extraction process at 80°C for 6 h using a Soxhlet apparatus and by employing 90% ethanol. A second cycle of extraction with the above-stated conditions was applied for 12 h. A coarse filtration through muslin cloth was followed by a fine filtration (using Whatman Grade No. 1 filter paper). Filtrates were then concentrated in a rotary evaporator at 40°C and stored in a refrigerator till further usage.

#### Extraction of lotus plant
100 g of sacred lotus (whole plant material) was macerated using 2000 ml of 70% methanol overnight. The macerated mixture was then heated at 60°C along with continuous stirring using overhead blade mixer (Eurostar, Germany) at 1000 rpm for 2 h. The resultant mixture was then cooled down and coarse-filtration was carried out using several layers of muslin cloth. Fine filtration was then carried out using Whatman Grade No. 1 filter paper. Filtrates were then concentrated in a rotary evaporator at 40°C and then stored in a refrigerator till further usage.

The yield (%) obtained from both the extractions was calculated by using the following Equation 1,

$$\text{Yield } (\%) = 100 \ (W_f/W_i) \tag{1}$$

Where $W_i$ is the initial weight of the plant material used for extraction and $W_f$ is the weight of dried extract after solvent evaporation.

### DPPH radical scavenging activity
To measure the DPPH radical scavenging activity of green tea and lotus extracts, the method as reported by Lee and Shibamoto [19] was used. For this, 0.1 mM solution of DPPH was prepared in ethanol. Then 5 μL of the extract was dissolved in DMSO and mixed with ethanolic DPPH solution (95 μL). The above mixture was then dispersed in a 96-well microplates reader (Spectra Max plus 384

Molecular Device, USA) and incubated at 37°C for 30 min and the absorbance was measured at 515 nm. From this, the radical scavenging activity was determined by using the following Equation 2:

$$\text{DPPH scavenging effect } (\%) = 1\text{-}(Ac\text{--}As/Ac)\ 100 \tag{2}$$

Where $A_c$ = absorbance of control and $A_s$ = absorbance of sample

## Preparation of multiple emulsions

Several pre-formulation studies were performed initially using different concentrations of oil and emulsifiers to arrive into a simple emulsion which was then further developed to fine multiple emulsions. Resultant simple emulsions with varying compositions of oil and emulsifier were stored at 40°C for a month and each sample was subjected to centrifugation and checked for stability at the end of one month. The most stable simple emulsion resistant to phase separation was then considered for the development of multiple emulsions. In the second stage of development of multiple emulsions, homogenization time and concentration of hydrophilic emulsifiers were varied to obtain a thicker creamy multiple emulsions. The concentration of HPMC was kept constant in all the experiments. The composition of multiple emulsion (MeC) formulations has been shown in Table 1.

All the multiple emulsions have been produced by a two-step emulsification strategy [20]. Briefly, simple emulsion was produced by emulsifying the oil with the lipophilic emulsifier and the mixture was preheated at 75°C before the emulsification. Green tea extract and conductometric tracer ($MgSO_4.7H_2O$) were then incorporated into the internal aqueous phase of the primary emulsion. Mixing of the aqueous phase with the

**Table 1 Composition of HPMC thickened green tea and lotus extracts loaded multiple emulsions (MeC) (% w/w)**

| Primary emulsion (W/O) | |
| --- | --- |
| Paraffin oil | 24 |
| Cetyl dimethicone copolyol | 4.25 |
| Green tea extract | 2.5 |
| Lotus extract | 2.5 |
| Magnesium sulfate | 0.7 |
| Deionized water (Q.S) | 100 |
| Multiple emulsion (W/O/W) | |
| Primary emulsion | 80 |
| Polyoxyethylene (20) cetyl ether | 3.75 |
| Cetomacrogol 1000® | 2.5 |
| Hydroxypropyl methylcellulose (HPMC) | 1.25 |
| Deionized water (Q.S) | 100 |

oil phase continued at 2000 rpm for 15 min and then at 1000 rpm for 10 min. Finally the obtained emulsion was cooled down to room temperature while maintaining a mixing speed of 500 rpm for a further 10 min. Mixing was accomplished by IKA Mixing Overhead Stirrer (Eurostar, Germany). For the second stage emulsification, the simple emulsion was added to the aqueous phase containing hydrophilic emulsifier at room temperature. A stirring speed of 700 rpm was maintained until the formation of multiple emulsion which was confirmed by microscopic analysis.

## Stability studies

In this study the focus is on the emulsion stability and not on the release behavior as the emulsion is designed to be spread on the skin and rubbing will aid in the release of active compounds. For the stability studies, multiple emulsions were weighed (50 g) and packed in glass containers with 100 g capacity and were kept in the incubation chambers at different storage temperature for a period of 30 days (accelerated storage period). Different storage conditions that were applied: room temperature (25±1°C), low temperature (8±1°C), high temperature (40±1°C) and high temperature with humidity (40±1°C with 75% relative humidity). At the pre-determined intervals (24 h, 48 h, 7 d, 15 d and 30 d, 8 months and 12 months), samples were removed from the storage and allowed to reach to room temperature (25°C) prior to evaluating their physico-chemical characteristics.

## Characterization of multiple emulsions (MeC)

After the preparation of multiple emulsions (MeC) they were characterized by different physico-chemical parameters such as microscopic analysis, conductivity, pH, phase separation and rheology for a follow-up period of 30 days and the influence of different storage conditions on these parameters were also determined. Initially the formulation samples (MeC) were stored at 8°C, 25°C, 40°C and at 40°C with 75% relative humidity. Also, one sample was kept in a plastic container and the stability was followed for 12 months at 25 ± 10°C with 40 ± 10% relative humidity.

## Microscopic analysis

The microscopic analysis of the generated multiple emulsions were examined using an optical microscope (Nikon E200, Nikon, Japan) with a camera (DCM-35 USB 2.0 and Minisee Image software). Observations were made at 100 X magnification after diluting the multiple emulsions. Measurements of 100 droplets per sample per storage condition were performed. The obtained images were analyzed using the software Digimizer (Version 4.1.1.0, MedCalc Software, Mariakerke, Belgium). After calculating the droplet diameter, the coefficient of variation (CV) was calculated by using the following Equation 3:

$$CV\ (\%) = [\text{standard deviation } (\mu m) \qquad (3)$$
$$/\text{mean droplet diameter } (\mu m)]\ 100$$

### Conductometric analysis

Conductometric analysis of the undiluted multiple emulsion was performed to examine the release of the electrolyte that has been initially entrapped in the internal aqueous phase. The specific conductivity of the emulsions was directly measured by using a digital conductivity meter (WTW- Tetracon®, Germany) at 25±2°C. Conductivity tests were performed for the multiple emulsion formulations immediately after their preparation and after 24 h, 7 d, 15 d, 30 d, 8 months and 12 months that have been kept at different storage conditions.

### pH determination

The pH of fresh multiple emulsion formulations and the formulations kept at different storage conditions was determined by using a digital pH meter (ProfiLine pH 197, WTW, Germany). The pH measurements were also taken for the formulations after 24 h, 7 d, 15 d, 30 d, 8 months and 12 months.

### Centrifugation

The generated multiple emulsions were centrifuged at 25°C (12) (Hettich EBA 20, Germany) and at 5000 rpm for 20 min. Centrifugation of each formulation was performed after 24 h, 7 d, 15 d, 30 d, 8 months and 12 months that have been kept at different storage conditions.

### Rheological examination

The rheological properties and viscosity measurements of multiple emulsions were determined using a Brookfield programmable rheometer (Model DV.III; Brookfield engineering laboratories Inc. USA). Rheocalc V 3.2 (Microsoft Corporation) software was used as a supporting program during the measurements. 0.5 g of each of the formulation was weighed and the viscosities were determined at 25°C with spindle speeds ranging from 100 to 200 rpm by using a spindle CP 41.

### Thermal stress test with repeated centrifugation

After one month of testing under accelerated conditions, one sample was packed in a plastic container and monitored for 12 months at varying conditions of 25±10°C with 40±10% relative humidity. The above-mentioned conditions are very close to the change in the environmental conditions of Pakistan. The samples after 8 and 12 months were then subjected to three elevated temperatures i.e. 50°C, 60°C and 80°C for 30 min in a thermostated water bath (Model H-4, China). In the first sequence, three 1 ml samples taken in plastic tubes (Eppendorf) were heated at three different temperatures i.e. 50°C, 60°C and 80°C for 30 min, and at the end they were centrifuged at 5000 rpm for 20 min and observed for the presence of any phase separation. In the second sequence, one sample filled in the tube was subjected to 50°C for 30 min and centrifuged again to observe the presence of any phase separation. The sample that was subjected to 50°C confirmed the absence of separation which was then subjected to the next elevated temperature i.e. 60°C and centrifuged to detect any phase separation. It showed no phase separation at 60°C, which was then finally subjected to 80°C to detect any phase separation. After each stage of heating and centrifugation, samples were observed under the microscope to confirm the integrity of the globules of multiple emulsions. Before subjecting to thermal stress testing, pH and conductivity measurements were performed for the samples of 8 and 12 months to investigate the effect of varying environmental conditions on the above parameters.

## Results and discussion
### Antioxidant activity

The extraction yield was first determined on the basis of dry plant material used for extraction and then the antioxidant potential of both the plant extracts was determined. From the results, a stronger antioxidant activity has been found out from the green tea extract while lotus extract has shown a less potent antioxidant activity as compared to green tea extract. Previous studies confirmed that the antioxidant activity of crude extract of green tea is higher than that of the standard antioxidant, ascorbic acid [21]. Also, earlier reports established that the methanolic extract of lotus has lower activity as compared to BHA and ascorbic acid [22]. Similar to above studies, the green tea in our study has shown a stronger antioxidant activity than lotus, irrespective of the extraction yield of green tea that has been found to be much lower than lotus extract. Table 2 shows the antioxidant activities and the extraction yield obtained from these two extracts.

### Microscopic analysis and centrifugation

Microscopic analysis is a useful and informative tool to understand the characteristics of multiple emulsions. Further, droplet size measurements indicate the stability of multiple emulsions as a faster increase in the droplet size with time is the indicator of lower stability of

**Table 2 Antioxidant activities and the extraction yield of green tea and lotus extracts**

| Extract | DPPH radical scavenging activity (%) | Yield (%) |
| --- | --- | --- |
| Green tea | 88 ± 0.2 | 2.1 |
| Lotus | 75 ± 0.41 | 4.15 |

the system [23]. The characteristics of multiple emulsions were investigated at the formulation stage as well as for the followed 30 days at different conditions of storage i.e. 8°C, 25°C, 40°C and at 40°C with 75% relative humidity. Furthermore, multiple emulsions were subjected to extensive centrifugation at 5000 rpm for 20 min to accelerate the phase separation of the system for the followed-up period of 12 months to understand the stability.

No phase separation was observed in any of the samples upon extensive centrifugation. Mean globule size was in the range of 8.41±3 µm to 10.81±7 µm (mean ± standard deviation). The sample that was kept at 8°C has shown smaller size of globules while an increase in the globule size was observed with an increase in the storage temperature. A maximum increase in the globule size was observed at 40°C. However this increase in the globule size with temperature does not seem to be a significant as the coefficient of variation (CV) was only varying in a narrow range at all the studied storage temperatures. A larger CV indicates an unacceptable limit of shrinkage or coalescence of the globules which indicate a lack of uniformity of globules. As reported earlier, an increase in the globule size is due to coalescence phenomena, while a decrease in the size is due to leakage of water from the internal to the external aqueous phase [24]. Results from the microscopic and centrifugation stability studies have been shown in Tables 3, 4 and 5. Photomicrographs of multiple emulsion formulations that have been kept at different storage conditions have been shown in Figure 1.

### Conductometric and pH analysis

Incorporating a conductometric tracer in the inner aqueous phase of the primary emulsion is necessary to detect any leakage from the internal to the external aqueous phase of W/O/W emulsion. Measurement of conductivity also provides the information about the entrapment of active substances in the primary emulsion. The amount of release is directly proportional to the amount of active substance that is available in the external aqueous phase and a rapid release therefore does not favor the slow release of the active substance. No significant variation in the conductivity was observed over 30 days of storage period and at any of the temperature that was subjected

**Table 3 Mean globule size and coefficient of variation (CV) of multiple emulsions followed for 30 days**

| Conditions | Multiple droplets (µm) | Inner droplets (nm) | CV (%) |
|---|---|---|---|
| 8°C | 8.05 ± 4.2 | 800 ± 20 | 0.53 |
| 25°C | 8.92 ± 4.2 | 1200 ± 60 | 0.47 |
| 40°C | 10.81 ± 7.0 | 1400 ± 50 | 0.65 |
| 40°C with 75% RH | 10.50 ± 5.5 | 1400 ± 80 | 0.53 |

**Table 4 Conductivity, pH and centrifugation stability of multiple emulsions kept at different storage conditions for 30 days**

|  | 8°C | 25°C | 40°C | 40°C with 75% RH |
|---|---|---|---|---|
| | | pH* | | |
| 24 h | 5.00 ± 0.09 | 5.19 ± 0.20 | 4.99 ± 0.18 | 4.99 ± 0.07 |
| 7 d | 4.97 ± 0.14 | 5.14 ± 0.07 | 4.91 ± 0.09 | 4.95 ± 0.10 |
| 15 d | 4.92 ± 0.15 | 5.12 ± 0.02 | 4.90 ± 0.10 | 4.93 ± 0.07 |
| 30 d | 5.35 ± 0.06 | 5.05 ± 0.04 | 5.12 ± 0.12 | 4.99 ± 0.12 |
| | | Conductivity** | | |
| 24 h | 76.40 ± 1.90 | 76.37 ± 1.19 | 66.50 ± 1.80 | 63.53 ± 1.51 |
| 7 d | 83.30 ± 1.37 | 85.90 ± 3.90 | 77.50 ± 1.71 | 69.83 ± 3.00 |
| 15 d | 84.27 ± 1.32 | 89.87 ± 1.90 | 79.20 ± 3.30 | 74.20 ± 2.08 |
| 30 d | 89.17 ± 2.80 | 57.73 ± 2.14 | 93.00 ± 1.20 | 60.30 ± 2.39 |
| | | Centrifugation stability*** | | |
| 24 h | S | S | S | S |
| 7 d | S | S | S | S |
| 15 d | S | S | S | S |
| 30 d | S | S | S | S |

*pH (Initial) = 5.04 ± 0.05, **Conductivity (Initial) = 77 ± 1.05 (µS/cm), ***Centrifugation at 5000 rpm for 20 min for each sample, S = Stable.

to. These results reveal that conductometric tracer has not diffused or rupturing of oil layer did not occur over the followed period of 30 days. In general it is believed that during the storage conductivity increases due to i) diffusion of an electrolyte, ii) the coalescence of globules iii) destruction of oil film because of the osmotic pressure and the leakage of internal aqueous phase [25].

**Table 5 Physicochemical characteristics of multiple emulsions shelf-stored at 8 and 12 months and subjected to thermal stress test with and without repeated centrifugation**

|  | 50°C | 60°C | 80°C |
|---|---|---|---|
| **After 8 months** | | | |
| *Centrifugation stability | | | |
|   Samples without repeated heating | S | S | S |
|   Sample with repeated heating | S | S | S |
| Microscopic globule integrity | | | |
|   Samples without repeated heating | S | S | S |
|   Sample with repeated heating | S | S | S |
| **After 12 months** | | | |
| *Centrifugation stability | | | |
|   Samples without repeated heating | S | S | S |
|   Sample with repeated heating | S | S | S |
| Microscopic globule integrity | | | |
|   Samples without repeated heating | S | S | S |
|   Sample with repeated heating | S | S | S |

*Centrifugation at 5000 rpm for 20 min for each sample, S = Stable.

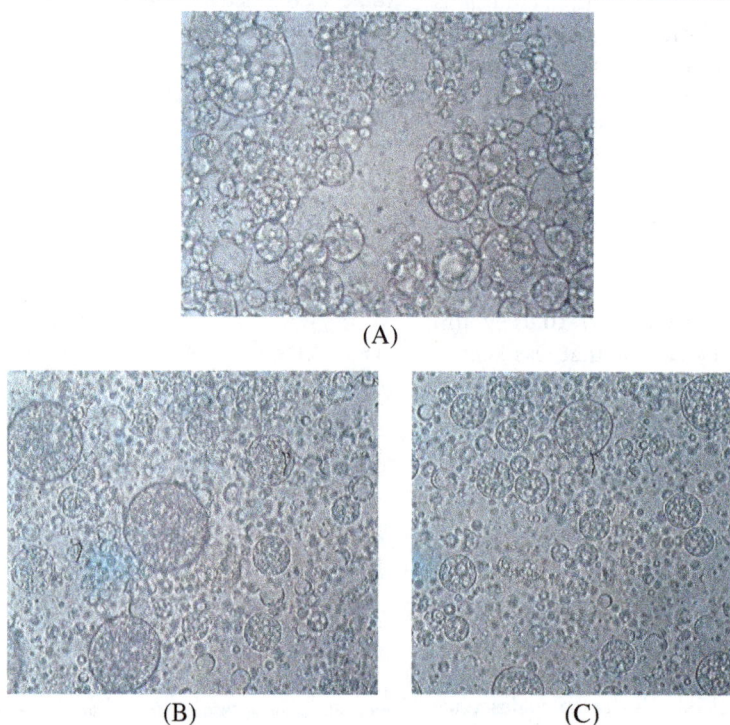

**Figure 1** Photomicrographs of multiple emulsion formulations kept at different storage conditions for 12 months. **A** = Fresh sample, **B** = After 8 months, **C** = After 12 months.

Conductivity values for the first 30 days decreased for the formulation that was kept at 40°C with 75% RH relative to the conductivity values of the fresh sample. This may be due to the formation of larger droplet size or phase separation. But no phase separation was observed at this condition. However, an increase in the droplet size was observed. A similar justification has been offered previously [26] where the reduction in the observed conductivities has been attributed to phase separation and larger droplet size. After 8 and 12 months of storage, conductivity values ($51.7 \pm 3.06$, $29.3 \pm 0.30$ respectively) tend to decrease but yet no phase separation occurred even after heating the samples at 80°C. Normal pH of skin is in between 5 and 6, and 5.5 is considered to be the average pH of skin [27]. Results of our pH analysis indicate that there is no variation in the pH of multiple emulsions kept at different conditions of storage and even after 8 and 12 months of storage ($5.4 \pm 0.01$, $5.6 \pm 0.01$ respectively) under varying temperature and humidity conditions. Results of this study (Table 4) indicate that different storage conditions did not have any influence on the pH and conductivity.

### Thermal stress test with repeated centrifugation

Thermal stress test with repeated centrifugation has been applied on multiple emulsions to study the stability of these emulsions under extensive stress conditions of temperature change and centrifugal force. The obtained results of this test have been shown in Table 5.

Obviously the generated multiple emulsions found to be extremely stable against elevated temperatures and the thermal stress could not produce any change in the globule integrity, phase inversion or phase separation. Even when the multiple emulsion was subjected to elevated temperatures and repeated centrifugation at 5000 rpm for 20 min, no phase separation was observed and globules have been found to be intact as observed through microscopic examinations. Hence, when the emulsion was tested after 12 months of storage at $25 \pm 10$°C with $40 \pm 10\%$ relative humidity it was found to be very stable and expected not to deteriorate soon. This resistance to phase separation appears due to the addition of Cetomacrogol 1000°, which acts as a film stabilizer and this has not been reported so far. The possible mechanism behind Cetomacrogol 1000° as film stabilizer is probably due to its self-bodying action in which rheological properties of the emulsions are related to gel networks formed in the continuous phases. This phenomenon has been presented previously when mixtures of emulsifiers of the surfactant-fatty alcohol type are used to stabilize oil in water emulsions [28,29].

### Rheological analysis

Consistency of cosmetic formulations is a very important aspect and is determined by the rheological methods.

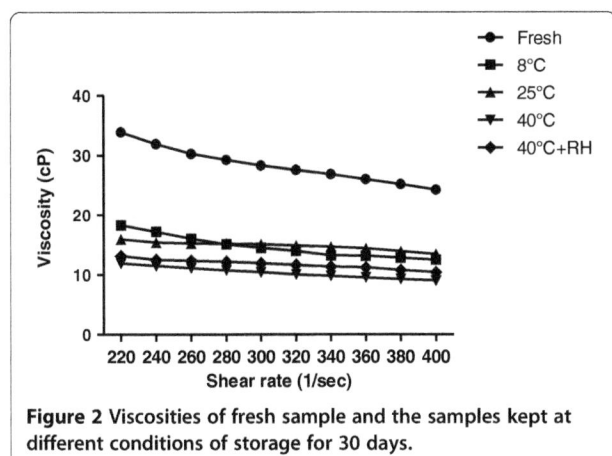

**Figure 2** Viscosities of fresh sample and the samples kept at different conditions of storage for 30 days.

Varying shear rates were applied to multiple emulsions to determine the flow parameters, while fitting the data in power law equation 4:

$$\tau = K\,\gamma^{n} \qquad (4)$$

Where $\tau$ is the shear stress, $\gamma$ is the shear rate, K is the consistency index and n is the flow behavior index.

Consistency index K is a measure of the system consistency and it is related to apparent viscosity. Flow behavior index n determines the degree of non-newtonian flow behavior and varies in the range between 0 and 1. The non–newtonian behavior of the investigated system is more pronounced for smaller values of "n". We measured the viscosities of multiple emulsions at a speed of 100 to 200 rpm while applying the shear rates from 200 to 400 on each of the samples. Varying shear rates were applied for the quality assurance of emulsions under stress conditions. Results for the rheology of fresh sample and for the samples that have been kept at different conditions of storage (followed for 30 days) have been shown in Figure 2.

Samples of multiple emulsions revealed non-newtonian flow and shear thinning behavior with different conditions of storage upon varying the shear rate. With an increase in shear rate and shear stress, a decrease in viscosity could be observed. Rheological analysis revealed excellent fits and were found to be from 98.6 to 99.6 and the obtained values of K and n have been shown in Table 6. It could be seen that all the samples exhibited pseudo-plastic

behavior with the flow behavior index (n) between 0.40 and 0.78. At higher temperatures, samples exhibited low consistency and thus a moderate shifting towards newtonian behavior.

## Conclusion

Fascination towards novel topical formulations loaded with functional actives having better antioxidant activity have emerged in the recent era. Encapsulation has been carried out with plant extracts (5%) in W/O/W multiple emulsion that have been fabricated using different emulsifiers. Promising stability characteristics have been observed in the generated multiple emulsions that were kept under different storage conditions. More importantly, multiple emulsions kept at different storage conditions were stable toward any phase separation followed for the 12 months i.e. accelerated and under shelf-storage. Globule size remained constant at different storage conditions. In addition, the coefficient of variation was lower enough to speculate that there was no internal deformation in the multiple emulsion characteristics. Conductivity analysis revealed that entrapment efficiency of multiple emulsions was excellent enough to offer sustained release of bio-functional agents. Rheological analysis revealed excellent curve fittings for all the formulations and the formulations subjected to elevated temperatures showed shifting toward newtonian behavior. Based on the above observations, it is expected that the green tea and lotus extracts loaded multiple emulsions could be excellent carrier of these powerful antioxidant substances ensuring long term stability of actives.

**Competing interests**
The authors declare that they have no competing interests.

**Authors' contributions**
All the authors have contribution in the design, planning and carrying out the experiments as well as drafting the manuscript. All the authors have read and approved the final manuscript.

**Acknowledgments**
The authors highly obliged for the financial support by Higher Education Commission of Pakistan.

**Author details**
[1]Department of Pharmacy, Faculty of Pharmacy and Alternative Medicine, The Islamia University of Bahawalpur, Bahawalpur 63100, Pakistan. [2]School of Pharmacy, The University of Faisalabad, Faisalabad 37610, Pakistan. [3]Manufacturing and Industrial Processes Research Division, Faculty of Engineering, University of Nottingham Malaysia Campus, Jalan Broga, Semenyih 43500, Malaysia.

**Table 6 Rheological analysis followed for 30 days**

|  | Initial | 8°C | 25°C | 40°C | 40°C RH* |
|---|---|---|---|---|---|
| Consistency index (cP) | 454.4 | 435.3 | 119.3 | 52.5 | 75.1 |
| Flow index | 0.51 | 0.40 | 0.57 | 0.78 | 0.67 |
| Confidence of fit (%) | 99.3 | 98.6 | 99.6 | 98.6 | 99 |

RH* = Relative humidity.

**References**
1. Patravale VB, Mandawgade SD: **Novel cosmetic delivery systems: an application update.** *Int J Cosmet Sci* 2008, **30**:19–33.
2. Hino T, Kawashima Y, Shimabayashi S: **Basic study for stabilization of w/o/w emulsion and its application to transcatheter arterial embolization therapy.** *Adv Drug Deliv Rev* 2000, **45**:27–45.

3. Rasul A, Akhtar N: **Formulation and in-vivo evaluation for anti-aging effects of an emulsion containing basil extract using non- invasive biophysical techniques.** *DARU* 2011, **19**(5):344–350.

4. Tang SY, Manickam S: **Design and evaluation of aspirin-loaded water-in-oil–water nano multiple emulsions prepared using two-step ultrasonic cavitational emulsification technique.** *Asia-Pac J Chem Eng* 2012, **7**(S1):S145–S156.

5. Tang SY, Manickam S, Ng AM, Shridharan P: **Anti-inflammatory and analgesic activity of novel oral aspirin-loaded nanoemulsion and nano multiple emulsion formulations generated using ultrasound cavitation.** *Int J Pharm* 2012, **430**:299–306.

6. Tang SY, Manickam S: **A novel and facile liquid whistle hydrodynamic cavitation reactor to produce submicron multiple emulsions.** *AIChEJ* 2013, **59**(1):155–167.

7. Tang SY, Manickam S, Billa N: **Impact of osmotic pressure and gelling in the generation of highly stable single core water-in-oil-in-water (W/O/W) nano multiple emulsions of aspirin assisted by two-stage ultrasonic cavitational emulsification.** *Colloids Surf B Biointerfaces* 2013, **102**:653–658.

8. Schmidts T, Dobler D, Nissing C, Garn H, Runkel F: **Development of multiple w/o/w emulsions as dermal carrier system for oligonucleotides: Effect of additives on emulsion stability.** *Int J Pharm* 2010, **398**:107–113.

9. Mishra B, Sahoo BL, Mishra M, Shukla D, Kumar V: **Design of a controlled release liquid formulation of lamotrigine.** *DARU* 2011, **19**(2):126–137.

10. Tang SY, Sivakumar M, Wei TK, Billa N: **Formulation development and optimization of a novel cremophore EL-based nanoemulsion using ultrasound cavitation.** *Ultrason Sonochem* 2012, **19**(2):330–45.

11. Wei TK, Sivakumar M: **Response surface methodology, an effective strategy in the optimisation of the generation of curcumin-loaded micelles.** *Asia-Pac J Chem Eng* 2012, **7**(S1):S125–S133.

12. Parthasarathy S, Tang SY, Sivakumar M: **Generation and optimization of palm oil based oil-in-water (O/W) nanoemulsions and nanoencapsulation of curcumin using liquid whistle hydrodynamic cavitation reactor (LWHCR).** *Ind Eng Chem Res* 2013, **52**(34):11829–11837.

13. Tang SY, Parthasarathy S, Sivakumar M: **Impact of process parameters in the generation of novel aspirin nanoemulsions – Comparative studies between ultrasound cavitation and microfluidiser.** *Ultrason Sonochem* 2013, **20**(1):485–497.

14. Schmidts T, Dobler D, Nissing C, Garn H, Runkel F: **Influence of hydrophilic surfactants on the properties of multiple W/O/W emulsions.** *J Colloid Interface Sci* 2009, **338**:184–192.

15. Preetha JP, Karthika K: **Cosmeceuticals – An evolution.** *Int J ChemTech Res* 2009, **1**(4):1217–1223.

16. Aburjai T, Natsheh FM: **Plants used in cosmetics.** *Phytother Res* 2003, **17**:987–1000.

17. Srichayanurak C, Phadungkit M: **Antityrosinase and antioxidant activities of selected Thai herbal extracts.** *KKU Res J* 2008, **13**(6):673–676.

18. Kim T, Kim HJ, Cho SK, Kang WY, Baek H, Jeon HY, Kim B, Kim D: **Nelumbo nucifera extract as skin whitening and anti-wrinkle cosmetic agent.** *Korean J of Chem Eng* 2011, **28**(1):424–427.

19. Lee K, Shibamoto T: **Antioxidant property of aroma extract isolated from clove bud [syzygiun aromaticum (L.) Merr. Et Perry].** *Food Chem* 2001, **74**:443–448.

20. Akhtar N, Ahmad M, Khan HMS, Akram J: **Gulfishan, Mahmood A, Uzair M. Formulation and characterization of a multiple emulsion containing 1% L-ascorbic acid.** *B Chem Soc Ethiopia* 2010, **24**:1–10.

21. Khalaf NA, Shakya AK, Al-Othman A, El-Agbar Z, Farah H: **Antioxidant activity of some common plants.** *Turk J Bio* 2008, **32**:51–55.

22. Wang L, Yen J-H, Liang H-L, Wu M-J: **Antioxidant effect of methanol extracts from lotus plumule and blossom (Nelumbo nucifera Gertn.).** *J Food Drug Anal* 2003, **11**:60–66.

23. Bernardi DS, Pereira TA, Maciel NR, Bortoloto J, Viera GS, Oliveira GC, Rocha-Filho PA: **Formation and stability of oil-in-water nanoemulsions containing rice bran oil: in-vitro and in-vivo assessments.** *J Nanobiotechnol* 2011, **9**(44):1–9.

24. Tirnaksiz F, Kalsin O: **A topical w/o/w multiple emulsions prepared with Tetronic 908 as a hydrophilic surfactant: Formulation, characterization and release study.** *J Pharm Pharm Sci* 2005, **8**:299–315.

25. Pays K, Giermanska-Kahn J, Pouligny B, Bibette J, Leal-Calderon F: **Double emulsions: How does release occur?** *J Control Release* 2002, **79**:193–200.

26. Kantarci G, Ozgüney I, Karasulu HY, Arzik S, Güneri T: **Comparison of different water/oil microemulsions containing diclofenac sodium: preparation, characterization, release rate, and skin irritation studies.** *AAPS PharmSciTech* 2007, **2**(8):E91.

27. Akhtar N, Ahmad M, Gulfishan, Masood MI: **Formulation and *in-vitro* evaluation of a cosmetic emulsion from almond oil.** *Pak J Pharm Sci* 2008, **21**:430–437.

28. Barry BW, Saunders GM: **Rheology of systems containing cetomacrogol 1000—cetostearyl alcohol. I. Self-bodying action.** *J Colloid Interface Sci* 1972, **38**(3):616–625.

29. Barry BW: **The control of oil-in-water emulsion consistency using mixed emulsifiers.** *J Pharma Pharmacol* 1969, **21**:533–540.

# Exploring cancer metastasis prevention strategy: interrupting adhesion of cancer cells to vascular endothelia of potential metastatic tissues by antibody-coated nanomaterial

Jingjing Xie[1], Haiyan Dong[1], Hongning Chen[1], Rongli Zhao[1], Patrick J Sinko[2], Weiyu Shen[1], Jichuang Wang[1], Yusheng Lu[1], Xiang Yang[1], Fangwei Xie[3] and Lee Jia[1*]

**Abstract**

**Background:** Cancer metastasis caused by circulating tumor cells (CTCs) accounts for 90% cancer-related death worldwide. Blocking the circulation of CTCs in bloodstream and their hetero-adhesion to vascular endothelia of the distant metastatic organs may prevent cancer metastasis. Nanomaterial-based intervention with adhesion between CTCs and endothelia has not been reported. Driven by the novel idea that multivalent conjugation of EpCAM and Slex antibodies to dendrimer surface may enhance the capacity and specificity of the nanomaterial conjugates for capturing and down-regulating colorectal CTCs, we conjugated the dendrimer nanomaterial with the EpCAM and Slex antibodies, and examined the capacity of the dual antibody-coated nanomaterial for their roles in interrupting CTCs-related cancer metastasis.

**Results:** The antibody-coated nanomaterial was synthesized and characterized. The conjugates specifically bound and captured colon cancer cells SW620. The conjugate inhibited the cells' viability and their adhesion to fibronectin (Fn)-coated substrate or human umbilical vein endothelial cells (HUVECs) in a concentration-dependent manner. In comparison with SW480 and LoVo cell lines, the activity and adhesion of SW620 to Fn-coated substrate and HUVECs were more specifically inhibited by the dual antibody conjugate because of the higher levels of EpCAM and Slex on SW620 cell surface. The hetero-adhesion between SW620 and Fn-coated substrate, or HUVECs was inhibited by about 60-70%. The dual conjugate showed the inhibition capacity more significant than its corresponding single antibody conjugates.

**Conclusions:** The present study provides the new evidence that coating nanomaterials with more than one antibody against CTCs may effectively interfere with the interaction between SW620 and HUVECs.

**Keywords:** Cancer metastasis prevention, Circulating tumor cells, Antibody conjugation, Multivalent binding, Cell adhesion

* Correspondence: cmapcjia1234@163.com
[1]Cancer Metastasis Alert and Prevention Center, and Biopharmaceutical Photocatalysis of State Key Laboratory of Photocatalysis on Energy and Environment, College of Chemistry, Fuzhou University, 523 Industry Road, Science Building, 3FL, Fuzhou, Fujian 350002, China
Full list of author information is available at the end of the article

## Background

Cancer is the second killer that leads people to death worldwide [1,2]. It was found that cancer metastasis was the principal cause of death among cancer patients [3-5]. The presence of circulating tumor cells (CTCs) [6-8], which are detached from primary tumor and enter the bloodstream [9], may contribute to initiate cancer metastasis. The progress of cancer metastasis usually depends on a series of consequential events, including the activation of dormant CTCs, the hetero-adhesion of CTCs to vascular endothelial bed of secondary organs, the continued survival and proliferation of CTCs after extravasation, and the formation of initial micrometastatic foci [10]. It seems that the effective prevention of cancer metastasis may be achieved by interrupting the circulation or activation of CTCs in blood and/or inhibiting the adhesion between CTCs and vascular endothelial cells.

CTCs as the hallmark of cancer metastasis have been paid more attention. To effectively interfere with the CTCs-related cancer metastasis, the residual CTCs should be preferentially captured and restrained with the enhanced specificity [11]. However, owing to the low number of CTCs in blood [12,13], capturing CTCs is a great technological challenge. Current chemotherapeutics and nanomaterials-based drug delivery system are designed to kill the malignant cancer cells, not CTCs *per se*. The serious adverse effects resulted in the damaged normal tissues and the decreased immunity [14,15]. Once CTCs in bloodstream were activated, cancer metastasis will be irrevocably initiated [16,17]. Some techniques were developed to capture CTCs, such as employing the epithelial cell adhesion molecule (EpCAM) antibody-coated three-dimensional nanostructured substrates [18,19], dendrimers [20], graphene oxide nanosheets [21] or immunomagnetic nanospheres [20,22,23]. However, these studies were only confined to functionalize nanomaterials with one targeting antibody against a single CTCs surface biomarker. Besides, the abundance of one biomarker varies dynamically with the cell cycle [24,25]. The level and activity of EpCAM expressed by CTCs not by hematologic cells was decreased with the epithelial-to-mesenchymal transition [26-28]. Therefore, the weak binding affinity of nanomaterials assembling one targeting antibody can't assure the extremely exact capture of CTCs. The unbound CTCs still made it possible to drive cancer metastasis.

Adhesion of CTCs to vascular endothelium was another crucial point of CTCs-derived cancer metastasis. Our previous studies demonstrated that chemopreventives such as S-nitrosocaptopril (CAP-NO) [10] and Metapristone (the metabolite of mifepristone) [29] with the low cytotoxicity had the intervention effects on the adhesion and invasion of colorectal CTCs to human umbilical vein endothelial cells (HUVECs). These related research laid the technical foundation for our study. Considering the fact that EpCAM

[30,31] and saliva acidifying louis oligosaccharides (Sialyl Lewis X, Slex) [32,33] are over-expressed on the surface of colorectal CTCs in circulation [32]. EpCAM antibody (anti-EpCAM) can directly interfere the adhesion process of CTCs [34] while Slex antibody (antiSlex) can indirectly block the adhesion between CTCs and endothelial cells through Slex/E-selection interaction [35,36]. Dendrimers-mediated multivalent binding effects were also exploited in previous studies [20]. Thus, we hypothesize that multivalent conjugation of both antiEpCAM and antiSlex to nanoscale polyamidoamine (PAMAM) dendrimers may significantly improve the anti-proliferation and anti-adhesion effects by enhancing the capture specificity, increasing the binding affinity, and avoiding the non-specific binding to similar cell subpopulations.

To test the hypothesis, we, herein, showed a novel strategy to realize the highly-specific binding, the restraint of colorectal CTCs, and the inhibition of adhesion of CTCs to vascular endothelial cells in vitro by using the bioconjugates that combine PAMAM dendrimers with dual targeting antibodies (antiEpCAM and antiSlex). Though attachment of both E-selectin and antiEpCAM to the functionalized glass substrates were previously reported [34], we, for the first time, showed conjugation of both antiEpCAM and antiSlex to dendrimers as one entity and its physicochemical characterization in this study. The dual roles of the bioconjugates in cancer metastasis prevention, including restraining the captured CTCs and inhibiting their adhesion, were also demonstrated here.

## Results

### Synthesis and physiochemical characterization of G6-5A-5S and PE-5A-G6-5S-FITC conjugates

Nanostructured PAMAM dendrimers with the functional group of 256 end amines were chosen as the good scaffolds to assemble dual antibodies, owing to their high payload and multivalent binding effect [20,37]. AntiEpCAM and/or antiSlex antibodies were sequentially conjugated onto the completely carboxylated G6 PAMAM (CC G6) dendrimer surface as previously reported [20]. Fluorescence-labeled dual antibody conjugate was similarly synthesized by using phycoerythrin (PE) linked antiEpCAM (antiEpCAM-PE) and fluorescein isothiocyanate (FITC) linked antiSlex (antiSlex-FITC, i.e., antiSlex and IgG/IgM-FITC antibodies were used together and abbreviated as antiSlex-FITC hereafter), instead (Figure 1a). The resultant antibody conjugates were used for the following cancer metastasis prevention assays including cell binding, cell activity regulation and cell adhesion (Figure 1a). For PE-5A-G6-5S-FITC conjugate in aqueous solution, fluorescence images taken at $\lambda_{ex}$ 488 and 543 nm demonstrated each antibody was coated on the modified dendrimer surface (Figure 1b). For G6-5A-5S conjugate, field emission

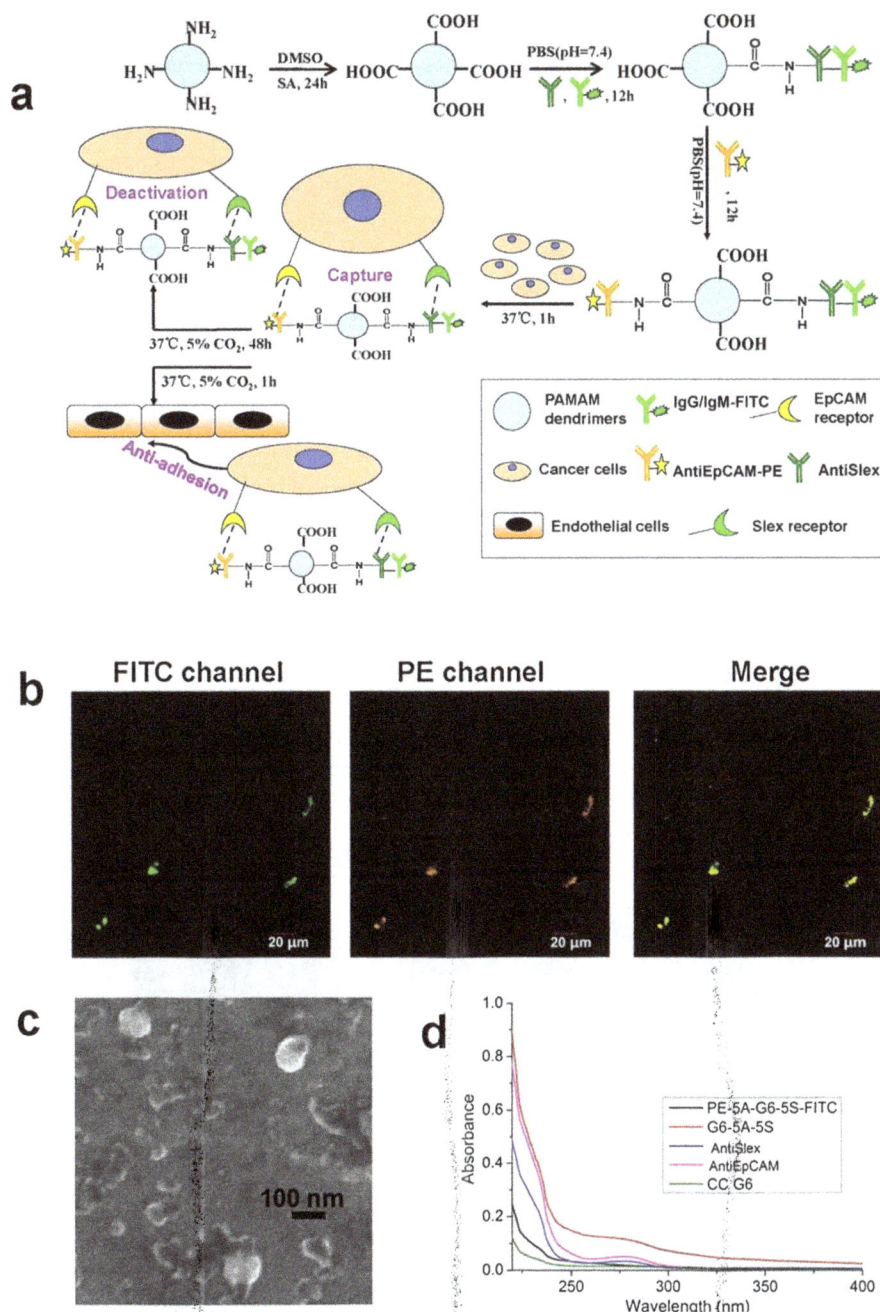

**Figure 1 Synthesis and physiochemical characterization of dendrimers assembling with antiEpCAM and antiSlex with or without fluorescence labeling. a**, Schematic illustration of the construction of dual antibody-conjugated dendrimers for exploring their biological functions in cancer metastasis through binding the target cancer cells. **b**, Images of PE-5A-G6-5S-FITC conjugate in aqueous solution under a laser confocal microscope. **c**, A FSEM image of the dry conjugate G6-5A-5S. **d**, The ultraviolet absorption spectra of CC G6 dendrimers, antibody and dual antibody conjugates in PBS (pH 7.4) solution.

scanning electron microscope (FSEM) measurement showed its characteristic morphology of round pie with the particle size of 100 nm (Figure 1c). UV spectra analysis at $\lambda_{220\ nm}$ indicated the successful coating of dendrimers with antibodies in comparison with the non-absorption of CC G6 dendrimers (Figure 1d).

## CTCs bound and captured by the conjugate

Considering colon cancer cell lines including SW480, SW620 and LoVo express different levels of biomarkers (e.g., EpCAM), which was confirmed by the flow cytometry (Additional file 1: Figure S1), the binding and capture capability of antiEpCAM and antiSlex sequentially-conjugated dendrimer conjugate (PE-5A-G6-5S-FITC) to the above EpCAM and Slex-expressing colon cancer cell lines was individually investigated by us.

### Specificity in recognizing and binding the adherent cells

Preliminary experiments indicated that G6-5A-5S conjugate could efficiently bind the target cells within 1 h (Additional file 1: Figure S2). To qualitatively evaluate the binding effects of G6-5A-5S conjugate at various concentrations on the three adherent colon cancer cell lines, laser confocal microscope analysis was performed. The cell nucleus was labeled with blue color to distinguish other cell components. Once cells were recognized and bound by PE-5A-G6-5S-FITC conjugate, the merged yellow-green color was displayed in cellular membrane. Fluorescence intensity was concentration-dependently increased with the conjugate increased from 10 to 20 μg mL$^{-1}$. Moreover, fluorescence intensity from the cytomembrane of SW620 cells was more stronger than that from LoVo and SW480 cell lines (Figure 2). It seemed that the conjugate was more internalized in SW620 cell than in SW480 and LoVo cell lines through the double specific antigen-antibody interactions.

### Efficiency in capturing the suspensory cells

The metastatic ability of non-adherent CTCs [38] may differ from that of adherent CTCs [39]. To further evaluate the capture capability of PE-5A-G6-5S-FITC conjugate (20 μg mL$^{-1}$) to the suspensory cell lines, fluorescence inverted microscope and flow cytometric analyses were both carried out. The cell lines have identical exposure time (1 h) and baseline of fluorescence intensity. Without the non-specific binding, the conjugate showed the specific receptor-mediated binding to both SW620 and LoVo cell lines, which was seen from the distinct yellow-green fluorescence on cytomembrane. The increased fluorescence intensity was displayed on SW620 cell than on LoVo cell. The number of captured SW620 cells seemed to be more than that of captured LoVo cells in any random visual field (Figure 3a). The capture efficiency of the conjugate was also quantitatively evaluated by the % FITC and PE-positive cells within Q2 quadrant analyzed by the flow cytometry. Relative to the isotype control, the capture efficiency for SW620 cells was 4-fold higher than that for LoVo cells based on the captured numbers (Figure 3b). The enhanced capture efficiency for SW620 cells might be up to the relatively higher expression levels of EpCAM and Slex (Additional file 1: Figure S1).

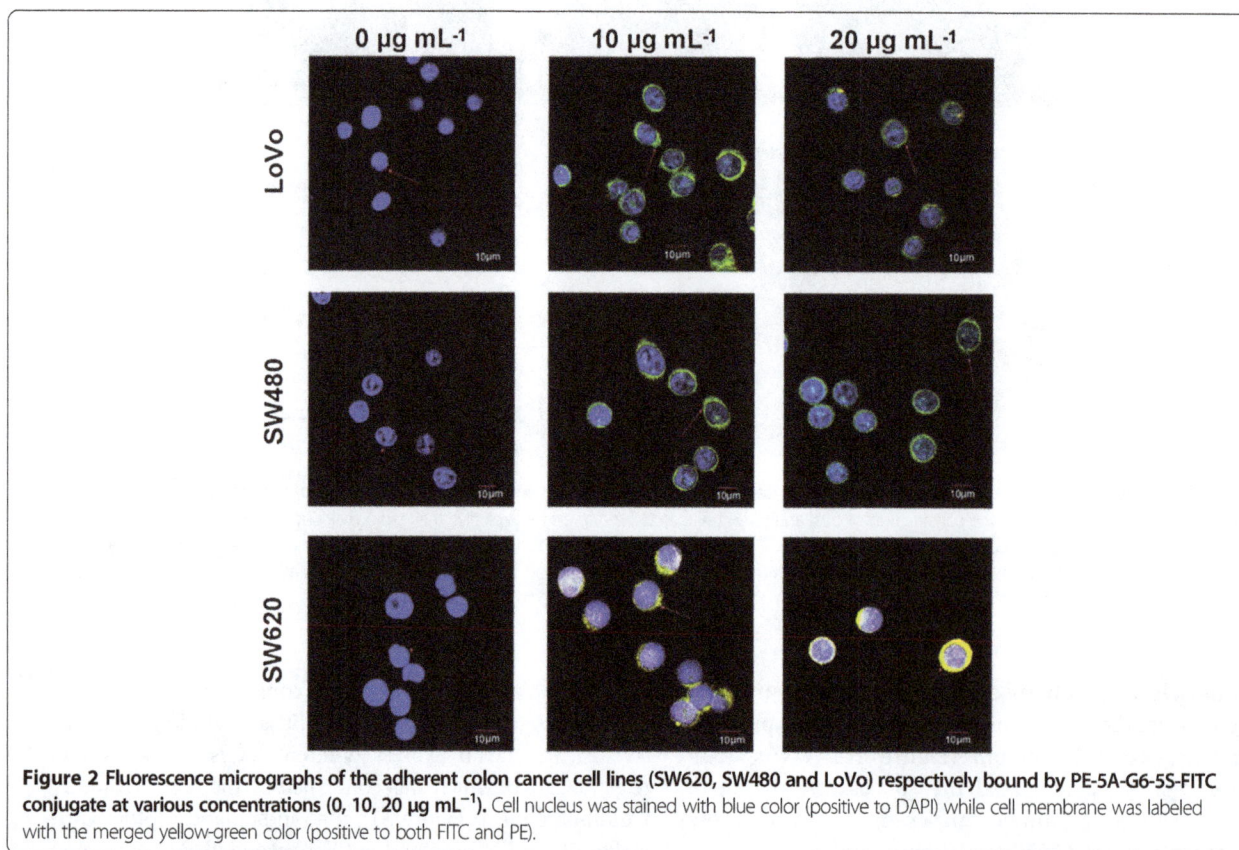

**Figure 2** Fluorescence micrographs of the adherent colon cancer cell lines (SW620, SW480 and LoVo) respectively bound by PE-5A-G6-5S-FITC conjugate at various concentrations (0, 10, 20 μg mL$^{-1}$). Cell nucleus was stained with blue color (positive to DAPI) while cell membrane was labeled with the merged yellow-green color (positive to both FITC and PE).

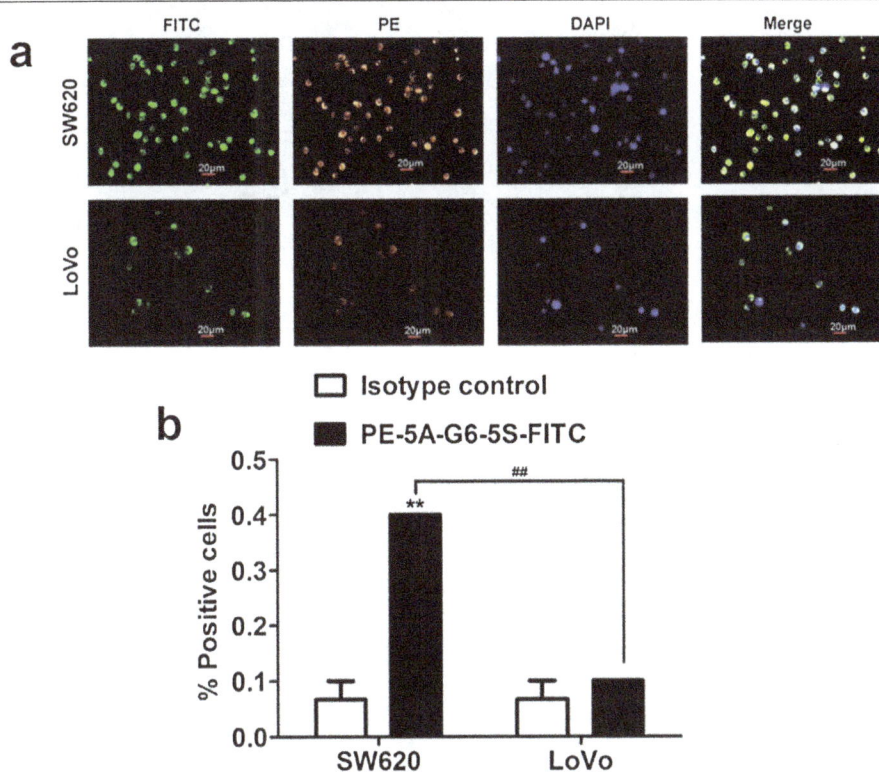

**Figure 3** Qualitative and quantitative analyses of the suspensory SW620 and LoVo cell lines respectively captured by PE-5A-G6-5S-FITC conjugate at the same concentration of 20 μg mL$^{-1}$. **a**, Representative fluorescence images of the captured cell lines at different excitation channels. **b**, Flow cytometric analysis of the capture efficiency of the conjugate in comparison with the isotype control.

## Down-regulation of the activity of captured cells by the conjugate

Once cancer cells were captured, the interactions between single or dual antibody conjugates and cell lines were investigated by us in parallel as follows.

### Inhibition of the proliferation of captured cells

The effects of single and dual antibody conjugates (G6-5A, G6-5S, G6-5A-5S) on the proliferation of captured cells were studied first of all. 48 h of binding treatment resulted in the decreased cell viability relative to control. With the concentration of conjugate increased from 1.25 to 20 μg mL$^{-1}$, the cell activity of each cell line was concentration-dependently restrained (Figure 4a-c). The cell viability caused by single and dual antibody conjugates was compared to that by CC G6 dendrimers. CC G6 dendrimers remained more than 80% of cell viability even at the concentration of 20 μg mL$^{-1}$. After single or dual antibody was conjugated onto dendrimer surface, the viability of each cancer cell line was obviously decreased. The single or dual antibody conjugate showed the stronger inhibitory effect on SW620 than on LoVo and SW480 cell lines. For example, the cell activity of SW620, LoVo and SW480 was reduced to 51.24%, 60.22% and 62.93%, respectively, by G6-5A conjugate (20 μg mL$^{-1}$) (Figure 4a-c). It

seemed that conjugates produced the inhibitory but not the lethal effects. The high levels of EpCAM and Slex might contribute to the selectivity of the conjugates for SW620 cells. G6-5A-5S and G6-5A conjugates had the stronger capability of restraining the activity of the same cells compared to G6-5S conjugate (Figure 4a-c). The decreased cell activity might be mainly attributed to the presence of EpCAM on cancer cells. The down-regulation of the captured colon cancer cell lines indicated that dual antibody-coated dendrimers may fit into a new class of therapeutic for preventing cancer metastasis by selectively restraining target CTCs rather than non-selectively killing normal and cancer cells.

### Flow cytometric analysis of cell cycle distribution

To further explore how the G6-5A-5S conjugate affected the cell activity, cell cycle distribution was analyzed by flow cytometry. Cell lines after individual incubation with the conjugate for 48 h were stained with propidium iodide (PI) staining to determine the cell population in every phases of G0/G1, S and G2/M. Flow cytometric images showed that the conjugate could cause a concentration-dependent increase in cell population of the G0/G1 phase and a decrease in cell population of the S phase without a significant increase in cell population of the G2/M phase

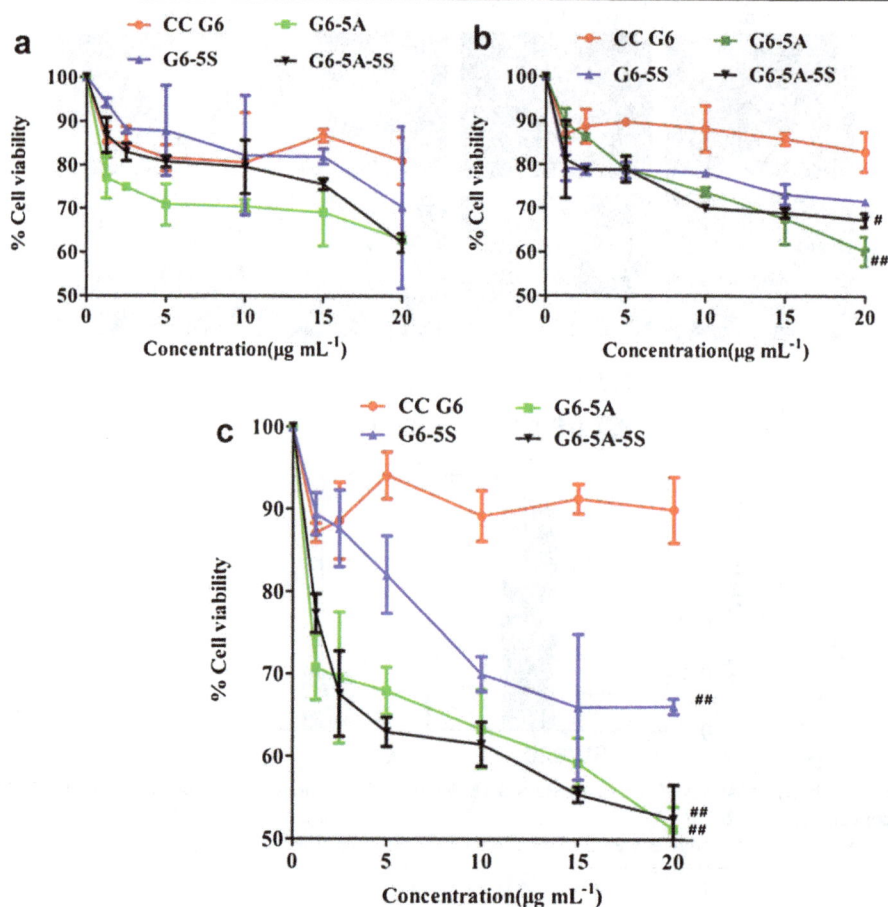

**Figure 4** Decrease in viability of colon cancer cell lines induced by single and dual antibody conjugates at concentrations ranging from 1.25 to 20 µg mL$^{-1}$. **a**, SW480 cells; **b**, LoVo cells; **c**, SW620 cells.

for SW620 cells (Figure 5a). Similar cell cycle distribution was found for SW480 cells. The significance between SW480 and SW620 cell lines indicated the increased inhibitory effect of conjugate on SW620 cells. However, for LoVo cells, the cell population in G2/M phase was concentration-dependently increased and that in S phase was decreased without a significant change in G0/G1 phase (Figure 5b), suggesting that dual antibody conjugate G6-5A-5S mainly arrested SW480 and SW620 cell lines at the G0/G1 stage and LoVo cells at the G2/M stage. The difference in cell cycle distribution might be attributed to the different interaction mechanism between dual antibody conjugate and each colon cancer cell line.

### Cellular mitochondrial membrane potential (MMP) evaluation

MMP ($\Delta\psi$m) depolarization is a prelude of cell apoptosis. The effects on cellular $\Delta\psi$m induced by various concentrations of G6-5A-5S conjugate (0, 10, 20 µg mL$^{-1}$) were

measured with DiOC6(3) (3,3'-Dihexyloxacarbocyanine iodide) staining. The increased fluorescence intensity predicts the decreased $\Delta\psi$m. Flow cytometric analysis showed that fluorescence intensity from the treated SW620 cells was stronger than that from the untreated ones. The cellular $\Delta\psi$m was decreased in a moderate concentration-dependent manner. However, the conjugate at the concentration up to 20 µg mL$^{-1}$ only produced 15% loss of $\Delta\psi$m. In contrast, the MMP of treated SW480 and LoVo cell lines wasn't significantly affected by the conjugate, indicating the mitochondrial function and electron transport chain activity are kept intact (Figure 5c). It seems like that the dual antibody-coated dendrimer conjugate could result in the change of cellular MMP.

### Inhibition of the adhesion of cancer cells by the conjugate

We performed the related adhesion assays as follows by using the conjugate at the safe concentrations ranging from 1.25, 2.5, 5 to 10 µg mL$^{-1}$ to determine whether the single and dual antibody conjugates could intervene

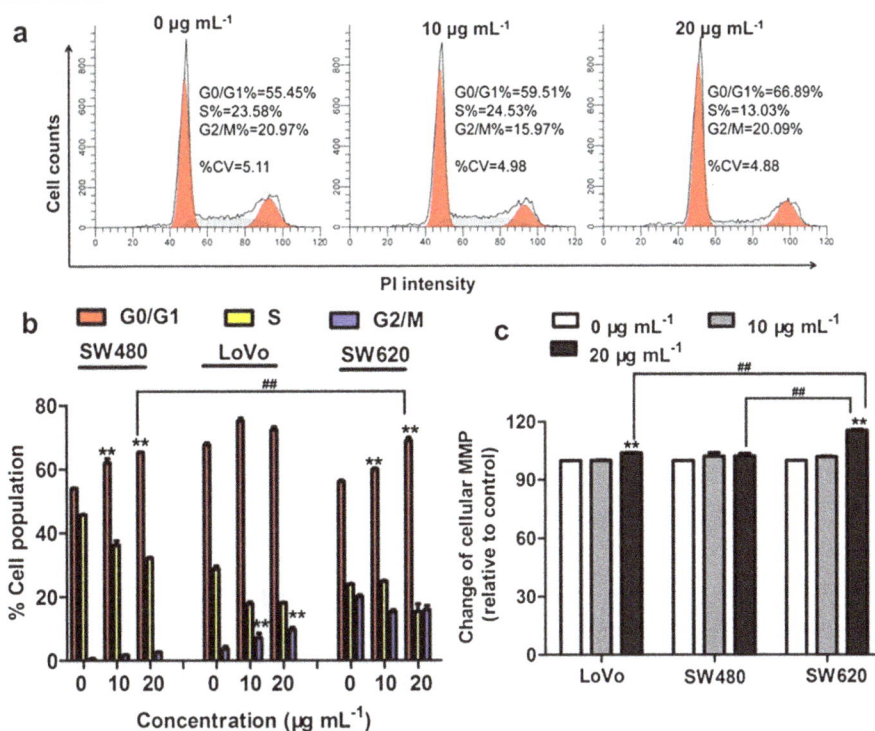

**Figure 5** Cell cycle distribution and cellular MMP of colon cancer cell lines after exposure to dual antibody conjugate G6-5A-5S at 10 and 20 μg mL$^{-1}$. **a**, DNA flow cytometric images of the treated SW620 cells with the conjugate. **b**, Percentage of cell population in every stage (G0/G1, S and G2/M). **c**, The influence of conjugate on the cellular MMP. The increased fluorescence intensity of DiOC6(3) usually indicated the decreased MMP.

the adhesion of cancer cells to endothelial cells for cancer metastasis prevention.

### Substrate adhesion analysis

EpCAM and Slex are two adhesion molecules expressed by CTCs not by hematologic cells [40-42]. E-selectin was mainly expressed on the activated endothelial cell surface of blood vessels. The interaction between Slex and E-selectin mediated the hetero-adhesion of CTCs to vascular endothelial cells and the continued survival and proliferation of CTCs [35,36]. Conjugation of antiEpCAM and/or antiSlex antibodies onto the dendrimer surface might effectively interfere the hetero-adhesion of cancer cells to basement membrane and endothelial cells. Similar MTT {[3-(4,5-dimethylthiazol-2-yl)-2,5-diphenyltetrasodium bromide] tetrazolium salt} assay was used to evaluate the inhibitory effects of G6-5A, G6-5S and G6-5A-5S conjugates on the adhesion of colon cancer cell lines to fibronectin (Fn)-coated artificial substrate membrane. CC G6 dendrimers used as the controls were also tested to demonstrate their low effects in adhesion process. In contrast, the antibody conjugates could lead to the reduced adhesion between cancer cells and Fn with concentrations increased from 1.25 to 10 μg mL$^{-1}$. The adhesive percentage of SW480, LoVo and SW620 cell lines was 74.20%, 34.30% and 32.93%, respectively, with the G6-5A-5S conjugate at 10 μg mL$^{-1}$ (Figure 6a-c). It was seen

that the conjugate was not effective in blocking the adhesion of SW480 cells but the adhesion of SW620 and LoVo cell lines to Fn. In comparison with single antibody conjugates G6-5A and G6-5S, dual antibody conjugate G6-5A-5S showed the stronger interference ability (Figure 6a-c). For LoVo cells, the mean anti-adhesion efficacy of G6-5S, G6-5A and G6-5A-5S were 44.90%, 61.09% and 65.70%, respectively (Figure 6b), suggesting that antiEpCAM and antiSlex antibodies might play the synergistic effects in inhibiting the adhesion of cancer cells to Fn-coated substrates, and the antiEpCAM was the key factor with this respect.

### HUVECs adhesion analysis

Hetero-adhesion of CTCs to local vascular endothelium initiates the irreversible cancer metastasis. Using the targeting antibodies-coated nanomaterials to interfere the adhesion process will be a new attempt. In this assay, fluorescence microscopic analysis was performed to assess the anti-adhesion effects of single and dual antibody conjugates (G6-5A, G6-5S and G6-5A-5S). The hetero-adhesion of three colon cancer cell lines to HUVECs was individually inhibited by the conjugates in a concentration-dependent manner (Figure 7a-c), for example, the number of SW480 cells with green fluorescence that adhered to HUVECs was concentration-dependently decreased by G6-5A-5S from 1.25 to 10 μg mL$^{-1}$ (Figure 7d). However, the

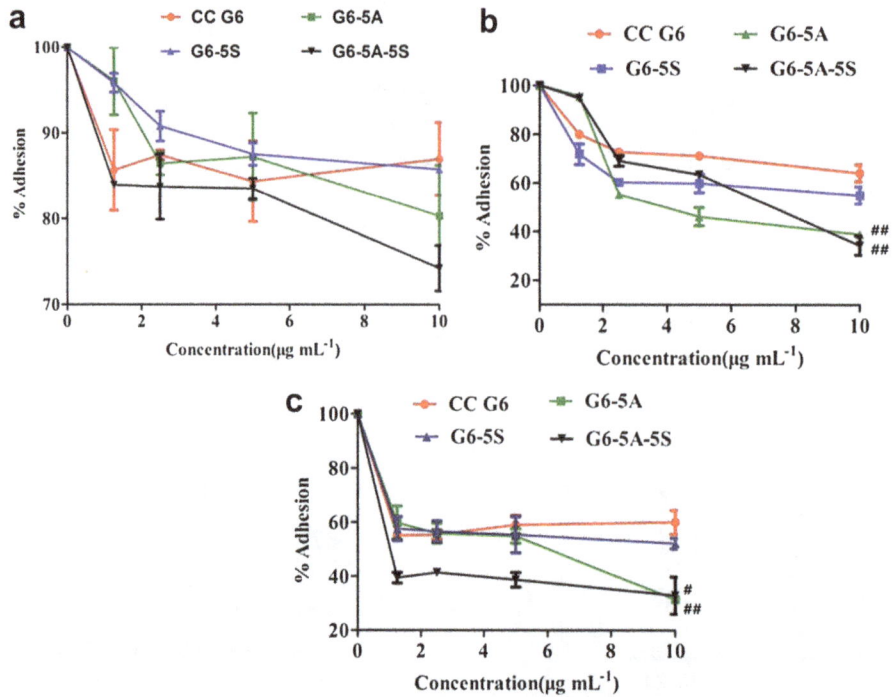

**Figure 6** The concentration-dependent inhibition of single and dual antibody conjugates (from 1.25 to 10 µg mL$^{-1}$) on the adhesion of colon cancer cell lines to Fn-coated substrate. **a**, SW480 cells; **b**, LoVo cells; **c**, SW620 cells.

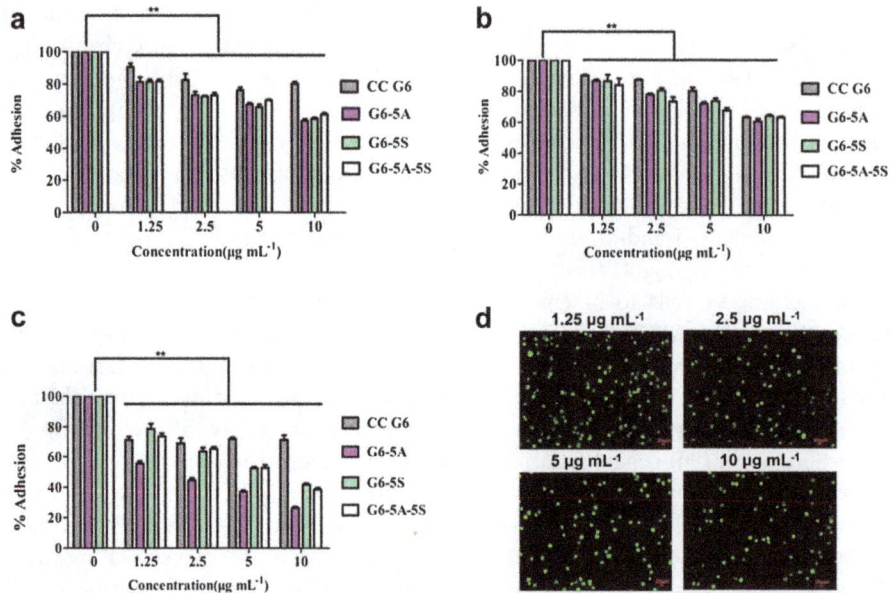

**Figure 7** Inhibition by single and dual antibody conjugates on the hetero-adhesion between colon cancer cell lines and HUVECs concentration-dependently (1.25 to 10 µg mL$^{-1}$). **a-c**, The conjugates showed the different capability in interfering with the adhesion of cancer cell lines to HUVECs. **a**, LoVo cells; **b**, SW480 cells; **c**, SW620 cells. **d**, Representative fluorescence images of Rhodamine 123-labeled SW480 cells that adhered to HUVECs when they were treated with different concentrations of G6-5A-5S conjugate.

anti-adhesion effects of the same conjugate varied with different colon cancer cell lines and the capability of single and dual antibody conjugates was also different. It seemed that conjugates displayed the stronger capability to interfere the adhesion between SW620 cells and HUVECs in comparison with the adhesion between other two cell lines and HUVECs. The maximum adhesion percentage of SW620, LoVo, and SW480 by the G6-5A-5S conjugate (10 μg mL$^{-1}$) was 38.60%, 61.11%, and 63.25%, respectively (Figure 7a-c). G6-5A and G6-5A-5S conjugates were superior to G6-5S conjugate in interfering the adhesion of colon cancer cells to HUVECs. The adhesion of SW620 cells to HUVECs was reduced by 73.68% (G6-5A), by 58.19% (G6-5S), and by 61.40% (G6-5A-5S), respectively, at the same concentration of 10 μg mL$^{-1}$ (Figure 7c). The different anti-adhesion capability of the conjugates to each cancer cell line might contribute to the selection of the appropriate conjugate as the specific target for preventing cancer metastasis.

## Discussion

AntiEpCAM and antiSlex collectively-coated dendrimer conjugates were synthesized, for the first time, by employing the surface coating technology [20,43]. The successful surface functionalization was demonstrated both by UV spectra and fluorescence images. FSEM measurement also showed the morphology and size of the antibody-conjugated dendrimers (Figure 1). The average size of G6 PAMAM dendrimer is about 7 nm. After conjugation with antibody, the size of the PAMAM dendrimer quickly grew to 100 nm. The increment in size may be attributed to the increased numbers of the dual antibodies conjugated. Moreover, two different antibodies with the different size, specie, character and charge may also result in the high variance in coating technique and particle size.

Compared to the reported single antibody-coated dendrimers [44,45], dual ones played the synergistic effects in biological functions [46]. After excluding the non-specific binding and cell autofluorescence with the isotype controls, whatever to the adherent or suspensory cells, PE-5A-G6-5S-FITC conjugate displayed the specific recognition and binding affinity (Figures 2 and 3), which might be attributed to the antiEpCAM/EpCAM and antiSlex/Slex double interactions. The conjugate could be internalized into cytomembrane and cytoplasm not organelles with the concentration increased or incubation time prolonged (Figure 2, Additional file 1: Figure S2), which might be caused by the cell endocytosis and the enhanced permeability and retention (EPR) effects of nanomaterials. The interactions between chemotherapeutics and cancer cells were studied by us [10,29,47], so we have a good understanding of the molecular characterization of cancer cells. Though dendrimers with carboxyl ending groups were reported to be less toxic and more compatible than dendrimers with amino terminal groups [48,49], whether

dendrimers coated with antiEpCAM and/or antiSlex affected the activity of captured cells was investigated. Our studies indicated that G6-5A and G6-5A-5S conjugates decreased the viability of colon cancer cell lines (especially SW620 cells) more significantly than G6-5S conjugate (Figure 4). We further explored the regulation mechanism by which G6-5A-5S conjugate blocked the cell cycle and reduced the cellular MMP in a modest concentration-dependent way (Figure 5). The result about cell activity regulation was in agreement with what we have previously reported [46,50]. Taken into account that the interaction between Slex and E-selectin mediated the adhesion of CTCs to endothelial cells, the intervention effects of the conjugates were further explored. Adhesion assays showed that both single and dual antibody conjugates effectively inhibited the hetero-adhesion between each colon cancer cell line and HUVECs/Fn-coated substrates (Figures 6 and 7). G6-5A and G6-5A-5S conjugates had the better anti-adhesion effects than G6-5S conjugate. In comparison with SW480 and LoVo cell lines, SW620 cells were more significantly affected by the conjugates, indicating that antiEpCAM/EpCAM interaction played the critical roles in the binding, regulation and adhesion processes (Additional file 1: Figure S1). Superior to the detection function, the dual biological functions of dual antibody-coated nanomaterials, including anti-proliferation and anti-adhesion effects, might interfere the critical points of initiating cancer metastasis.

## Conclusions

In conclusion, the present study firstly synthesized the dual antibody-coated dendrimers with or without fluorescence labeling, and fully characterized their physicochemical properties. The dual antibody conjugates bound or captured the colon cancer cell lines with the enhanced affinity and specificity, and exhibited the superiority to their single counterparts in the restraint of cell activity and in the inhibition of the hetero-adhesion of cancer cells to endothelial cells. These newly-found biological functions of the re-engineered nanomaterials with antibodies may aid in designing new strategy to effectively prevent cancer metastasis by targeting the biomarkers-abundant cancer cells.

## Materials and methods
### Synthesis and characterization of antiEpCAM- and antiSlex- coated dendrimer conjugates with or without fluorescence labeling
To investigate the dual roles of dendrimer-antibody conjugates in cancer metastasis prevention, PAMAM dendrimers with the ethylenediamine core [generation 6 (G6), theoretical MW 624,00 Da] provided by Shandong Weihai Chenyuan New Silicone Materials, Co. Ltd were firstly surface-modified and sequentially conjugated with two fluorescence or non fluorescence-labeled

antibodies. Briefly, G6 PAMAM dendrimers with amine ends of 256 (100 mg, 1.60 μmol) were dissolved in 2 mL DMSO, and respectively reacted with 410 mg succinic anhydride (SA) (4.1 mmol, ten molar excess) under vigorous stirring overnight. The obtained CC G6 dendrimers were dialyzed against DDI water to remove the unreacted molecules as well as organic solvents before lyophilization.

The dual antibody-conjugated dendrimer conjugate was synthesized by employing the 1-ethyl-3-(3-dimethylaminopropyl) carbodiimide (EDC) catalytic method and designated as G6-5A-5S based on the reaction molar ratio of 1 dendrimers to 5 antiEpCAM to 5 antiSlex. AntiSlex (MW150KDa), IgG/IgM-FITC, antiEpCAM-PE were provided by BD company, and antiEpCAM (MW150 KDa) was purchased from Sigma-Aldrich and Abcam (Hong Kong) Ltd. CC G6 dendrimers (0.55 μg, 7.9 pmol) dissolved in 2 mL phosphate-buffered saline (PBS) were activated using EDC (75.8 ng, 395.3 pmol, 50 molar excess) and N-hydroxysuccinimide (NHS) (45.5 ng, 395.3 pmol, 50 molar excess) at room temperature for 1 h. The activated dendrimers were reacted with the combined antiEpCAM (39.5 pmol, 5 molar excess) and antiSlex (39.5 pmol, 5 molar excess) under vigorous stirring overnight. The single antibody-coated dendrimer conjugates were similarly synthesized and denoted as G6-5A and G6-5S. There were approximately two aEpCAM molecules in one G6-5A conjugate and six aSlex in one G6-5S conjugate according to the UV analysis. The fluorescence-labeled dual antibody conjugate PE-5A-G6-5S-FITC was also synthesized and designated following the similar procedure just by using antiEpCAM-PE and antiSlex-FITC (antiSlex and IgG/IgM-FITC were used together) antibodies, instead. The activated CC G6 dendrimers were reacted with antiSlex-FITC for 12 h, then with antiEpCAM-PE for another 12 h. All the reactions were conducted under vigorous stirring overnight in dark. Finally, all the conjugates were purified via dialysis (10,000 MWCO) against DDI water overnight before lyophilization. Dendrimers with large surface functional groups were almost able to assemble with all of the antibodies as one entity at the above designed molar ratios. Transmembrane dialysis was used to remove the intermediates and small molecules.

The presence of antibody or fluorescence-labeled antibody onto the dendrimer surface was respectively confirmed by the UV absorption value at $\lambda_{220\,nm}$ (Quawell 5000 UV–vis Spectrophotometer, America) and the merged fluorescence intensity at FITC $\lambda_{ex}$ 488 nm, $\lambda_{em}$ 500–535 nm and PE $\lambda_{ex}$ 568 nm, $\lambda_{em}$ 560–660 nm (Olympus FluoView 1000). The morphological property and particle size of G6-5A-5S conjugate were determined by FSEM measurement.

## Cell culture
Human colorectal carcinoma cell lines including SW480, SW620 and LoVo were purchased from the Type Culture

Collection of the Chinese Academy of Sciences, Shanghai, China and kept in a minimal number of passages, then cultured in RPMI 1640 medium supplemented with 10% heat-inactivated fetal calf serum (FCS) and 1% penicillin/streptomycin (P/S). HUVECs were obtained from the fresh human umbilical cords of new-born babies with 1 mg mL$^{-1}$ of collagenase in PBS and cultured by us with M199 medium supplemented with 20% fetal bovine serum (FBS), 100 μg mL$^{-1}$ endothelial cell growth supplement (ECGS), 50 μg mL$^{-1}$ heparin sodium and 1% P/S in a culture flask coated with 0.2% gelatin after some necessary pretreatment [10,51]. All the cell lines above were grown in a humidified atmosphere of 5% $CO_2$ at 37°C for the subsequent experiments. HUVECs were used for no more than six passages.

## Flow cytometric procedures
A Becton Dickinson (BD) multiparametric fluorescence-activated cell sorting (FACS) Aria III with laser excitation set at 488 was used for flow cytometric analysis. According to the forward versus side scatter histograms, gating strategy was used to set P1 gate for determining the target colon cancer cell lines. Fluorescence signals derived from PI (or PE) and DiOC6(3) (or FITC) were respectively detected through 585 and 530 nm bandpass filters. Side angle scattered light (SSC) versus PI histogram displayed the cell cycle distribution, SSC versus PE (or FITC) histogram showed the expression levels of biomarkers, SSC versus DiOC6(3) histogram revealed the cellular MMP, and PE versus FITC dot plots showed the captured cell numbers by the synthesized conjugate. All the data were acquired based on the collected 10,000 cells satisfying the light scatter criteria and analyzed using the BD FACS Diva software provided with the system.

## CTCs binding and capture assays
To explore the binding and capture capability of fluorescence-labeled dual antibody conjugate PE-5A-G6-5S-FITC at various concentrations (0, 10, 20 μg mL$^{-1}$) to the adherent and suspensory colon cancer cell lines, the optimal incubation time was in advance determined according to the analysis of time-response cell capture assay shown in Additional file 1. The preliminary result showed that 1 h-binding time was *sine qua non* for the quick and efficient cell capture.

### Binding to the adherent cells
Cell lines at the density of $10^5$/mL were cultivated on 35 mm dishes with glass coverslips in the bottom, and individually treated with PBS containing 1% bovine serum albumin (BSA) (1% PBSA) for 30 min. After 1 h of co-incubation with PE-5A-G6-5S-FITC conjugate at various concentrations (0, 10, 20 μg mL$^{-1}$) in a humidified

atmosphere of 5% $CO_2$ at 37°C, cell lines were washed with PBS to remove the unbound conjugate, and fixed with stationary liquid ($V_{methanol}:V_{acetone} = 7:3$) for 1 min, then stained with 10 µg mL$^{-1}$ of nuclei stain dihydrochloride (DAPI) solution for 15 min. Finally, cell lines were covered with serum-free medium for images taken by an Olympus Fluo-View 1000 laser confocal microscope respectively in the channel of DAPI, Alex Fluor 488 and 568.

### Capturing the suspensory cells

To evaluate the efficiency of PE-5A-G6-5S-FITC conjugate at capturing the colon cancer cell lines, SW620 and LoVo cell lines at the density of $10^6$/mL were suspended in each tube. Cell lines were treated with 1% PBSA, then with 20 µg mL$^{-1}$ of PE-5A-G6-5S-FITC conjugate for 1 h at 37°C water bath. Cell lines without the treatment of conjugate were incubated with immunoglobulins labeled with PE or FITC in the similar way as isotype controls. After washing and centrifugation, the unbound conjugates or antibodies were abandoned. Cell lines suspended with PBS buffer were directly analyzed on a BD FACS Aria III analyzer with laser excitation set at 488 nm or further stained with Hoechst 33258 (labeling the nucleus) for analysis with a fluorescence inverted microscope (Axio Observer A1, Zeiss, Germany).

### Restraining the captured CTCs for preventing cancer metastasis

#### Cell viability

To investigate how the single and dual antibody conjugates (G6-5A, G6-5S and G6-5A-5S) affected the cell proliferation, MTT analysis was conducted as we previously described. The effect of completely-carboxylated G6 dendrimers on cell activity was also tested. Cell lines at the density of $5 \times 10^3$-$1 \times 10^4$ cells/mL were cultivated on the 96-well plates with 1640 medium. When grew in the confluence of 70%-80%, cell lines were individually exposed to the conjugates at various concentrations (0, 1.25, 2.5, 5, 10, 15, 20 µg mL$^{-1}$) for 48 h. Then, 100 µL of serum-free medium containing 1 mg mL$^{-1}$ MTT solution was added to incubate for another 4 h. Finally, the supernatant was aspirated and 150 µL of DMSO was added to each well to dissolve the water-insoluble blue formazan. The viability of each cell line induced by the conjugates was determined based on the optical absorption value at the wavelength of 570 nm ($A_{570\ nm}$) and expressed as $A_{570\ nm}$ of the treated group divided by that of the control group.

#### Cell cycle distribution

To further discuss the effects of the antibody conjugates (e.g., G6-5A-5S) on the cell population distribution in every phases (G0/G1, S, and G2/M), PI staining experiment was performed at 37°C as the kit instructions. Cell lines were cultivated in 6-well plates overnight, and incubated with various concentrations of G6-5A-5S conjugate (0, 10, 20 µg mL$^{-1}$) for 48 h. Then cell lines were trypsinised and washed with ice-cold PBS for three times. After fixed with 70% ice-cold ethanol overnight at −20°C, cell lines were washed and stained with PI solution at 37°C for 15 min. Finally, data acquisition and analysis were performed on a BD FACS Aria III flow cytometer and DNA integration software mflt32, respectively.

### Cellular MMP

Depolarization of cellular MMP usually predicts the starting of cell apoptosis. In this assay, DiOC6(3) (a lipotropy cationic fluorescent dye) staining was used to determine the change of MMP in colon cancer cell lines. Increment of fluorescence intensity with the accumulation of DiOC6(3) in mitochondria was accompanied with the descent of MMP. After exposure to the antibody conjugates (e.g., G6-5A-5S) at various concentrations (0, 10, 20 µg mL$^{-1}$) for 48 h, cell lines were trypsinized and collected after centrifugation. 500 µL of DiOC6(3) (2 nM) working solution was individually added into each tube, and kept at 37°C water bath for 20 min. The cellular MMP ($\Delta\psi m$) was finally assessed according to the fluorescence intensity of DiOC6(3) examined by the flow cytometry (BD FACS Aria III).

### Inhibiting the adhesion of captured CTCs to endothelial cells for preventing cancer metastasis

#### Blocking the adhesion of cancer cells to Fn-coated substrates

Adhesion of CTCs to extracellular matrix (ECM) was a critical step in the process of cancer metastasis. Colon cancer cell lines were usually used as CTC models. In cell adhesion assays, CC G6 dendrimers were used as the control. Fn-coated substrates were prepared as the ECM to test the capability of single and dual antibody conjugates (G6-5A, G6-5S, G6-5A-5S) in interfering the adhesion of colon cancer cell lines to ECM. First of all, 10 ng mL$^{-1}$ of Fn was pre-coated on the substrates of 96-well plate overnight, then discarded and sealed up with 2% PBSA for 30 min. 100 µL of the mixtures of cell lines and conjugate at various concentrations (0, 1.25, 2.5, 5, 10 µg mL$^{-1}$) were added onto each well for 1 h of incubation. The post-processing was similar to that described in MTT assay. 100 µL of serum-free medium containing 1 mg mL$^{-1}$ MTT solution was used for another 4 h. After the supernatant was aspirated, 100 µL of DMSO was added to each well to dissolve the water-insoluble blue formazan. The optical density was read on an ELISA reader at a wavelength of 570 nm to determine the abilities of the conjugates to interfere with the adhesion. The relative adhesion (%) was finally evaluated by the $A_{570\ nm}$ in the treated group compared to that in the control group.

## *Blocking the adhesion of cancer cells to HUVECs*

Adhesion of CTCs to vascular endothelium was another crucial starting point of cancer metastasis. HUVECs instead of vascular endothelial cells were used for adhesion assay in vitro. After grew in the confluence of 100% on the 24-well plates, HUVECs were pre-treated with 1 ng mL$^{-1}$ cytokine IL-1β for 4 h followed by individually incubated with the mixture of rhodamine 123-labeled colon cancer cell lines and single or dual antibody conjugates (G6-5A, G6-5S, G6-5A-5S) at various concentrations (0, 1.25, 2.5, 5, 10 μg mL$^{-1}$) for 1 h. After removal of the non-adhered cells with PBS washing, ten visual fields were randomly selected and taken images by a fluorescence inverted microscope (Axio Observer A1, Zeiss, Germany). The capability of the conjugates in inhibiting the adherence of cancer cell lines to HUVECs was determined by counting the numbers of fluorescence-labeled cells that adhered to HUVECs in sample groups relative to those in the control group.

## Statistical analysis

Every experiment was performed independently and repeated at least three times. Data were expressed as the means ± standard deviations (SD). Statistical analysis was done by Student's t-test and one-way analysis of variance (One-ANOVA). Multiple comparisons of the means were made through One-ANOVA analysis and demonstrated by the least significance difference (LSD) test (IBM SPSS Statistics 19.0). The symbol of * and ** represented the comparison between sample and control, while # and ## represented the comparison between any two samples. A probability value of <0.05 was considered significant (* and #), and <0.01 was considered extremely significant (** and ##).

## Additional file

Additional file 1: Biomarkers expression and Time-response cell capture.

## Abbreviations

CTCs: Circulating tumor cells; Fn: Fibronectin; HUVECs: Human umbilical vein endothelial cells; CAP-NO: S-nitrosocaptopril; EpCAM: Epithelial cell adhesion molecule; AntiEpCAM: EpCAM antibody; PE: Phycoerythrin; AntiEpCAM-PE: PE linked antiEpCAM; Sialyl Lewis X (Slex): Saliva acidifying louis oligosaccharides; AntiSlex: Slex antibody; FITC: Fluorescein isothiocyanate; AntiSlex-FITC: FITC linked antiSlex; PAMAM: Polyamidoamine; CC G6: Completely carboxylated G6 PAMAM; SA: Succinic anhydride; EDC: 1-ethyl-3-(3-dimethylaminopropyl) carbodiimide; NHS: N-hydroxysuccinimide; FSEM: Field emission scanning electron microscope; FCS: Fetal calf serum; P/S: Penicillin/ streptomycin; FBS: Fetal bovine serum; ECGS: Endothelial cell growth supplement; SSC: Side angle scattered light; BD: Becton Dickinson; FACS: Fluorescence-activated cell sorting; PBS: Phosphate-buffered saline; BSA: Bovine serum albumin; 1% PBSA: PBS containing 1% BSA; PI: Propidium iodide; DAPI: Dihydrochloride; MTT: [3-(4,5-dimethylthiazol-2-yl)-2,5-diphenyltetrasodium bromide] tetrazolium salt; DiOC6(3): 3,3'-Dihexyloxacarbocyanine iodide; MMP: Mitochondrial membrane potential; ECM: Extracellular matrix; EPR: Enhanced permeability and retention; SD: Standard deviations; One-ANOVA: One-way analysis of variance; LSD: Least significance difference.

## Competing interests

The authors declare that they have no competing financial interest.

## Authors' contributions

XJJ and JL conceived the study. XJJ designed the experiments. XJJ, CHN and ZRL performed most and others did some of the experiments. XJJ, DHY and ZRL analyzed and interpreted the data. XJJ and JL co-wrote the paper. All authors discussed the results and commented on the manuscript. All authors read and approved the final manuscript.

## Acknowledgements

This work was supported by grants from the National Science Foundation of China Nos. 81273548 and the Ministry of Science and Technology of China Nos. 2015CB931804.

## Author details

[1]Cancer Metastasis Alert and Prevention Center, and Biopharmaceutical Photocatalysis of State Key Laboratory of Photocatalysis on Energy and Environment, College of Chemistry, Fuzhou University, 523 Industry Road, Science Building, 3FL, Fuzhou, Fujian 350002, China. [2]Rutgers, The State University of New Jersey, 160 Frelinghuysen Road, Piscataway, NJ 08854, USA. [3]Department of Medicine Oncology, East Hospital of Xiamen University, Fuzhou 350004, China.

## References

1. Ferlay J, Parkin DM, Steliarova-Foucher E. Estimates of cancer incidence and mortality in Europe in 2008. Eur J Cancer. 2010;46:765–81.
2. Bach PB. Costs of cancer care: a view from the centers for Medicare and Medicaid services. J Clin Oncol. 2007;25:187–90.
3. Sethi N, Kang Y. Unravelling the complexity of metastasis - molecular understanding and targeted therapies. Nat Rev Cancer. 2011;11:735–48.
4. Klein CA. Cancer. The metastasis cascade. Science. 2008;321:1785–7.
5. Steeg PS. Tumor metastasis: mechanistic insights and clinical challenges. Nat Med. 2006;12:895–904.
6. Bernards R, Weinberg RA. A progression puzzle. Nature. 2002;418:823.
7. Kaiser J. Medicine. Cancer's circulation problem. Science. 2010;327:1072–4.
8. Plaks V, Koopman CD, Werb Z. Cancer. Circulating tumor cells. Science. 2013;341:1186–8.
9. Kling J. Beyond counting tumor cells. Nat Biotechnol. 2012;30:578–80.
10. Lu Y, Yu T, Liang H, Wang J, Xie J, Shao J, et al. Nitric oxide inhibits heteroadhesion of cancer cells to endothelial cells: restraining circulating tumor cells from initiating metastatic cascade. Sci Rep. 2014;4:4344.
11. Cristofanilli M, Budd GT, Ellis MJ, Stopeck A, Matera J, Miller MC, et al. Circulating tumor cells, disease progression, and survival in metastatic breast cancer. N Engl J Med. 2004;351:781–91.
12. Zieglschmid V, Hollmann C, Bocher O. Detection of disseminated tumor cells in peripheral blood. Crit Rev Clin Lab Sci. 2005;42:155–96.
13. Pantel K, Alix-Panabieres C. Circulating tumour cells in cancer patients: challenges and perspectives. Trends Mol Med. 2010;16:398–406.
14. Jia L, Schweikart K, Tomaszewski J, Page JG, Noker PE, Buhrow SA, et al. Toxicology and pharmacokinetics of 1-methyl-d-tryptophan: absence of toxicity due to saturating absorption. Food Chem Toxicol. 2008;46:203–11.
15. Morgan G, Ward R, Barton M. The contribution of cytotoxic chemotherapy to 5-year survival in adult malignancies. Clin Oncol (R Coll Radiol). 2004;16:549–60.
16. Ellis P, Barrett-Lee P, Johnson L, Cameron D, Wardley A, O'Reilly S, et al. Sequential docetaxel as adjuvant chemotherapy for early breast cancer (TACT): an open-label, phase III, randomised controlled trial. Lancet. 2009;373:1681–92.
17. Ramsey SD, Moinpour CM, Lovato LC, Crowley JJ, Grevstad P, Presant CA, et al. Economic analysis of vinorelbine plus cisplatin versus paclitaxel plus carboplatin for advanced non-small-cell lung cancer. J Natl Cancer Inst. 2002;94:291–7.
18. Wang S, Wang H, Jiao J, Chen KJ, Owens GE, Kamei KI, et al. Three-Dimensional Nanostructured Substrates toward Efficient Capture of Circulating Tumor Cells. Angewandte Chemie. 2009;121:9132–5.
19. Wang S, Liu K, Liu J, Yu ZTF, Xu X, Zhao L, et al. Highly efficient capture of circulating tumor cells by using nanostructured silicon substrates with integrated chaotic micromixers. Angew Chem Int Ed. 2011;50:3084–8.

20. Myung JH, Gajjar KA, Saric J, Eddington DT, Hong S. Dendrimer-mediated multivalent binding for the enhanced capture of tumor cells. Angew Chem Int Ed Engl. 2011;50:11769–72.

21. Yoon HJ, Kim TH, Zhang Z, Azizi E, Pham TM, Paoletti C, et al. Sensitive capture of circulating tumour cells by functionalized graphene oxide nanosheets. Nat Nanotechnol. 2013;8:881.

22. Yoon HJ, Kim TH, Zhang Z, Azizi E, Pham TM, Paoletti C, et al. Sensitive capture of circulating tumour cells by functionalized graphene oxide nanosheets. Nat Nanotechnol. 2013;8:735–41.

23. Wen CY, Wu LL, Zhang ZL, Liu YL, Wei SZ, Hu J, et al. Quick-response magnetic nanospheres for rapid, efficient capture and sensitive detection of circulating tumor cells. ACS Nano. 2014;8:941–9.

24. Dobrovolskaia MA, McNeil SE. Immunological properties of engineered nanomaterials. Nat Nanotechnol. 2007;2:469–78.

25. Balic M, Williams A, Lin H, Datar R, Cote RJ. Circulating tumor cells: from bench to bedside. Annu Rev Med. 2013;64:31–44.

26. Kalluri R, Weinberg RA. The basics of epithelial-mesenchymal transition. J Clin Invest. 2009;119:1420–8.

27. Rhim AD, Mirek ET, Aiello NM, Maitra A, Bailey JM, McAllister F, et al. EMT and dissemination precede pancreatic tumor formation. Cell. 2012;148:349–61.

28. Yu M, Bardia A, Wittner BS, Stott SL, Smas ME, Ting DT, et al. Circulating breast tumor cells exhibit dynamic changes in epithelial and mesenchymal composition. Science. 2013;339:580–4.

29. Wang J, Chen J, Wan L, Shao J, Lu Y, Zhu Y, et al. Synthesis, Spectral Characterization, and In Vitro Cellular Activities of Metapristone, a Potential Cancer Metastatic Chemopreventive Agent Derived from Mifepristone (RU486). AAPS J. 2014;16:289–98.

30. Baccelli I, Schneeweiss A, Riethdorf S, Stenzinger A, Schillert A, Vogel V, et al. Identification of a population of blood circulating tumor cells from breast cancer patients that initiates metastasis in a xenograft assay. Nat Biotechnol. 2013;31:539–44.

31. Balzar M, Winter MJ, de Boer CJ, Litvinov SV. The biology of the 17-1A antigen (Ep-CAM). J Mol Med (Berl). 1999;77:699–712.

32. Haier J, Nasralla M, Nicolson GL. Cell surface molecules and their prognostic values in assessing colorectal carcinomas. Ann Surg. 2000;231:11–24.

33. Matsushita Y, Hoff S, Nudelman E, Otaka M, Hakomori S, Ota D, et al. Metastatic behavior and cell surface properties of HT-29 human colon carcinoma variant cells selected for their differential expression of sialyl-dimeric Lex antigen. Clin Exp Metastasis. 1991;9:283–99.

34. Myung JH, Launiere CA, Eddington DT, Hong S. Enhanced tumor cell isolation by a biomimetic combination of E-selectin and anti-EpCAM: implications for the effective separation of circulating tumor cells (CTCs). Langmuir. 2010;26:8589–96.

35. Berg EL, Magnani J, Warnock RA, Robinson MK, Butcher EC. Comparison of L-selectin and E-selectin ligand specificities: the L-selectin can bind the E-selectin ligands sialyl Le(x) and sialyl Le(a). Biochem Biophys Res Commun. 1992;184:1048–55.

36. Ohyama C, Tsuboi S, Fukuda M. Dual roles of sialyl Lewis X oligosaccharides in tumor metastasis and rejection by natural killer cells. EMBO J. 1999;18:1516–25.

37. Hong S, Leroueil PR, Majoros IJ, Orr BG, Baker Jr JR, Banaszak Holl MM. The binding avidity of a nanoparticle-based multivalent targeted drug delivery platform. Chem Biol. 2007;14:107–15.

38. Yu M, Bardia A, Aceto N, Bersani F, Madden MW, Donaldson MC, et al. Cancer therapy. Ex vivo culture of circulating breast tumor cells for individualized testing of drug susceptibility. Science. 2014;345:216–20.

39. Zhang L, Ridgway LD, Wetzel MD, Ngo J, Yin W, Kumar D, et al. The identification and characterization of breast cancer CTCs competent for brain metastasis. Sci Transl Med. 2013;5:180ra148.

40. Thomas TP, Majoros IJ, Kotlyar A, Kukowska-Latallo JF, Bielinska A, Myc A, et al. Targeting and inhibition of cell growth by an engineered dendritic nanodevice. J Med Chem. 2005;48:3729–35.

41. Quintana A, Raczka E, Piehler L, Lee I, Myc A, Majoros I, et al. Design and function of a dendrimer-based therapeutic nanodevice targeted to tumor cells through the folate receptor. Pharm Res. 2002;19:1310–6.

42. Wang S, Wang H, Jiao J, Chen KJ, Owens GE, Kamei K, et al. Three-dimensional nanostructured substrates toward efficient capture of circulating tumor cells. Angew Chem Int Ed Engl. 2009;48:8970–3.

43. Li W, Zhao H, Qian W, Li H, Zhang L, Ye Z, et al. Chemotherapy for gastric cancer by finely tailoring anti-Her2 anchored dual targeting immunomicelles. Biomaterials. 2012;33:5349–62.

44. Thomas TP, Patri AK, Myc A, Myaing MT, Ye JY, Norris TB, et al. In vitro targeting of synthesized antibody-conjugated dendrimer nanoparticles. Biomacromolecules. 2004;5:2269–74.

45. Shukla R, Thomas TP, Peters JL, Desai AM, Kukowska-Latallo J, Patri AK, et al. HER2 specific tumor targeting with dendrimer conjugated anti-HER2 mAb. Bioconjug Chem. 2006;17:1109–15.

46. Xie J, Zhao R, Gu S, Dong H, Wang J, Lu Y, et al. The Architecture and Biological Function of Dual Antibody-Coated Dendrimers: Enhanced Control of Circulating Tumor cells and Their Hetero-Adhesion to Endothelial Cells for Metastasis Prevention. Theranostics. 2014;4:1250–63.

47. Shao J, Dai Y, Zhao W, Xie J, Xue J, Ye J, et al. Intracellular distribution and mechanisms of actions of photosensitizer Zinc(II)-phthalocyanine solubilized in Cremophor EL against human hepatocellular carcinoma HepG2 cells. Cancer Lett. 2013;330:49–56.

48. Duncan R, Izzo L. Dendrimer biocompatibility and toxicity. Adv Drug Deliv Rev. 2005;57:2215–37.

49. Malik AN, Wiwattanapatapee R, Klopsch R, Lorenz K, Frey H, Weener J, et al. Dendrimers: Relationship between structure and biocompatibility in vitro, and preliminary studies on the biodistribution of 125I-labelled polyamidoamine dendrimers in vivo. J Control Release. 2000;65:133–48.

50. Xie J, Zhao R, Lu Y, Wang J, Gu S, Jia L. Re-engineering PAMAM dendrimer for its multivalent binding to colon cancer cells HT29. In: Ministry of Science and Technology (MOST) and Chinese Academy of Sciences, The 450th Xiangshan-Science Conference and the 3th US-China Symposium on Cancer Nanotechnology and Nanomedicine. Beijing, China. 2012.

51. Jaffe EA, Nachman RL, Becker CG, Minick CR. Culture of human endothelial cells derived from umbilical veins. Identification by morphologic and immunologic criteria. J Clin Invest. 1973;52:2745–56.

# Invertase-nanogold clusters decorated plant membranes for fluorescence-based sucrose sensor

Dipali Bagal-Kestwal, Rakesh Mohan Kestwal and Been-Huang Chiang[*]

## Abstract

In the present study, invertase-mediated nanogold clusters were synthesized on onion membranes, and their application for sucrose biosensor fabrication was investigated. Transmission electron microscopy revealed free nanoparticles of various sizes (diameter ~5 to 50 nm) along with clusters of nanogold (~95 to 200 nm) on the surface of inner epidermal membranes of onions (Allium cepa L.). Most of the polydispersed nanoparticles were spherical, although some were square shaped, triangular, hexagonal or rod-shaped. Ultraviolet–visible spectrophotometric observations showed the characteristic peak for nanoparticles decorated invertase-onion membrane at approximately 301 nm. When excited at 320 nm in the presence of sucrose, the membranes exhibited a photoemission peak at 348 nm. The fluorescence lifetime of this nanogold modified onion membrane was 6.20 ns, compared to 2.47 ns for invertase-onion membrane without nanogold. Therefore, a sucrose detection scheme comprised of an invertase/nanogold decorated onion membrane was successfully developed. This fluorescent nanogold-embedded onion membrane drop-test sensor exhibited wide acidic to neutral working pH range (4.0-7.0) with a response time 30 seconds (<1 min). The fabricated quenching-based probe had a low detection limit ($2 \times 10^{-9}$ M) with a linear dynamic range of $2.25 \times 10^{-9}$ to $4.25 \times 10^{-8}$ M for sensing sucrose. A microplate designed with an enzyme-nanomaterial-based sensor platform exhibited a high compliance, with acceptable percentage error for the detection of sucrose in green tea samples in comparison to a traditional method. With some further, modifications, this fabricated enzyme-nanogold onion membrane sensor probe could be used to estimate glucose concentrations for a variety of analytical samples.

**Keywords:** Nanogold clusters, Gold nanoparticles, Invertase, Onion membrane, Sucrose, Glucose, Analyte, Fluorescence, Quenching-based biosensor

## Introduction

Metal nanoparticles are outstanding building blocks for fabrication of biosensors, due to their surface plasmon resonance shifts in response to a biorecognition event. Properties of nanoparticles vary in accordance with their size and composition, which facilitates diverse applications in various areas including catalysis, sensors and medicine. Production of nanoparticles can be achieved through chemical, physical or biological methods. Among them, there has been considerable attention focused on biological methods for synthesis of metallic nanoparticles, as there is a vast array of biological resources available in nature [1,2]. Furthermore, the biological approach to nanoparticles synthesis is also a low cost, non-toxic, biocompatible and environmentally friendly process.

Enzymes can work as nanoreactors that allow generation of nanostructures often of controlled size by limiting the rate of nucleation of nascent nanocrystals [3]. For example, enzyme-guided nanoparticles have been used for fabrication of biosensors that can detect prostate-specific antigen (a biomarker of prostate cancer), with outstanding sensitivity. The development of optical sensors using enzyme-stimulated synthesis of metallic nanoparticles has been also reported by Willner et al. [4]. According to their study, in the presence of bovine serum albumin, bacterial protease not only

* Correspondence: bhchiang@ntu.edu.tw
Institute of Food Science and Technology, National Taiwan University, No.1, Roosevelt Road, section 4, Taipei, Taiwan

mediated biosynthesis of gold (Au) nanoparticles, but also acted as reducing and shape-directing agents. *In vitro* enzymatic synthesis of Au nanoparticles using alpha-NADPH-dependent sulfite reductase and phytochelatin was reported by Kumar et al. [5]. Kalishwaralal et al., investigated biosynthesis of gold, silver, and gold-silver alloy nanoparticles by harnessing free and exposed thiol groups of α-amylase [1]. Furthermore, the native reducing properties of plant proteins have also been harnessed for synthesis of Au nanoclusters [6,7]. Mittal et al., has reviewed the methods for making nanoparticles using plant extracts and the potential applications of these nanoparticles for various applications [8]. Moreover, the biosynthesis of Au nanoparticles by many plants such as *Medicago sativa* [9], *Azadirachta indica* [10], *Aloe vera* [11], *Cinnamomum camphora* [12], *Magnolia kobus, Diopyros kaki* [13], *Szyygium aromaticum* [14], *Putranjiva Roxburghii* [15], *Cassia auriculata* [16], among others, has been well documented.

Consumption of sugar added beverages and numerous other foods have increased across the globe. Therefore, quantitative determination of sugar content in food is an important issue. In particular, there is need for a fast, simple, and reproducible method for determination of sucrose content [17]. Fluorescent-sensing systems have become increasingly popular, due to their versatility, ease of use and low cost. Several analytical techniques have been developed, to exploit changes in fluorescence properties of a molecule in different environments, including quenching, Forster resonance energy transfer, and surface-modified fluorescence (FL) [18]. Invertase (INV; β-fructofuranosidase) is an enzyme with a high rate of enzymatic turnover. Under ambient conditions, nanomolar levels of INV are capable of converting a millimolar concentration of sucrose into glucose, making it an ideal catalyst for amplification of 'turn-on' signals in sucrose sensors [19]. Consequently, optical nanoprobe utilizing electrospun polyamide meshes containing gold salts and invertase have been reported to be useful for sugar-sensing [20]. Blue and pink colorimetric assays based on sugar and glucose oxidase-assisted synthesis of nanoparticles for sugar detection, have also been reported [17]. However, reports of fluorometric biosensors for sucrose estimation are very few.

The inner epidermal membrane of the onion bulk scales is a good bio-platform to immobilize enzymes, as it has excellent gas and water permeability for substrates and products. Onion membranes (Oms) mainly consist of elongated tubular cells, with blunt or tapered ends, along with numerous guard cells. For biosensors fabrication, this natural membrane is mechanically stronger than other natural membranes, due to its microfibrillar cellulosic elongated tubular structure. Thus, it could be ideal as a biocompatible platform for enzyme immobilization

[21,22]. Kumar and Pundir reported immobilization of lipase on onion membrane and its possible commercial application in food-processing industries [23]. A glucose biosensor comprising glucose oxidase/O-(2-hydroxyl) propyl-3-trimethylammonium chitosan chloride nanoparticle-immobilized on the inner membrane of onion and a dissolved oxygen sensor have also been reported [24]. Furthermore, a glucose biosensor based on onion primary cuticula that immobilized glucose oxidase was reported for determining glucose concentrations in human serum [25].

The objective of our study was to synthesis and characterize invertase-nanogold clusters (INV-NAuCs) embedded in plant membranes and investigate their application in designing fluorescent probes for sucrose detection. The novel feature of our proposed method is employment of a new biomaterial along with enzyme for gold nanomaterial synthesis and biosensor development. Various factors that might influence the sensor performance have been investigated. The fabricated drop-test sensor was then used to detect sucrose in various green tea samples to demonstrate its high sensitivity and specificity.

## Results and discussion
### Invertase- mediated nanogold synthesis: UV-Visible studies

The UV-Visible absorption spectra of invertase, blank (untreated with invertase) onion membrane, hydrogen tetrachloroaurate (HAuCl$_4$) solution and lastly, onion membrane with invertase and HAuCl$_4$ for 96 h in acetate buffer (20 mM, pH 5.0) are shown in Figure 1a. The HAuCl$_4$ solution had no obvious absorption peak, whereas the spectrum of INV had an absorption peak at 260 nm and blank onion membrane in assay buffer had an absorption peak at 275 nm. After incubating the onion membrane with invertase and gold chloride solution for 96 h, the absorption peak shifted from 275 to 301 nm, which indicated that there was a direct reaction among gold chloride, invertase and onion membrane to form nanogold clusters (NAuCs). Moreover, one minor peak was observed at 540 nm, which is also a characteristic of gold nanoparticles (AuNPs).

Formation of gold nanomaterials was further confirmed by UV–vis spectroscopy. Periodic UV–vis absorption spectra of onion membranes embedded with invertase-nanogold clusters (INV-NAuCs-Om) are shown in Figure 1b. All of spectra displayed the same plasmonic band for major dominant peak I at 301 nm (inset I) and minor surface plasmon peak II at 540 nm (Inset II), which intensified with time. The nanogold biosynthesis process was monitored continuously for 96 h. Absorption intensity increased with duration of incubation, reaching a plateau at 72 h, indicating saturation of

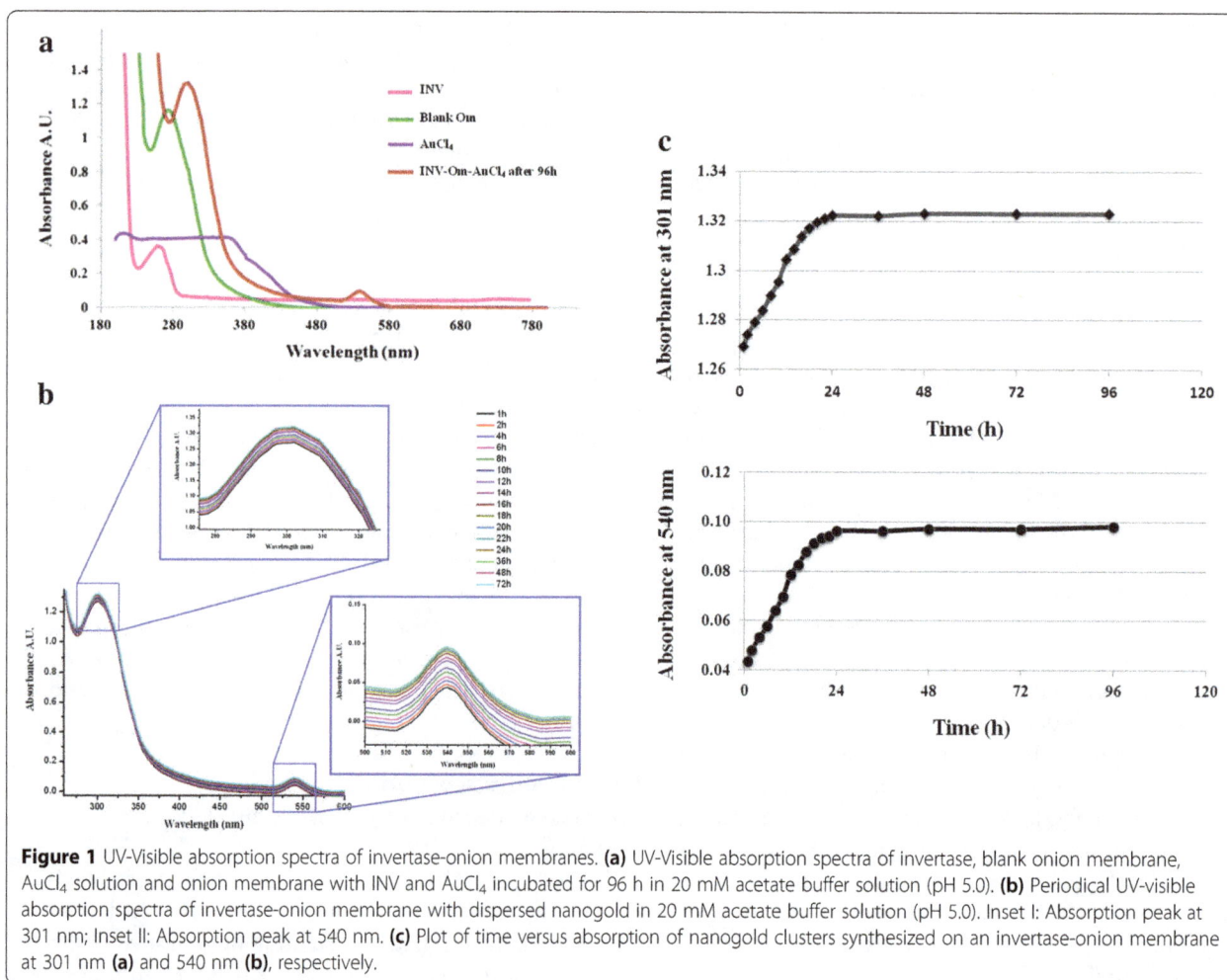

**Figure 1** UV-Visible absorption spectra of invertase-onion membranes. **(a)** UV-Visible absorption spectra of invertase, blank onion membrane, AuCl₄ solution and onion membrane with INV and AuCl₄ incubated for 96 h in 20 mM acetate buffer solution (pH 5.0). **(b)** Periodical UV-visible absorption spectra of invertase-onion membrane with dispersed nanogold in 20 mM acetate buffer solution (pH 5.0). Inset I: Absorption peak at 301 nm; Inset II: Absorption peak at 540 nm. **(c)** Plot of time versus absorption of nanogold clusters synthesized on an invertase-onion membrane at 301 nm **(a)** and 540 nm **(b)**, respectively.

nanogold formation on onion membrane after 72 h (Figure 1c). These results are similar to those reported by Parida et al. [6]. Biological activity of INV after nanogold formation was confirmed by DNSA method [26-30].

Nanogold synthesis on the invertase-immobilized onion membrane was consistent with the properties of onion membrane as a reducing and stabilizing agent. A similar approach was used for synthesis of silver nanoparticles, with onion (*Allium cepa*) extract acting as both a reducing as well as capping agent [31]. Parida et al., stated that reduction of gold nanoparticles occurred in onion extract due to the presence of ample vitamin C, citric acid, ascorbic acid, flavonoids and extracellular electron shutters, etc. [6,32]. However, the specific role of the invertase in the synthesis of nanogold has not been well established. One hypothesis is that a high content of vitamin C, flavonoids, thiosulphonates and other organosulfur in onion membranes are directly involved in the gold reduction mechanism. The exposed S–H groups of invertase may allow enzyme binding to the gold ions via gold–S bond without jeopardizing INV structure. The increased rate of NAuC production by

the enzyme indicated a rapid reduction of $Au^{3+}$ to $Au^0$ by the exposed functional groups of reducing amino acids (*e.g.* the thiol group of cysteine and the tertiary amine group of histamine) [33]. Furthermore, it is possible that intrinsic enzymatic generation of reactive sulfur species develops surface plasmon resonance at gold nanostructures which may turn them into tiny fluorophores. A more in-depth investigate is needed to understand the internal mechanism responsible for the formation of nanoparticles or clusters in the presence of invertase.

## Topological investigation of nanogold membrane

We also analyzed the topography of nanogold using both scanning and transmission electron microscopy (SEM and TEM, respectively). In the SEM images nanoclusters were found aligned with onion epidermal cell walls (Figure 2a-d). The biosynthesized nanogold clusters, ranging in size from 95 to 200 nm, either adhered to or embedded in the membrane. Most of the poly-dispersed nanoparticles were spherical, although other shapes were also visible, e.g. square-shaped,

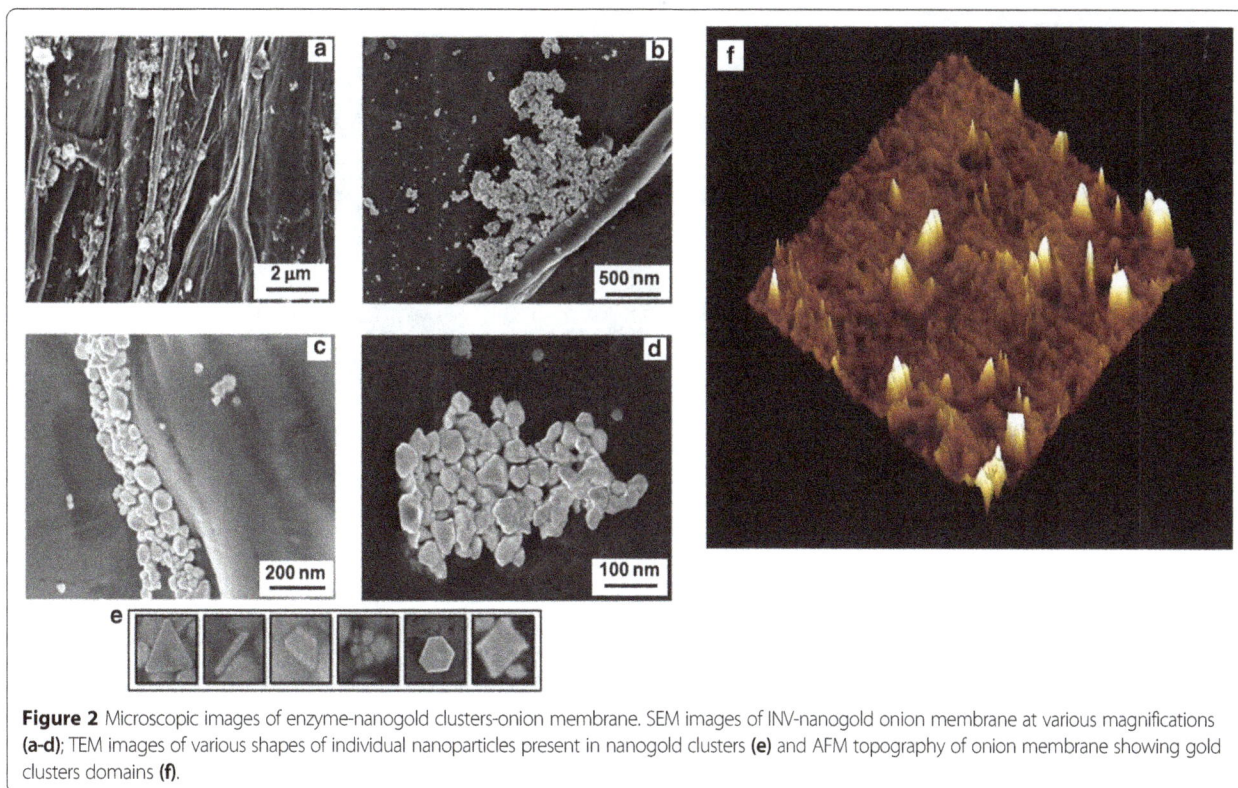

**Figure 2** Microscopic images of enzyme-nanogold clusters-onion membrane. SEM images of INV-nanogold onion membrane at various magnifications (**a-d**); TEM images of various shapes of individual nanoparticles present in nanogold clusters (**e**) and AFM topography of onion membrane showing gold clusters domains (**f**).

triangular, rectangular, hexagonal and cylindrical, but to a lesser extent. Based on transmission electron microscopy (TEM), these nanoparticles were ~5 to 50 nm in size (Figure 2e). Three-dimensional atomic force microscopic (AFM) images of onion membrane also showed prominent domain impressions of gold nanoclusters into the membrane surface of the onion (Figure 2f).

## Characteristic fluorescence spectra of INV- NAuCs- Om

Invertase-onion membranes, both with and without nanoparticles, were excited at 320 nm and the spectra were recorded in an emission range of 330 to 700 nm. The image of INV-Om, indicate that the inner epidermal onion membranes with invertase only did not possess fluorescence, whereas onion membrane decorated with gold clusters did have a fluorescent image (Figure 3a). Furthermore, invertase-immobilized onion membrane had a small peak at 337 nm (Figure 3b). Similarly, Hou et al., reported that the free invertase had an emission peak at 335 nm when excited at the same wavelength [34]. The slight red shift in emission peak (~2 nm) may have been due to the immobilization process. The possibility of emissions arising from the reagents (*e.g.* HAuCl$_4$, sucrose and the mixture of HAuCl$_4$–sucrose), were also examined. Blank onion membranes incubated with reagents showed very weak fluorescence which was similar to that of the background signal. However, INV-

NAuCs-Oms in 25 µL acetate buffer (20 mM, pH 5.0) had a photoemission peak at 346 nm. Perhaps the native reducing property of invertase was also harnessed for synthesis, capping and aggregation of gold particles into stable nanogold clusters. The electrostatic bonding and steric protection due to the bulkiness of the protein may also be responsible for stable INV-Om-scaffolds. Similar observations were reported for photoluminescent BSA-protected nanoparticles, with excitation and emission maxima at 320 and 404 nm, respectively [35].

## Fluorescence lifetime and quantum yield of the membrane

The fluorescence lifetime for INV-NAuCs-Om was 6.20 ns and $x^2 = 1.150$ (Figure 4). In addition, the lifetime for blank Om and INV-Om without nanogold, were 1.23 and 2.47 ns respectively. Therefore, the association of invertase and other proteins with nanoparticles could increase the fluorescence lifetime. The fluorescence average lifetime of INV-nanogold-onion membranes was similar to bovine serum albumin modified gold nanoparticles (BSA-GNPs) previously reported [36].

The quantum yield of gold nanoparticles was conservatively estimated to be $0.065 \pm 0.0050$ (P = 0.90), approximately eight orders of magnitude greater than that of gold films. Based on the INV-AuNPs-Om high quantum yield ($\Phi = 0.17 \pm 0.004$), we may inferred that the membrane is a useful element for fluorescent

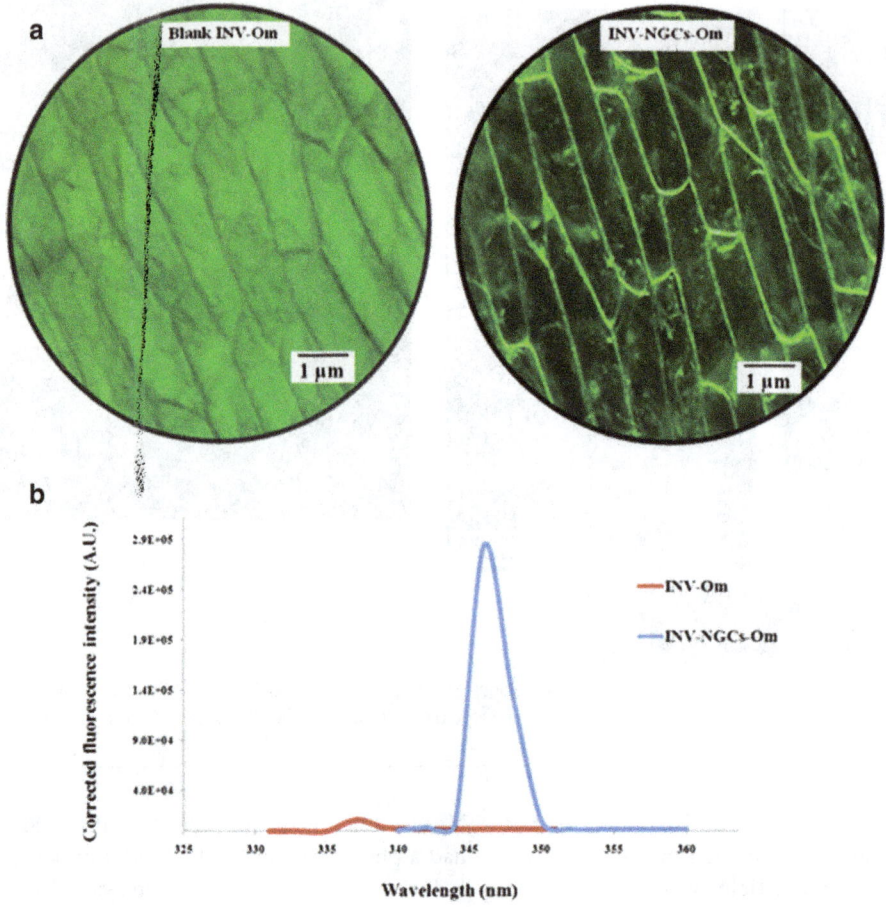

**Figure 3** Fluorescence studies of INV-Om and INV-NAuCs-Om. **(a)** Fluorescence images of INV-OM and INV-NAuCs-Om **(b)** Fluorescence spectra of INV-Om and INV-NAuCs-Om at excitation wavelength 320 nm.

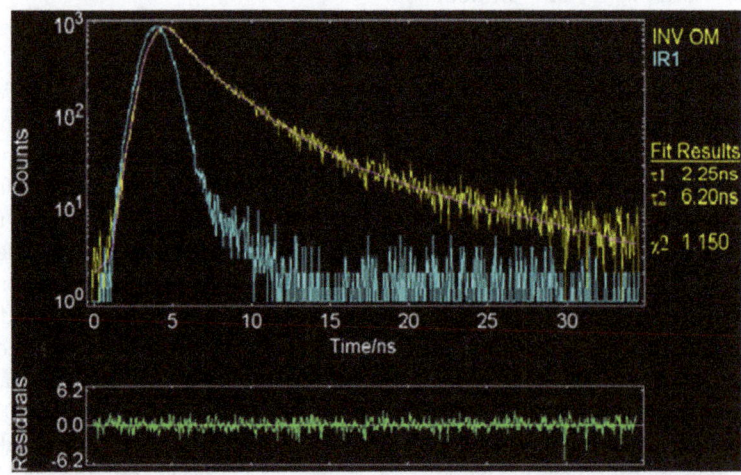

**Figure 4** Fluorescence lifetime for INV-NAuCs-Om in aqueous assay buffer solution (yellow line) whereas pink line corresponds to the non-linear least square fit value ($x^2 = 1.150$). Blue line represents IR spectrum. The lower panel represents the residual plot of the fit.

sensors. In addition, as the invertase-conjugated nano gold particles individually acted as embedded fluorophores in the membranes, it can be concluded that the intrinsic fluorescence of invertase tryptophan was scarcely used. Furthermore, the conservatively estimated quantum yield of INV-gold nanoparticles was found to be approximately seven and twelve orders of magnitude greater than that of blank INV-Om and blank Om, respectively.

## Influence of pH on nanogold clusters synthesis and fluorescence property

The influence of pH on the synthesis of nanogold clusters and their morphological properties was studied and observations are provided in Additional file 1 (S1.1 section). The fluorescence of the INV-NAuCs-onion membranes at various pH levels was also checked. For this purpose, membranes in respective pH solutions were excited in the range of 340 to 360 nm and pH range of 3.0 to 11.0. There were emission peaks at 344, 346 and 348 nm for pH 3.0, 4.0 and 5.0, respectively (Additional file 1: Figure S3), whereas emission peaks for pH 6.0 to 9.0 were all observed at 348 nm. For pH 10.0 and 11.0 the fluorescence peaks were observed at 350 and 353 nm, respectively (Additional file 1: Figure S3 inset). The fluorescence intensity from pH 3.0 to 5.0 increased in a linear fashion, with the maxima at pH 5.0, which could be attributed to the invertase pH optima. Nanogold decorated onion membranes were non-fluorescent when they were excited in the range of 540–600 nm.

## Effect of invertase and HAuCl$_4$ concentration on nanogold clusters synthesis

The influence of enzyme concentration on nanogold formation and fluorescence intensity is shown and discussed in Additional file 1: Figure S1.2. The chloroauric acid concentration affected the size of nanogold and assemblies formation. For the concentration of HAuCl$_4$ from 0.25 to 1.0 mM, fluorescent gold nanoparticles were formed, with diameters from 2-50 nm (Table 1). As the concentration increased to 1.5 mM, nanoparticles approaching 70 nm (± 32%) in diameter were produced.

At 2.0 mM HAuCl$_4$, spherical gold particles (diameter ~90 nm) were formed. These gold nanoparticles had surface plasmon ~543 nm without fluorescence properties. However, further increasing the concentration above 2.0 mM had no major effect on particle growth. The surface plasmon band in the gold nanoparticles solution remained close to 543 nm throughout the reaction period. Therefore, we inferred that nanoparticles were dispersed in the aqueous solution, leaving no evidence of aggregation in UV-Vis absorption spectrum. However, at a high gold salt concentration, steric hindrance and salt crowding on the enzyme surface may change protein structure, eventually causing enzyme precipitation with diminished invertase activity. Surface plasmon resonance peaks and fluorescence intensities with respect to gold salt concentration are shown in Table 1. These results were in good agreement with a previous report, in which effects of gold nanoparticle morphology on adsorbed protein structure and function were thoroughly studied [37].

## Application of the INV-NAuCs-Om sensor for sucrose sensing

Fluorescence measurement allows direct background subtraction strategy while colorimetric assays suffer background interference problem when onion membranes were used directly. Moreover, FL technique is extremely sensitive and fast. FL measurements also provide structure and micro-environment of molecules which help to understand the detailed reaction mechanism. All these special features are important for sensing applications and therefore fluorescent sensors are more attractive as compare to colorimetric sensor.

A schematic representation of the sucrose sensing mechanism behind a microplate sensor modified with a nanoparticle-decorated invertase-onion membrane is shown in Figure 5a. The sucrose-sensing performance of INV-NAuCs-Om was evaluated with fluorometric measurements, which showed a slight blue shift (emission at 348 nm) after sucrose addition at excitation wavelength 320 nm. Furthermore, sucrose was a strong quencher for INV-NAuCs in the UV region. A similar quenching

**Table 1 Effect of gold concentration on nanogold formation and fluorescence intensity**

| HAuCl$_4$ (mM) | Size distribution range (nm) | SPR peak I (nm) | SPR peak II (nm) | Normalized fluorescent intensity (A.U.) ($\lambda_{ex} = 320$ nm, $\lambda_{em} = 348$ nm) |
|---|---|---|---|---|
| 0.25 | 2↔10.8 | 301.3 | 517 | 219291 |
| 0.50 | 13.4↔27.2 | 301.0 | 538 | 330387 |
| 0.75 | 15.8↔34.2 | 301.1 | 543 | 302343 |
| 1.00 | 31.1↔47.2 | 301.3 | 542 | 95327 |
| 1.25 | 50↔65.7 | 301.3 | 543 | 53299 |
| 1.50 | 70.4↔81.2 | 301.5 | 543 | 39397 |
| 2.00 | 91.4↔91.2 | 302.0 | 544 | 37159 |

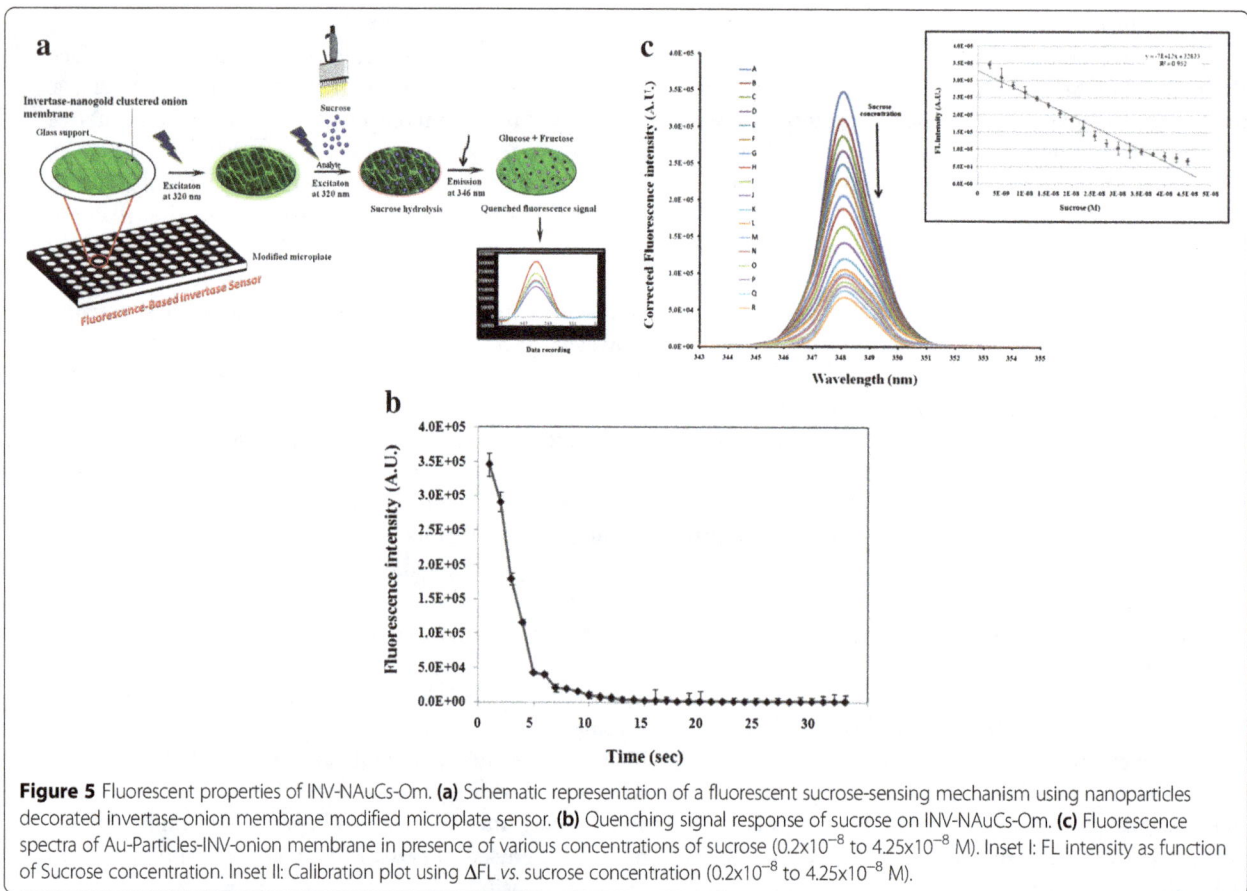

**Figure 5** Fluorescent properties of INV-NAuCs-Om. **(a)** Schematic representation of a fluorescent sucrose-sensing mechanism using nanoparticles decorated invertase-onion membrane modified microplate sensor. **(b)** Quenching signal response of sucrose on INV-NAuCs-Om. **(c)** Fluorescence spectra of Au-Particles-INV-onion membrane in presence of various concentrations of sucrose ($0.2 \times 10^{-8}$ to $4.25 \times 10^{-8}$ M). Inset I: FL intensity as function of Sucrose concentration. Inset II: Calibration plot using ΔFL $vs.$ sucrose concentration ($0.2 \times 10^{-8}$ to $4.25 \times 10^{-8}$ M).

effect was reported by many researchers measuring monosaccharides; therefore, this property is exploited for analyte sensing [38,39]. Consequently, this enzyme-based onion membrane assembly was used as a fluorescence-based optical biosensor. The fluorescence behavior of the biosensor membrane was recorded at room temperature and $\lambda_{ex} = 320$ nm wavelength excitation. Fluorescence intensities of the biosensor membrane steadily decreased with increased sucrose concentrations, with no effect on spectral position and shape. In the present study, glucose, the product of sucrose hydrolysis, not only acted as a quencher, but also as a reducing agent for gold produced in the vicinity of nano gold clusters [39-41]. The same principle was used by Scampicchio et al., where the reaction of glucose (produced by an invertase) with gold salt in alkaline media was used for sucrose sensing [42]. Sensor output was expressed by the change in fluorescence intensity relative to the sucrose concentration (ΔFL/ΔSuc) [39]. The quenching reaction progress was observed for 5 min after sucrose addition for this sensor. However, a typical fast quenching response due to invertase action was observed, within 30 seconds (less than1 min) as shown in Figure 5b. Therefore, the response time for the current

sensor was superior to previously reported absorbance-based sucrose sensors [28,43,44]. An additional advantage to note is that these fluorescent biosensor membranes retained invertase activity for one week when stored at 4°C in acetate buffer.

The linear dynamic range for a sucrose standard obtained was $2.25 \times 10^{-9}$ to $4.25 \times 10^{-8}$ M, and the limit of detection was $2 \times 10^{-9}$ M (Figure 5c). The $R^2$ value was 0.952 which indicates a strong positive relationship of the calibration. Sucrose concentrations $< 2 \times 10^{-9}$ M were not differentiated from the reference spectrum. Therefore, this was designated as a cut-off value and limit of detection. This threshold was attributed to the limited amount of invertase-nanogold conjugates that can react with sucrose in an onion membrane. The presently designed microplate readout sensor is found to be faster with less sample volume compared to other transducer-based biosensors [28,43-45]. Table 2 summarizes the sensor analysis times, dynamic range and sensitivities of the various sucrose biosensors reported previously.

## Spiked samples testing
The feasibility of a quenching biosensor for sucrose detection was evaluated by analysis of green tea samples.

**Table 2 Comparison of the sensor readout time, dynamic range and limit of detection for different types of sucrose biosensors**

| Type of sensor | Readout time | Dynamic linear range | Lower detection limit | Reference |
|---|---|---|---|---|
| Amperometric Sucrose-fructose biosensor | 8 s | $1\times10^{-4}$ to $5\times10^{-3}$ M. | $2\times10^{-6}$ M | [46] |
| Microbial sensor based on *E. coli* strain K-802 | Real time | $5\times10^{-5}$ to $5\times10^{-4}$ M | $5\times10^{-5}$ M | [45] |
| Microbial sensor based on *B. subtilis* strain VKM B-434 | Real time | $5\times10^{-6}$ to $5\times10^{-5}$ M | $5\times10^{-6}$ M | [45] |
| Conductometric tri-enzyme biosensor | 1–2 min | $2\times10^{-6}$ to $5\times10^{-3}$ M | $2\times10^{-6}$ M | [43] |
| Electrochemical tri-enzyme based sensor | 1 min | $4\times10^{-6}$ to $8\times10^{-6}$ M | $4.5\times10^{-6}$ M | [44] |
| Quenching based on INV-NAuCs-Om sensor | 30 s | $2.25\times10^{-9}$ to $4.25\times10^{-8}$ M | $2\times10^{-9}$ M | [Present biosensor] |

Measurements with the fluorescence biosensor were further validated against a standard analytical dinitrosalicylic (DNSA) method. Green tea samples spiked with various sucrose concentrations were prepared for testing, while for blank correction for fluorescence spectra, green tea without sucrose was used. Samples were analyzed using the current biosensor at room temperature (after appropriate dilutions). The comparison of testing results between INV-NAuCs-Om biosensor and DNSA analysis of the sucrose-spiked samples are shown (Table 3). These two methods showed a high compliance, with an acceptable error. Satisfactory recoveries ranging from 94 to 108% were obtained, indicating acceptable accuracy of the proposed detection sensor for sucrose in green tea samples.

## Conclusion

Our present work is apparently the first to use a fluorometric optical onion membrane-based sensor for detection of sucrose. The sensor was based on formation of invertase-induced nanogold clusters and particles within the membrane. Sucrose was hydrolysed by invertase to glucose, which in turn quenched fluorescence. The microplate-based biosensor yielded comparable results with a traditional method for quantifying sucrose in green tea, providing evidence of reliable sensitivity. Therefore, the proposed fluorescent biosensor has potential as a sensitive one-step measurement of sucrose. Furthermore, after some modifications and future investigations, we expect that this technology can be

used to estimate glucose concentrations in various sugar-sweetened beverages and other food products. Likewise, this application can be easily adapted in pharmaceutical research where routine screening of glucose is mandatory.

## Methods
### Materials

Invertase from baker's yeast (*Saccharomyces cerevisiae*; EC 3.2.1.26) was purchased from Fluka (Milwaukee, WI, USA). Albumin from bovine serum, glutaraldehyde, sucrose, hydrogen tetrachloroaurate, trisodium citrate, gold chloride and glucose were purchased from Sigma-Aldrich (St. Louis, MO, USA). Ultra-pure water filtered by Millipore SAS 67120, Molsheim, France, was used for all experiments. All other chemicals were of the highest purity and used without further purification.

### Invertase assay and immobilization

Large yellow onion bulb (*Allium cepa* L.) with significantly high natural organosulfur compounds which are also flavor precursors of onion were used for the present study. Fully mature onions were purchased from a local vegetable market (Taipei, Taiwan). The onions were cut into halves, bulb scales separated and inner epidermis was stripped from the outer fleshy scales [22]. This thin bulb epidermal cell wall was used as support for invertase immobilization and matrix for nano gold synthesis. The onion membranes (diameter, 6.0 mm) were thoroughly washed and then incubated with glutaraldehyde

**Table 3 Spike sample testing with a fabricated INV-NGC-Om sensor**

| Green tea* | Spiked concentration (ng mL$^{-1}$) | DNSA method (ng mL$^{-1}$) | | INV-NAuCs-Om Sensor Output (ng mL$^{-1}$) | | |
|---|---|---|---|---|---|---|
| | | Detected concentration | % Error | Detected concentration | % Error | Recovery (%) |
| Spike A | 25 | 26 | 4.0 | 24 | −4.0 | 96.0 |
| Spike B | 50 | 51 | 2.0 | 47 | −6.0 | 94.0 |
| Spike C | 75 | 75 | 0.0 | 77 | 2.6 | 102.67 |
| Spike D | 100 | 108 | 8.0 | 108 | 8.0 | 108.0 |
| Spike E | 125 | 127 | 1.6 | 119 | −4.8 | 95.20 |
| Spike F | 150 | 155 | 3.3 | 160 | 6.6 | 106.67 |

*Data represented an average of three independent experiments.

(0.01%) in acetate buffer (pH 4.5) at 4°C in the dark, for 1 h. Activated membranes were washed gently three times to remove excess glutaraldehyde and treated with a mixture of invertase (500 μL, 220 U mL$^{-1}$) and bovine serum albumin (1 mL, 1 mg mL$^{-1}$) at 4°C for 12 h under gentle stirring. The onion membranes with immobilized invertase were washed twice and tested for enzyme activity using the DNSA method with sucrose ($2.5 \times 10^{-8}$ M) as substrate [26-28]. Invertase-immobilized onion membranes were then stored at 4°C until use. One unit of invertase activity was defined as the amount of enzyme that hydrolyzed 1 μmole of sucrose in 1 min at 30°C in sodium acetate buffer (20 mM, pH 4.5).

### Invertase-mediated nano gold synthesis

Invertase-immobilized onion membranes were immersed in a mixture of 1.0 mL hydrogen tetrachloroaurate (0.5 mM) and 1.0 mL of assay buffer for 24 h at 55°C in a shaking incubator. The transparent, thin onion membranes changed from colorless to a slight yellowish pink color, indicating nanogold synthesis during incubation. Resulting membranes were stored at 4°C in acetate buffer (pH 5.0).

### UV–vis absorbance spectroscopy

UV-Visible spectra analysis was used to confirm reduction of hydrogen tetrachloroaurate (HAuCl$_4$) and formation of nanogold on the invertase-bound onion membranes. Biosynthesis of invertase-assisted nanogold clusters and nanoparticles on the Oms were monitored periodically for 72 h. The Om samples were scanned from 300 to 600 nm wavelengths using a dual beam UV-Visible spectrophotometer (1 nm resolution). The UV–vis spectra of the immobilized membranes in assay buffer solution was measured and compared to blank onion membranes.

### Fluorescent imaging

The invertase-immobilized onion membranes (1.0 × 1.0 cm$^2$) both with and without nanogold were analyzed under fluorescence imaging using a Leica MZ16F fluorescence stereomicroscope equipped with a DFC 500 camera having GFP filter from Leica Microsystems (Switzerland) Ltd. A magnification range 7.1× to 115×, with a 10× eyepiece, was used to obtain optical images (exposure time of 10.41 s and gain of 80.4%). Images were analyzed with Leica image manager 50, V1.20 software.

### SEM images of the onion membrane

A scanning electron microscope (Model JEOL JSM-6300 F, Tokyo, Japan; 2–5 kV with Auto Fine Coater, JEOL-JFC-1600E Ion Sputtering Device) was used to study modified invertase-onion membranes. During SEM analysis, onion membrane(s) were mounted on stubs and coated with Au/Pd. SEM micrographs of both the invertase immobilized and those that were blank were taken at various magnifications.

### TEM study of onion membrane

The invertase-onion membranes with nanogold clusters were also analyzed by transmission electron microscopy to identify the effects of pH on nanogold synthesis. The INV-NAuCs-Om was cut into circles (~ Φ3 mm) using a razor at room temperature. Membrane pieces were supported on a conventional Φ3 mm Cu mesh with a carbon micro-grid. The TEM observations were performed using a JEOL JEM-3000 F transmission electron microscope (Topcon Co., Ltd., Japan) operated at an accelerating voltage of 300 kV.

### Fluorescence lifetime of INV-NAuCs-Om

Fluorescence lifetime data for light-emitting NAuCs-onion membranes were obtained with an FLS920 combined steady-state lifetime fluorescence spectrometer (Hitachi, Japan). Decay curves were analyzed with a multi-exponential iterative fitting program provided with the instrument. The quantum yield (QY) of INV-NAuCs-Om and INV-Om were determined using l-tryptophan as a criterion (QY = 0.14) at room temperature.

### Sensor fabrication and sucrose measurement

To develop a simple read-out and highly sensitive biosensor system, we used a 96-well fluorescence-compatible microplate as a convenient platform. There are many reports of innovative optical and electrochemical biosensors using a microplate as a reusable component of biosensor [22,29,30]. Black polystyrene FluoroNunc™/LumiNunc™ plates (with minimum back-scattered light and background fluorescence) were used for the measurements. Tecan Infinite® 200 PRO microplate reader with Tecan i-control software was used for microplate analysis. The INV-NAuCs-Om disc (Φ 5 mm) was prepared and adhered on a cover glass disc, without any adhesive, with the hydrophobic side of the onion membrane downward. These modified glass sensor chips were placed at the bottom of each well of the microplate cassette. The INV-NAuCs-Om microplate sensor was calibrated using a standard sucrose solution and fluorescence measurements were recorded. After sensor characterization, sucrose-spiked real samples were also tested. An INV-NAuCs-Om modified microplate was used as a transducer tool to evaluate performance of bioconjugated membranes. For this, 25 μL buffered sucrose solution ($2.5 \times 10^{-8}$ M) was added to the sensing zone and fluorescence was analyzed by exciting the probe at 320 nm. Thereafter, sucrose concentration was increased to $4.25 \times 10^{-8}$ M, and fluorescence intensity at 348 nm of the INV-NAuCs-Om sensor was measured.

Experiments were repeated at least three times, and similar results were obtained. One representative set of data was used for analysis as described.

## Preparation and determination of sucrose in spiked samples

Spike recovery is important to investigate the accuracy of an analytical method. The applicability of the fabricated sensor was evaluated using green tea samples obtained from a local market (Taipei, Taiwan). Tea samples were filtered through a 0.22 μm membrane prior to dilution with assay buffer (acetate buffer, 20 mM, pH 5.0). Spiked tea samples were prepared with 25–150 ng mL$^{-1}$ sucrose concentrations. For testing, 25 μL buffered spiked sample solution was drop-tested on the sensor zone and fluorescence analysis was performed by exciting the INV-NAuCs-Om probe at 320 nm. The quenched fluorescence signal was monitored for all tea samples.

## Additional file

**Additional file 1: Invertase-nanogold based quenching sucrose sensor.**

## Abbreviations

INV: Invertase; Om: Onion membrane; AuNPs: Gold nanoparticles; NAuCs: Nanogold clusters; DNSA: Dinitrosalisilic acid; TEM: Transmission electron microscopy; SEM: Scanning Electron Microscopy; AFM: Atomic force microscopy; FL: Fluorescence.

## Competing interests

The authors declare that they have no competing interests.

## Authors' contributions

DK gathered the research data. DK and RK contributed to the project design, data analysis and data interpretation and manuscript preparation. All authors read and approved the final manuscript.

## Authors' information

Dr. Dipali Bagal-Kestwal is Post-Doctoral research fellow, working in the biosensor research for last 12 years. She completed her PhD in enzyme-based biosensors and is currently working on nanomaterial-based biosensors under the guidance of Prof. Been-Huang Chiang. Dr. Rakesh Mohan Kestwal has also been involved in biosensor research for the last 7 years and now he is working on the pesticide biosensors with Prof. Chiang. Professor Been-Huang Chiang is a distinguished Professor at the Institute of Food Science and Technology, National Taiwan University, Taipei, Taiwan, ROC. He is actively engaged in food science and food safety research. His expertise includes food packaging, separation technology, functional foods and fabrication of different enzyme based sensors for food application.

## Acknowledgment

The authors gratefully acknowledge all financial assistance from the National Science Council, Taipei, Taiwan (NSC-100-2221-E-002-032) and National Taiwan University (Project No. 101R4000). We also are grateful to Dr. Senthil Kumar and Prof. Kai-Wun Yeh for providing facilities and guidance during this research.

## References

1. Kalishwaralal K, Gopalram S, Vaidyanathan R, Deepak V, Pandian SRK, Gurunathan S. Optimization of α-amylase production for the green synthesis of gold nanoparticles. Colloids Surf B Biointerfaces. 2010;77:174–80.
2. Thakkar NK, Mhatre SS, Parikh RY. Biological synthesis of metallic nanoparticles. Nanomed Nanotechnol Biol Med. 2010;6:257–62.
3. Rodríguez-Lorenzo L, Rica R, Álvarez-Puebla RA, Liz-Marzán LM, Stevens MM. Ultrasensitive detection of PSA in serum. Nature Mater. 2012;11:604–7.
4. Willner I, Baron R, Willner B. Growing metal nanoparticles by enzymes. Adv Mater. 2006;18:1109–20.
5. Kumar SA, Abyaneh MK, Gosavi SW, Ahmad A, Khan MI. Sulfite reductase-mediated synthesis of gold nanoparticles capped with phytochelatin. Biotechnol Appl Biochem. 2007;47:191–5.
6. Parida UK, Bindhani BK, Nayak P. Plant extract mediated synthesis of silver and gold nanoparticles and its antibacterial activity against clinically isolated pathogens. World J Nanosci Eng. 2011;1:93–8.
7. MubarakAli D, Thajuddin N, Jeganathan K, Gunasekaran M. Plant extract mediated synthesis of silver and gold nanoparticles and its antibacterial activity against clinically isolated pathogens. Colloids Surf B Biointerfaces. 2011;85:360–5.
8. Mittal AM, Chisti Y, Banerjee UM. Synthesis of metallic nanoparticles using plant extracts. Biotechnol Adv. 2013;31:346–56.
9. Gardea-Torresdey JL, Parsons JG, Gomez E, Peralta-Videa J, Troiani HE, Santiago P. Formation and growth of Au nanoparticles inside live alfalfa plants. Nano Lett. 2002;2:397–401.
10. Shankar SS, Rai A, Ankamwar B, Singh A, Ahmad A, Sastry M. Biological synthesis of triangular gold nanoprisms. Nature Mater. 2004;3:482–8.
11. Chandran PA, Chaudhary M, Pasricha R, Ahmad A, Sastry M. Synthesis of gold nanotriangles and silver nanoparticles using Aloe vera plant extract. Biotechnol Prog. 2006;22:577–83.
12. Huang J, Li Q, Sun D, Lu Y, Su Y, Yang X, et al. Biosynthesis of silver and gold nanoparticles by novel sundried cinnamomum camphora leaf. Nanotechnol. 2007;18:105104–11.
13. Song JY, Jang HK, Kim BS. Biological synthesis of gold nanoparticles using Magnolia kobusand Diopyros kaki leaf extracts. Process Biochem. 2009;44:1133–8.
14. Singh AK, Talat M, Singh DP, Srivastava ON. Biosynthesis of gold and silver nanoparticles by natural precursor clove and their functionalization with amine group. J Nanopart Res. 2010;12:1667–75.
15. Badole MR, Dighe VV. Synthesis of gold nano particles using Putranjiva roxburghii wall leaves extract. Int J Drug Discovery Herbal Res. 2012;44:275–8.
16. Venkatachalam M, Govindaraju K, Mohamed Sadiq A, Tamilselvan S, Ganesh Kumar V, Singaravelu G. Functionalization of gold nanoparticles as antidiabetic nanomaterial. Spectrochim Acta Mol. 2013;116:331–8.
17. Plazzo G, Facchini L, Mallardi A. Colorimetric detection of sugars based on gold nanoparticle formation. Sensors Actu B. 2012;161:366–71.
18. Simonian AL, Good TA, Wang SS, Wild JR. Nanoparticle-based optical biosensors for the direct detection of organophosphate chemical warfare agents and pesticides. Anal Chim Acta. 2005;534:69–77.
19. Xiang Y, Lu Y. Using personal glucose meters and functional DNA sensors to quantify a variety of analytical targets. Nature Chem. 2011;3:697–703.
20. Scampicchio M, Fuenmayor CA, Mannino S. Sugar determination via the homogeneous reduction of Au salts: a novel optical measurement. Talanta. 2009;79:211–5.
21. Kumar J, D'Souza SF. Inner epidermis of onion bulb scale: as natural support for immobilization of glucose oxidase and its application in dissolved oxygen based biosensor. Biosens Bioelectron. 2009;24:1792–5.
22. Kumar J, D'Souza SF. Immobilization of microbial cells on inner epidermis of onion bulb scale for biosensor application. Biosens Bioelectron. 2011;26:4399–404.
23. Kumar VR, Pundir CS. Covalent immobilization of lipase onto onion membrane affixed on plastic surface: Kinetic properties and application in milk fat hydrolysis. Indian J Biotechnol. 2007;6:479–84.
24. Wang F, Yao J, Russel M, Chen H, Chen K, Zhou Y, et al. Development and analytical application of a glucose biosensor based on glucose oxidase/O-(2-hydroxyl)propyl-3-trimethylammonium chitosan chloride nanoparticle-immobilized onion inner epidermis. Biosens Bioelectron. 2010;25:2238–43.
25. Jia W, Liu W, Zhang Y, Cui M, Shuang S, Dong C, et al. Determination of glucose in human serum based on an onion primary cuticula biosensor immobilized glucose oxidase. Anal Methods. 2012;4:1432–7.
26. Miller GL. Use of dinitrosalicylic acid reagent for determination of reducing sugar. Anal Chem. 1959;31:426–8.
27. Bagal DS, Vijayan A, Aiyer RC, Karekar RN, Karve MS. Fabrication of sucrose biosensor based on single mode planar optical waveguide using co-immobilized plant invertase and GOD. Biosens Bioelectron. 2007;22:3072–9.

28. Bagal-Kestwal D, Kestwal R, Chiang BH, Karve MS. Development of dip-strip sucrose sensors: application of plant invertase immobilized in chitosan-guar gum, gelatin and poly-acrylamide films. Sensors Actuat B. 2011;160:1026–33.

29. Kalyuzhny AE. Protein blotting and detection. Methods Mol Biol. 2009;536:355–65.

30. Liu Y, Jia S, Guo LH. Synthesis of chitosan-Prussian blue-graphene composite nanosheets for electrochemical detection of glucose based on pseudobienzyme channeling. Senso Actuators. 2012;161:334–40.

31. Saxena A, Tripathi RM, Singh RP. Synthesis of silver nanoparticles using onion (Allium cepa) extract and their antibacterial activity. Dig J Nanomater Bios. 2010;5:427–32.

32. Pandey S, Oza G, Gupta A, Shah R, Sharon M. Green synthesis of highly stable gold nanoparticles using *momordica charantia* as nano fabricator. Euro J Exp Bio. 2012;2:475–83.

33. Sanghi R, Verma P, Puri S. Enzymatic formation of gold nanoparticles using *Phanerochaete Chrysosporium*. Adv Chem Engg Sci. 2011;1:154–62.

34. Hou Y, Hansen TB, Staby A, Cramer SM. Effect of urea induced protein conformational changes on ion exchange chromatographic behaviour. J Chromatograph. 2010;1217:7393–400.

35. Liu L, Zheng HZ, Zhang ZJ, Huang YM, Chen SM, Hu YF. Photoluminescence from water-soluble BSA-protected gold nanoparticles. Spectrochim Acta A. 2008;69:701–5.

36. Chakraborty S, Joshi P, Shanker V, Ansari ZA, Singh SP, Chakrabarti P. Contrasting effect of gold nanoparticles and nanorods with different surface modifications on the structure and activity of bovine serum albumin. Langmuir. 2011;27:7722–31.

37. Gagner JE, Lopeze MD, Dordickb JS, Siegel RW. Effect of gold nanoparticle morphology on adsorbed protein structure and function. Biomaterials. 2011;32:7241–52.

38. Panigrahi S, Kundu S, Ghosh SK, Nath S, Pal T. Sugar assisted evolution of mono- and bimetallic nanoparticles. Colloids Surf A Physicochem Eng Aspects. 2005;264:133–8.

39. Anker JN, Hall WP, Lyaders O, Shah NC, Zhao J, Van Duyne RP. Biosensing with plasmonic nanosensors. Nat Mater. 2008;7:442–53.

40. Huang X, Neretina S, El-Sayed MA. Gold nanorods: from synthesis and properties to biological and biomedical applications. Adv Mater. 2009;21:4880–910.

41. Hussain AMP, Sarangi SN, Kesarwani JA, Sahu SN. Au-nanocluster emission based glucose sensing. Biosens Bioelectron. 2011;29:60–5.

42. Scampicchio M, Arecchi A, Mannino S. Optical nanoprobes based on gold nanoparticles for sugar sensing. Nanotechnol. 2009;20:135501–10.

43. Soldatkin OO, Peshkova VM, Dzyadevych SV, Soldatkin AP, Jaffrezic-Renault N, El'skaya AV. Novel sucrose three-enzyme conductometric biosensor. Mater Sci Eng C. 2008;28:959–64.

44. Haghighi B, Varma S, Alizadeh SFM, Yigzaw Y, Gorton L. Prussian blue modified glassy carbon electrodes-study on operational stability and its application as a sucrose biosensor. Talanta. 2004;64:3–12.

45. Kitova A, Reshetilov A, Ponamoreva O, Leathers T. Microbial biosensors for selective detection of disaccharides. Internet J Microbiol. 2009;8:2.

46. Antiochia R, Tasca F, Mannina L. Osmium-polymer modified carbon nanotube paste electrode for detection of sucrose and fructose. Mater Sci Appl. 2013;4:15–22.

# Multifunctional polymeric nanoparticles doubly loaded with SPION and ceftiofur retain their physical and biological properties

Paula Solar[1,2], Guillermo González[3,4], Cristian Vilos[1,3], Natalia Herrera[1], Natalia Juica[1], Mabel Moreno[3], Felipe Simon[5,6,7] and Luis Velásquez[1,3]*

## Abstract

**Background:** Advances in nanostructure materials are leading to novel strategies for drug delivery and targeting, contrast media for magnetic resonance imaging (MRI), agents for hyperthermia and nanocarriers. Superparamagnetic iron oxide nanoparticles (SPIONs) are useful for all of these applications, and in drug-release systems, SPIONs allow for the localization, direction and concentration of drugs, providing a broad range of therapeutic applications. In this work, we developed and characterized polymeric nanoparticles based on poly (3-hydroxybutyric acid-co-hydroxyvaleric acid) (PHBV) functionalized with SPIONs and/or the antibiotic ceftiofur. These nanoparticles can be used in multiple biomedical applications, and the hybrid SPION–ceftiofur nanoparticles (PHBV/SPION/CEF) can serve as a multifunctional platform for the diagnosis and treatment of cancer and its associated bacterial infections.

**Results:** Morphological examination using transmission electron microscopy (TEM) showed nanoparticles with a spherical shape and a core-shell structure. The particle size was evaluated using dynamic light scattering (DLS), which revealed a diameter of 243.0 ± 17 nm. The efficiency of encapsulation (45.5 ± 0.6% w/v) of these polymeric nanoparticles was high, and their components were evaluated using spectroscopy. UV–VIS, FTIR and DSC showed that all of the nanoparticles contained the desired components, and these compounds interacted to form a nanocomposite. Using the agar diffusion method and live/dead bacterial viability assays, we demonstrated that these nanoparticles have antimicrobial properties against *Escherichia coli*, and they retain their magnetic properties as measured using a vibrating sample magnetometer (VSM). Cytotoxicity was assessed in HepG2 cells using live/dead viability assays and MTS, and these assays showed low cytotoxicity with $IC_{50} > 10$ mg/mL nanoparticles.

**Conclusions:** Our results indicate that hybrid and multifunctional PHBV/SPION/CEF nanoparticles are suitable as a superparamagnetic drug delivery system that can guide, concentrate and site–specifically release drugs with antibacterial activity.

**Keywords:** PHBV, SPION, Ceftiofur, Polymeric nanoparticles, Drug delivery, Superparamagnetic nanoparticles

## Background

Superparamagnetic nanoparticles are used in multiple biomedical applications, such as for contrast medium in magnetic resonance imaging (MRI)-based diagnosis, in hyperthermia applications and for tissue-specific drug delivery using an external magnetic field in cancer treatment [1]. Hyperthermia is a treatment were the target tissue is exposed to temperatures slightly higher than physiological to damage and kill cells.

There are many different superparamagnetic nanoparticles; however, superparamagnetic iron oxide nanoparticles (SPIONs) are currently the best option for biomedicine because they are biocompatible with the body. Actually, for diagnostics, gadolinium salts (Gd-DTPA) are the gold standard contrast medium for MRI, but their resolution and retention time in vivo remains low (2 mm for MRI) [2-4]. However, Gd-DTPA have been linked to

* Correspondence: luis.velasquez@unab.cl
[1]Universidad Andres Bello, Facultad de Medicina, Center for Integrative Medicine and Innovative Science, Echaurren 183, Santiago, Chile
[3]Center for the Development of Nanoscience and Nanotechnology, CEDENNA, 9170124, Av. Ecuador 3493, Estación Central, Santiago, Chile
Full list of author information is available at the end of the article

anaphylactic reactions and toxicity at the hepatic and renal levels. Therefore, it is important to find alternative contrast media that have low toxicity and better resolution for MRI. In this field, SPIONs function as a more sensitive contrast medium; the obtained images have better resolution, and the particles have a better retention time *in vivo* and are more biocompatible than Gd-DTPA [4]. The use of a SPION-like contrast medium for MRI is not the only potential application of SPIONs. Theoretically, SPIONs can be used for multiple actions, thus generating multifunctional nanoparticles. SPIONs can be used for hyperthermia applications and drug delivery [5], allowing for the localized and controlled release of drugs.

The application of nanoparticles *in vivo* depends on their physicochemical properties such as their hydrophobicity, net surface charge and size. These properties have a direct impact on their biocompatibility, biostability and toxicity. The use of SPIONs is limited because SPIONs are not stable in physiological environments due to their very reactive surface. SPIONs liberate $Fe^{2+}$ ions, produce oxidative stress via the release of reactive oxygen species (ROS), and alter ion transport [6]. Lastly, SPIONs are not stable in aqueous media because they tend to agglomerate and precipitate [4]. To overcome these limitations, we have proposed coating SPIONs with biocompatible materials to preserve all of their potential uses while preventing their toxicity [7].

The development of nanotechnologies that utilize biocompatible and biodegradable polymers has provided new tools for diagnostic and therapeutic strategies in biomedicine [8], wherein one of the most important areas is drug delivery. Current novel therapeutic strategies include the delivery of cell tissue-specific drugs, the use of agents that allow for the imaging of release sites [9], delivery systems that can cross epithelial and endothelial barriers [10], the intracellular delivery of macromolecules, improvements in drugs with poor water solubility, the co-release of therapeutic agents, and an improvement in effective therapies [11,12]. For these applications, a biocompatible and biodegradable polymer is the best candidate for a SPION coating because this type of coating may decrease their natural reactivity and maintain their physical properties. Poly (3-hydroxybutyrate-co-3-hydroxyvalerate) (PHBV), a polymer that has recently been used for drug delivery and tissue engineering [13], is cost effective and has physicochemical properties similar to those of the most widely used polymers such as poly(L-lactide) [14], poly(D,L-lactide-co-glycolide) [15-18], poly-sebacic anhydride [19] and poly-ε-caprolactone [20-22].

One aim in drug delivery is to transport antibiotics. In this area, we determined that PHBV can be used to encapsulate and deliver the antibiotic ceftiofur in microparticles [23,24] and PHBV can interact with β-lactams to form a new microcomposite with specific physicochemical properties. We analyzed the same antibiotic that was previously assayed (ceftiofur), and we incorporated SPIONs into PHBV nanoparticles to generate a new nanocomposite. Furthermore, we analyzed the molecular structure and the cytotoxicity of PHBV nanoparticles loaded with SPIONs and ceftiofur (PHBV/CEF/SPION); for controls, we used each component and nanoparticles with different combinations of their components: PHBV nanoparticles with only ceftiofur (PHBV/CEF) and PHBV nanoparticles with only SPIONs (PHBV/SPION).

Due to the potential uses of SPIONs and the advantages of polymeric nanoparticles, PHBV/CEF/SPION nanoparticles could (i) be used as a nanocarrier for tissue-specific drug delivery, (ii) allow for MRI-based diagnostics and for therapy using their antibacterial activity or hyperthermia using only one injection, and (iii) be used as a multifunctional treatment for hyperthermia and for the release of antibiotics at an infection site.

## Results and discussion

TEM images (Figure 1) show PHBV/CEF/SPION nanoparticles with a core-shell structure, a spherical shape, a smooth surface and a moderately uniform size distribution. This image shows nanoparticles with dense black spots inside. The mean size (diameter, nm) of the formulated PHBV/CEF/SPION measured using DLS was $243.0 \pm 17$.

To determine the amount of ceftiofur that was able to incorporate into the PHBV nanoparticles, we analyzed the individual components of these lyophilized nanoparticles using UV-visible (UV–VIS) spectroscopy (Figure 2). The UV–VIS spectra of nanoparticles with different combinations of components are shown: PHBV nanoparticles with ceftiofur, PHBV nanoparticles with SPION and PHBV nanoparticles with SPION and ceftiofur. For controls, we used empty PHBV nanoparticles and free ceftiofur.

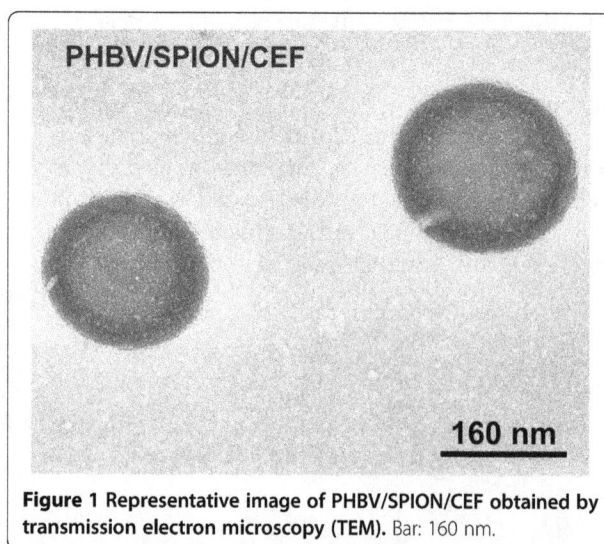

**Figure 1 Representative image of PHBV/SPION/CEF obtained by transmission electron microscopy (TEM).** Bar: 160 nm.

Multifunctional polymeric nanoparticles doubly loaded with SPION and ceftiofur retain their...

191

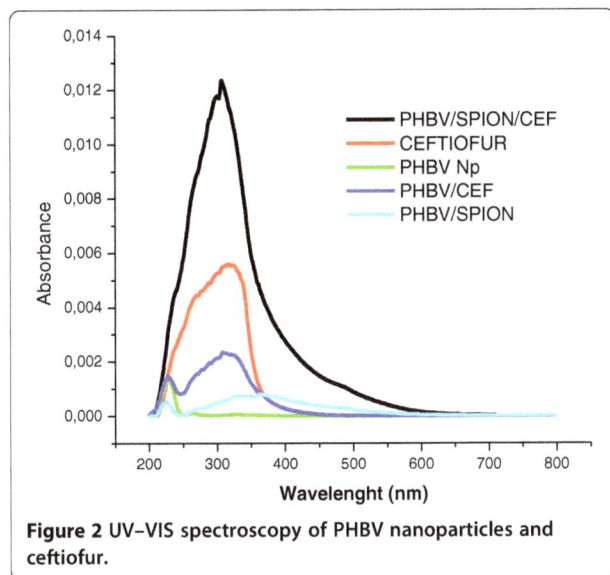

**Figure 2** UV–VIS spectroscopy of PHBV nanoparticles and ceftiofur.

Analysis of the ceftiofur spectrum (Figure 2, red line) shows an intense broad band centered at approximately 320 nm. The deconvolution of the signal at 320 nm showed that it mainly corresponds to the superposition of four peaks, namely, two low-intensity absorption peaks centered at 235 and 294 nm and two more intense absorption peaks centered at 265 and 325 nm. The lower energy and higher intensity of the latter absorbances correspond to π-π* transitions for a highly delocalized system. Because of the molecular structure of ceftiofur, this absorption band could be assigned to the polyenone arising from the alkyloximino C = N- chromophore modified by the thiazolyl moiety. The peak observed at 294 nm with lower intensity could correspond to the C = O chromophore from the amide group, which is also near the thiazolyl group. The high intensity band at 265 nm could also be assigned to π-π* transitions centered in the furan-carboxylic-thioester, whereas the lower intensity peak at higher energy (or lower wavelength) most likely arises from the C = O lactam moiety. Absorption by the carboxylic acid-dihydrothiazine chromophore is expected to occur at higher energies, and thus it was not observed in the spectrum.

PHBV nanoparticles could potentially act as a nanocarrier that incorporates ceftiofur; thus, we evaluated the encapsulation efficiency (EE%) of ceftiofur. The EE% of lyophilized nanoparticles, measured using Ultra performance liquid chromatography (UPLC) and UV–VIS absorbance at 302 nm, was compared with a previously set calibration curve. The PHBV/CEF nanoparticles contain 11.41 wt% ceftiofur, and the PHBV/CEF/SPION nanoparticles contain 11.43 wt% ceftiofur. As the SPIONs produce high levels of noise in the UV–VIS spectrum, we corroborated this result using UPLC with three

independent samples. The %EE of ceftiofur in PHBV nanoparticles was 45.5 ± 0.6.

To assess the composition and molecular architecture of the new PHBV/CEF/SPION nanoparticles and their precursors, FT-IR spectra were analyzed (Figure 3A-D). It is well-known that the ceftiofur molecule contains different functional groups that can be identified in an infrared spectrum; therefore, the most representative signals of the drug could be assigned.

From the IR spectrum, it is possible to assign signals corresponding to the carbonyl, thioester, amide and amine groups of ceftiofur at the bond-stretching frequencies of 1771, 1709, 1661 and 1661 $cm^{-1}$, respectively. The presence of these signals helps to corroborate the formation of the drug-polymer nanoconjugate.

The spectra of the precursors of crystalline PHBV and SPION correspond to those expected for both (Additional file 1). However, some significant changes are observed in the spectrum of PHBV when the polymer is prepared in the form of nanoparticles. The most prominent bands in the spectrum of the polymer, one band centered at approximately 1530 $cm^{-1}$ in the resonance region of N-H and O-H oscillators and another band at 1637 $cm^{-1}$ that is assignable to the carbonyl ester stretching strongly associated with hydrogen bonding, are severely decreased. However, the normal absorption expected for the ester carbonyl group at 1737 $cm^{-1}$ became clearly detectable in the NPs. Both features clearly indicate that the confinement of the polymer in such a small volume appears to produce a drastic loss of water.

The PHBV/CEF composite has a strong band centered at 1732 $cm^{-1}$, possibly due to the superposition of the absorption of the ester with the carboxylic acid group of ceftiofur. Moreover, the band corresponding to N-H absorption in ceftiofur (Additional file 1) is broadened and shifted to a lower energy in the presence of the polymer; thus, this band shows a relative intensity as large as or even larger than that of the PHBV crystals. Meanwhile, the width of the band region corresponding to C = O stretching vibrations remains practically unaltered. Furthermore, the addition of SPIONs induces a decrease in the intensity of the spectrum and a well-known red shift of the bands of the C = O oscillators, whereas the band of the N-H absorption remains practically unaltered. In the PHBV/SPION/CEF nanocomposite (Figure 3C), the spectrum in these regions is better resolved in general, although the intensity is lower. Nevertheless, there is still much overlapping of vibrational bands. However, the particular frequency values of the absorption band components may at least be partially attained by analyzing the second derivative of the spectrum. Thus, an analysis of the spectral changes of ceftiofur that are associated with the formation of the nanocomposite is discussed below.

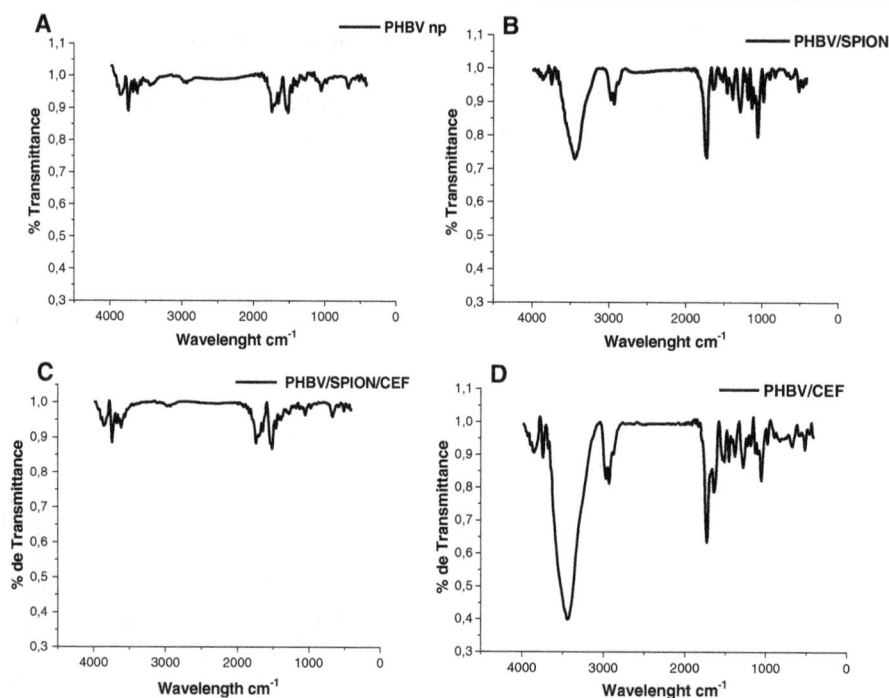

**Figure 3** FTIR spectroscopy of PHBV nanoparticles. **A)** empty PHBV nanoparticles, **B)** PHBV/SPION nanoparticles, **C)** PHBV/SPION/CEF nanoparticles and **D)** PHBV/CEF nanoparticles. The controls, FTIR of ceftiofur, SPION and PHBV crystal are in Additional file 1.

To investigate possible interactions between ceftiofur, the polymer and the metal oxide, this analysis of vibrations has been limited to the functional groups that are potentially more reactive: electrophilic or nucleophilic centers that can interact with the polymer and the metal oxide. From such a perspective, the behavior of electron donor centers (principally the carbonyl groups), as well as that of acceptor centers (e.g., the hydrogen atoms in the N-H and -O-H groups), is particularly interesting. In the polymer, the donor functionality of the carbonyl groups in the polyester is certainly the most important. However, the activity of the -COOH groups at the end of the polymer chains should also be considered. The SPION surface itself is rich in acceptor centers, coordinately unsaturated iron atoms and O-H groups, which are normally found in stabilized oxide surfaces.

In the PHBV/CEF (Figure 3D) and PHBV/SPION/CEF composites (Figure 3C), the amine and amide N-H protons of ceftiofur strongly interact with the polymer. Thus, the bands centered at 3522 and 3393 cm$^{-1}$ in ceftiofur, which we have assigned to asymmetric and symmetric N-H stretching, are shifted to lower energy in the composite: 3421 and 3250 cm$^{-1}$, respectively. The stretching vibration of the secondary amide N-H is found at a relatively low wavenumber, 3289 cm$^{-1}$, in ceftiofur, indicating an association in the pristine compound; thus, the red shift observed in the composites is rather moderate. The addition of polymer affects the acidic centers of ceftiofur in a

manner that does not change after the addition of SPIONs. As mentioned above, the polymer also contains some acidic centers; thus, it is expected that the addition of PHBV should also affect the carbonyl oscillators in ceftiofur. Such an effect is clearly observed in the stretching vibration of the C = O amide bond. Indeed, a red shift of approximately 14 cm$^{-1}$ is detected in both composites.

To clarify the molecular structure of the PHBV nanoparticles, we analyzed these nanoparticles using DSC (Figure 4). The thermograms show a change in the crystalline behavior of the nanostructured PHBV. This fact should be attributed to the loss of degrees of freedom of the polymer due to its nanostructure; this loss could be favored by an increase in the amorphous fraction in the PHBV nanoparticles. Additionally, the PHBV-CEF nanoconjugate thermogram shows behavior that differs from that of the PHBV NP precursor because ceftiofur is incorporated inside of the cavities generated by the polymeric chains, preventing further ordering of the chains. Thus, the disorder in the polymeric chains and the amorphous fraction increases relative to the empty PHBV nanoparticles. Assays of PHBV/CEF showed that it is more amorphous than PHBV NP because ceftiofur inserted between the polymeric chains via hydrogen bridges and because ceftiofur broke some of the polymeric chains that form aldehyde groups in the valerate group of PHBV. The conformation of PHBV/SPION is similar to that of PHBV/CEF. However, PHBV/CEF/SPION has a

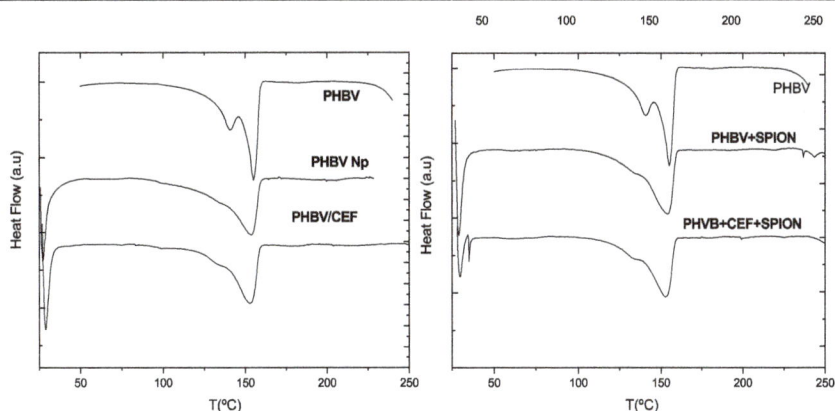

**Figure 4 DSC assay for PHBV nanoparticles.** The melting temperature for each sample is shown at the bottom.

| Sample | Tg | Tm1 | Tm2 |
|---|---|---|---|
| PHBV crystal | - | 140 °C | 155 °C |
| PHBV NP | 90,66 °C | 133,67 °C | 152,67 °C |
| PHBV/CEF | 86,67 °C | 132,67 °C | 151,67 °C |
| PHBV/SPION | 64,83 °C | 133 °C | 152,67 °C |
| PHBV /SPION/ CEF | 35 °C | 132,67 °C | 152 °C |

completely different conformation (Figures 4 and 5). The crystalline fraction of PHBV/CEF/SPION is dramatically diminished; it is possible that ceftiofur and SPION interleave between polymeric chains, preventing any interaction, and this fact may explain the huge difference at 3500 cm$^{-1}$ between Figure 3B/D and 3C. All of these results suggest a possible molecular structure for the PHBV/SPION/CEF nanoparticles. The interactions between ceftiofur and the polymer are shown in Figure 5. We have demonstrated that SPIONs and ceftiofur are

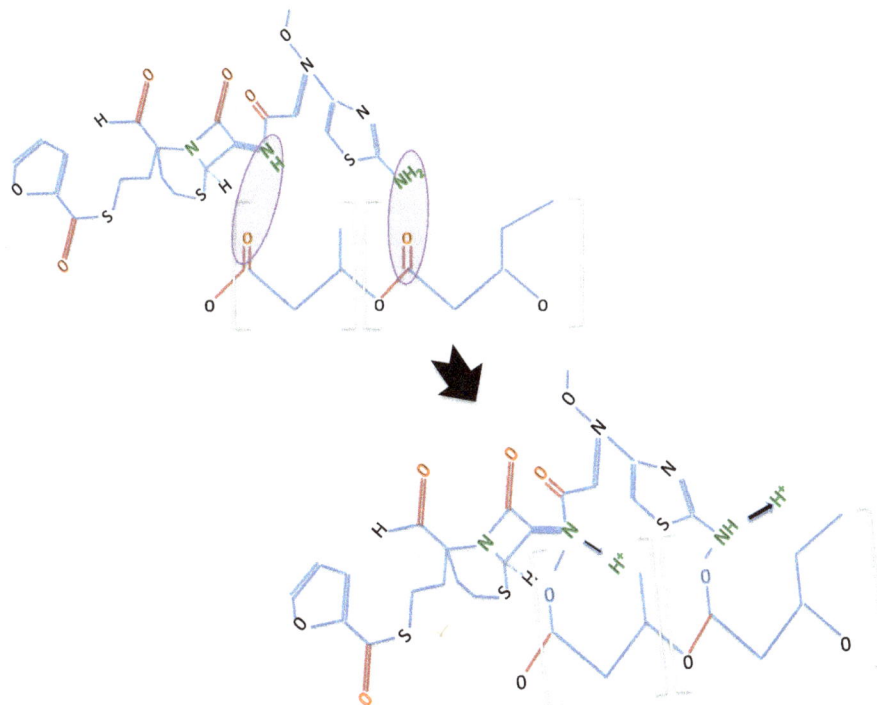

**Figure 5** Possible interactions between ceftiofur and the polymer in the nanoparticles according to data obtained through FTIR, UV–VIS and DSC assays.

incorporated into the nanoparticles and that both interact with the polymeric chains; thus, we need to further demonstrate that the SPIONs and ceftiofur do not lose their physicochemical and biological properties. Magnetic characterization using VSM measurements demonstrated the superparamagnetic behavior of the PHBV/SPION and PHBV/CEF/SPION nanoparticles (Figure 6). This fact proves that PHBV/SPION and PHBV/CEF/SPION nanoparticles have biomedical applications in MRI, hyperthermia and drug delivery to specific tissues.

The antimicrobial activity of PHBV/CEF/SPION nanoparticles against *Escherichia coli* was measured using the agar diffusion method. These nanoparticles showed positive antibacterial activity. The inhibition halo of PHBV/CEF/SPION nanoparticles was 29 mm and that of free ceftiofur was 36 mm. PHBV nano particles without CEF did not show an inhibition halo (Figure 7). Live/dead bacterial viability assays further showed that the viability of bacteria decreased in a dose-dependent manner in samples treated with PHBV/CEF/SPION (Figure 8A-E). The positive control (30 µg/mL ceftiofur, Figure 8F) showed results similar to those of 20 µg/mL PHBV/CEF/SPION (Figure 8E).

We lastly assayed the toxicological effects of 0.1 µg/mL to 10,000 µg/mL of polymeric nanoparticles in a HepG2 cell line using fluorescence microscopy (Figures 9 and 10). We used the HepG2 cell line because PHBV/CEF/SPION have potential applications in biomedicine by intravenous administration. Thus it is very important determine the toxicology in hepatic cells. Figure 9 shows a statistical analysis of the number of live and dead cells per field from a double-blind analysis of 1200 images. We show an example of these images in Figure 10 (live cells in green and dead cells in red). The background fluorescence intensity for the 1,000 µg/mL and 10,000 µg/mL samples was high because at high concentrations, the

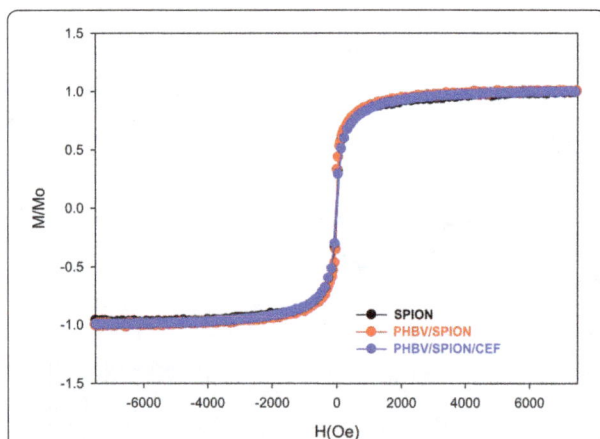

**Figure 7** Determination of the antimicrobial activity of PHBV nanoparticles, ceftiofur and SPIONs against *Escherichia coli* (ATCC 25922) measured using the agar diffusion method at 24 hours of incubation.

nanoparticles tended to deposit on the sample, hindering the washing. For these samples, we used fluorescence intensity plots to quantify the number of live and dead cells per field (bottom, Figure 10). Treatment with PHBV, PHBV/CEF, PHBV/SPION or PHBV/CEF/SPION nanoparticles at 0.1 µl/mL to 10,000 µg/mL did not cause significant cell death in the HepG2 cells (ANOVA p-value >0.05) (Figure 9). A more detailed analysis shows that the PHBV NP, PHBV/SPION and PHBV/CEF/SPION nanoparticles have $IC_{50} > 10,000$ µg/mL and that the PHBV/CEF nanoparticles have $IC_{50} = 10,000$ µg/ml. Statistical analysis shows that only PHBV/CEF caused significant differences in cell viability, but at doses much higher than physiological doses (10,000 µg/ml) (ANOVA and T-test < 0,05). The toxicity of PHBV/CEF is primarily caused by the high concentration of ceftiofur in the external shell of the nanoparticles.

Cell viability was measured using an MTS assay (Figure 11). The MTS assay showed that the $IC_{50}$ of the empty PHBV, PHBV/SPION and PHBV/CEF/SPION nanoparticles was higher than 10,000 µg/mL and that the $IC_{50}$ of PHBV/CEF nanoparticles was 10,000 µg/mL. Although the viability of the HepG2 cells decreased after treatment with PHBV/CEF nanoparticles at 10,000 µg/ml, the number of dead cells did not increase. This finding is very important because it indicates that the % viability decreased but not enough to kill the cells.

In this work, we report on the use of SPIONs in a formulation of PHBV nanoparticles loaded with ceftiofur and on the use of SPIONs as a carrier with a superparamagnetic

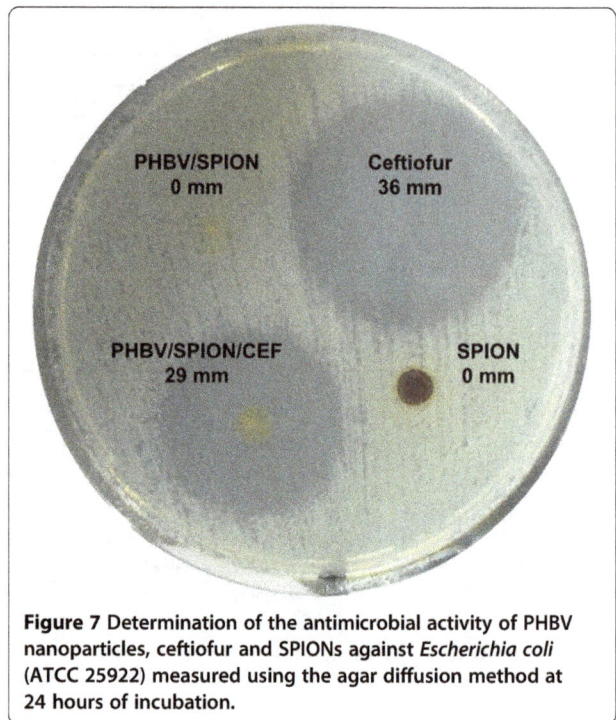

**Figure 6** Hysteresis loop of PHBV/SPION/CEF nanoparticles measured using a vibrating sample magnetometer (VSM). All samples showed superparamagnetic characteristics.

Multifunctional polymeric nanoparticles doubly loaded with SPION and ceftiofur retain their...

195

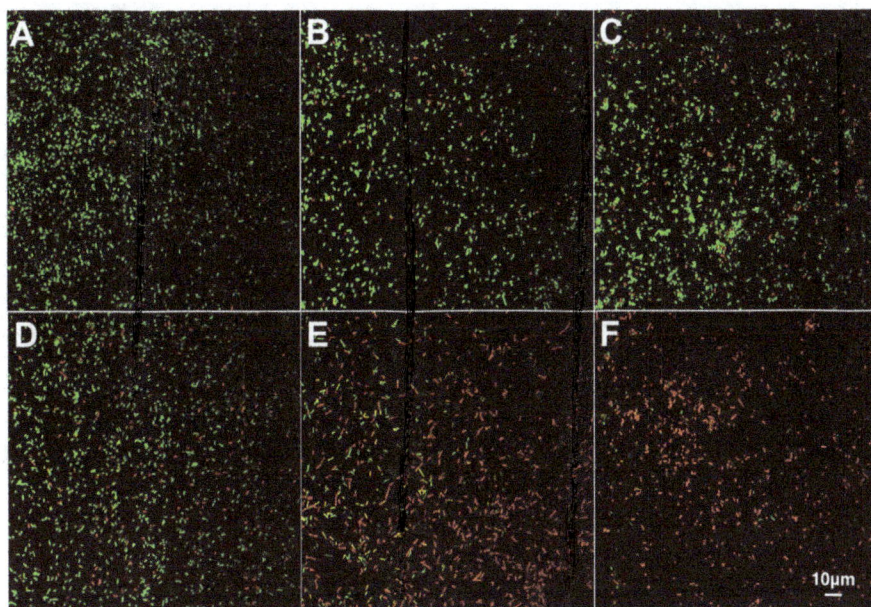

(A) control; (B) 0.1 ug/mL; (C) 1 ug/mL; (D) 10 ug/mL; (E) 20 ug/mL; (F) ceftiofur 30 ug

**Figure 8** Determination of the antibacterial activity of PHBV/SPION/CEF against *Escherichia coli* (ATCC 25922) at 24 hours of incubation using LIVE/DEAD viability assays. **A)** control (without treatment), **B-E)** treatment with 0.1, 1, 10 and 20 μg/ml of PHBV/SPION/CEF nanoparticles, **F)** treatment with free ceftiofur 30 μg.

response. The PHBV/CEF/SPION nanoparticles exhibited a spherical shape with a core-shell structure and a smooth surface. A TEM image (Figure 1) shows nanoparticles with dense black spots inside; these spots show that the SPIONs were well incorporated into the PHBV nanoparticles. The PHBV/SPION/CEF size (243.0 ± 17 nm) is optimal to have low cytotoxicity and high retention in vivo [6]. The results also show that interactions between PHBV, SPIONs and ceftiofur can change the conformation of the polymeric chains and their properties to favor the entrapment of the drug molecule. In addition, the structure of the polymer and its functionality suggest that the release process of the

drug should be controlled by hydrolysis of the polymeric chains. The changes in the spectrum of ceftiofur mentioned above can be explained by considering a model in which the PHBV/CEF nanocomposite contains a SPION core. Ceftiofur is anchored to this core by a Lewis donor acceptor interaction through the amide and carboxylic C = O moieties, whereas its amine groups simultaneously interact with the polymer, forming the external shell of the nanoparticles.

Given the nature of the interactions in this nanocomposite, the system should be sensitive to changes in the pH of medium, a feature that is interesting from the perspective of the encapsulation and release of the drug.

FT-IR analysis of the products described in this work (Figure 3A-D) demonstrated the efficiency of the method used to prepare the PHBV nanoparticles and also contributed to a better understanding of the phenomena involved in this process. Indeed, the results described above show that an essential variable in this procedure is the creation of conditions that produce polymer self-aggregation. The dependence on these conditions, along with other factors, forces the polymers to a lower energy by saturating, at least partially, the ester donor moieties; this saturation occurs because the ends of the chains displace water molecules from these sites, and thus, the volume can decrease to the desired nanometer scale. Such a condition appears to fail in the presence of ceftiofur, where hydration also appears to be necessary to stabilize the excess of donor

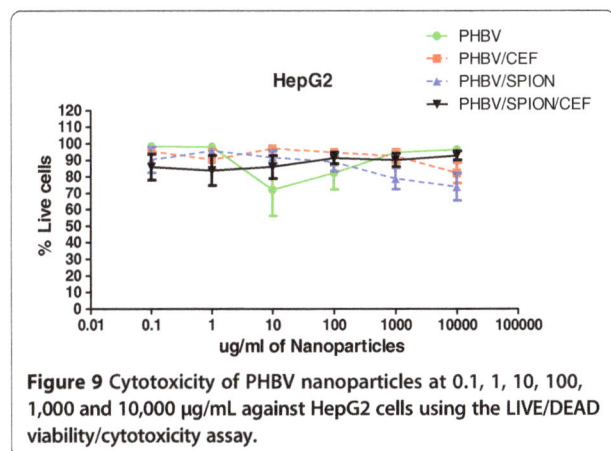

**Figure 9** Cytotoxicity of PHBV nanoparticles at 0.1, 1, 10, 100, 1,000 and 10,000 μg/mL against HepG2 cells using the LIVE/DEAD viability/cytotoxicity assay.

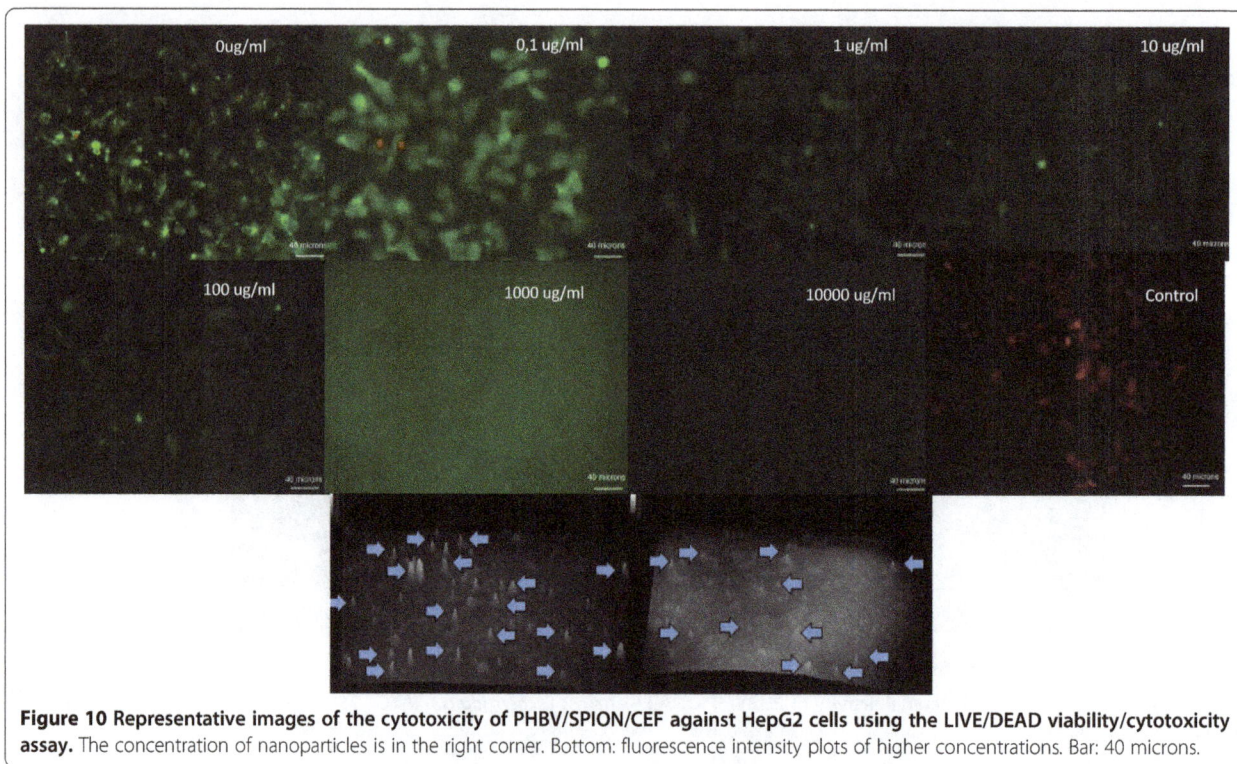

**Figure 10** Representative images of the cytotoxicity of PHBV/SPION/CEF against HepG2 cells using the LIVE/DEAD viability/cytotoxicity assay. The concentration of nanoparticles is in the right corner. Bottom: fluorescence intensity plots of higher concentrations. Bar: 40 microns.

centers in the composite. The shape of the band for the PHBV/CEF nanoparticles (Figure 3D) is different and a bit more complex than that previously described for isolated ceftiofur (Additional file 1). The deconvolution of this spectrum showed a sharp band at 227 nm, which is characteristic of PHBV nanoparticles (Figure 3A), and an absorbance at approximately 337 nm that apparently corresponds to a wavelength shift in the drug-polymer ceftiofur interactions. The band assigned to the imide chromophore undergoes a relatively strong blue shift, most likely due to an inductive effect between the $NH_2$ group of ceftiofur and the ester carbonyls groups of the polymer. These observations are in good agreement with the infrared spectra described below. Indeed, the observed shifts of the electronic absorption bands indicate that the polymer provides the drug with a microenvironment that differs from the pristine bulk state. In this case, these shifts are caused not only by dielectric effects but also by rather specific interactions between the donor and acceptor chromophore centers of ceftiofur and the respective groups. Along with hydration water, the electrophilic and nucleophilic groups of ceftiofur, which are available within the polymer backbone, could be involved. Due to the numerous donor centers

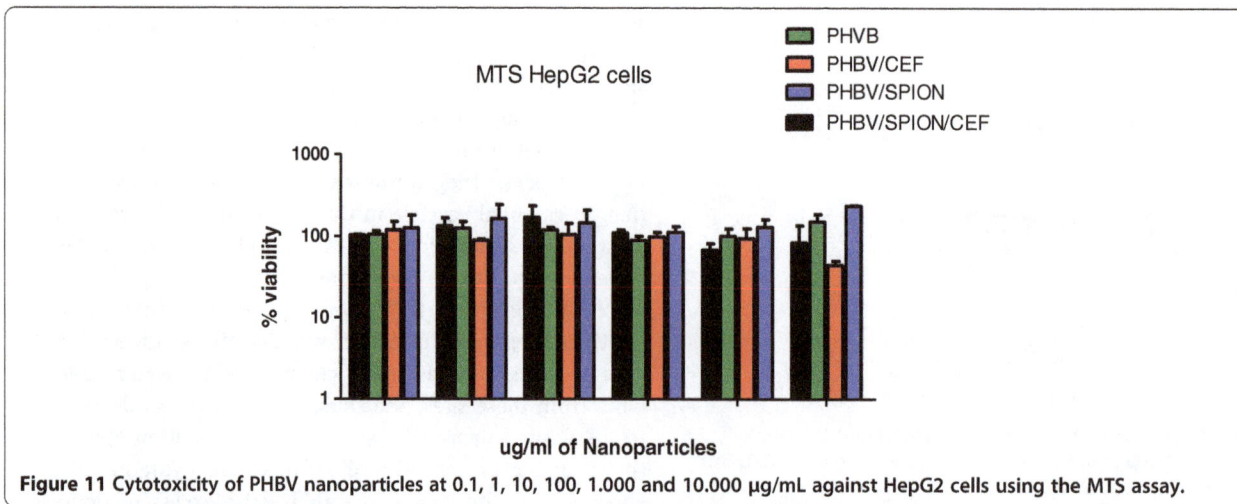

**Figure 11** Cytotoxicity of PHBV nanoparticles at 0.1, 1, 10, 100, 1.000 and 10.000 μg/mL against HepG2 cells using the MTS assay.

available in the polyester (acceptor sites are restricted to only the carboxylic groups at the polymer-chain ends), it is expected that the principal contribution to the modification of the electronic configuration of ceftiofur should be the formation of hydrogen bonds between the ester carboxylic residues within the polymer backbone and the protons from the thiazolyl-amine groups. Such interactions seem to mainly be caused by the hypsochromic effects produced in both the imide and amide chromophores.

The presence of SPIONs in the nanocomposite provides a high concentration of new strong electrophilic sites. These sites are able to compete favorably to stabilize the excess of charge on the $C = O$ groups of ceftiofur and polymer. Indeed, these groups often appear to prefer to interact with SPIONs, thus replacing polymer chain terminals as well as exogenous water molecules from the electrophilic sites in ceftiofur and the polyester. In the case of $\pi$-$\pi^*$ transitions in the polyenone chromophores, these types of interactions are expected to induce bathochromic effects. Of the two bands in the PHBV/CEF nanocomposite assigned above to the polymer, only the low energy broad band is detected at approximately 370 nm in the spectrum of PHBV/SPION/CEF. This red shift is also the most likely reason that the characteristic band of the polymer nanoparticles at 227 nm is not detected in the nanocomposite with SPIONs. In the case of ceftiofur confined in the nanoparticle, the amide-CO-based chromophore is supposed to be more active in the interaction with SPIONs than the chromophores based on the imide-C = N. Thus, the fact that the addition of SPIONs to the nanoparticles only moderately affects the position of the lower energy band of ceftiofur in the electronic spectrum of the PHBV/CEF nanocomposite corroborates the assignment of this absorption to the imide unit proposed above. In contrast, in the case of the amide band, the presence of SPIONs not only reverses the blue shift caused by the hydrogen bonding of the thiazolyl moiety of ceftiofur with the polymer but also further induces a red shift of approximately 15 nm with respect to the position of this band in the pristine drug. A similar (but more moderate) effect of approximately 7 nm is observed in the band assigned to the ceftiofur thiolester carbonyl, whereas the band proposed for the lactam carbonyl remains practically unaltered.

Lastly, the PHBV nanoparticles showed low cytotoxicity at high concentrations in HepG2 cells. These results indicate that PHBV is a suitable agent for entrapment antibiotics and further demonstrate its biocompatible properties.

## Conclusions

Our results indicate that these nanoparticles have potential for use in hyperthermia and could be used for drug delivery. PHBV/CEF/SPION have potential applications as a multifunctional platform because they could be used for MRI, hyperthermia applications and the release of antibiotics at an infection site with only one injection. The encapsulation of ceftiofur in PHBV nanoparticles is only one example of cephalosporin encapsulation. In theory, any cephalosporin with the same physicochemical properties as ceftiofur will behave similarly.

The size and shape of the PHBV/CEF/SPION nanoparticles can be seen in the TEM image (Figure 1). UV–VIS spectroscopy of these nanoparticles (Figure 2) shows the presence of all the desired components, and the efficiency of encapsulation was calculated using a calibration curve. Using FT-IR spectroscopy (Figure 3), we analyzed the interactions between ceftiofur, SPION and the polymer. These data were correlated with the DSC results (Figure 4) and were also used to do the model in Figure 5. The SPION activity of the PHBV nanoparticles was measured by VSM (Figure 6), and the antimicrobial activity of ceftiofur was measured by Kirby-Bauer assay (Figure 7) and microscopy (Figure 8). The toxicity of these nanoparticles was measured by microscopy in HepG2 cells. The percentage of live cells was plotted in Figure 9, and in Figure 10 an example of the microscopy photos is shown. Finally we measured the percentage of viability of the HepG2 cells by MTS (Figure 11).

The characteristics of PHBV/CEF/SPION nanoparticles give them many possible applications in biomedicine. Their superparamagnetic and antibacterial properties are very useful for treating infections in low irrigation sites, for example bone infections, because these nanoparticles allow the antibacterial dose to be localized to the site of infection. This ability can be used to reduce the doses of antibiotics, which at the same time would decrease associated adverse drug reactions. On the other hand, PHBV/CEF/SPION nanoparticles can be used in hyperthermia at infection sites, and this capability can be used to treat patients with bacterial infections associated with cancer. Vulvar, prostate, lung, gallbladder, colon, and stomach cancer have all been associated with different bacterial infections in these organs. Eliminating the infection in the cancer site can be critical to achieving tumor resection and decreasing cancer recurrence. For this reason, therapy with PHBV/CEF/SPION can be very useful for killing cancer cells and eliminating infection in the target tissue, all at the same time and with very low doses of antibiotics.

## Methods

### Preparation of PHBV nanoparticles

PHBV/CEF/SPION nanoparticles were synthesized using the water–oil-water ($w_1/o_1/w_2$) double emulsion with solvent-evaporation method. This protocol was described previously [7,23], but in this case it was slightly modified. For the experiments, 100 μL of ceftiofur (kindly provided by Centrovet (Santiago, Chile) dissolved in methanol (100 mg/mL) (Merck, Germany) was mixed with 300 μL of a SPION suspension (magnetite, BioPAL Inc, MS, USA) (1.65 mg/mL) and was added to 1 mL of PHBV (6.25 mg/mL) (Sigma-Aldrich, St. Louis, MO, USA) dissolved in dichloromethane (DCM) (Merck, Germany). A first $w_1/o_1$ emulsion was prepared by sonication for 40 s at 100% amplitude (VCX130 ultrasonic processor, Sonics & Materials, CT, USA) over an ice bath. The water-in-oil emulsion was further emulsified under the same conditions in 2 mL of an aqueous solution of polyvinyl alcohol (PVA) (5 mg/mL) (Sigma-Aldrich, St. Louis, MO, USA). The $w_1/o_1/w_2$ emulsion was immediately poured into a beaker that contained 30 mL of PVA solution (0.5 mg/mL) and was stirred in a hood under an exhaust fan for 16 h, allowing the solvent to evaporate. The solidified nanoparticles were harvested by centrifugation and washed with distilled water three times using an Amicon Ultra-4 centrifugal filter (Millipore, Billerica, MA, USA) with a molecular weight cut-off of 10 kDa. The PHBV/CEF/SPION were either stored at 4°C for immediate use or freeze-dried in liquid nitrogen and lyophilized for storage at −80°C for later use. We carefully synthesized empty PHBV nanoparticles (PHBV NP) using the same protocol without SPIONs or ceftiofur. PHBV/SPION nanoparticles were prepared using the same protocol but without ceftiofur, and PHBV/CEF nanoparticles were prepared using the same protocol but without SPIONs.

### Transmission electron microscopy (TEM)
The morphological examinations of the nanoparticles were performed as we described previously [23,24], using a transmission electron microscope (Phillips-TECNAI 12 BIOTWIN EM Microscope, FEI Company, Hillsboro, OR) at an acceleration voltage of 80 kV. The TEM sample was prepared by depositing 0.5 mL of the nanoparticle suspension (1.0 mg/mL) onto a 300-mesh carbon-coated copper grid (Electron Microscopy Sciences, PA, USA) that had been previously hydrophilized under UV light (Electron Microscopy Sciences, Hatfield, PA). The samples were blotted away after 20 min of incubation, and the grids were negatively stained for 5 min at room temperature with freshly prepared and sterile-filtered 2% (w/v) uranyl acetate aqueous solution (Electron Microscopy Sciences, PA, USA). Then, the grids were washed twice with distilled water and air-dried prior to imaging.

### Dynamic light scattering (DLS)
The diameter (nm) and ζ potential (mV) of the nanoparticles were analyzed by dynamic light scattering (DLS) in the Zetasizer Nano ZSP 3000 (Malvern Instruments, UK)

as we described previously [12,23,24]. Each preparation was dissolved in 1 mL phosphate buffered saline (Merck, Germany) at pH 7.4, and the measurements were carried out on 3 independent formulations (batches).

### UV–VIS spectroscopy
UV–VIS spectroscopy was used to measure the lyophilized nanoparticles. The nanoparticles were analyzed using a Shimadzu spectrophotometer (UV-2450, Columbia, USA). We analyzed PHBV crystals (PHBV), empty PHBV nanoparticles (PHBV NP), PHBV/SPION nanoparticles, PHBV/CEF nanoparticles and PHBV/CEF/SPION nanoparticles. The data were analyzed using Origin software (6.0, Northampton, USA).

### FTIR spectroscopy
We lyophilized PHBV, PHBV NP, PHBV/SPION and PHBV/CEF/SPION nanoparticles. Each lyophilized nanoparticle was mixed with KBr (Sigma-Aldrich, St. Louis, MO, USA). These samples were used to make pills that were 1 cm in diameter. Each pill was analyzed using Bruker equipment (Vector 22, Hardtstrabe, Karlsruhe, Germany), and our data were collected using OPUS software (Optical user software, Hardtstrabe, Karlsruhe, Germany). Thereafter, the data were analyzed using Origin software (6.0, Massachusetts, USA).

### Differential scanning calorimetry
We lyophilized PHBV, PHBV NP, PHBV/SPION and PHBV/CEF/SPION nanoparticles. Each nanoparticle sample was weighed and then analyzed using a Mettler Toledo instrument (DSC822-E module, Barcelona, Spain) from 25°C to 250°C. The data were analyzed using Origin software (6.0, Massachusetts, USA).

### Ceftiofur entrapment efficiency
The ceftiofur entrapment efficiency was analyzed using UV–VIS spectroscopy. We obtained a calibration curve for ceftiofur and measured different concentrations of ceftiofur at 302 nm using a Shimadzu spectrophotometer (UV-2450, Columbia, USA). We then measured the lyophilized PHBV/CEF/SPION nanoparticles at 302 nm using the same equipment and calculated the amount of ceftiofur in the nanoparticles.

Alternatively, the ceftiofur entrapment efficiency was analyzed by an extraction method described previously [25]. Experimentally, 10 mg of PHBV/CEF/SPION was dissolved in 1 mL of DCM followed by the addition of 5 mL PBS buffer (pH 7.4), and it was then agitated in an orbital shaker maintained at 37°C for 24 h at 100 rpm.

The ceftiofur concentration was determined by ultra-performance liquid chromatography (UPLC). Each sample was measured in triplicate, and the actual drug loading and drug encapsulation efficiency (EE) were calculated using the following equations:

$$Theoretical\ drug\ loading = drug\ total/(drug\ total+polymer)$$
$$Actual\ drug\ loading = drug\ encapsulated/(drug\ total+polymer)$$
$$Encapsulation\ efficiency=(actual\ drug\ loading/theoretical\ drug\ loading)\times100\%$$

## Cell cultures
HepG2 cells (ATCC NUMBER HB-8065) were obtained from ATCC at ampule passage N° 74. These cells were cultured in MEM (Gibco, Invitrogen, Life Technologies, Grand Island, NY, USA) supplemented with 10 v/v% FBS Hyclone (Thermo Scientific, Utah, USA), 100 U/mL penicillin Hyclone (Thermo Scientific, Utah, USA), and 100 µg/mL streptomycin Hyclone (Thermo Scientific, Utah, USA) at 37°C under a humidified atmosphere of 5% $CO_2$ to passage N° 76. These cells were used in all experiments.

## Antibacterial activity
Antimicrobial susceptibility was measured using the agar diffusion method as we performed previously [23] and with live/dead bacterial viability assays (Invitrogen Corporation, Carlsbad, USA). The tests were performed in triplicate against Escherichia coli (ATCC 25922), and we analyzed the antibacterial activity of PHBV/CEF/SPION. The agar diffusion method was conducted in accordance with the National Committee for Clinical Laboratory Standards (NCCLS) [26]. The inoculum was prepared from a Mueller-Hinton plate that had been streaked with a single colony from an initial subculture plate and incubated for 18 to 24 h. The test involved inoculating Mueller-Hinton medium and adjusting the inoculum to $1.5 \times 10^8$ CFU/mL (equal to a 0.5 McFarland turbidity standard). SPIONs and PHBV/SPIONs (20 µg) were used as negative controls, and ceftiofur (30 µg) was used as a positive control. The diameter of each zone of inhibition was determined to the nearest millimeter after 24 h of incubation. In the live/dead bacterial viability assays, 1 mL of $1.5 \times 10^8$ CFU/mL Escherichia coli and PHBV/CEF/SPION (0.1, 1, 10 and 20 µg/mL) were added to tubes with 3 mL of Mueller-Hinton liquid medium. As a negative control, we used 0.9% NaCl (Sigma-Aldrich, St. Louis, MO, USA), and as a positive control, we used ceftiofur (30 µg/mL). The live/dead bacterial viability assays were conducted according to the manufacturer's protocol. After 24 h of incubation, each sample was mounted on a cover slip and visualized using laser scanning confocal microscopy (Axiovert 100 M Microscope, Carl Zeiss, Germany).

## Vibrating sample magnetometer (VSM)
The magnetic properties of the magnetite, PHBV/SPION and PHBV/CEF/SPION nanoparticles were performed as we described previously [7] using a vibrating sample magnetometer (VSM). The hysteresis loop was measured at 300 K as a function of an external applied field.

## Cytotoxicity assays for polymeric nanoparticles
Cell viability was examined using the MTS CellTiter 96 AQ Non-Radioactive Cell Proliferation assay (Promega, Madison, USA) as we described previously [23]. Alternatively, we used the LIVE/DEAD Viability/Cytotoxicity kit for mammalian cells (Invitrogen Corporation, Carlsbad, USA) following the manufacturer's protocols. Cells were treated for 24 h with the following: PHBV NP, PHBV/SPION or PHBV/CEF/SPION at 0.1, 1, 10, 100, 1,000 and 10,000 µg/mL. As a negative control, the cells were treated with just vehicle (MEM medium without serum).

In the LIVE/DEAD Viability/Cytotoxicity assay, the cells were cultured in CultureSlides from BD Falcon (8 wells) with MEM supplemented 10 v/v% FBS at 37°C under a humidified atmosphere of 5% $CO_2$. The seeding was 60000 cells per well, and the cells were used after 24 hours and 90% confluence. Each sample was tested in triplicate with three independent experiments, and 10 images were examined at 40× (Olympus U-RLF-T microscopy) to count the live and dead cells for each treatment. As a cytotoxicity control (100% cell death), the cells were treated for 30 minutes with 70% methanol (Merck, Germany). The statistical analyses (two-way ANOVA and T-test) were conducted using GraphPad Prism 5.0 software (California, USA).

In the MTS cell proliferation assay, the seeding was 75000 cells per well in a 96-well plate. Each sample was measured in triplicate using three independent experiments. Before the addition of MTS, the cells were washed with Hank's medium. The cells were incubated with MTS for 1 h at 37°C under a humidified atmosphere of 5% $CO_2$. The supernatant was collected in an Eppendorf tube and was centrifuged at 13000 rpm for 15 min. The measurement was performed using 90 µL of supernatant per well in a 96-well plate in an ELISA reader at 450 nm (Thermo Scientific, USA). SPIONs can interfere with the spectrometry readings. To correct these data we used the same concentration of nanoparticles in a 96-well-plate without cells. As controls we used untreated cells for 100% viability and only medium for 0% viability. The statistical analyses (two-way ANOVA and t-test) were performed using GraphPad Prism 5.0 software (California USA).

## Statistical analysis
We evaluated the significance of differences in viable and dead cells between the treatments with PHBV NP,

PHBV/SPION, PHBV/CEF/SPION and without nano-particles. We used two-way ANOVA to evaluate the differences between the treatments and the t-test to assess the differences between the treatments and the negative control. We calculated all of the statistics using Graph-Pad Prism 5.0 software (California, USA).

## Additional file

Additional file 1: FT-IR of the components of PHBV nanoparticles. FT-IR spectra of ceftiofur, SPION and PHBV crystal. These data were used as controls to analyze the spectra in Figure 3.

## Competing interests

The authors declare that they have no competing interests.

## Authors' contributions

PS carried out the physicochemical characterization by UV–VIS and FT-IR spectroscopic studies, efficiency of encapsulation, DSC analysis, and the interaction model between ceftiofur, SPION and the polymer. PS carried out the toxicological analysis and drafted the manuscript. GG analyzed the spectroscopic studies and performed the deconvolutions of the UV–VIS and FT-IR spectra. CV carried out the TEM imaging and analyzed the SPION and ceftiofur activities of the PHBV nanoparticles. NH carried out the UPLC assays to measure % efficiency of encapsulation. NJ carried out the double blind analysis of the microscopy in the toxicological assays. MM did the analysis of the DSC assay. FS performed the statistical analysis. LAVC conceived of the study, participated in its design and coordination and helped to draft the manuscript. All authors read and approved the final manuscript.

## Acknowledgements

This work was supported by the Center for the Development of Nanoscience and Nanotechnology (CEDENNA), CONICYT—PCHA/ Doctorado nacional/ 2013-21130869, Grant Number: Anillo ACT 1107. CV acknowledges the support of the Grant TPI06 and the Postdoctoral Program of Becas-Chile/CONICYT, FONDECYT 1120712 and Convenio de Desempeño de Apoyo a la Educación Superior MINEDUC-UNAB, PMI UAB1301.

## Author details

[1]Universidad Andres Bello, Facultad de Medicina, Center for Integrative Medicine and Innovative Science, Echaurren 183, Santiago, Chile. [2]Departamento de Ciencias y Tecnología Farmacéutica, Universidad de Chile, Facultad de Ciencias Químicas y Farmacéuticas, Santos Dumont 964, Independencia, Santiago, Chile. [3]Center for the Development of Nanoscience and Nanotechnology, CEDENNA, 9170124, Av. Ecuador 3493, Estación Central, Santiago, Chile. [4]Departamento de Química, Laboratorio de Síntesis Inorgánica y Electroquímica, Universidad de Chile, Facultad de Ciencias, Las Palmeras 3425, Nuñoa, Santiago, Chile. [5]Departamento de Ciencias Biológicas, Facultad de Ciencias Biológicas, Universidad Andres Bello, República 252, Santiago, Chile. [6]Facultad de Medicina, Universidad Andres Bello, República 590, Santiago, Chile. [7]Millennium Institute on Immunology and Immunotherapy, Avenida Libertador Bernardo O'Higgins 340, Santiago, Chile.

## References

1. Gao J, Gu H, Xu B. Multifunctional magnetic nanoparticles: design, synthesis, and biomedical applications. Acc Chem Res. 2009;42:1097–107.
2. LaConte L, Nitin N, Bao G. Magnetic nanoparticle probes. Nanotoday. 2005;8:32–8.
3. Dousset V, Ballarino L, Delalande C, Coussemacq M, Canioni P, Petry K, et al. Comparison of Ultrasmall Particles of Iron Oxide (USPIO)-Enhanced T2-Weighted, Conventional T2-Weighted, and Gadolinium-Enhanced T1-Weighted MR Images in Rats with Experimental Autoimmune Encephalomyelitis. AJNR Am J Neuroradiol. 1999;20:223–7.
4. Wang A, Bagalkot V, Vasilliou C, Gu F, Alexis F, Zhang L, et al. Superparamagnetic iron oxide nanoparticle-aptamer bioconjugated for combined prostate cancer imaging and therapy. Chem Med Chem. 2008;3:311–1315.
5. Gupta A, Naregalkar R, Vaidya V, Gupta M. Recent advances on surface engineering of magnetic iron oxide nanoparticles and their biomedical applications. Nanomedicine. 2007;2:23–39.
6. Nel A, Mädler L, Velegol D, Xia T, Hoek E, Somasundaran P, et al. Understanding biophysicochemical interactions at the nano-bio interface. Nat Mater. 2009;8:543–57.
7. Vilos C, Gutiérrez M, Escobar R, Morales F, Denardin J, Velasquez L, et al. Superparamagnetic Poly (3-hydroxybutyrate-co-3 hydroxyvalerate) (PHBV) nanoparticles for biomedical applications. Electron J Biotechnol. 2013;16:5. http://dx.doi.org/10.2225/vol16-issue5-fulltext-8.
8. Vilos C, Velásquez LA. Therapeutic strategies based on polymeric microparticles. J Biomed Biotechnol. 2012;2012:Article ID 672760.
9. Lee JH, Lee K, Moon SH, Lee Y, Park TG, Cheon J. All-in-one target-cell-specific magnetic nanoparticles for simultaneous molecular imaging and siRNA delivery. Angew Chem Int Ed Engl. 2009;48:4174–9.
10. Barbu E, Molnar E, Tsibouklis J, Gorecki DC. The potential for nanoparticle-based drug delivery to the brain: overcoming the blood–brain barrier. Expert Opin Drug Deliv. 2009;6:553–65.
11. Farokhzad OC, Langer R. Impact of nanotechnology on drug delivery. ACS Nano. 2009;3:16–20.
12. Vilos C, Morales FA, Solar PA, Herrera NS, Gonzalez-Nilo FD, Aguayo DA, et al. Paclitaxel-PHBV nanoparticles and their toxicity to endometrial and primary ovarian cancer cells. Biomaterials. 2013;34(16):4098–108.
13. Luzier W. Materials derived from biomass/biodegradable materials. Proc Natl Acad Sci USA. 1999;89:839–42.
14. Liu Q, Cai C, Dong CM. Poly(L-lactide)-b-poly(ethylene oxide) copolymers with different arms: hydrophilicity, biodegradable nanoparticles, in vitro degradation, and drug-release behavior. J Biomed Mater Res A. 2009;88:990–9.
15. Tsukada Y, Hara K, Bando Y, Huang CC, Kousaka Y, Kawashima Y, et al. Particle size control of poly(dl-lactide-co-glycolide) nanospheres for sterile applications. Int J Pharm. 2009;370:196–201.
16. Bertram JP, Jay SM, Hynes SR, Robinson R, Criscione JM, Lavik EB. Functionalized poly(lactic-co-glycolic acid) enhances drug delivery and provides chemical moieties for surface engineering while preserving biocompatibility. Acta Biomater. 2009;5(8):2860-2871.
17. Miao LF, Yang J, Huang CL, Song CX, Zeng YJ, Chen LF, et al. Rapamycin-loaded poly (lactic-co-glycolic) acid nanoparticles for intraarterial local drug delivery: preparation, characterization, and in vitro/in vivo release. Zhongguo Yi Xue Ke Xue Yuan Xue Bao. 2008;30:491–7.
18. Xu P, Gullotti E, Tong L, Highley CB, Errabelli DR, Hasan T, et al. Intracellular drug delivery by poly(lactic-co-glycolic acid) nanoparticles, revisited. Mol Pharm. 2009;6:190–201.
19. Furtado S, Abramson D, Burrill R, Olivier G, Gourd C, Bubbers E, et al. Oral delivery of insulin loaded poly(fumaric-co-sebacic) anhydride microspheres. Int J Pharm. 2008;347:149–55.
20. Park EK, Kim SY, Lee SB, Lee YM. Folate-conjugated methoxy poly(ethylene glycol)/poly(epsilon-caprolactone) amphiphilic block copolymeric micelles for tumor-targeted drug delivery. J Control Release. 2005;109:158–68.
21. Wang YC, Liu XQ, Sun TM, Xiong MH, Wang J. Functionalized micelles from block copolymer of polyphosphoester and poly(epsilon-caprolactone) for receptor-mediated drug delivery. J Control Release. 2008;128:32–40.
22. Balmayor ER, Tuzlakoglu K, Azevedo HS, Reis RL. Preparation and characterization of starch-poly-epsilon-caprolactone microparticles incorporating bioactive agents for drug delivery and tissue engineering applications. Acta Biomater. 2009;5:1035–45.
23. Vilos C, Constandil L, Herrera N, Solar P, Escobar-Fica J, Velásquez L. Ceftiofur-loaded PHBV microparticles: a potential formulation for a long-acting antibiotic to treat animal infections. Electron J Biotechnol. 2012;15:4.
24. Vilos C, Constandil L, Rodas PI, Cantin M, Zepeda K, Herrera A, et al. Evaluation of ceftiofur-PHBV microparticles in rats. Drug Des Devel Ther. 2014;29(8):651–66.
25. Mao S, Xu J, Cai C, Germershaus O, Schaper A, Kissel T. Effect of WOW process parameters on morphology and burst release of FITC-dextran loaded PLGA microspheres. Int J Pharm. 2007;334:137–48.
26. Katz OT, Peled N, Yagupsky P. Evaluation of the current National Committee for Clinical Laboratory Standards guidelines for screening and confirming extended-spectrum beta-lactamase production in isolates of Escherichia coli and Klebsiella species from bacteremic patients. Eur J Clin Microbiol Infect Dis. 2004;23:813–7.

# Permissions

# List of Contributors

**George Seghal Kiran**
Department of Food Science and Technology, Pondicherry University, Puducherry 605014, India

**Asha Dhasayan**
Department of Microbiology, Bharathidasan University, Tiruchirappalli 620 024, India

**Anuj Nishanth Lipton**
Department of Microbiology, Pondicherry University, Puducherry 605014, India

**Joseph Selvin**
Department of Microbiology, Pondicherry University, Puducherry 605014, India

**Mariadhas Valan Arasu**
Department of Botany and Microbiology, Addiriyah Chair for Environmental Studies, College of Science, King Saud University, P. O. Box 2455, Riyadh 11451, Saudi Arabia

**Naif Abdullah Al-Dhabi**
Department of Botany and Microbiology, Addiriyah Chair for Environmental Studies, College of Science, King Saud University, P. O. Box 2455, Riyadh 11451, Saudi Arabia

**Bahareh Salehi**
Young Researchers and Elites Club, North Tehran Branch of Islamic Azad University, Tehran, Iran

**Sedigheh Mehrabian**
Microbiology Group, Biological Sciences Faculty, North Tehran Branch of Islamic Azad University, Tehran, Iran

**Mehdi Ahmadi**
Institute of Biochemistry and Biophysics, University of Tehran, Tehran, Iran

**Michal Kolitz-Domb**
Department of Chemistry, The Institute of Nanotechnology and Advanced Materials, Bar-Ilan University, Ramat-Gan 52900, Israel

**Igor Grinberg**
Department of Chemistry, The Institute of Nanotechnology and Advanced Materials, Bar-Ilan University, Ramat-Gan 52900, Israel

**Enav Corem-Salkmon**
Department of Chemistry, The Institute of Nanotechnology and Advanced Materials, Bar-Ilan University, Ramat-Gan 52900, Israel

**Shlomo Margel**
Department of Chemistry, The Institute of Nanotechnology and Advanced Materials, Bar-Ilan University, Ramat-Gan 52900, Israel

**Caio Pinho Fernandes**
Programa de Pós, Graduação em Biotecnologia Vegetal, Centro de Ciências da Saúde, Universidade Federal do Rio de Janeiro – UFRJ, Bloco K, 2° andar – sala 032, Av. Brigadeiro Trompowski s/n, CEP: 21941-590 Ilha do Fundão, RJ, Brazil
Laboratório de Farmacotécnica, Colegiado de Ciências Farmacêuticas, Universidade Federal do Amapá, Campus Universitário Marco Zero do Equador, Rodovia Juscelino Kubitschek – KM – 02-Jardim Marco Zero, CEP: 68903-419 Macapá, AP, Brazil

**Fernanda Borges de Almeida**
Programa de Pós, Graduação em Biotecnologia Vegetal, Centro de Ciências da Saúde, Universidade Federal do Rio de Janeiro – UFRJ, Bloco K, 2° andar – sala 032, Av. Brigadeiro Trompowski s/n, CEP: 21941-590 Ilha do Fundão, RJ, Brazil

**Amanda Nunes Silveira**
Laboratório de Tecnologia Farmacêutica I, Faculdade de Farmácia, Universidade Federal Fluminense, Rua: Mario Viana, 523, Santa Rosa, CEP: 24241-000 Niterói RJ, Brazil

**Marcelo Salabert Gonzalez**
Laboratório de Biologia de Insetos – LABI, Departamento de Biologia Geral (GBG), Universidade Federal Fluminense, Morro do Valonguinho S/No, CEP 24001-970 Niterói, RJ, Brazil

**Cicero Brasileiro Mello**
Laboratório de Biologia de Insetos – LABI, Departamento de Biologia Geral (GBG), Universidade Federal Fluminense, Morro do Valonguinho S/No, CEP 24001-970 Niterói, RJ, Brazil

**Denise Feder**
Laboratório de Biologia de Insetos – LABI, Departamento de Biologia Geral (GBG), Universidade Federal Fluminense, Morro do Valonguinho S/No, CEP 24001-970 Niterói, RJ, Brazil

**Raul Apolinário**
Laboratório de Biologia de Insetos – LABI, Departamento de Biologia Geral (GBG), Universidade Federal Fluminense, Morro do Valonguinho S/No, CEP 24001-970 Niterói, RJ, Brazil

**Marcelo Guerra Santos**
Faculdade de Formação de Professores, UERJ, Rua: Dr. Francisco Portela, 1470 – Patronato, CEP: 24435-005 São Gonçalo, Rio de Janeiro, Brazil

**José Carlos Tavares Carvalho**
Laboratório de Pesquisa em Fármacos, Colegiado de Ciências Farmacêuticas, Universidade Federal do Amapá, Rodovia Juscelino Kubitschek – KM – 02 – Jardim Marco Zero, CEP: 68903-419 Macapá, AP, Brazil

**Luis Armando Cândido Tietbohl**
Laboratório de Tecnologia de Produtos Naturais – LTPN, Departamento e Tecnologia Farmacêutica, Faculdade de Farmácia, Universidade Federal Fluminense – UFF Rua, Mario Viana, 523, CEP: 24241-000, Santa Rosa, Niterói, RJ, Brazil

**Leandro Rocha**
Laboratório de Tecnologia de Produtos Naturais – LTPN, Departamento e Tecnologia Farmacêutica, Faculdade de Farmácia, Universidade Federal Fluminense – UFF Rua, Mario Viana, 523, CEP: 24241-000, Santa Rosa, Niterói, RJ, Brazil

**Deborah Quintanilha Falcão**
Laboratório de Tecnologia Farmacêutica I, Faculdade de Farmácia, Universidade Federal Fluminense, Rua: Mario Viana, 523, Santa Rosa, CEP: 24241-000 Niterói RJ, Brazil

**Maria Luisa Bondì**
Istituto per lo Studio dei Materiali Nanostrutturati- U.O.S. di Palermo-Consiglio Nazionale delle Ricerche-via Ugo La Malfa, 153 90146 Palermo, Italy

**Maria Ferraro**
Istituto di Biomedicina e Immunologia Molecolare- Consiglio Nazionale delle Ricerche – via Ugo La Malfa, 153 90146 Palermo, Italy

**Serena Di Vincenzo**
Istituto di Biomedicina e Immunologia Molecolare- Consiglio Nazionale delle Ricerche – via Ugo La Malfa, 153 90146 Palermo, Italy

**Stefania Gerbino**
Istituto di Biomedicina e Immunologia Molecolare- Consiglio Nazionale delle Ricerche – via Ugo La Malfa, 153 90146 Palermo, Italy

**Gennara Cavallaro**
Laboratory of Biocompatible Polymers-Dipartimento di Scienze e Tecnologie, Biologiche, Chimiche e Farmaceutiche (STEBICEF), Università di Palermo -via Archirafi, 32-90123 Palermo, Italy

**Gaetano Giammona**
Laboratory of Biocompatible Polymers-Dipartimento di Scienze e Tecnologie, Biologiche, Chimiche e Farmaceutiche (STEBICEF), Università di Palermo -via Archirafi, 32-90123 Palermo, Italy

**Chiara Botto**
Laboratory of Biocompatible Polymers-Dipartimento di Scienze e Tecnologie, Biologiche, Chimiche e Farmaceutiche (STEBICEF), Università di Palermo -via Archirafi, 32-90123 Palermo, Italy

**Mark Gjomarkaj**
Istituto di Biomedicina e Immunologia Molecolare- Consiglio Nazionale delle Ricerche – via Ugo La Malfa, 153 90146 Palermo, Italy

**Elisabetta Pace**
Istituto di Biomedicina e Immunologia Molecolare- Consiglio Nazionale delle Ricerche – via Ugo La Malfa, 153 90146 Palermo, Italy

**Anamika Singh**
BMRL, Department of Chemistry, Government Model Science College, Jabalpur, India

**Jaya Bajpai**
BMRL, Department of Chemistry, Government Model Science College, Jabalpur, India

**Anil Kumar Bajpai**
BMRL, Department of Chemistry, Government Model Science College, Jabalpur, India

**Moraima Morales-Cruz**
Department of Biology, University of Puerto Rico, Río Piedras Campus, San Juan, PR 00931, USA

**Cindy M Figueroa**
Departments of Chemistry, University of Puerto Rico, Río Piedras Campus, San Juan, PR 00931, USA

**Tania González-Robles**
Departments of Chemistry, University of Puerto Rico, Río Piedras Campus, San Juan, PR 00931, USA

**Yamixa Delgado**
Department of Biology, University of Puerto Rico, Río Piedras Campus, San Juan, PR 00931, USA

**Anna Molina**
Departments of Chemistry, University of Puerto Rico, Río Piedras Campus, San Juan, PR 00931, USA

**Jessica Méndez**
Departments of Chemistry, University of Puerto Rico, Río Piedras Campus, San Juan, PR 00931, USA

**Myraida Morales**
Department of Graduate Studies, University of Puerto Rico, Río Piedras Campus, Río Piedras Campus, San Juan, PR 00931, USA

**Kai Griebenow**
Departments of Chemistry, University of Puerto Rico, Río Piedras Campus, San Juan, PR 00931, USA

**Kirstie Salinas**
UTSA Neurosciences Institute, The University of Texas at San Antonio, San Antonio, Texas 78249, USA

**Zurab Kereselidze**
Department of Physics and Astronomy, The University of Texas at San Antonio, San Antonio, Texas 78249, USA

**Frank DeLuna**
UTSA Neurosciences Institute, The University of Texas at San Antonio, San Antonio, Texas 78249, USA

**Xomalin G Peralta**
Department of Physics and Astronomy, The University of Texas at San Antonio, San Antonio, Texas 78249, USA

**Fidel Santamaria**
UTSA Neurosciences Institute, The University of Texas at San Antonio, San Antonio, Texas 78249, USA

**Yong Ho Kim**
Curriculum in Toxicology, University of North Carolina at Chapel Hill, Chapel Hill, NC, USA

**Elizabeth Boykin**
Environmental Public Health Division, National Health and Environmental Effects Research Laboratory, United States Environmental Protection Agency, Research Triangle Park, NC, USA

**Tina Stevens**
Research Triangle Park Division, National Center for Environmental Assessment, United States Environmental Protection Agency, Research Triangle Park, NC, USA

**Katelyn Lavrich**
Curriculum in Toxicology, University of North Carolina at Chapel Hill, Chapel Hill, NC, USA

**M Ian Gilmour**
Environmental Public Health Division, National Health and Environmental Effects Research Laboratory, United States Environmental Protection Agency, Research Triangle Park, NC, USA

**Sangiliyandi Gurunathan**
Department of Animal Biotechnology, Konkuk University, 1 Hwayang-Dong, Gwangin-gu, Seoul 143-701, South Korea
GS Institute of Bio and Nanotechnology, Coimbatore, Tamil Nadu 641024, India

**Jae Woong Han**
Department of Animal Biotechnology, Konkuk University, 1 Hwayang-Dong, Gwangin-gu, Seoul 143-701, South Korea

**Eunsu Kim**
Department of Animal Biotechnology, Konkuk University, 1 Hwayang-Dong, Gwangin-gu, Seoul 143-701, South Korea

**Deug-Nam Kwon**
Department of Animal Biotechnology, Konkuk University, 1 Hwayang-Dong, Gwangin-gu, Seoul 143-701, South Korea

**Jin-Ki Park**
Animal Biotechnology Division, National Institute of Animal Science, Suwon 441-350, Korea

**Jin-Hoi Kim**
Department of Animal Biotechnology, Konkuk University, 1 Hwayang-Dong, Gwangin-gu, Seoul 143-701, South Korea

**Azamal Husen**
Department of Biology, College of Natural and Computational Sciences, University of Gondar, P.O. Box 196, Gondar, Ethiopia

**Khwaja Salahuddin Siddiqi**
Department of Chemistry, College of Natural and Computational Sciences, University of Gondar, P.O. Box 196, Gondar, Ethiopia

**Wei Ge**
State Key Lab of Bioelectronics (Chien-Shiung Wu Lab), Department of Biological Science and Medical Engineering, Southeast University, Nanjing 210096, China

**Yuanyuan Zhang**
State Key Lab of Bioelectronics (Chien-Shiung Wu Lab), Department of Biological Science and Medical Engineering, Southeast University, Nanjing 210096, China

**Jing Ye**
State Key Lab of Bioelectronics (Chien-Shiung Wu Lab), Department of Biological Science and Medical Engineering, Southeast University, Nanjing 210096, China

**Donghua Chen**
School of Chemistry and Chemical Engineering, Southeast University, Nanjing 211189, China

**Fawad Ur Rehman**
State Key Lab of Bioelectronics (Chien-Shiung Wu Lab), Department of Biological Science and Medical Engineering, Southeast University, Nanjing 210096, China

**Qiwei Li**
State Key Lab of Bioelectronics (Chien-Shiung Wu Lab), Department of Biological Science and Medical Engineering, Southeast University, Nanjing 210096, China

**Yun Chen**
State Key Lab of Bioelectronics (Chien-Shiung Wu Lab), Department of Biological Science and Medical Engineering, Southeast University, Nanjing 210096, China

**Hui Jiang**
State Key Lab of Bioelectronics (Chien-Shiung Wu Lab), Department of Biological Science and Medical Engineering, Southeast University, Nanjing 210096, China

**Xuemei Wang**
State Key Lab of Bioelectronics (Chien-Shiung Wu Lab), Department of Biological Science and Medical Engineering, Southeast University, Nanjing 210096, China

**Tomoko Sugiyama**
Department of Dentistry, Oral and Maxillofacial Surgery, Jichi Medical University, 3311-1 Yakushiji, Shimotsuke, Tochigi 329-0498, Japan

**Motohiro Uo**
Advanced Biomaterials Department, Graduate School of Medical and Dental Sciences, Tokyo Medical and Dental University, 1-5-45 Yushima, Bunkyo-ku, Tokyo 113-8549, Japan

**Takahiro Wada**
Advanced Biomaterials Department, Graduate School of Medical and Dental Sciences, Tokyo Medical and Dental University, 1-5-45 Yushima, Bunkyo-ku, Tokyo 113-8549, Japan

**Toshio Hongo**
Advanced Biomaterials Department, Graduate School of Medical and Dental Sciences, Tokyo Medical and Dental University, 1-5-45 Yushima, Bunkyo-ku, Tokyo 113-8549, Japan

**Daisuke Omagari**
Department of Pathology, Nihon University School of Dentistry, 1-8-13 Kanda-Surugadai, Chiyoda-ku, Tokyo 131-8310, Japan

**Kazuo Komiyama**
Department of Pathology, Nihon University School of Dentistry, 1-8-13 Kanda-Surugadai, Chiyoda-ku, Tokyo 131-8310, Japan

**Hitoshi Sasaki**
Nakayamagumi Co. Ltd., North 19, East 1, Higashi-ku, Sapporo, Hokkaido 065-8610, Japan

**Heishichiro Takahashi**
Graduate School of Engineering, Hokkaido University, North 13, West 8, Sapporo, Hokkaido 060-8628, Japan

**Mikio Kusama**
Department of Dentistry, Oral and Maxillofacial Surgery, Jichi Medical University, 3311-1 Yakushiji, Shimotsuke, Tochigi 329-0498, Japan

**Yoshiyuki Mori**
Department of Dentistry, Oral and Maxillofacial Surgery, Jichi Medical University, 3311-1 Yakushiji, Shimotsuke, Tochigi 329-0498, Japan

**Vinh Quang Nguyen**
Faculty of System Design, Tokyo Metropolitan University, 6-6 Asahigaoka, Hino, Tokyo 191-0065, Japan
Research Institute, National Defense Medical College, 3-2 Namiki, Tokorozawa, Saitama 359-1324, Japan

**Masayuki Ishihara**
Research Institute, National Defense Medical College, 3-2 Namiki, Tokorozawa, Saitama 359-1324, Japan

**Jun Kinoda**
Department of Oral and Maxillofacial Surgery, National Defense Medical College, 3-2 Namiki, Tokorozawa, Saitama 359-8513, Japan

**Hidemi Hattori**
Research Institute, National Defense Medical College, 3-2 Namiki, Tokorozawa, Saitama 359-1324, Japan

**Shingo Nakamura**
Research Institute, National Defense Medical College, 3-2 Namiki, Tokorozawa, Saitama 359-1324, Japan

**Takeshi Ono**
Department of Global Infectious Diseases and Tropical Medicine, National Defense Medical College, 3-2 Namiki, Tokorozawa, Saitama 359-8513, Japan

**Yasushi Miyahira**
Department of Global Infectious Diseases and Tropical Medicine, National Defense Medical College, 3-2 Namiki, Tokorozawa, Saitama 359-8513, Japan

**Takemi Matsui**
Faculty of System Design, Tokyo Metropolitan University, 6-6 Asahigaoka, Hino, Tokyo 191-0065, Japan

**Muhammad Ajaz Hussain**
Department of Chemistry, University of Sargodha, Sargodha 40100, Pakistan

**Abdullah Shah**
Department of Chemistry, University of Sargodha, Sargodha 40100, Pakistan

**Ibrahim Jantan**
Drug and Herbal Research Centre, Faculty of Pharmacy, Universiti Kebangsaan Malaysia, Jalan Raja Muda Abdul Aziz, Kuala Lumpur 50300, Malaysia

**Muhammad Nawaz Tahir**
Institute of Inorganic and Analytical Chemistry, Johannes Guttenberg University of Mainz, Duesbergweg 10-14, Mainz 55128, Germany

**Muhammad Raza Shah**
International Center for Chemical and Biological Sciences, University of Karachi, Karachi 75270, Pakistan

**Riaz Ahmed**
Centre for Advanced Studies in Physics (CASP), GC University, Lahore 54000, Pakistan

**Syed Nasir Abbas Bukhari**
Drug and Herbal Research Centre, Faculty of Pharmacy, Universiti Kebangsaan Malaysia, Jalan Raja Muda Abdul Aziz, Kuala Lumpur 50300, Malaysia

**Tariq Mahmood**
Department of Pharmacy, Faculty of Pharmacy and Alternative Medicine, The Islamia University of Bahawalpur, Bahawalpur 63100, Pakistan
School of Pharmacy, The University of Faisalabad, Faisalabad 37610, Pakistan

**Naveed Akhtar**
Department of Pharmacy, Faculty of Pharmacy and Alternative Medicine, The Islamia University of Bahawalpur, Bahawalpur 63100, Pakistan

**Sivakumar Manickam**
Manufacturing and Industrial Processes Research Division, Faculty of Engineering, University of Nottingham Malaysia Campus, Jalan Broga, Semenyih 43500, Malaysia

**Jingjing Xie**
Cancer Metastasis Alert and Prevention Center, and Biopharmaceutical Photocatalysis of State Key Laboratory of Photocatalysis on Energy and Environment, College of Chemistry, Fuzhou University, 523 Industry Road, Science Building, 3FL, Fuzhou, Fujian 350002, China

**Haiyan Dong**
Cancer Metastasis Alert and Prevention Center, and Biopharmaceutical Photocatalysis of State Key Laboratory of Photocatalysis on Energy and Environment, College of Chemistry, Fuzhou University, 523 Industry Road, Science Building, 3FL, Fuzhou, Fujian 350002, China

**Hongning Chen**
Cancer Metastasis Alert and Prevention Center, and Biopharmaceutical Photocatalysis of State Key Laboratory of Photocatalysis on Energy and Environment, College of Chemistry, Fuzhou University, 523 Industry Road, Science Building, 3FL, Fuzhou, Fujian 350002, China

**Rongli Zhao**
Cancer Metastasis Alert and Prevention Center, and Biopharmaceutical Photocatalysis of State Key Laboratory of Photocatalysis on Energy and Environment, College of Chemistry, Fuzhou University, 523 Industry Road, Science Building, 3FL, Fuzhou, Fujian 350002, China

**Patrick J Sinko**
Rutgers, The State University of New Jersey, 160 Frelinghuysen Road, Piscataway, NJ 08854, USA

**Weiyu Shen**
Cancer Metastasis Alert and Prevention Center, and Biopharmaceutical Photocatalysis of State Key Laboratory of Photocatalysis on Energy and Environment, College of Chemistry, Fuzhou University, 523 Industry Road, Science Building, 3FL, Fuzhou, Fujian 350002, China

**Jichuang Wang**
Cancer Metastasis Alert and Prevention Center, and Biopharmaceutical Photocatalysis of State Key Laboratory of Photocatalysis on Energy and Environment, College of Chemistry, Fuzhou University, 523 Industry Road, Science Building, 3FL, Fuzhou, Fujian 350002, China

**Yusheng Lu**
Cancer Metastasis Alert and Prevention Center, and Biopharmaceutical Photocatalysis of State Key Laboratory of Photocatalysis on Energy and Environment, College of Chemistry, Fuzhou University, 523 Industry Road, Science Building, 3FL, Fuzhou, Fujian 350002, China

**Xiang Yang**
Cancer Metastasis Alert and Prevention Center, and Biopharmaceutical Photocatalysis of State Key Laboratory of Photocatalysis on Energy and Environment, College of Chemistry, Fuzhou University, 523 Industry Road, Science Building, 3FL, Fuzhou, Fujian 350002, China

**Fangwei Xie**
Department of Medicine Oncology, East Hospital of Xiamen University, Fuzhou 350004, China

**Lee Jia**
Cancer Metastasis Alert and Prevention Center, and Biopharmaceutical Photocatalysis of State Key Laboratory of Photocatalysis on Energy and Environment, College of Chemistry, Fuzhou University, 523 Industry Road, Science Building, 3FL, Fuzhou, Fujian 350002, China

**Dipali Bagal-Kestwal**
Institute of Food Science and Technology, National Taiwan University, No.1, Roosevelt Road, section 4, Taipei, Taiwan

**Rakesh Mohan Kestwal**
Institute of Food Science and Technology, National Taiwan University, No.1, Roosevelt Road, section 4, Taipei, Taiwan

**Been-Huang Chiang**
Institute of Food Science and Technology, National Taiwan University, No.1, Roosevelt Road, section 4, Taipei, Taiwan

**Paula Solar**
Universidad Andres Bello, Facultad de Medicina, Center for Integrative Medicine and Innovative Science, Echaurren 183, Santiago, Chile
Departamento de Ciencias y Tecnología Farmacéutica, Universidad de Chile, Facultad de Ciencias Químicas y Farmacéuticas, Santos Dumont 964, Independencia, Santiago, Chile

**Guillermo González**
Center for the Development of Nanoscience and Nanotechnology, CEDENNA, 9170124, Av. Ecuador 3493, Estación Central, Santiago, Chile
Departamento de Química, Laboratorio de Síntesis Inorgánica y Electroquímica, Universidad de Chile, Facultad de Ciencias, Las Palmeras 3425, Nuñoa, Santiago, Chile

**Cristian Vilos**
Universidad Andres Bello, Facultad de Medicina, Center for Integrative Medicine and Innovative Science, Echaurren 183, Santiago, Chile
Center for the Development of Nanoscience and Nanotechnology, CEDENNA, 9170124, Av. Ecuador 3493, Estación Central, Santiago, Chile

**Natalia Herrera**
Universidad Andres Bello, Facultad de Medicina, Center for Integrative Medicine and Innovative Science, Echaurren 183, Santiago, Chile

**Natalia Juica**
Universidad Andres Bello, Facultad de Medicina, Center for Integrative Medicine and Innovative Science, Echaurren 183, Santiago, Chile

**Mabel Moreno**
Center for the Development of Nanoscience and Nanotechnology, CEDENNA, 9170124, Av. Ecuador 3493, Estación Central, Santiago, Chile

**Felipe Simon**
Departamento de Ciencias Biológicas, Facultad de Ciencias Biológicas, Universidad Andres Bello, República 252, Santiago, Chile
Facultad de Medicina, Universidad Andres Bello, República 590, Santiago, Chile
Millennium Institute on Immunology and Immunotherapy, Avenida Libertador Bernardo O'Higgins 340, Santiago, Chile

**Luis Velásquez**
Universidad Andres Bello, Facultad de Medicina, Center for Integrative Medicine and Innovative Science, Echaurren 183, Santiago, Chile
Center for the Development of Nanoscience and Nanotechnology, CEDENNA, 9170124, Av. Ecuador 3493, Estación Central, Santiago, Chile

www.ingramcontent.com/pod-product-compliance
Lightning Source LLC
Chambersburg PA
CBHW080656200326
41458CB00013B/4876